ROCK MECHANICS AND ENGINEERING GEOLOGY IN VOLCANIC FIELDS

Rock Mechanics and Engineering Geology in Volcanic Fields includes keynote lectures and papers from the 5th International Workshop on Rock Mechanics and Engineering Geology in Volcanic Fields (RMEGV2021, Fukuoka, Japan, 9-10 September 2021). This book deals with challenging studies related to solving engineering issues around volcanic fields, including:

- Volcanic geology, disasters and their mitigation
- Resources and energy in volcanic fields
- Mechanical behavior of volcanic rocks and soils
- Groundwater and environmental problems in volcanic fields
- Geotechnical engineering in volcanic fields

Rock Mechanics and Engineering Geology in Volcanic Fields is of great interest to civil engineers and engineering geologists working in the areas of rock and soil mechanics, geotechnical engineering, geothermal energy, engineering geology, and environmental science.

T0172407

PROCEEDINGS OF THE 5TH INTERNATIONAL WORKSHOP ON ROCK MECHANICS AND ENGINEERING GEOLOGY IN VOLCANIC FIELDS (RMEGV 2021), FUKUOKA, JAPAN, 9–10 SEPTEMBER 2021

Rock Mechanics and Engineering Geology in Volcanic Fields

Editors

Takehiro Ohta
Yamaguchi University, Yamaguchi, Japan

Takatoshi Ito
Tohoku University, Sendai, Japan

Masahiko Osada
Saitama University, Saitama, Japan

CRC Press
Taylor & Francis Group
Boca Raton London New York

CRC Press is an imprint of the
Taylor & Francis Group, an **informa** business

A BALKEMA BOOK

Cover image: Takehiro Ohta

First published 2023
by CRC Press/Balkema
4 Park Square, Milton Park, Abingdon, Oxon OX14 4RN, United Kingdom
e-mail: enquiries@taylorandfrancis.com
www.routledge.com – www.taylorandfrancis.com

CRC Press/Balkema is an imprint of the Taylor & Francis Group, an informa business

Library of Congress Cataloging-in-Publication Data

A catalog record has been requested for this book

ISBN: 978-1-032-27656-4 (hbk)
ISBN: 978-1-032-27657-1 (pbk)
ISBN: 978-1-003-29359-0 (ebk)
DOI: 10.1201/9781003293590

Table of Contents

Resources and energy in volcanic fields

Mechanical behavior of volcanic rocks and soils

Groundwater and environmental problems in volcanic fields

Geoengineering and infrastructures in volcanic fields

Preface

Numerous Quaternary volcanos and older pre-Neogene volcanic rocks are found throughout the world. In particular, active volcanos are located along subduction zones and mid-ocean ridges and in hot spots. Huge plateau basalts are distributed on many continents, e.g., Deccan Trap in India, Siberian Trap in Russia, Colombia River Basalts in America, etc. Numerous resources can be found in the vicinity of these volcanic fields, such as hydro-geothermal energy and groundwater. On the other hand, many natural disasters and other risks to infrastructure construction also exist due to the geoscientific and geotechnical features in these areas. Thus, civil engineers and engineering geologists strive to solve these problems based on the existing data on volcanic fields.

The International Society for Rock Mechanics (ISRM) has held four workshops on volcanic rocks in Madeira, Azores, Tenerife, and Ischia. The Japanese Society for Rock Mechanics and the Japan Society of Engineering Geology co-organized "RMEGV2021: 5th International Workshop on Rock Mechanics and Engineering Geology in Volcanic Fields" as the fifth edition of this event on a volcanic field on Kyushu Island. This workshop was held as an ISRM Specialized Conference and an official conference of the International Association for Engineering Geology and Environment (IAEG).

The goals of this workshop were to promote the exchange of ideas and information among civil engineers and geologists regarding serious issues related to volcanic fields, and to find possible solutions to them, as well as to aid in the fusion of Civil Engineering and Geology in terms of applied volcanic science, including construction, environment, groundwater, disasters, and geothermal energy.

The themes of this workshop were as follows:

✓ Volcanic geology, disasters and their mitigation
✓ Resources and energy in volcanic fields
✓ Mechanical behavior of volcanic rocks and soils
✓ Groundwater and environmental problems in volcanic fields
✓ Geotechnical engineering in volcanic fields.

The Organizing Committee accepted 55 proceedings papers from eight countries for the above themes. Keynote lectures were delivered by six prominent researchers.

On behalf of the Organizing Committee, we would like to thank all the attendees and to express our gratitude to the sponsors and exhibitors for their support. We are also very grateful for the assistance of the International Advisory Committee, the hard work of the Steering Committee, and the dedication of the chairpersons of the sessions. The workshop could not have happened without the efforts of everyone. Our thanks are also extended to the Japan Society for the Promotion of Science.

Committees

Organizing Committee

Takatoshi Ito: Chair, Japan
Takehiro Ohta: Secretary General, Japan
Atsushi Hasegawa: Japan
Nobusuke Hasegawa: Japan
Hiroaki Ikemi: Japan
Hisatoshi Ito: Japan
Kiyoshi Kishida: Japan
Shinichiro Nakashima: Japan
Mitsuru Okuno: Japan
Masahiko Osada: Japan
Akira Sato: Japan
Koji Shimizu: Japan
Norikazu Shimizu: Japan

Steering Committee

Takehiro Ohta: Chair, Leader of General Affairs Working Group, Japan
Masahiko Osada: Leader of Science Working Group, Japan
Mitsuru Okuno: Leader of Venue and Events Working Group, Japan
Hisatoshi Ito: Core-Member, Japan
Hideki Kosaka: Core-Member, Japan
Shinichiro Nakashima: Core-Member, Japan
Akira Sato: Core-Member, Japan
Tomohiro Tsuji, Core-Member, Japan
Hideaki Yasuhara: Core-Member, Japan
Shuichi Hasegawa: Japan
Ryota Hashimoto: Japan
Shigetaka Ishihama: Japan
Takashi Inokuchi: Japan
Kiyoshi Kanematsu: Japan
Junko Kiyosaki: Japan
Tomofumi Koyama: Japan
Tomoya Miyahara: Japan
Akira Momoshima: Japan
Tatsuaki Nakazuka: Japan
Tomohisa Otsuka: Japan
Shiro Tanaka: Japan
Yota Togashi: Japan
Mitsuhiro Tsuchiya: Japan
Yuji Uehara: Japan
Koji Umeda: Japan
Tomohiro Yasuda: Japan

International Advisory Committee

Őmer Aydan: Japan
Rafig Azzam: Germany
Faquan Wu: China
Vojkan Jovičić; Srovenia
Suseno Kramadibrata: Indonesia
Vassilis Marinos: Greece
Asnawir Nasution: Indonesia
Resat Ulusay: Turkey
Eugene A. Voznesensky: Russia

Geology, disaster and its mitigation in volcanic fields

Rock Mechanics and Engineering Geology in Volcanic Fields – Ohta, Ito & Osada (eds)
© 2023 copyright the Author(s), ISBN 978-1-032-27657-1

Volcanic events and magma interactions before, during and after the caldera-forming eruption of Aso, Kyushu island, Japan

Toshiaki Hasenaka*
Center for Water Cycle, Marine Environment and Disaster management, Kumamoto University Kumamoto, Japan

Masataka Kawaguchi
Graduate school of Science and Technology, Kumamoto University, Kumamoto, Japan
Laboratoire Magmas et Volcans, Université Clermont Auvergne, Aubière, France

Masayuki Torii
Center for Water Cycle, Marine Environment and Disaster management, Kumamoto University Kumamoto, Japan

Atsushi Yasuda & Natsumi Hokanishi
Earthquake Research Institute, University of Tokyo, Tokyo, Japan

ABSTRACT: Melt inclusions in minerals contained in tephra from different stages of Aso caldera volcano show distinct patterns of compositional plots. They provide information of different state of magma plumbing system. Melt inclusions in Aso-4 pumice representing caldera-forming eruption, those of Aso ABCD pumice and Omine scoria, which were precursory to caldera-forming eruptions, and those of post-caldera central cone pumice (ACP-2 and ACP-4) all show narrow regions in the compositional plots. They indicate existence of homogeneous magma reservoir, probably relatively large in scale. The compositions of melt inclusions are often the same as those of regional tephra or groundmass glass of pumice and scoria, and are quite similar irrespective of host minerals. ACP-1 melt inclusion exceptionally shows a very wide compositional range. It corresponds to the initiation of Holocene basaltic eruptions including that of currently active Nakadake. It may indicate the mixing of magma of different origin, or the intrusion of magma into an inhomogeneous crust initiating partial melting. Melt inclusions of ACP3/4 scoria show distinct compositional range reflecting difference depending on host minerals. The composition becomes SiO_2-rich from Olivine-hosted inclusions, to pyroxene-hosted, and to plagioclase-hosted ones. Those melts were probably tapped by host minerals at different portions of the magma supplying system. The compositions of melt inclusions from Holocene basaltic activities of Aso volcano are similar to those of ACP3/4. When all the compositional data of melt inclusions are taken into consideration, development of relatively large homogeneous silicic magma reservoir at present seems unlikely.

Keywords: Aso-4 tephra, caldera-forming eruption, precursory event, magma plumbing system, melt inclusions

*Corresponding author: hasenaka@kumamoto-u.ac.jp

DOI: 10.1201/9781003293590-1

1 INTRODUCTION

Silicate-melt inclusions in tephra provide information on the composition and evolution of complex magma supply systems (Wallace et al., 1999; Kent, 2008). We sampled melt inclusions from different stages of Aso volcanic system, and try to find the melt composition of the magma reservoir before eruption. Selected samples for this study include tephras representing precursory event of caldera-forming eruptions, those of caldera-forming eruptions, and those of post-caldera eruptions.

Aso area of central Kyushu island has three contrasting stages of volcanism: (1) pre-caldera stage, (2) caldera-forming stage, and (3) post-caldera stage (Miyoshi et al., 2011; Miyoshi et al., 2013). (1) Pre-caldera volcanism is represented by occurrence of lava flows and lava domes, and by abundant tuff breccia. They occur mainly at Aso caldera walls or outside caldera area. The peak activities were between 400 ka and 800 ka (Furukawa et al., 2009; Miyoshi et al., 2009). (2) Caldera-forming stage is represented by four large-scale pyroclastic eruptions, called as Aso-1, Aso-2, Aso-3 and Aso-4 with K-Ar ages of 266 ka, 141 ka, 123 ka, and 89 ka, respectively (Ono, 1965; Watanabe, 1978, 1979; Matsumoto et al., 1991). The last Aso-4 caldera eruption was the largest, with the recent updated volume estimate of volcanic products exceeding 1,000 km^3 (Takarada and Hoshizumi, 2020). Large-scale pyroclastic eruptions were intervened by intra-caldera volcanism of lava flows and moderate explosive eruptions producing tephra. Precursory lava effusion event occurred before large-scale eruptions of Aso-1, Aso-2 and Aso-4 (Ono, 1965; Ono and Watanabe, 1969). (3) Post-caldera volcanism is a collection of cones, lava flows and a sequence of tephra to form central cone complex inside the caldera. The compositions of volcanic products show a wide variation from basalt to rhyolite with some alkaline rocks (Miyoshi et al., 2005).

Whether post-caldera stage (89 ka to the present) is a real post-caldera or an intra-caldera stage is a crucial question to be answered. Geophysical monitoring is one of the key methods to find the present conditions of crustal magma reservoirs which may lead to caldera-forming eruptions. Petrological approach is useful for estimating physical and chemical conditions of the past and present magma system. Comparison of the information of melts between caldera-forming eruptions and other minor eruptions will give us an important clue to detect the conditions of crustal magma reservoirs of different types of eruptions.

We review our melt inclusion data from different stages of volcanism from Aso region, including precursory activities to caldera-forming eruption, main caldera-forming eruption, and post-caldera eruptions. We will add several unpublished melt inclusion data to find the relationship between distinct types of eruptions and magmatic melts which existed in the shallow crust prior to eruptions.

2 ANALYTICAL METHODS

Separate grains of olivine, clinopyroxene, orthopyroxene, and plagioclase were prepared from pumice and scoria, representing precursory stage, caldera-forming stage and post-caldera stage of Aso volcano. Crystals containing melt inclusions were selected under binoculars. Several of them were then mounted on a microscope slide using Petro-Poxy cement and only one side was polished to the roughness of ~1 μm. They were then analyzed by JEOL JXA-8800R electron probe micro-analyzer of Earthquake Research Institute (ERI), University of Tokyo, and JEOL JSM-7001F FE-SEM EDS of Department of Geological and Environmental Science, Kumamoto University.

3 RESULTS

We obtained melt compositions including major elements, sulfur and chlorine from the electron probe analyses. Total of the major element content was recalculated as 100 % to compensate different analytical conditions. In this paper, only major element content of melt inclusions, especially SiO_2 and K_2O contents are examined to see the different stages of activities at Aso caldera volcano.

(1) Aso ABCD tephra

Aso A, Aso B, Aso C and Aso D tephras, the youngest ones among which erupted between Aso-3 and Aso-4 caldera-forming eruptions, are collectively called as Aso ABCD tephra here. They were defined by Ono et al. (1977) who showed that they originated from the southern part of the Aso caldera. The age is 98 ka, and the volume is 3.5 km^3 (Nagahashi et al., 2007; Machida and Arai, 2003).

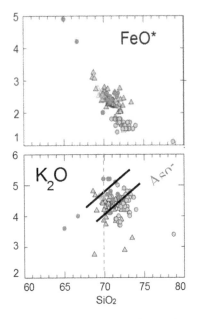
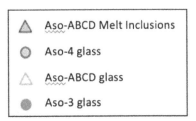

Figure 1. Compositions of melt inclusions in pyroxenes and plagioclase from Aso-A,C,D pumice. Compositions of volcanic glass from tephras of Aso-4, Aso-ABCD, and Aso-3 are also plotted for comparison. FeO*: Fe total as FeO.

Compositions of melt inclusions in plagioclase and pyroxenes of Aso ABCD tephra show some scatters (Sugiyama et al., 2017), but includes those of regional Aso ABCD tephra glass as reported by Nagahashi et al. (2007) (Figure 1). They are similar to those of Aso-3 glass (Kaneko et al., 2015), which shows distinctively high FeO*. The composition of Aso ABCD tephra are apparently different from those of Aso-4 glass (Kaneko et al., 2007), although they erupted 9,000 years before Aso-4 eruption.

(2) Omine scoria

Omine volcano erupted just before Aso-4 caldera-forming eruption, forming Omine scoria cone and associated Takayubaru lava flow plateau (Matsumoto et al., 1991; Watanabe and Ono, 1969). The age difference between Omine and Aso-4 is probably several hundreds of years. The volume of Omine cone and Takayubaru lava flow is 2.0 km^3 (Watanabe and Ono, 1969). The southern part of lava flow shows vertical offset of 100 m by Futagawa active fault.

The compositions of melt inclusions in plagioclase, orthopyroxene, clinopyroxne of Omine scoria are almost the same, and are plotted in a closely spaced area (Figure 2; Shiihara et al., 2017). They are similar to the groundmass glass in Omine scoria. A few exceptions of SiO_2-rich composition are found among groundmass glass and melt inclusions hosted by orthopyroxene.

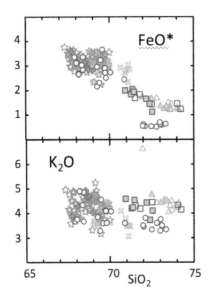
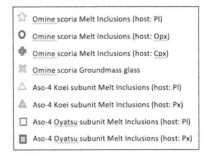

Figure 2. Compositions of melt inclusions in pyroxenes and plagioclase and groundmass glass from Omine scoria. Compositions of volcanic glass from pumice of Aso-4 Koei subunit and Oyatsu subunit are also plotted for comparison. Two populations of groundmass glass and Opx-hosted melt inclusion compositions are found.

(3) Aso-4 tephra: Koei and Oyatsu subunits

Watanabe (1978) divided Aso-4 pyroclastic flow deposits into eight stratigraphical subunits in the west of Aso caldera. We selected the earlier two subunits of Aso-4, i.e. Koei and Oyatsu for analyses. Melt inclusion compositions in Aso-4 pumice in pyroclastic flow deposit are plotted in Figure 2 in addition to Omine tephra. Accepting some scatter, they are plotted in a narrow region. Koei melt inclusions are more SiO_2-rich than Oyatsu melt inclusions, however, a few Oyatsu pumice glass show the same composition as Koei pumice (Yamasaki et al., 2015). Two population of pumice glass compositions are exactly as reported by Machida and Arai (2003). Omine and Aso-4 tephras obviously show different compositions, as were also confirmed by bulk composition plots (not shown here).

(4) Aso central cone pumice (ACP1, ACP2, ACP3 and ACP4)

Aso central cone pumice represents large-scale silicic eruptions among post-caldera activities. It includes ACP1 through ACP6 (Miyabuchi et al., 2004; Miyabuchi, 2009). We used ACP1 to ACP4 for our melt inclusion study. ACP2 is correlated with eruption of Kusasenri-ga-hama pumice cone at 30 ka, ACP3, Takano-obane rhyolite lava flow at 51 ka, ACP4, Tateno lava flow at 54 ka. ACP2 produced the largest volume of tephra amounting to 2.5 km^3 (Miyabuchi, 2009).

Pumice glass of ACP2, ACP3, and ACP4 shows narrow area in compositional plot (Figure 3; Nagaishi et al., 2019). Melt inclusions in pyroxenes and plagioclase of ACP4 and ACP3 show a wider compositional range than pumice glass counterpart. They both are more SiO_2-rich than the host pumice. ACP3 melt inclusions in pyroxenes show quite different composition from the host pumice. ACP1 is unique in that melt inclusion composition displays quite a large scatter (Brouille et al., 2018). Host minerals includes olivine, pyroxenes, plagioclase, and biotite. We discovered a scoria layer between ACP3 and ACP4 and name it as ACP3/4. Minerals of this ACP3/4 scoria have different compositions of melt inclusion depending on the host, i.e. olivine, pyroxenes, and plagioclase. Holocene basalt of Kishimadake, Kamikomezuka, and Nakadake show similar but slightly higher K_2O content (not plotted in Figure 3; Kawaguchi et al., 2021).

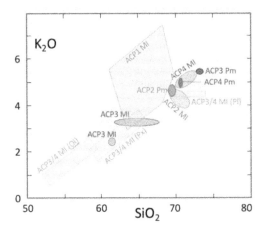

Figure 3. Compositions of melt inclusions and pumice glass from Aso Central Cone Pumice (ACP1, ACP2, ACP3, and ACP4). Compositions of melt inclusions in olivine, pyroxene, and plagioclase from newly discovered scoria (ACP3/4) are also plotted.

4 SUMMARY

General trend of melt inclusion composition at Aso volcano from 98 ka to the present indicate dis-appearance of relatively large and stable shallow magma reservoir. The last existence of stable shallow reservoir probably was at 30 ka when ACP2 or Kusasenri-ga-hama was active. The current image of magma reservoirs depicted by geophysical and petrological data does not indicate an existence of shallow large silicic magma body (Hata et al, 2016; Ushioda et al., 2020; Kawaguchi et al., 2021).

ACKNOWLEDGEMENTS

The first author is grateful to the organizing committee of the RMEGV2021 symposium for the honorable chance of giving keynote lecture. He especially thanks Prof. Ohta, the secretary general, for a thoughtful handling of the manuscript. We appreciate Ms. F. Sugiyama, Mr. K. Shiihara, Mr. H. Yamasaki, Mr. F. Brouille for kind permission to use their melt inclusion data. Part of this work was supported by JSPS Grants-in-Aid for Scientific Research awarded to T. Hasenaka (17K05682), joint research program of Earthquake Research Institute, University of Tokyo, and Integrated Program for Next Generation Volcano Research and Human Resource Development (Theme C-1: A. Yasuda, leader).

REFERENCES

Brouille, F., Hasenaka, T., Kawaguchi, M. and Nishiyama, T., 2018, Disequilibrium features found among Janoo scoria, Akamizu lava and ACP1 tephra, precursory event of Holocene basaltic volcanism of Aso Volcano, SW Japan, *Program and Abstracts Volcanological Society Japan, 2017 fall meeting*, P071.

Furukawa, K., Miyoshi, M., Shinmura, T., Shibata, T. and Arakawa, Y., 2009, Geology and petrology of the Pre-Aso volcanic rocks distributed in the NW wall of Aso caldera: the eruption and magma plumbing system of the pre-caldera volcanism, *J. Geological Society Japan*, 115, 658–671 (in Japanese with English abstract).

Hata, M., Takakura, S., Matsushima, N., Hashimoto, T. and Utsugi, M., 2016, Crustal magma pathway beneath Aso caldera inferred from three-dimensional electrical resistivity structure, *Geophys. Res. Lett.*, 43, doi:10.1002/ 2016GL070315.

Kaneko, K., Kamata, H., Koyaguchi, T., Yoshikawa, M. and Furukawa, K., 2007, Repeated large-scale eruptions from a single compositionally stratified magma chamber: An example from Aso volcano, Southwest Japan, *J. Volcanol. Geotherm. Res.*, 167, 160–180.

Kaneko, K., Koyaguchi, T., Yoshikawa, M., Shibata, T., Takahashi, T. and Furukawa, K., 2015, Magma plumbing system of the Aso-3 large pyroclastic eruption cycle at Aso volcano, Southwest Japan: Petrological constraint on the formation of a compositionally stratified magma chamber, *J. Volcanol. Geotherm. Res.*, 303, 41–58.

Kawaguchi, M., Hasenaka, T., Koga, K., Rose-Koga, E., Yasuda, A., Hokanishi, N., Mori, Y., Shimizu, K., Ushikubo, T., 2021, Persistent gas emission originating from a deep basaltic magma reservoir of an active volcano: the case of Aso volcano, Japan. *Contrib. Mineral. Petrol.*, 176, 6, 10.1007/s00410-020-01761-6.

Kent, A. J., 2008, Melt inclusions in basaltic and related volcanic rocks, *Reviews in Mineralogy and Geochemistry*, 69 (1): 273–331.

Machida, H. and Arai, F., 2003, Atlas of Tephra in and Around Japan, rev. ed. *University of Tokyo Press*, Tokyo, 336 p.

Matsumoto, A., Uto, K., Ono, K. and Watanabe, K., 1991, K–Ar age determinations for Aso volcanic rocks—concordance with volcanostratigraphy and application to pyroclastic flows, *Program and Abstracts Volcanol. Society Japan*, 2, 73 (in Japanese).

Miyabuchi, Y., Hoshizumi, H. and Watanabe, K., 2004, Late-Pleistocene tephrostratigraphy of Aso Volcano, southwestern Japan, after deposition of AT ash, *Bulletin Volcanol. Society Japan*, 49, 51–64.

Miyabuchi, Y., 2009, A 90,000-year tephrostratigraphic framework of Aso Volcano, Japan. *Sedimentary Geology*, 220, 169–189.

Miyoshi, M., Hasenaka, T., Sano, T., 2005, Genetic relationships of the compositionally diverse magmas from Aso post-caldera volcanism, *Bulletin Volcanol. Society Japan*, 50, 269–283 (in Japanese with English abstract).

Miyoshi, M., Furukawa, K., Shinmura, T., Shimono, M. and Hasenaka, T., 2009, Petrography and Whole-rock Geochemistry of Pre-Aso Lavas from the Caldera Wall of Aso Volcano, Central Kyusyu, *J. Geological Society Japan*, 115, 672–687 (in Japanese with English abstract).

Miyoshi, M., Shibata, T., Yoshikawa, M., Sano, T., Shinmura, T. and Hasenaka, T., 2011, Genetic relationship between post-caldera and caldera-forming magmas from Aso volcano, SW Japan: Constraints from Sr isotope and trace element compositions, *J. Mineralogical Petrological Sciences*, 106, 114–119.

Miyoshi, M., Miyabuchi, Y., Shinmura, T. and Sumino, H., 2013, Long-term volcanic activity in Aso area, central Kyushu, Japan, *Chikyu Monthly Special Issue*, No. 62: 168–176, (in Japanese).

Nagahashi, Y., Sato, T., Takeshita, Y., Tawara, T. and Kumon, F., 2007, Stratigraphy and chronology of widespread tephra beds intercalated in the TKN-2004 core sediment obtained from the Takano Formation, central Japan, *Quater. Res*, 46: 305–325.

Nagaishi, R., Hasenaka, T., Torii, M., Yasuda, A. and Hokanishi, N., 2019, Transition of magma plumbing system deduced from the melt inclusions contained in tephra minerals originated from Aso central cones, *Japan Geoscience Union Meeting 2019 Abstract*, SVC36–03.

Ono, K., 1965, Geology of the eastern part of Aso Caldera, central Kyushu, southwest Japan. *J. Geological Society Japan*, 71, 541–553 (in Japanese with English abstract).

Ono, K. and Watanabe, K., 1969, Geology of the vicinity of Omine on the western flank of the Aso caldera, *J. Geological Society Japan*, 75 (7), 365–374 (in Japanese with English abstract).

Ono, K., Matsumoto, Y., Miyahisa, M., Teraoka, Y. and Kanbe, N., 1977, Geology of the Taketa district. *Quadrangle series, scale 1: 50,000. Geol. Surv. Japan*, 145 p.

Shiihara, K., Hasenaka, T., Yasuda, A., Hokanishi, N. and Mori, Y., 2017, Omine volcano erupted just before Aso-4 pyroclastic flow –Transition of magma supply system inferred from composition of melt inclusions–, *Chikyu Monthly Special Issue*, No. 68, 80–85 (in Japanese).

Sugiyama, F., Hasenaka, T., Yasuda, A., Hokanishi, N. and Mori, Y., 2017, Melt inclusions in minerals from tephra between Aso-3 and Aso-4: estimation of magma plumbing system before caldera-forming eruption, *Chikyu Monthly Special Issue*, No. 68, 74–79, (in Japanese).

Takarada, S. and Hoshizumi, H., 2020, Distribution and Eruptive Volume of Aso-4 Pyroclastic Density Current and Tephra Fall Deposits, Japan: A M8 Super-Eruption. *Frontiers 8*, 170, 10.3389/feart.2020.00170.

Ushioda, M., Miyagi, I., Suzuki, T., Takahashi, E. and Hoshizumi, H., 2020, Preeruptive P-T conditions and H_2O concentration of the Aso-4 silicic end-member magma based on high-pressure experiments. *J. Geophys. Res.: Solid Earth*, 125: e2019JB018481, 10.1029/2019JB018481.

Wallace, P. J., Anderson Jr, A. T. and Davis, A. M., 1999, Gradients in H_2O, CO_2, and exsolved gas in a large-volume silicic magma system: Interpreting the record preserved in melt inclusions from the Bishop Tuff, *J. Geophys. Res.: Solid Earth*, 104 (B9), 20097–20122.

Watanabe, K., 1978, Studies on the Aso pyroclastic flow deposits in the region to the west of Aso caldera, southwest Japan. I: geology, *Mem. Fac. Educ, (Natural Science) Kumamoto Univ.*, No. 27, 97–120.

Watanabe, K., 1979, Studies on the Aso pyroclastic flow deposits in the region to the West of Aso caldera, southwest Japan, II: Petrology of the Aso-4 pyroclastic flow deposits, *Mem. Fac. Educ. (Natural Science) Kumamoto Univ.*, No. 28, 75–112.

Watanabe, K. and Ono, K., 1969, Geology of the vicinity of Omine on the western flank of the Aso caldera, *J. Geological Society Japan*, 75, 365–374 (in Japanese with English abstract).

Yamasaki, H., Hasenaka, T. and Yasuda, A., 2015, Zonal structure of Aso-4 magma reservoir as estimated from compositions of plagioclase and melt inclusions, *Japan Geoscience Union Meeting 2015 Abstract*, SVC47–06.

Rock Mechanics and Engineering Geology in Volcanic Fields – Ohta, Ito & Osada (eds)
© 2023 copyright the Author(s), ISBN 978-1-032-27657-1

Landform development process of a volcanic fan at the eastern foot of Mt. Sakurajima

Takahito Kuroki*
Kansai University, Osaka, Japan

Hiroaki Ikemi
Nippon Bunri University, Oita, Japan

Kensuke Goto
Osaka Kyoiku University, Osaka, Japan

Tatsuroh Soh
Shigakukan University, Kagoshima, Japan

ABSTRACT: This study aims to clarify the landform development process ongoing since 1946 at the eastern foot of Mt. Sakurajima. It is the most active volcano in Japan. Pyroclastic materials such as volcanic ash and rock, related to volcanic activities, are removed to downstream areas through debris flow or traction, forming a volcanic fan at the foot of the mountain. By analyzing aerial photographs, satellite images, and laser profiler (LP) data, we confirmed that a large number of sediments were deposited on the Showa lava. Based on a field survey, the characteristics of landforms and vegetation differences depending on them were confirmed. Because of partial erosion at the same time as sedimentation being predominant, the volcanic fan area can be classified into four categories of landforms in the descending order of relative elevation: residual hill, Terrace I, Terrace II, and transportation surface. This order is also the order of the emergence of landform surfaces. The volcanic fan area expansion and landform development process were interpreted using aerial orthophotographs. The landform distribution could be mapped based on the interpretation of the land cover classification using satellite images. The distribution of elevation change was mapped using the LP data for different measurement periods. Because the sedimentation effect was larger in the upstream area, the elevation of the volcanic fan gradually increased. Alternatively, because the sediment erosion effect was larger in the downstream area, earlier-emerging landforms were found. This development process was closely related to the amount of volcanic ash associated with volcanic activities. These landform features were very similar to ordinary dissected fans that take a long time to develop.

Keywords: Volcanic Fan, Landform, Aerial Photograph, Satellite Data, Laser Profiler Data

1   INTRODUCTION

Volcaniclastic materials, such as volcanic ash and lava, ejected by volcanic eruptions are removed to downstream areas through debris flow or traction, and a volcanic fan composed of these materials is formed at the foot of a mountain (Moriya, 1975) when the volcano is

*Corresponding author: kuroki0@kansai-u.ac.jp

DOI: 10.1201/9781003293590-2

active. The new landforms composed of these materials cover underlying deposits from debris avalanches, pyroclastic flows, and lava flows. However, when the volcano is dormant, the landform undergoes erosion and terraces are developed in downstream areas. These long-term landform changes have been studied using tephrochronology (Kuroki, 1995). Alternatively, extremely short-term landform changes have also been discussed based on the interpretation of aerial photographs (Tomiyama et al., 2011) and field measurements (Yamakoshi and Suwa, 2000; Iso and Kuroki, 2017). Recent progress in measurement technologies has facilitated the discussion of landform changes in volcanic fans at high spatial and temporal resolutions. For disaster prevention, the investigation of these extremely short-term landform changes for identifying safe areas at the foot of a volcano is essential. In this study, we investigated a volcanic fan forming by remarkable sedimentation at the eastern foot of Mt. Sakurajima, which is currently experiencing volcanic activities. Landform classification maps and elevation change maps were created using aerial photographs, satellite images, and laser profiler (LP) data, and the landform development process was explored.

2 METHODOLOGY

2.1 *Study area*

Mt. Sakurajima, located in the central part of Kagoshima Prefecture in western Japan, is an active volcano with a height of 1117 m. A volcanic fan in the Kurokami area for research is located at the eastern foot of the volcano (Figure 1). In this area, the Showa lava flowed from southwest to northeast in 1946.

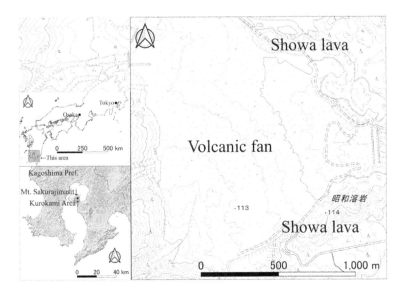

Figure 1. Study area.

2.2 *Elevation data*

The aerial photographs used in this study comprised data collected by the US military in 1947 and the Geospatial Information Authority of Japan in 1966, 1975, 1984, 1996, 2005, and 2016. We also captured aerial photographs using an unmanned aerial vehicle. These photographs were analyzed using the technique of structure from motion, and aerial orthophotographs and a digital elevation model (DEM) were created. The GCP (Ground Control Point) coordinate data acquired using a global navigation satellite system receiver were employed in the analysis. The LP data with a resolution of 1 m and obtained using the measurements performed by the Kyushu Regional Development Bureau, MLIT (the Ministry of Land, Infrastructure, Transport, and Tourism) in 2006, 2007, 2012, 2018, and 2019, were also used.

2.3 Landform classification map

Land cover classification maps were created using the Google Earth satellite images obtained in March 2012, February 2016, and January 2019 by employing the maximum likelihood classification tool of QGIS. In our field survey, the landform, geology, vegetation, and their relationships were investigated. The land cover categories were classified as the landform categories based on the field survey.

2.4 Discussion on landform development

Data on the amount of volcanic ashfall in the Kurokami area since 1978 were obtained from Kagoshima city. The annual and maximum daily precipitation data of this city since 1945 were obtained from the Japan Meteorological Agency. The elevation changes in some periods were estimated using the DEM developed using the aerial photographs and obtained from the LP data. We investigated the landform development process of the volcanic fan based on the landform classification, elevation changes, and ashfall and precipitation data.

3 RESULTS OF THE FIELD SURVEY AND ANALYSIS

3.1 Vegetation closely related to the landform classification

The volcanic fan area can be divided into four categories of landforms in the descending order of relative elevation, namely, residual hill, Terrace I, Terrace II, and transportation surface. The residual hill, covered with thriving pine trees, is an aggregate of huge lava boulders

Figure 2. Landscapes showing landform features.

(Figure 2(1)). Terrace I mainly comprises cobble- to pebble-sized sediments and exhibits undulating surfaces (Figure 2(2)). On this terrace, many Japanese pampas grasses grow and small pine trees are scattered. Moreover, almost no remnants of channels can be observed on the landform and the landscape shows that the landform is stable. Terrace II mainly comprises pebble- to sand-sized sediments and exhibits flat surfaces (Figure 2(3)). However, some shallow gullies with flat bottoms are observed in this landform. Small natural levees composed of gravels and trashes can be observed on the terrace adjacent to the gullies, indicating the occurrence of overflow from the gullies to the terrace and the progression of sedimentation during floods. This landform shows a scattering of Japanese pampas grass. Because the soil is captured by their roots and not easily eroded, the ground surface where the plants grow is slightly elevated than its surroundings. Based on the landscape, some parts of Terrace II show predominant weak lateral erosion and slightly decreased elevation. The transportation surface comprises boulder- to sand-sized sediments and exhibits flat surfaces (Figure 2(4)). Many sand or gravel bars and remnants of braided channels are distributed on this landform, and no vegetation can be observed. The landscape reflects predominant active sedimentation shaping the geomorphology. Considering the relationship between landform and vegetation, the descending order of relative elevation represents the order of the emergence of landform surfaces.

3.2 Observation of aerial orthophotographs

In 1947, the Showa lava was covered with sediments around the dissected valley exit (Figure 3 (1)). Since then, the sedimentation area has expanded over the foothill of the Showa lava. Many lava corrugations forming the regularly spaced folds observed at the surface, developed in the downstream area of the lava. In 1975, the upstream area of the Showa lava was mostly covered

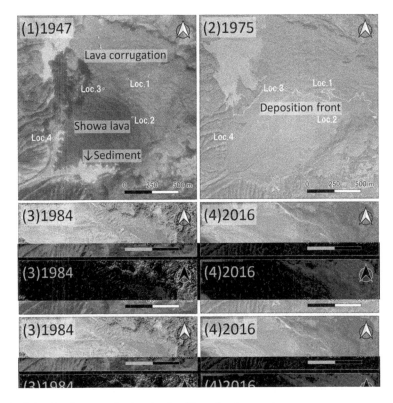

Figure 3. Aerial orthophotographs showing landform development.

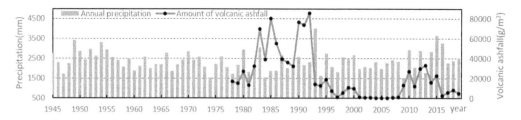

Figure 4.   Precipitation and volcanic ashfall.

with sediments, which intruded the recesses of the lava corrugations at the sedimentation front (Figure 3(2)). Figure 4 shows the annual precipitation and the amount of volcanic ashfall in each year. The average precipitation is about 2400 mm, and the fluctuation range is not large. In 1984, when the amount of ashfall increased, the lava area was almost entirely covered with sediments and a volcanic fan developed (Figure 3(3)). Since 1996, when the amount of ashfall was low, many lineaments indicating erosion can be identified based on the colors of the aerial photographs in the downstream area of the volcanic fan. In the colored photograph obtained in 2016 (Figure 3(4)), the difference in vegetation related to the landforms can be clearly observed. The development process indicates that the Showa lava forms the basement of the fan, sediments covering the basement are thick in the upstream area and thin in the downstream area, and the landform tends to emerge first in the downstream area where erosion is dominant. The residual hill is a portion of the Showa lava not covered by sediments. The terraces and transportation surface comprise sediments covering the lava. The terraces have become easier to interpret since around 2000, when the amount of ashfall decreased. On the transportation surface, sediment transportation is active and sedimentation is predominant.

3.3   *Landform classification based on land cover classification*

The land cover classification was performed with respect to the colors of the vegetation and surface geology in the satellite images. Based on the field survey, the land cover categories in 2019 were classified as the vegetation of residual hill, the vegetation of Terrace I, the vegetation of Terrace II, and the no vegetation of transportation surface. In the landform classification map (Figure 5), the residual hill is located at the edge of the distal fan, Terrace I mainly lies at the distal fan, Terrace II lies between the proximal fan and midfan, and transportation surface is throughout the volcanic fan area. The Terrace II and transportation surface landforms have elongated features oriented in the longitudinal direction of the volcanic fan area. This distribution fits well with the field survey results.

4   DISCUSSION OF THE LANDFORM DEVELOPMENT PROCESS

We divided the trends of elevation changes related to the volcanic fan development into four periods: Period P1 from 1947 to 1984, Period P2 from 1975 to 2016, Period 3 from 2006 to 2012, and Period 4 from 2012 to 2019. Based on Figure 3, in Period P1, no elevation changes were observed on the residual hill, slight elevation increments were observed on Terrace I, and large elevation increments were observed on Terrace II and transportation surface. Elevation changes were mapped using photographic measurements in Period P2 and via LP data analysis in Periods 3 and 4 (Figure 6). Table 1 summarizes the trends of elevation changes for each landform during the four periods. Considering the accuracy of measurement, changes from −1 to 1 m in Period P2 and those from −0.25 to 0.25 m in Periods 3 and 4 showed no elevation changes. In Period P2, no elevation changes were observed on in the residual hill, small elevation increments were observed on Terraces I and II, and large elevation increments were observed on the transportation surface (Figure 6(1)). In Period 3, no elevation changes were found on the residual hill and Terrace I, small elevation increments were found on

14

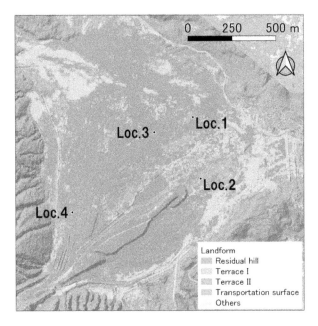

Figure 5.   Landform classification map based on an image captured in 2019.

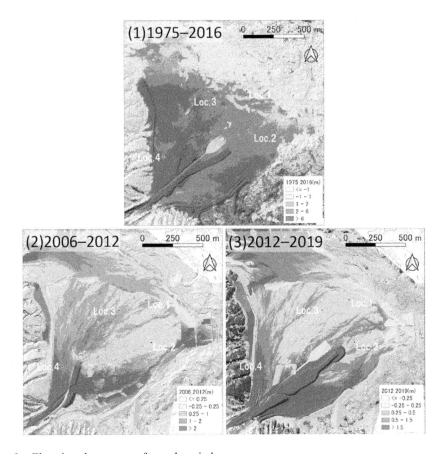

Figure 6.   Elevation change maps for each period.

Terrace II, and large elevation increments were found on the transportation surface (Figure 6 (2)). In Period 4, no elevation changes were observed on the residual hill and Terraces I and II, whereas large elevation increments occurred on the transportation surface (Figure 6(3)). These patterns of elevation changes suggest that Terrace I emerged before 2006 and Terrace II emerged before 2012.

Table 1. Landform development process for each period.

Periods (year_year)	Period P1 (1947_1984)	Period P2 (1975_2016)	Period 3 (2006_2012)	Period 4 (2012_2019)
Data	Photograph interpretation	Phtopographic measurements	LP data analysis	LP data analysis
Landforms	Trend of elevation change			
Residual hill	no change	no change	no change	no change
Terrace I	slight increment	small increment	no change	no change
Terrace II	large increment	small increment	small increment	no change
Transportation surface	large increment	large increment	large increment	large increment

5 CONCLUSIONS

The landform development process of the volcanic fan at the eastern foot of Mt. Sakurajima is discussed based on the results of a field survey and the analysis of aerial photographs, satellite images, and LP data. The primary results are summarized as follows.

1) The increasing volcanic fan area and landform development are associated with changes in the amount of volcanic ashfall.
2) The landforms are classified based on the descending order of relative elevation, namely, residual hill, Terrace I, Terrace II, and transportation surface, and this order is also the order of emergence of landform surfaces.
3) The residual hill comprises the Showa lava uncovered by sediments, and Terraces I and II and transportation surface comprise sediments covering the lava.
4) The landform classification can be performed based on the land cover classification using satellite images.
5) The residual hill is located at the edge of the distal fan, Terrace I is mainly located at the distal fan, Terrace II is located between the proximal fan and midfan, and the transportation surface is found throughout the volcanic fan area.
6) Based on the analysis of elevation changes, Terrace I is found to emerge before 2006, Terrace II is found to emerge before 2012, and the transportation surface with active sediment movements did not emerge at this time.

ACKNOWLEDGMENTS

We profoundly thank the staff of the Osumi Office of River and National Highway, Kyushu Regional Development Bureau, MLIT for the opportunity of using the LP data. This study was funded by the Japan Society for the Promotion of Science (Grant-in-Aid for Scientific Research (C) 18K01121).

REFERENCES

Moriya, I., 1975, Volcanic fan and stratified tuff breccia, *J. of the Hokkaido Komazawa Univ.*, 9 10, 107–126.

Kuroki, T., 1995, Volcanic-fan formation in the northern foot of Mt. Iwaki, Northeast Japan, *Quarterly Journal of Geography*, 47, 285–301.

Tomiyama, N., Koike, K., Oomura, M. and Iguchi, M., 2011, Analysis of topographic change at Mount Sakurajima, South Kyushu, Japan, using JERS-1 SAR interferometry, *Geoinformatics*, 22, 17–24.

Yamakoshi, T. and Suwa, H., 2000, Post-eruption characteristics of surface runoff and sediment discharge on the slopes of pyroclastic-flow deposits, Mount Unzen, Japan, *Trans. Jpn. Geomorphological Union*, 21, 469–497.

Iso, N. and Kuroki, T., 2017, Transportation of tephra fall deposits on a valley slope and topographic changes of a valley bottom after the 2011 Shinmoe-dake eruption in the Kirishima Volcanoes, *Trans. Jpn. Geomorphological Union*, 38, 27–40.

Rock Mechanics and Engineering Geology in Volcanic Fields – Ohta, Ito & Osada (eds)
© 2023 copyright the Author(s), ISBN 978-1-032-27657-1

Tephra from large eruption in Kyushu area observed in UT-iwk core from Uwa basin, western Shikoku

Masashi Ushioda*
Shikoku Research Institute Inc., Takamatsu, Japan

Chisato Nakamura
NAIBA Co. Ltd., Takamatsu, Japan

Tomohiro Tsuji
Yamaguchi University, Yamaguchi, Japan

Michiharu Ikeda
Shikoku Research Institute Inc., Takamatsu, Japan

Kozo Ohnishi & Naoki Nishizaka
Shikoku Electric Power Co. Ltd., Takamatsu, Japan

ABSTRACT: Large eruptions from caldera volcanoes in the Kyushu region during the Quaternary periods caused large amount of tephra over a wide area of the Japanese archipelago. Because the Shikoku region is close to the Kyushu region, many tephras were deposited relatively thickly, and it is important to identify these tephras and measure tephra thickness for disaster prevention. The UT core (Tsuji et al. 2018), collected in the Uwa Basin, western Ehime Prefecture, Japan, preserves tephra over the past 600,000 years, and Uwa basin is a good field for comparison of tephra thickness and magma storage conditions. Using UT-iwk core collected from 500 m north of the UT core site, we attempted to determine the precise layer thickness of tephra derived from the large eruption in Shikoku region, identify cryptotephra, and understand the magma storage conditions based on the glass and mineral chemical compositions for the visible tephra. In the UT-iwk core, visible widespread tephras are K-Ah, AT, Aso-4, K-Tz, Aso-ABCD, Ata, Aso-2, Ata-Th, Aso-1, Kkt, Oda resemble. For example, net thicknesses of Aso-1 and Kkt are about 15 cm and 20 cm, respectively, based on grain-size and component analyses. AT is 65 cm thick including the reworked layer. The net thickness of Aso-4 was estimated to be 20 to 30 cm (Nakamura et al. 2020). These results are expected to be useful for the estimation of the volume of eruptive products and the probability evaluation of the impact of a large eruption in western Shikoku.

Keywords: Thickness of widespread tephra, borehole core, Uwa basin, Aso, Kuju

1 INTRODUCTION

Volcanic ash from huge caldera-forming eruptions is able to be transported far from the source volcano and affect human society. In order to assess the probabilistic impact of volcanic ash for human society, it is essential to obtain actual data on the number and thickness of volcanic ash

*Corresponding author: m-ushioda@ssken.co.jp

DOI: 10.1201/9781003293590-3

layers from geological surveys. In particular, Kyushu region in Japan has experienced a number of large eruptions in the Quaternary period which are revealed by detailed geological surveys. If a huge eruption occurs in Kyushu, the damage caused by volcanic ash in the downwind Shikoku region is enormous. Major VEI (Volcanic Explosive Index) 7 and over VEI 7 eruptions on Kyushu region in the last 400 ka are listed as follows: K-Ah and K-Tz (Kikai caldera), AT (Aira caldera), Aso-4, 3, 2, and 1 (Aso caldera), Ata and Ata-Th (Ata caldera) and Kkt (Kakuto caldera). These tephra are widely distributed in the Japanese archipelago. On the other hand, eruptions of VEI 6 or less are also important because tephra may be transported far from Kyushu depending on the eruptive scale and wind direction. The tephrostratigraphy in the Uwa Basin, western Shikoku, has been established by Tsuji et al. (2018), but problems of tephra thickness evaluation and identification of trace tephra (cryptotephra) remain. The purpose of this study is to accumulate basic data for the probabilistic assessment of the impact of volcanic eruptions by correlation between visible tephra and widespread known tephra and evaluation of the tephra thickness using borehole cores as continuous samples for the past 600,000 years in western Shikoku area.

2 METHODOLOGY

2.1 UT-iwk and UT core

The UT-iwk core which was collected in the Uwa Basin (Figure 1), western Ehime Prefecture, Japan, was used to determine the thicknesses of tephra layers from a huge eruption in the western Shikoku region. One visible tephra (Aso-4) of the UT core (Tsuji et al. 2018) was also used for a comparison. The UT-iwk core was collected at about 500 m north of the UT core collection point in the Uwa Basin. Length of this core is 95 m long with penetrating the sedimentary layers to the basement rock. The UT core stores tephras from K-Ah to Yfg (600ka) as visibility tephra. In this study, we mainly used the newly drilled UT-iwk core to improve the accuracy of thickness evaluation of tephra and to identify the cryptotephras. Products

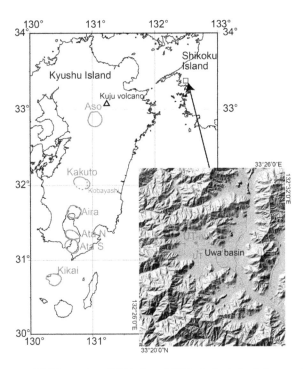

Figure 1. Locality map of drilling points (UT-iwk and UT). A shaded relief map released by Geospatial Information Authority of Japan was used.

from Kuju and Aira volcanoes, which are candidates for tephra correlation, were also collected near the volcanoes. Handa pyroclastic flow (53.52±1.14 ka: Okuno et al. 2013), Kj-P1, Shimosakata pyroclastic flow (110±40 ka: Kamata et al. 1998), and Miyagi pyroclastic flow (150±40 ka: Kamata et al. 1998) erupted from Kuju volcano were collected from Shirani, Ariuji, Shimosakata, and Kariono in Taketa city, Oita Prefecture, respectively. Pre-AT samples (Kenashino ash, Fukaminato pumice fall, Otsuka pumice fall (31, 32~31, 33 ka, respectively: Geshi et al. 2020), and Iwato pyroclastic flow (60 ka: Nagaoka et al. 2001)) were also collected at Hareyamabaru, Kirishima city, Kagoshima Prefecture.

2.2 *Analytical procedures*

Tephra samples of UT-iwk core were used for the chemical composition analysis of volcanic glass for comparison of visible tephra, observation of lithology, component analysis, and grain size analysis. X-ray CT analysis of the UT-iwk core samples was also performed, and the relative change in density of the core samples was evaluated from the CT number. The CT number was measured at the center line of the core sample. The chemical composition of the major elements was analyzed using a scanning electron microscope and energy dispersive X-ray spectrometer installed at Furusawa Geological Survey Co., Ltd. Grain size analysis for visible tephra in UT-iwk core and part of UT core was performed by dry sieving. In the UT-iwk core, tephra-derived particles were collected from the tephra-mixed layer, and chemical compositions of volcanic glasses and amphibole crystals were analyzed to compare with the known tephra reported in previous studies.

Table 1. Visible tephra of UT-iwk core. "Gross thickness" means summation of net and reworked tephra thicknesses.

	UT-iwk depth (m)	gross thickness (cm)
K-Ah	1.940-2.055	11.5
AT(including Osumi pumice)	2.44-3.09	65
Aso-4	8.17-8.565	39.5
K-Tz	8.79-8.95	16
Aso-ABCD	9.77-9.87	10
Ata	9.88-10.49	61
Aso-2	14.915-15.09	17.5
Ata-Th	28.71-29.25	54
Aso-1 resemble	34.65-35.35	70
Aso-1	37.115-37.375	26
Kkt	47.340-47.545	20.5
Oda resemble	54.185-54.53	34.5

3 CORRELATION OF VISIBLE TEPHRA IN UT-IWK CORE WITH WIDESPREAD KNOWN TEPHRA

Based on the stratigraphic relationships, component ratio, and chemical compositions of the core samples, the visible tephra shown in Table 1 were correlated with the widespread tephra from Kyushu region. In Table 1, the summation of net and reworked tephra thickness is shown as the "gross" thickness. The net layer thickness is under analysis, but it is not significantly different from the net layer thickness of the UT core (Tsuji et al. 2018). In the UT core, the visible tephras were correlated with Aso-3, Tky-Ng1 (Takayama Ng1), Hwk (Hiwaki), Yfg (Yufugawa), and correlated tephras of UT-iwk (Tsuji et al. 2018).

4 RESULTS OF GRAIN SIZE ANALYSIS FOR VISIBLE TEPHRA

Figure 2 shows an example of the results of the grain size analysis by dry sieving (Aso-4) in order to qualify the net layer of visible tephra. The gross tephra layer of Aso-4 is a range of 8.17-8.565 m in the UT-iwk core and a net layer of Aso-4 in the UT core is a range of 6.90-7.21 m (Tsuji et al. 2018). Overall, the UT core tends to be richer in fine-grained components than the UT-iwk core. The difference in the fine-grained component between the UT and UT-iwk core is comparable to the difference in the "other" (accretional material) component. The tephra layers in the lower part of Aso-4 of both the UT and UT-iwk cores are crystal-rich, and these grain sizes are clearly larger than those in upper layer. The Aso-4 layers of UT and UT-iwk tends to become finer with decreasing these depths. Amounts of fine-grained particles below the silt level increases at 8.30-8.32m. At the depth of 8.26-8.28 m, Aso-4 of the UT-iwk core shows contamination of altered rock fragments. Even in the UT core, the fine grain component increases at 6.96-6.98 m, which is the upper part of Aso-4. The amounts of fine particles below silt level in Aso-4 layers are different between UT-iwk and UT cores, but the other components are similar. Therefore, most of Aso-4 layers both UT-iwk and UT core are possible to show primary lithology.

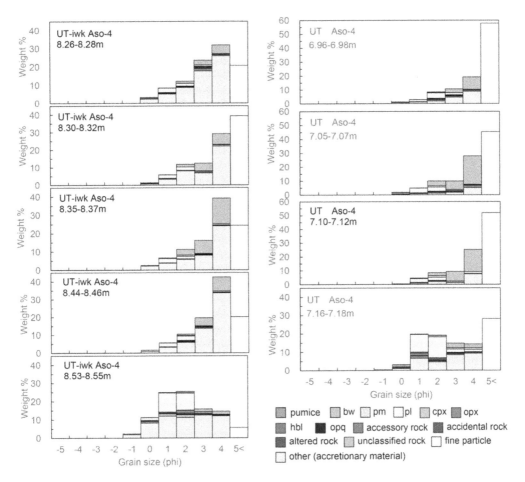

Figure 2. Grain size distribution with dry sieving of Aso-4 in UT-iwk and UT core. pumice: pumice fragment, bw: bubble wall type glass, pm: pumice type glass, pl: plagioclase, cpx: clinopyroxene, opx: orthopyroxene, hbl: hornblende, opq: opaque mineral.

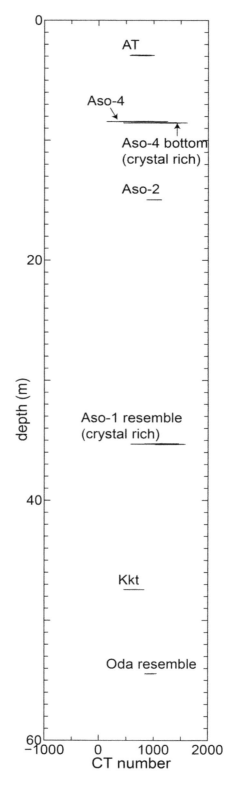

Figure 3. Relation between Collection depth and CT numbers of undisturbed visible tephra samples. Measurement ranges of CT number are listed in Table 2. Vertical measurement interval is 0.5mm.

5 CT NUMBER OF VISIBLE TEPHRA IN UT-IWK CORE

For the visible tephra layer of the UT-iwk core, we selected some points without core turbulence, muddy water contamination, and voids from CT photographs and observation of half section. Table 2 and Figure 3 show the results of measuring the CT numbers of the core centers of such selected layers. The CT numbers of the visible tephra were obtained for AT, Aso-4, Aso-2, Aso-1 resemble, Kkt, resemble Oda. Because the components and grain size of Aso-4 layer are clearly different between the bottom and the other part from Figure 2, the CT numbers were measured for the bottom of Aso-4 and the central part which mainly contains volcanic glass. Kkt had the smallest CT number, AT, Aso-4, Aso-2, and Oda resemble had intermediate CT numbers, and Aso-4 bottom and Aso-1 resemble had high CT numbers. These may be derived from the fact that Kkt has the smallest central grain size among these tephras and is mostly composed of volcanic glass, while Aso-4 bottom and Aso-1 resemble are rich in crystals, and the others have intermediate characteristics. Changes in CT values are largely reflected in the grain size and type of components. Systematic changes such as a monotone increasing in CT number (density) with increasing depth are not observed from this data.

6 IDENTIFICATION OF CRYPTOTEPHRA

An example of a tephra-mixed sample (UT-iwk 5.45-5.465m) from between AT and Aso-4 is described in this chapter. This sample has a tephra particle (volcanic glass and amphibole crystals) in a sand layer sandwiched in clay and silt layers. Results of major element measurements for volcanic glass and amphibole crystals are shown in Figure 4. The composition of this volcanic glass is similar to the composition of proximal products (AT and pre-AT) from Aira volcano. In this study, the following proximal products of pre-AT were used for correlations: Kenashino ash, Fukaminato pumice fall, Otsuka pumice fall, and Iwato pyroclastic flow deposits (Nagaoka et al., 2001). Comparing the volcanic glass composition of AT and pre-AT products, K_2O contents of volcanic glasses of pre-AT tend to be slightly higher than those of AT, while the volcanic glass of cryptotephra (UT-iwk 5.45-5.465m) has an intermediate composition. Therefore, we are not able to correlate volcanic glass at 5.45-5.465m to specific tephra of Aira products using volcanic glass compositions. The amphibole composition of cryptotephra at 5.45-5.465m is similar to that of Kuju volcano, not to that of Aso-4 or Aira (lowest part of AT). In the composition of amphibole, a detailed correlation to the specific tephra (Handa pyroclastic flow, Kj-P1, Shimosakata pyroclastic flow and Miyagi pyroclastic flow) from Kuju volcano has not been clear. Based on the stratigraphic relationship of this cryptotephra (5.45-5.465m) sandwiched between AT and Aso4, these amphibole crystals were correlated to Kj-P1 or Handa from Kuju volcano, and these volcanic glasses are correlated to those of pre-AT products (Kenashino. Fukaminato, Otsuka, Iwato). Because this cryptotephra is sandwiched in continuous sediments between thick AT and Aso-4 layers, the primary thickness of this cryptotephra may be thin, relatively. It is expected that these results help us draw an isopach map. In a borehole core sample (MD012422) which was dirilled in the Pacific Ocean off Shikoku Island, the composition of volcanic glass in a layer correlated to that from the Kuju volcano (Shimosakata pyroclastic flow or Miyagi pyroclastic flow)

Table 2. Collection depth and CT numbers of undisturbed visible tephra samples.

depth(m)	tephra	Average CT number
2.90-2.95	AT	872
8.42-8.45	Aso-4	781
8.523-8.565	Aso-4 bottom	1131
14.95-14.96	Aso-2	1070
35.26-35.34	Aso-1 resemble	1215
47.42-47.43	Kkt	648
54.44-54.46	Oda resemble	949

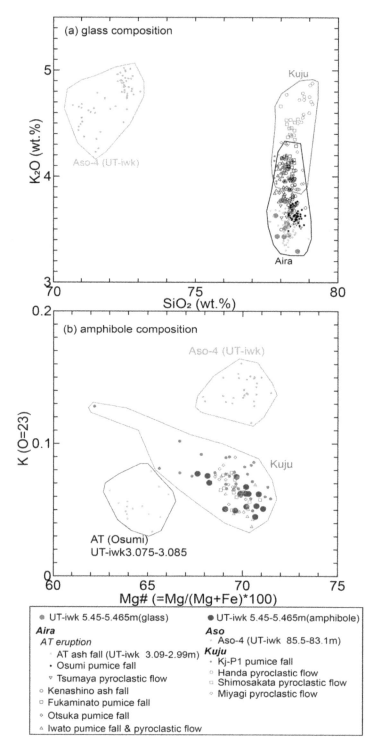

Figure 4. Comparison of volcanic glass and amphibole composition between UT-iwk 5.45-5.465m sample and proximal volcanic products.

(Matsu'ura et al. 2021). For pre-AT tephras from Aira volcano, the following tephras were observed in the MD012422 core: the Kenashino, Fukaminato (?), Iwato, Fukuyama tephras (Ikehara et al. 2006; Matsu'ura et al. 2021). While the depositional environment of MD012422 marine core differs from that of the Uwa Basin core in this study, tephras of VEI 6 or less from Kyushu region that reached Shikoku island and off Shikoku was discovered. In order to evaluate the probabilistic impact of volcanic eruptions, it is necessary to examine the frequency of incoming tephra, including relatively small eruptions.

In this study, we revealed that the amphibole in the UT-iwk 5.45-5.465m is originated from Kuju volcano. Variations in the major element composition of silicic volcanic glass samples from proximal Kuju volcano products show that glasses from Handa pyroclastic flow are richest in K_2O, next is that of Simosakata pyroclastic flow, and then the Miyagi pyroclastic flow and Kj-P1 (Figure 4). Therefore, it is possible to make a correlation by using not only amphibole but also volcanic glass composition. The chemical composition of glass inclusions in amphibole may be useful to correlate a tephra (e.g. tephra from Daisen volcano: Furusawa 2008).

7   CONCLUSIONS

We identified 12 visible tephras deposited in the Uwa Basin by lithological observation, chemical composition analysis, and grain size analysis of UT-iwk core samples. We also found pre-AT tephra and tephra from Kuju volcano by chemical composition analysis of cryptotephras. These results are expected to be useful for the probabilistic impact assessment of the volcanic eruptions from Kyushu in western Shikoku.

ACKNOWLEDGEMENTS

Dr. Tatsuyuki Ueno helped us a lot with outcrop information and outcrop observation during the geological survey. Mr. Hideto Uchida analyzed the X-ray CT scan data for us.

REFERENCES

Furusawa, A., 2008, Characterizing tephras by major-element analysis of glass inclusions in plagioclase phenocrysts: Preliminary results from the DKP tephra of Daisen Volcano, Japan, *J. Geol. Soc. Japan*, 114(12). 618–631.

Geshi, N., Yamada, I., Matsumoto, K., Nishihara, A. and Miyagi, I., 2020, Accumulation of rhyolite magma and triggers for a caldera-forming eruption of the Aira Caldera, Japan, *Bull. Volcanol*, 82, 44.

Ikehara, M., Murayama, M., Tadai, O., Hokanishi, N., Daido, N., Kawahata, H. and Yasuda, H., 2006, Late Quaternary tephrostratigraphy of two IMAGES cores taken from the off Shikoku in the Northwest Pacific, *Fossils*, 79, 60–76.

Kamata, H., Danhara, T., Itoh, J., Hoshizumi, H. and Kawanabe, Y., 1998, Fission-track Ages of Zircons in the Miyagi, Shimosakata and Handa Pyroclastic-Flow Deposits Erupted from Kuju Volcano in Central Kyushu, Japan, Bull. Volcanol. Soc. Jpn, 43(2), 69–73.

Matsu'ura, T., Ikehara, M. and Ueno, T., 2021, Late Quaternary tephrostratigraphy and cryptotephrostratigraphy of core MD012422: Improving marine tephrostratigraphy of the NW Pacific, *Quat. Sci. Rev.*, 257, 106808

Nagaoka, S., Okuno, M. and Arai, F., 2001, Tephrostratigraphy and eruptive history of the Aira caldera volcano during 100-30 ka, Kyushu, Japan, *J. Geol, Soc. Japan.*, 107(7), 432–450.

Nakamura, C., Tsuji, T., Ikeda, M., Nishizaka, N., Ohnishi, K. and Sakakibara, M., 2020, Validity of thickness of Aso-4 tephra fall deposit of Uwa Basin, western Shikoku, *JpGU-AGU joint Meeting 2020*, SVC-46, P03.

Okuno, M., Nagaoka, S., Saito-Kokubo, Y., Nakamura, T. and Kobayashi, T., 2017, AMS Radiocarbon Dating of Pyroclastic-Flows Deposits on the Southern Slope of Kuju Volcanic Group, Kyushu, Japan, *Radiocarbon*, 59(2), 483–488.

Tsuji, T., Ikeda, M., Furusawa, A., Nakamura, C., Ichikawa, K., Yanagida, M., Nishizaka, N., Ohnishi, K. and Ohno, Y., 2018, High resolution record of Quaternary explosive volcanism recorded in fluvio-lacustrine sediments of the Uwa basin, southwest Japan, *Quat. Int.*, 471(B), 278–297.

Rock Mechanics and Engineering Geology in Volcanic Fields – Ohta, Ito & Osada (eds)
© 2023 copyright the Author(s), ISBN 978-1-032-27657-1

Ash transport and deposition from 2019-2020 eruption of Aso volcano, Japan

Tomohiro Tsuji*
Graduate School of Science and Technology for Innovation, Yamaguchi University, Yamauchi, Japan

Karin Suzuki
SYSTEM'S Co., Ltd, Ibaraki, Japan

ABSTRACT: Tephra fallout from the eruption causes social damages (electric power facilities, agricultural products, health, aircraft, etc.), environmental impact (air pollution, temperature decreasing, etc.) and secondary disasters (removing by wind, lahar, etc.). To prevent and mitigate these disasters, it is necessary to understand the tephra transport and deposition (TTD) process. However, TTD process is still on debate despite more than four decade of research due to various complications (diversity of eruptions, uncertainty of eruption parameters, aggregation, etc.). Here, in order to discuss tephra transport and deposition process from the plumes on Dec. 15, 2019 (plume Dec15) and on Feb. 21, 2020 (plume Feb21) eruption of Aso volcano (Japan), we conducted the observation of eruption, direct sampling of ash particles within 3 km from the vent, as well as grain-size analyses, SEM observation and tephra simulation analyses using Tephra2. The white-gray colored plume Dec15 gently rose up to 1300 m above crater rim and horizontally flowed to outside of the caldera rim by wind, releasing aggregated ash. The gray colored plume Feb21 gently rose diagonally up to 800 m and started advection not horizontally but descended gently, releasing non-aggregated ash almost within the caldera. The aggregated particles were 100–250 μm in diameters and consist of mainly 5–7 ϕ particles. The aggregated fine particles gained terminal velocities up to several orders of magnitude greater than their individual fall velocities. The deposits include 26.8 wt.% of aggregated particles that is notable and affect to the tephra transport and deposition. Based on the observation, grain-size analyses and Tephra2 simulation, the fine particles segregated from lower plume at early stage and spreading current at Dec. 15, 2019. The spreading plume Feb21 subsided gently possibly due to the depletion of fine particles. The production of fine particles attribute to segregation from plume at early stage, aggregation within plume and ash dispersal to distal area.

Keywords: Ash transport, Weak plume, Segregation, Aggregation, Aso volcano

1   INTRODUCTION

Tephra fallout from the eruption causes social damages, environmental impact (air pollution, temperature decreasing, etc.) and secondary disasters (removing by wind, lahar, etc.). To prevent and mitigate these disasters, it is necessary to understand the tephra transport and deposition (TTD) process. However, TTD process is still on debate despite more than four decade of research due to various complications (diversity of eruptions, uncertainty of eruption parameters, aggregation,

*Corresponding author: t-tsuji@yamaguchi-u.ac.jp

DOI: 10.1201/9781003293590-4

etc.). Here, in order to discuss tephra transport and deposition process from the 2019-2020 eruption of Aso volcano (Japan), we conducted the observation of eruption, direct sampling of ash particles within 3 km from the vent, as well as grain-size analyses, SEM observation and tephra simulation analyses using Tephra2 that is a tephra diffusion-advection equation model.

2 GEOLOGIC SETTING AND OUTLINE OF ERUPTION

Japan's 24-km-wide Aso caldera in Kyushu Island has been active since its first caldera-forming eruption (270 ka; Japan Meteorological Agency (JMA) and Volcanological Society of Japan (VSJ), 2013). Nakadake is an active basaltic to basaltic-andesitic stratocone and has been most active of the 17 central cones during the past 2000 years (Global Volcanism Program, 2013). Its main eruption style has been continuous ash emission, which is referred as "ash emission" (Ono et al., 1995). Only First crater (its rim is around 1280 m in altitude) in Nakadake has been active during the last 80 years (JMA and VSJ, 2013). The first eruption began at First crater at April 16, 2019 and long-lasting ash eruption from July 26, 2019 to June 2020.

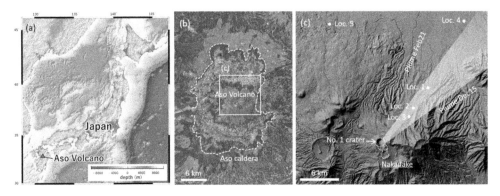

Figure 1. Location of Aso volcano (a, b) and sampling localities (c). The traces of plume Dec15 and plume Feb21 are also shown. Base images (b, c) are from GSI Maps.

3 METHODOLOGY

3.1 Observation of eruptions and direct sampling of fallen ash

We observed the eruptions at Nov. 30, 2019, Dec. 15, 2019 and Feb. 21, 2020. The crater lake was not detected on the days (Fukuoka Regional Volcanic Observation and Warning Center, 2020). The plume observed at Dec. 15, 2019 (plume Dec15) was blown by wind at ground level (2.6 m s^{-1} from the south at ~1,100 m in altitude) and wind at higher altitude (2.3–6.4 m s^{-1} from SW–WSW at ~1500–3000 m in altitude) around Mt. Aso (Figure 1c). We carried out direct sampling of falling ash using umbrella at the downwind region of 1.9 km (loc. 3), 2.3 km (loc. 2), and 3.5 km (loc. 1), 6.2 km (loc. 5) and 7.2 km (loc. 4) from the vent (Table1, Figure 1c).

3.2 Microscopic observation and grain size analysis

Ash samples were dried in an oven at 60 °C for 1 day before being dry-sieved down to 4 ϕ ($\phi = -\log 2D$, where D is the particle diameter in millimeters) at 1 ϕ intervals. Wet-sieved ash fractions finer than were analyzed down to 13 ϕ using a laser scattering particle size distribution analyzer (HORIBA LA-300) at Yamaguchi Prefectural Industrial Technology Institute, assuming refractive index of 1.50 and dispersed with water. To avoid redundancy, we herein refer to the combination of wet-sieving and laser diffraction analyses as "wet-sieving". We also performed microscopic observation of dry- and wet-sieved samples and component analysis of wet-sieved samples at 1.9 km from the crater collected at Dec. 15, 2019 and Feb. 21, 2020. We identified particle clusters (ash clusters, PC1, and coated particles, PC2; Brown et al., 2012) and poorly structured pellets, AP1; Brown et al., 2012) using light microscopy, and observed AP1 using scanning electron microscopy (SEM).

3.3 *Simulation for tephra dispersal*

To discuss the influence of segregated ash particles from lower part of plume Dec15 on the tephra distribution, we carried out two cases of tephra dispersal simulations, using Tephra2 that is a tephra diffusion-advection equation model (Bonadonna et al., 2005). In case-1, plume ratio (= height of plume bottom / height of plume top) is 0.42 that assumes the ash particles do not segregated from lower part but only from upper part of the plume. In case-2, plume ratio is 0 that assumes the particles segregated from the lower to upper part of the plume. Other parameters are shown in Table 2.

Table 1.   Data of sampling, depositional rate and grain size.

Sample name	Locality	Latitude	Longitude	Distance (direction) from vent (km)	Sampling time (min)	Depositional rate (g m^{-2} min^{-1})	Dry-sieving Mdϕ	$\sigma\phi$	Wet-sieving Mdϕ	$\sigma\phi$
Dec15-1	loc. 1	131.110208	32.908646	3.55 (NE)	10:11~13:31 (200)	9.15E-02	3.76	0.70	4.24	1.32
Dec15-2	loc. 2	131.102780	32.898725	2.27 (NE)	10:23~12:59 (156)	5.07E-03	3.52	0.84	4.03	0.89
Dec15-3	loc. 3	131.100268	32.896054	1.90 (NE)	10:55~11:59 (64)	1.08E-03	2.85	1.40	3.10	1.24
Feb21-1	loc. 4	131.059226	32.935323	6.23 (NNW)	14:12~17:10 (178)	1.63E-04	3.47	0.81	-	-
Feb21-2	loc. 5	131.129825	32.936638	7.16 (NE)	15:03~6:00 (897)	-	-	-	3.65	1.07
Feb21-3	loc. 2	131.102780	32.898725	2.27 (NE)	15:31~18:26 (175)	3.15E-03	2.80	0.69	2.86	0.81
Feb21-4	loc. 3	131.100268	32.896054	1.90 (NE)	15:57~18:04 (127)	3.60E-03	2.78	0.71	2.84	0.77
Feb21-5	loc. 2	131.102706	32.898691	2.27 (NE)	16:33~18:27 (114)	3.73E-03	2.97	0.72	3.05	0.74

Table 2.   Input parameters for the simulation.

Cases	Plume ratio	Plume height (m)	Vent elevation (m)	Eruption mass (m^3)	Grain size distribution Mdϕ	$\sigma\phi$	DC (m^2 s^{-1})	FTT (s)	Density (kg m^{-3}) Lithic	Pumice	Wind direction	Wind speed (m s^{-1})
Case-1	0.42	2480	1280	300000	3.56	1.06	200	200	2600	1000	SW	2~10*
Case-2	0	2480	1280	300000	3.56	1.06	200	200	2600	1000	SW	2~10*
Sources	Plume Dec15		GSI Maps	Comparison with measured mass	Average of dry-sieving of Dec15-2		Mannen (2013)		Bonadonna et al. (2005)			Comparison with the actual distribution

* Wind speed increases from 2 m s^{-1} at altitude of 600 m to 10 m s^{-1} at 1900 m.

4   RESULTS

4.1   *Eruption plume*

Gray ash emitted at Nov. 30 and Dec. 15, 2019. The white gray, probably water-dominant eruption plume at Dec. 15, 2019 (plume Dec15) rose and overshot up to 1,200 m above crater rim and started advection horizontally flowed to NE outside the caldera rim by the wind from SW (Figure 2a, b). The top and bottom heights of the plume were 900 m and 500 m above crater rim. Very weak gray-colored plume Feb. 21, 2020 (plume Feb21) rose diagonally up to 800 m and started advection not horizontally but descending gradually to NNE direction (Figure 2c). Most of volcanic ash fallen within Aso caldera, indicating that deposited within approximately 10 km. The crater lake does not existed at Dec. 15, 2019 and Feb. 21, 2020 (JMA, 2020).

We observed slightly deposited very fine ash particles, and aggregated particles dropped on the umbrella from the plume Dec15 (Figure 2b). Some aggregates were broken during impact (Figure 2b).

4.2   *Ash sample and composition*

Sample from plume Dec15 were gray-colored and mostly consists of particles coated by fine ash particles (PC2). Ash clusters (PC1) and subspherical aggregates (AP1) were included (Figure 3a). In contrast, the sample from plume Feb21 were black-colored and not coated by fine ash particles (Figure 3b). Washed sample at loc. 2 from plume Dec15 and Feb21 are rich in lithic fragments (Figure 3c, d, e, f). The sample at loc. 2 from plume Feb21 includes much more fresh glass shards (8 wt.%) than that from plume Dec15 (2 wt.%) (Figure 3e, f).

4.3   *Grain-size between plumes*

The dry-sieved results represent the grain-size distribution (GSDs) of particles that fell on the umbrella, although some particles were broken or disrupted during impact (Figure 2b), sampling and sieving (Tsuji et al., 2020). Since wet-sieving can disrupt aggregated particles, the wet-sieved results represent the initial GSDs of tephra within the plume (before aggregation and deposition).

At loc. 2, 2.3 km NE from the vent, grain-size distributions from plume Dec15 indicate a peak shifted from 3–4 ϕ for dry-sieved to 4–5 ϕ for wet-sieved sample (Figure 3g). Fine ash accounted for only 24.5 wt.% of the dry-sieved sample and 51.3 wt.% of the wet-sieved sample. In contrast, the dry- and wet-sieved results from plume Feb21 are similar indicating a peak at 2–3 ϕ and well-sorted.

4.4   *Aggregates*

The subspherical ash pellets (AP1) included in sample from plume Dec15 were 100–250 μm in diameters and fragile that they could be broken by a drop of water. Based on SEM observation, they were poorly structured and consist of mainly 5–7 ϕ particles (Figure 4).

4.5   *Simulation*

As a result of simulation, ash dispersed to NE direction from Nakadake first crater that is consistent with the observation (Figure 5). In case-1, ash particles did not deposited at loc. 3 (Figure 5). Only 1–2 ϕ particles deposited and finer particles did not deposited at loc. 1 and 2 (Table 3) that is not consistent with measured grain-size distributions (Table 1, Figure 2). In case-2, although the ash particles did not deposited at loc. 3, particles finer than 1–2 ϕ increased at loc. 1 and 2 (Table 3) that is relatively close to the measured grain size distributions.

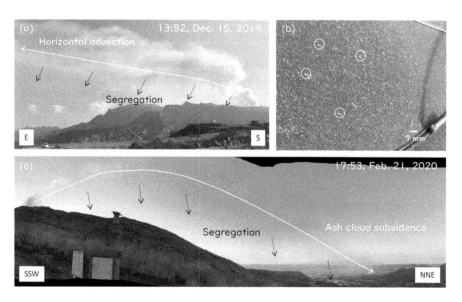

Figure 2.   Ash cloud advection (a) and ash fallen on umbrella (b) at loc. 3 at Dec. 15, 2019. Collision tracks of aggregated particles are indicated as white circles. Ash cloud subsidence at Feb. 21, 2020 (c).

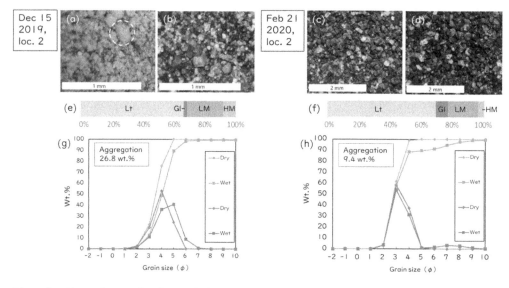

Figure 3.   Photomicrographs of unwashed (a, c) and washed samples (b, d) from loc. 2. AP1 is shown by white broken circle. Components of washed samples (e, f). Grain-size distribution of dry-sieved and wet-sieved samples (g, h). Percentages of aggregation are also shown. Gl: glass shards, HM: heavy minerals, LM: light minerals, Lt: lithic fragments.

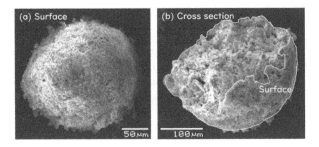

Figure 4. SEM-BSE image showing surface (a) and cross section (b) of accretionary pellets (AC1) included in the ash from the eruption at Dec 15, 2019.

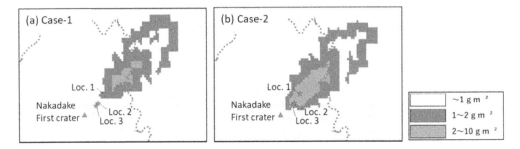

Figure 5. Simulation results of case-1 (a) and case-2 (b). Sample localities are also shown.

Table 3. Calculated accumulation by each grain size.

		Grain size (ϕ)	0~1	1~2	2~3	3~4	4~5	Grain size (ϕ)	0~1	1~2	2~3	3~4	4~5
Calculated	Case-1	Loc. 1	0	1	0	0	0	Case-2	Loc. 1	0	1	2	0	0
deposits (g m^{-2})		Loc. 2	0	0	0	0	0		Loc. 2	0	2	1	0	0
		Loc. 3	0	0	0	0	0		Loc. 3	0	0	0	0	0

5 DISCUSSION

5.1 *Ash aggregation*

Aggregation was observed in the ash eruption of Aso volcano. This findings suggest that aggregation is probably more common than previously thought. Even in a series of continuous ash eruptions, there were eruptions that were rich in aggregation and eruptions that were poor.

The subspherical ash pellets (AP1) suggesting the surface tension of water on particle surfaces (hydrostatic bonds) (Durant and Brown, 2016). AP1 were observed only in moist ash plume Dec15 that indicates the aggregation were formed due to interaction between ash particles and vapor within plum. The origin of vapor could be water included in magma, groundwater and/or vapor from atmosphere. Aggregation does not occur in plume Feb21 probably due to the lack of vapor within the plume.

Calculated terminal velocities of 5 ϕ and 2 ϕ particles is 0.07 m s^{-1} and 2 m s^{-1} at sea level (Walker et al., 1971). Grain-size of AP1 suggests that the aggregated fine ash particles have terminal velocities up to several orders of magnitude greater than their individual fall velocities. The deposits include 26.8 wt.% of aggregated particle that is notable and affect to the tephra transport and deposition model.

5.2 Ash segregation from plumes

The fine particles ubiquitously deposited at Loc. 1–3 from plume Dec15 suggests that they fall individually. The simulation results indicate that the fine particles segregated from upper part of the plume could not deposited at Loc. 1 and 2 but the fines from the lower part could be deposited at the localities. This could be attributed to that the fine particles at Loc. 1–3 were segregated from the lower part of plume Dec15. This is consistent to observation of segregation from plume Dec15 (Figure 2a).

This findings that the fine particles were separated from lower rising plume and upper spreading current at Dec 15, 2019 (Figure 6a).

5.3 Subsidence of fine-deplete ash cloud

Plume Feb21 did not spread horizontally but subsided to the NE direction that is because the ash cloud was deplete on fine ash (Figure 6b). This is obvious difference with plume Dec15. The production of fine particles attributed to segregation from plume at early stage and aggregation within uprising and migrating plume. Thus the production of fine ash particles influence on ash dispersal, accumulation mass and grain-size distribution.

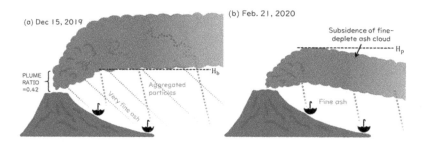

Figure 6. Ash transportation and deposition model of eruption at Dec 15, 2019 (a) and Feb 21, 2020 (b).

6 CONCLUSIONS

To discuss tephra transport and deposition process from the plumes on Dec. 15, 2019 (plume Dec15) and on Feb. 21, 2020 (plume Feb21) eruption of Aso volcano, we conducted the observation of eruption, direct sampling of ash particles within 3 km from the vent, as well as grain-size analyses, SEM observation and tephra simulation analyses using Tephra2. The white-gray colored plume Dec15 gently rose up to 1300 m above crater rim and horizontally flowed to outside of the caldera rim by wind, releasing aggregated ash. The gray colored plume Feb21 gently rose diagonally up to 800 m and started advection not horizontally but descended gently, releasing non-aggregated ash almost within the caldera. The aggregated particles were 100–250 μm in diameters and consist of mainly 5–7 ϕ particles. The aggregated fine particles gained terminal velocities up to several orders of magnitude greater than their individual fall velocities. The deposits include 26.8 wt.% of aggregated particles that is notable and affect to the tephra transport and deposition. Based on the observation, grain-size analyses and Tephra2 simulation, the fine particles segregated from lower plume at early stage and spreading current at Dec. 15, 2019. The spreading plume Feb21 subsided gently possibly due to the depletion of fine particles. The production of fine particles attribute to segregation from plume at early stage, aggregation within plume and ash dispersal to distal area.

ACKNOWLEDGEMENTS

The authors greatly appreciates Dr. Mannen K. for support on Tephra2 analyses and Dr. Ohkura T. for support on ash sampling.

REFERENCES

Bonadonna, C., Connor, C. B., Houghton, B. F., Connor, L., Byrne, M., Laing, A. and Hincks, T. K., 2005, Probabilistic modeling of tephra dispersal: Hazard assessment of a multiphase rhyolitic eruption at Tarawera, New Zealand, *J. Geophysical Research*, 110, B03203, doi:10.1029/2003JB002896.

Brown, R. J., Bonadonna, C. and Durant, A. J., 2012, A review of volcanic ash aggregation, *J. Physics Chemical Earth*, 45–46, 65–78

Durant, A. J. and Brown, R. J., 2016, Ash Aggregation in Volcanic Clouds, *In* Mackie, S., Cashman, K., Ricketts, H., Rust, A. and Watson, M., eds *VOLCANIC ASH HAZARD OBSERVATION, ELSEVIER*, 53–65.

Fukuoka Regional Volcanic Observation and Warning Center, 2020, Volcanic activity of Asosan, 2020.

Global Volcanism Program, 2013, Asosan, in Volcanoes of the World, v. 4.7.4. Venzke E (ed.) Smithsonian Institution (https://volcano.si.edu/volcano.cfm?vn=282110).https://doi.org/10.5479/si.GVP.VOTW4-2013

Japan Meteorological Agency and Volcanological Society of Japan, 2013, National catalogue of the active volcanoes in Japan. http://www.data.jma.go.jp/svd/vois/data/tokyo/STOCK/souran_eng/menu.htm.

Mannen, K., 2013, Theoretical background and recent progress of the pyroclastic fall simulation code Tephra2: essence for application in Quaternary research, *The Quaternary Research*, 52, 173–187 (Japanese with English abstract).

Ono, K, Watanabe, K, Hoshizumi, H. and Ikebe, S., 1995, Ash eruption of the Naka-dake crater, Aso volcano, southwestern Japan, *J. Volcanol Geotherm Res.*, 66, 137–148.

Tsuji, T., Nishizaka, N. and Ohnishi, K., 2020, Influence of particle aggregation on the tephra dispersal and sedimentation from the October 8, 2016, eruption of Aso volcano, *Earth, Planets and Space*, 72, 104, https://doi.org/10.1186/s40623-020-01233-y

Walker, G. P. L., Wilson, L. and Bowell, E. L. G., 1971, Explosive volcanic eruptions, I, The rate of fall of pyroclasts. Geophys. J. Res. Astron. Soc., 22, 377–383.

Rock Mechanics and Engineering Geology in Volcanic Fields – Ohta, Ito & Osada (eds)
© 2023 copyright the Author(s), ISBN 978-1-032-27657-1

Resistivity of volcanic ash and its relation to chemical composition

Takuya Urakoshi*, Yuichiro Nishikane & Shoichi Kawamura
Railway Technical Research Institute, Japan

ABSTRACT: Electric resistivity of volcanic ash is one of important physical property to estimate influences of volcanic ash fall on social activities such as railway operation. Therefore, to clarify factors that determines electric resistivity of volcanic ash, we developed a new equipment to measure the electric resistivity of volcanic ash, and applied it to the six samples of volcanic ash collected in Japan. According to the test result, the resistivity ranges from 10^6 Ωm to 10^8 Ωm or more in dry condition and 10 to 100 Ωm in the wet condition with the water content of 20%. We have reconfirmed that the resistivity of volcanic ash strongly depends on the water content.

Furthermore, to understand why the resistivity differs even at the same water content, we compared the obtained resistivity to chemical composition of the volcanic ashes and the ionic composition of the suspension of 10 g volcanic ashes mixed with 100 ml of pure water. As a result, sulfur content of the volcanic ashes ranges from 0.00 to 7.64 wt.%, and pH and total ion equivalent concentration of the suspension are 4 - 6.5 and 0.1 - 100 meq/L, respectively. We have revealed that the volcanic ashes with higher sulfur content produces the suspension with lower pH and the higher total ion equivalent concentration, showing the lower resistivity in the wet condition.

These results implicate that the sulfur in volcanic ash is oxidized to generate hydrogen ions, which promote the dissolution of ionic compounds contained in volcanic ash, resulting in a higher total ion equivalent concentration of suspension and lower resistivity of volcanic ash in the wet condition. Thus, we concluded that the influence of volcanic ash on electric apparatuses depends on the sulfur content and the water content of the volcanic ash.

Keywords: Electric Resistivity, Volcanic ash, Chemical Composition

1 INTRODUCTION

Phenomena associated with volcanic activity, such as, ash fall, pyroclastic flow, lahar, and ground movement, cause damage to social infrastructure. Among the phenomena, especially, ash fall affect extensively because volcanic ash in atmosphere extends by advective diffusion. To estimate the influences of ash fall on social activities such as railway operation, electric resistivity of volcanic ash is one of the important features (Urakoshi et al., 2015; Nishikane et al., 2021). There are two sources of the problem: one is volcanic ash with low resistivity and the other is volcanic ash with high resistivity. An example of the former is the eruption of Mt. St. Helens in 1980. The volcanic ash from the eruption caused the malfunctions in the train signal system (Kumagai and Suto, 2004). The reason of the malfunctions was that the volcanic ash caused the short-circuit due to its low resistivity. Volcanic ash with low resistivity also

*Corresponding author: urakoshi.takuya.38@rtri.or.jp

DOI: 10.1201/9781003293590-5

causes the short-circuit of insulators for power supply lines, leading leakage of electric current (Konishi et al., 2018).

An example of the latter is the eruption of Mt. Unzen in 1991. The volcanic ash deposited on rails caused the malfunction of railroad crossings (Ashi-Shobo (ed), 1998). The reason was that the signal system using signal current was interfered by the volcanic ash. The signal current is the electric current used for the train detection, which flows rails separately from the power supply for train drive. If the signal current flows from one of rails to the other rail via wheels, then the signal system judges a train be on rails. The volcanic ash with high resistivity on rails interrupted the signal current between rails and wheels, leading failures of signal system and railroad crossing (Nishikane et al., 2021). Same phenomena were reported at Mt. Sakurajima (Uchikura and Sasaki, 2012), and Mt. Shinmoe (Iguchi, 2011).

These examples show that mechanisms of malfunctions caused by volcanic ash deposited on railway equipment depend on whether its electric resistivity high or low. Thus, the electric resistivity is desirable information to predict what volcanic ash would cause and to plan countermeasures.

In this study we aimed to clarify factors that determines electric resistivity of volcanic ash. Firstly, we sampled volcanic ashes from active volcanos, Sakurajima, Mt. Aso, Mt. Kirishima and Mt.Fuji, in Japan, and analyzed their chemical composition and mineral composition. Secondly, we measured the resistivity of the volcanic ashes using a newly developed equipment, considering the water content. Additionally, we measured the properties of suspension of volcanic ashes, including the electrical conductivity, pH, and dissolved ion. Finally, we discussed the relation between the chemical composition and the electric resistivity.

2 EXPERIMENTS

2.1 *Volcanic ash for the experiments*

(1) Sampling of volcanic ash

We collected six volcanic ash samples from four active volcanos in Japan and prepared Toyoura sand for comparison (Table 1 and Figure 1). The samples Sa and Sb are the volcanic ashes collected around Mt. Sakurajima. The sample Sa was collected immediately after the eruption in Mt. Sakurajima, and the sample Sb was collected from a volcanic ash pile, which was consist of ash fall gathered from roads near Mt. Sakurajima and was kept outside over one year. The samples Aa and Ab from Mt. Aso were collected on the day of the eruption and five days later, respectively. The sample K from Mt. Kirishima (Mt. Shinmoe) was collected on the next day of the eruption. The sample F was collected from the scoria layer determined to be the Hoei eruption of Mt. Fuji. In addition, commercially available Toyoura sand, sample T, collected from Yamaguchi Prefecture, Japan, was prepared for comparison. It mainly consists of quartz with a particle distribution between 0.1mm and 0.2mm .

The median grain size diameter of the samples Aa and Ab are 6.106mm and 0.062 mm, respectively, as the result of the difference of the distances between the crater and the sampling points although both samples were collected from volcanic ash of the same eruption. The median grain size diameter of the other samples ranges from 0.190 mm to 0.311mm. The particle density of the samples ranges from 2.617 g/cm^3 to 2.786 g/cm^3.

(2) Chemical composition and mineral composition

The major chemicals of the chemical composition of the volcanic ash samples and Toyoura sand were analyzed by wavelength dispersive X-ray fluorescence (XRF). The samples were air-dried and milled into powder, and glass beads were prepared using $Li_2B_4O_7$ flux, which mass was 15 times of the mass of the powder sample. Then the glass beads were analyzed with a Rigaku Primus II. The applied calibration curve for this analysis was composed of volcanic rocks from the geochemical reference samples distributed by the Geological Survey of Japan. The analysis yields that the Silica (SiO_2) content of the volcanic ash samples ranges from 51.92 wt.% to 60.32 wt.%, while the Silica content of Toyoura sand is 91.48 wt.% (Table 2).

Table 1.   Description of collected volcanic ash samples and Toyoura sand.

ID	Volcano	Sampling Date	Explanation	Median grain size diameter (mm)	Particle Density (g/cm^3)
Sa	Sakurajima	Dec. 18, 2014	Ash of an eruption on Dec. 18, 2014. It was collected at a point about 3 km from the crater in Sakurajima Island immediately after the eruption.	0.210	2.716
Sb	Sakurajima	Dec. 18, 2014	Ash from a pile at a volcanic ash storage site in Kagoshima City, about 12km from Sakurajima. The ash pile had kept outdoors for over a year.	0.186	2.726
Aa	Mt. Aso	Oct. 8, 2016	Ash of an eruption on Oct. 8, 2016. It was collected at a point about 4.4 km from the crater during the same day.	6.106	2.689
Ab	Mt. Aso	Oct. 13, 2016	Ash of an eruption on Oct. 8, 2016. It was collected at a point 6.3 km from the crater five days later. There was rainfall between the eruption and the collection.	0.062	2.617
K	Mt. Kirishima (Mt. Shinmoe)	Mar. 7, 2018	Ash of an eruption on Mar. 6, 2018. It was collected at a point 3.5 km from the crater on the next day.	0.278	2.719
F	Mt. Fuji	Oct. 14, 2016	Ash of the Hoei eruption in 1707. It was collected from scoria layer in Gotenba City, Shizuoka Prefecture.	0.311	2.786
T	-	-	Toyoura sand. As a comparison. It was purchased.	0.190	2.648

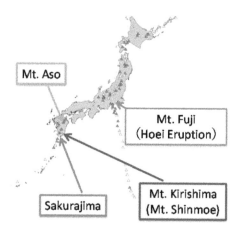

Figure 1.   Volcanic ash sampling sites. (after JMA, 2021).

The sulfur content of the samples was analyzed three times a sample by frontal chromatograph method of combustion gas with a Perkin Elmer 2400 II CHNS/O elemental analyzer. The sulfur contents of the samples Aa and Ab are 4.16 wt.% and 7.64 wt.% respectively, which are higher than those of the other samples (Table 2). That of the samples F and T are 0.003 wt.% or less.

The mineral composition was analyzed using the random mounts, the oriented mounts and the ethylene glycol treated oriented mounts of the samples by the powder X-ray diffraction (XRD) with a Rigaku RINT 2000. This analysis showed that all samples contained feldspar (Table 3). The samples Aa and Ab contain gypsum, anhydrite and pyrite, which is consistent with their high content of sulfur. The sample T is mainly composed of quartz, which is consistent with its high content of silica.

Table 2.　Chemical compositions of volcanic ash samples and Toyoura sand.

Sample	SiO_2	TiO_2	Al_2O_3	Fe_2O_3	MnO	MgO	CaO	Na_2O	K_2O	P_2O_5	S
Sa	58.91	0.84	16.11	8.55	0.17	3.66	6.87	3.00	1.42	0.15	0.02
Sb	58.79	0.81	16.37	8.30	0.16	3.47	6.98	3.08	1.41	0.16	0.05
Aa	52.22	0.77	19.25	7.39	0.09	3.24	7.69	2.33	1.83	0.23	4.16
Ab	51.92	0.87	13.15	7.19	0.07	2.14	8.27	1.76	1.43	0.22	7.64
K	60.32	0.62	18.58	8.14	0.13	5.18	7.69	2.91	1.50	0.10	0.03
F	52.65	1.34	17.00	11.51	0.19	4.89	9.00	2.86	0.85	0.28	0.003
T	91.48	0.33	3.26	1.17	0.02	0.13	0.18	0.53	1.59	0.01	<0.003

Unit: wt.%

Table 3.　Mineral compositions of volcanic ash samples and Toyoura sand.

Sample	Quartz	Feldspar	Mica	Pyroxene	Amphibole	Gypsum	Anhydrite	Pyrite	Cristbarite	Smectite
Sa		++								
Sb		++								
Aa		+++	tr		tr	+	++	+		tr
Ab		++	tr		tr	+	++	++	++	tr
K		+++		tr						tr
F		+++								
T	+++	+								

Relative Intensity of XRD: <u>Strong</u> +++ > ++ > + > tr <u>Weak</u>

2.2　Electric resistivity measurement

Electric resistivity of the samples was measured by using a newly developed equipment which has electrodes arranged as the quadrupole (Figure 2). The samples are air-dried and crushed to pass the one-millimeter sieve to normalize the grain diameter regardless of the samples. Water content, a ratio of the weight of water to the weight of solid of a sample, is adjusted to 2.5%, 5%, 10%, and 20% with pure water. Hereinafter, the sample without adding pure water is referred to a dry condition sample, and the others are wet condition samples. Note the water content of the dry condition sample is not 0% because the sample is only air-dried. Then, a sample is filled in the measurement vessel, and the voltage between the measuring electrodes, V_S, and the voltage between the electrodes of the resistor, V_R, are measured. The electric resistivity ρ of the sample is determined as Eqn. (1).

$$\rho = \frac{A}{L} \frac{V_s}{V_R} R \tag{1}$$

Here, A is the cross-sectional area of the filled sample in the measurement vessel, L is the distance between the measuring electrodes, and R is the resistance of the resistor.

With DC power source, it is conceivable that the sample is polarized, and measured electrical resistivity changes with time. Therefore, we conducted a preliminary test and confirmed that relative error of measured electrical resistivity is within 10% from the start of the power supply until ten minutes later. On the other hand, in some cases, it taken several tens of seconds for the electric current to stabilize. Therefore, we decided to measure V_S, and V_R at one minute after the start of the power supply, and thus the relative error of measured electric resistivity is about 10%.

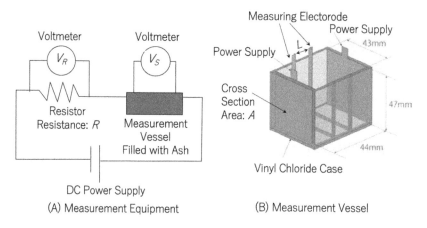

Figure 2.   Electric resistivity measurement equipment.

2.3   Electrical properties of the suspension

We analyzed the electrical properties of suspension of the volcanic ash samples and Toyoura sand, based on three methods: "Test Method for Electric Conductivity of Suspended Soils" (JGS 0212-2009), "Test Method for pH of Suspended Soils" (JGS 0211-2009), and "Test Method for Water-Soluble Components in Soils" (JGS 0241-2009).

For the electric conductivity test and pH test of the soil suspension, an air-dried sample is mixed with five times the weight of pure water. Next, we measure the electrical conductivity and pH of the supernatant, 30 min. after the mixing.

For the water-soluble component analysis, the air-dried sample is mixed with ten times the weight of pure water using a shaker at 200 rpm for six hours. Then, after filtration, the dissolved ions are quantified by an ion chromatography or an atomic absorption spectroscopy.

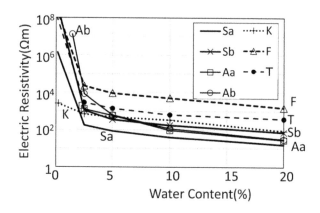

Figure 3.   Electric resistivity of volcanic ash samples and Toyoura sand.

3 RESULTS AND DISCUSSIONS

3.1 *Electric resistivity*

The electric resistivity is as high as 10^6 Ωm to 10^8 Ωm or more except for the sample K, in the dry condition (Figure 3). Note that 10^8 Ωm is the measurement limit, and the electric resistivity of the samples Sb, F, and T in the dry condition are above this limit. In contrast, the electrical resistivity is 10^2 Ωm to 10^4 Ωm when the water content is 2.5%, and it decreases gently as the water content increases. At the water content of 20%, the electrical resistivity of the samples ranges from 10 to 100 Ωm except for the samples T and F. It is found that the electrical resistivity of volcanic ashes is strongly depends on the water content. For instance, if the water

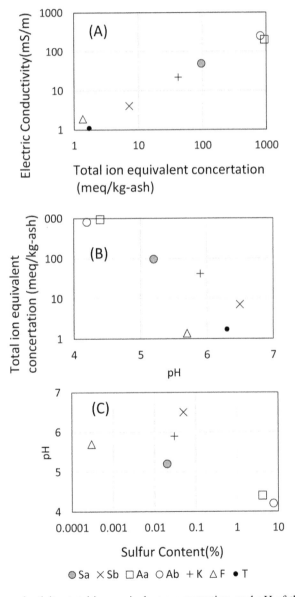

Figure 4.   Electrical conductivity, total ion equivalent concentration, and pH of the suspension of volcanic ash samples and Toyoura sand.

content changes from less than 2% to 20%, the electric resistivity of volcanic ashes changes by about 10^6 times.

For the sample K, the tendency of decreasing in the electrical resistivity with increasing the water content is the same as for the other samples. However, the electric resistivity is 10^5 Ωm when the water content is 0.2%.

3.2 *Electrical properties of the suspension*

The results of electrical conductivity measurement, pH measurement and water-soluble component analysis of the suspension are shown in Figure 4. The results of the water-soluble component analysis are shown as the total ion equivalent concentration eluted from one kilogram of dried volcanic ash. The electrical conductivity of the suspension is positively correlated with the total ion equivalent concentration (Figure 4 (A)). Furthermore, the total ion equivalent concentration is high in acidic suspension, and low in neutral one (Figure 4 (B)). Generally, an acidic solution dissolves mineral components more easily than the neutral solution. Thus, it is considered that the same phenomenon occurs in volcanic ash.

In addition, from the relationship between pH and sulfur content, it is found that suspension tends to be acidic when the sulfur content is high (Figure 4 (C)). This is because the sulfur and pyrite contained in the volcanic ashes react with water and oxygen to generate hydrogen ions. Thus, if volcanic ash has high sulfur content, then the pH decreases, total ion equivalent concentration increase, and electric conductivity also increase. Here, we consider gypsum and anhydrite in the samples Aa and Ab. Gypsum and anhydrite are easily to dissolve, so that total ion equivalent concentration increases. Thus, high sulfur content also contributes to high electric conductivity by dissolving gypsum and anhydrite.

Although both samples Sa and Sb are volcanic ash collected around Mt. Sakurajima, the suspension of sample Sa shows higher electrical conductivity and total ion equivalent concentration than that of Sb (Figure 4 (A)). This will be because the sample Sb was collected from the pile exposed to rainfall for more than a year after the eruption, and water-soluble components have already flowed out. In addition, the samples F and T shows low electrical conductivity and low total ion equivalent concentration (Figure 4 (A)). The reason is that the sample F of Mt. Fuji was collected from the stratum and the water-soluble components dissolved due to rainfall and groundwater flow. Further, the sample T, which is sand sample as comparison, is mainly composed of quartz.

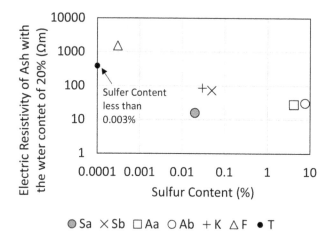

Figure 5.   Comparison between sulfur content and electric resistivity of samples at a water content of 20%.

3.3 Relation between chemical composition and electric resistivity

From Figure 3, there are difference of electric resistivity at the same water content. The difference is about 100 times. Figure 5 shows the cross plot between the electric resistivity of ashes at the water content of 20% and the sulfur content. It is found that Figure 5 shows a negative correlation. Thus, the difference in the electrical resistivity of the volcanic ashes is considered to depend on the sulfur content as the following mechanism: Native sulfur or pyrite generate acidic water, and enhance dissolving the mineral of ash. Then, the total ion equivalent concentration of suspension increases, and the electric conductivity of suspension also increases. Hence, the electric resistivity of ashes in wet condition decreases. Also, dissolution of gypsum and anhydrite contributes to the decrease of the electric resistivity of ashes.

This discussion is based on the obtained samples in this research. We need to increase samples and see if this discussion can be applied more generally.

4 CONCLUSIONS

We developed a new equipment to measure the electric resistivity of volcanic ash, and successfully determined the resistivity of the six samples of volcanic ash collected in Japan. The obtained value ranges from 10^6 Ωm to 10^8 Ωm or more in the dry condition and from 10 Ωm to 100 Ωm in the wet condition with the water content of 20%. We found the resistivity of volcanic ashes strongly depended on the water content.

Furthermore, we revealed that the volcanic ashes with higher sulfur content produced the suspension with lower pH, enhancing the dissolving the minerals of volcanic ashes, and leads the higher total ion equivalent concentration of the suspension, and showed the lower electric resistivity in the wet condition. Here, gypsum and anhydrite in high sulfur content volcanic ashes also contributed to decreasing the electric resistivity.

These results suggest that the extent of the electric influence of volcanic ash on electric apparatuses to depend on the sulfur content and the water content of the volcanic ash.

REFERENCES

Ashi-shobo (ed), 1998, *Funkato Tatakatta Shimabara Tetsudo* (Shimabara Railway to have confronted the eruption), Ashi shobo, Fukuoka, Japan, p.145 (in Japanese).

Iguchi, T., 2011, Shinmoedake funkano testudo denki setsubieno eikyoto sono taio (Impact of eruption of Mt. Shinmoedake on railway electrical facilities and correspondence on the eruption), *Railway and Electrical Technology*, 22, 33–37 (in Japanese).

JMA; Japan Meteorological Agency, *Katsukazantoha* (What is the active volcano), Retrieved from https://www.data.jma.go.jp/svd/vois/data/tokyo/STOCK/kaisetsu/katsukazan_toha/katsukazan_toha. html (Accessed: 11 May 2021) (in Japanese).

Konishi, T., Hirakawa, S., and Nishikane, Y., 2018, Experimental study of deterioration for the insulator with volcanic ash in electrified railway, *2018 Annual Meeting Record IEEJ*, pp. "9-3-19" – "9-3-20" (in Japanese).

Kumagai and Suto, 2004, *Impact analysis on the volcanic ash-fall in the metropolitan area*, 546p (in Japanese with English abstract).

Nishikane, Y., Terada, N., Urakoshi, T., and Kawamura, S., in press, volcanic ash impacts on railway signal facilities and utilizing of the volcanic ash fall forecast, *The 5th International Workshop on Rock Mechanics and Engineering Geology in Volcanic Fields*, Fukuoka, Japan.

Uchikura, K., and Sasaki, K., 2012, Kohai kenchi sochide jikobousi (Prevent accidents by using volcanic ash fall detection system), *25th Railway Electrical Technical Forum*, 271–271 (in Japanese).

Urakoshi, T., Nishikane, Y., and Kawagoe, T., 2016, A case study of disasters due to volcanic activities and countermeasures against them in railway of Japan, *RTRI report*, 29, 1, 47–52 (in Japanese with English abstract).

Rock Mechanics and Engineering Geology in Volcanic Fields – Ohta, Ito & Osada (eds)
© 2023 copyright the Author(s), ISBN 978-1-032-27657-1

An experimental study on the dynamic stability of rock slopes with hexagonal discontinuity pattern

Yuki Murayama*, Ömer Aydan, Naohiko Tokashiki & Takashi Ito
Department of Civil Engineering, University of the Ryukyus, Japan

ABSTRACT: Various discontinuity patterns are observed in rock masses as a result of their geological past. One of the common discontinuity patterns is the hexagonal pattern. This pattern is mostly observed in all extrusive volcanic rocks such as basalt, andesite, rhyolite and welded tuffs as well as in some sedimentary rocks subjected to desiccation or freezing-thawing processes. Such patterns are even observed in igneous and sedimentary rocks on Mars. In this study, the authors investigate the dynamic stability of rock slope consisting of hexagonal blocks through model tests on shaking table. The outcomes of these experimental studies are presented and their implications in practice are discussed.

Keywords: Hexagonal Discontinuity Pattern, Rock Slope, Model Test, Slope Stability, Earthquake

1 INTRODUCTION

Rock masses irrespective of its type contain discontinuities associated with their geological formation. In nature, rock masses consisting of hexagonal blocks receive a great attention of ordinary people due to their beautiful geometrical patterns. These patterns are commonly observed in basic extrusive volcanic rocks such as basalt and andesite. However, the rock masses having hexagonal discontinuity pattern often observed in nature and they constitute some spectacular features, which often constitute geo-parks in many countries all over the World. Such structures are even observed in other planets such as Mars (e.g. Aydan 2019) (Figure 1). Figure 2 show some examples of rock masses having hexagonal discontinuity pattern. There is a great concern on the both

(a) Orhun River (Central Asia)   (b) Kelbecer (Azerbaijan)   (c) Marte Vallis (Mars)

Figure 1. Some examples of columnar jointing on Earth and Mars.

*Corresponding author: k208482@eve.u-ryukyu.ac.jp

DOI: 10.1201/9781003293590-6

stability of slopes, underground openings and seepage and bearing capacity of rock formations having hexagonal joint sets in recent years. As the hexagonal discontinuity sets result in columnar structures, there is a big concern on rockfalls from steep rock slopes and tunnel portals (Aydan et al. 2021).

This study is concerned with the dynamic stability of rock mass models having hexagonal discontinuity pattern. The dynamic stability of slopes of rock mass models consisting of hexagonal blocks was investigated through model tests on shaking table. The authors present the outcomes of these studies and discuss their implications in practice.

   (a) Kızılcahamam (Turkey)   (b) Tachijami (Ryukyu)   (c) Tsumekizaki (Izu)

Figure 2.   Some examples of rock slopes consisting of rock masses with hexagonal discontinuity pattern.

2   DYNAMIC FAILURES OF ROCK SLOPES WITH ROCK MASSES HAVING HEXAGONAL DISCONTINUITY PATTERN

Earthquakes may induce induce failures of rock slopes (Aydan 2016, 2019). This issue is less studied in literature. As rock mass in nature contains numerous discontinuities, usually in the form of sets, the stability of such slopes depends upon the spatial orientations of discontinuity

Figure 3.   Views of earthquake induced rock slope failures having hexagonal discontinuity pattern.

43

sets with respect to slope geometry, their continuity and their mechanical properties. As long as the rock material itself does not break up under induced state of stress, and two sets of discontinuity, whose strikes are parallel or nearly parallel to the axis of the slope, exist, the possible forms of slope instability are sliding failure, toppling failure and combined sliding and toppling failure. These instability forms may appear depending upon the discontinuity pattern, their inclination, frictional properties and the geometry of slopes. More general discussion on the stability of rock slopes is given in Aydan (1989) and his follow-up publications. Observations on failure modes also indicated that the passive modes of failure forms can also occur in nature (Aydan 2016, 2019). Figure 3 shows some actual rock slope failure observed in 2008 Iwate-Miyagi, 2011 Christchurch, 2011 Mt. Fuji and 2016 Kumamoto earthquakes. Most of rock slopes failures are dominated by toppling or combined toppling and sliding modes.

3 MODEL SLOPES

3.1 *Model materials and their frictional properties*

Aluminum hexagonal blocks were selected to create rock slope models. Block were 50mm long with side length of 6 mm. The physical and mechanical properties of aluminum are given in Table 1. In these particular experiments, the blocks were selected such that they will remain

Table 1.   Physical and elastic constants of Aluminum.

Elastic Modulus(GPa)	68×10^4
Poisson's ratio	0.33
Unit weight (kN/mm^3)	26.89

Table 2.   Frictional characteristics of aluminum block interfaces.

Test No	$\varphi_s{}^\circ$	$\varphi_d{}^\circ$	A (cm/s^2)
①	15.0	14.1	16
②	15.9	14.54	24
③	15.7	14.23	26

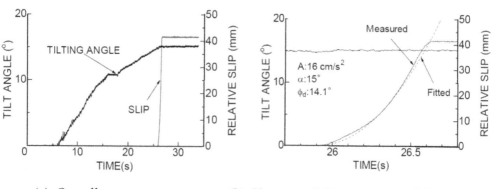

(a) Overall response   (b) Close-up of slip response and fitting

Figure 4.   An example of tilting test and determination of kinetic friction angle.

elastic while the movements can take place in the form of sliding and/or separation along block interfaces. Therefore, the friction properties are necessary. As the height of slopes is selected to be about 250mm, any friction test should consider this normal stress levels. Although direct shear tests are possible, the applicable normal stresses would be too high. Taking into account this fact, tilting tests were carried out on hexagonal aluminum blocks. In tests, the motion of blocks is also recorded using laser-transducers so that it was possible to determine both static and kinetic friction angles (Aydan 2017, 2019) (Figure 4). Table 2 summarizes experimental results. As noted from the table, static friction angle ranges between 15 to 15.9 degrees while the kinetic (dynamic) friction angle ranges between 14.1 and 14.54 degrees. Figure 4 shows an example of slip response of the block during a tilting experiment. Kinetic friction angle is determined from the measured slip response, which is described by Aydan (2019).

3.2 *Discontinuity pattern of model slopes*

Fundamentally, the discontinuity patterns used in model slope tests are denoted as Pattern A and Pattern B as illustrated in Figure 5. The 90-degree rotation of Pattern-A results in Pattern-B. In other words, the variety of model slopes is quite restricted to two patterns. If the friction angle of block interfaces is less than 30 degrees, the slope angle cannot be greater than 60 degrees under gravitational conditional (Figure 6a). For Discontinuity Pattern-A, the anticipated slope failure modes would always be passive type (Aydan 2016, 2019). As for Discontinuity Pattern-B, it is possible to build-up slopes with an angle of 90 degrees (Figure 6b) and the slope failure modes would be always active.

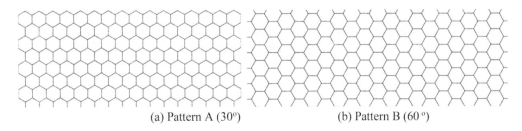

(a) Pattern A (30°)   (b) Pattern B (60°)

Figure 5.   Selected discontinuity Patterns.

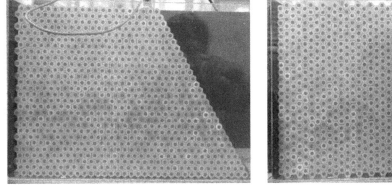
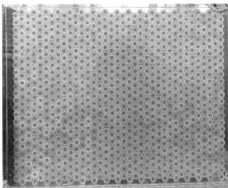

(a) Discontinuity Pattern-A   (b) Discontinuity Pattern-B

Figure 6.   Views of model slopes.

4 SHAKING TABLE MODEL TESTS

Model slopes for each discontinuity pattern were subjected to shaking using the shaking table test device at the University of the Ryukyus. During experiments, the accelerations at the top of the model slope and on the shaking table and displacements at the slope crest were monitored, simultaneously. Accelerations were recorded by using Tokyo Sokki 10G accelerometer and displacements were measured using the laser displacement transducers. The sampling was set at 10 ms and YOKOGAWA SL1000 Dynamic acquisition system is utilized together with recordings on laptop computers. In addition, experiments were recorded by the high-speed camera. At least three experiments were carried out for each discontinuity pattern. However, only one experiment for each discontinuity pattern is explained herein.

4.1 Model slopes with hexagonal discontinuity pattern A

The slope angle was 60 degrees, which is the stable slope angle under given discontinuity pattern. Figure 7 shows views of the model slope at the before and during shaking. As noted from the figure, a columnar motion of the passive toppling mode is recognized. The column consists of hexagonal blocks tend to be in motion like an equivalent monolithic column. Figure 8 shows the base acceleration and displacement of the top of the slope. The columnar behavior is noted at the acceleration level of 720 gals and the model slope become totally unstable when the acceleration reached the level of 790 gals.

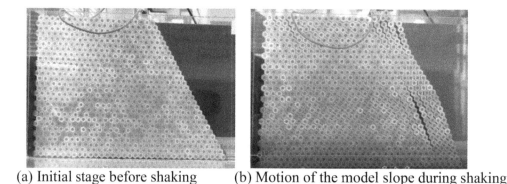

(a) Initial stage before shaking (b) Motion of the model slope during shaking

Figure 7. Views of the model slope before shaking (a) and during shaking (b).

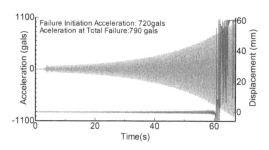

Figure 8. Recorded base acceleration and the response of the top of the model slope.

4.2 Model slopes with hexagonal discontinuity pattern B

The slope angle was 90 degrees. As pointed out previously in the previous sub-section, the column consisting of hexagonal blocks is anticipated to be quite vulnerable to failure under even very slight vibration (24-40 gals) or slight tilting of the base in the range of 1.37 – 3.17

degrees inclination. Figure 9 shows views of the model slope at the before shaking and during shaking. As noted from the figure, a columnar motion of the active toppling mode is recognized. The column consists of hexagonal blocks above the plane inclined at an angle of 30 degrees tend to be in motion like equivalent monolithic columns. Figure 10 shows the base acceleration and displacement of the top of the slope. The columnar behavior is noted at the acceleration level of 20 gals and the model slope become totally unstable when the acceleration reached the level of 40 gals as estimated (Aydan et al. 2021).

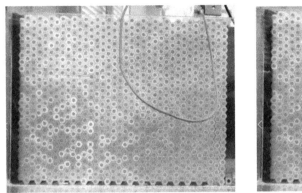 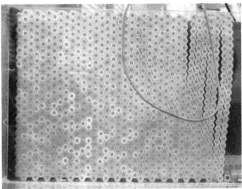

(a) Initial stage before shaking   (b) Motion of the model slope during shaking

Figure 9.   Views of the model slope before shaking (a) and during shaking (b).

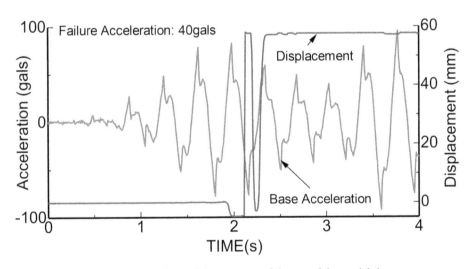

Figure 10.   Recorded base acceleration and the response of the top of the model slope.

5   DISCUSSION OF FAILURE MODES IN SHAKING TABLE MODEL TESTS

As for Discontinuity Pattern A, the failure regions within the slopes can be denoted as shown in Figure 11(a), depending upon the slope geometry and frictional characteristics of discontinuities. Region III is the first most likely failure form of the slope. The region denoted by III will fail in the active sliding mode and/or active toppling mode under gravity. If the slope angle such that the region III does not exist, the region denoted as II, can only fail in the form of passive toppling and/or passive sliding mode when the slope is subjected shaking. When slope height is relatively high, the passive toppling mode would be dominant. On the other hand, the passive sliding mode may be dominant depending upon the frictional properties of discontinuities and dimensions of hexagonal blocks with respect to slope geometry (Aydan et al. 1989, 2021; Aydan 2017). The stable denoted by Region I may also become unstable if the passive sliding mode is prevailing.

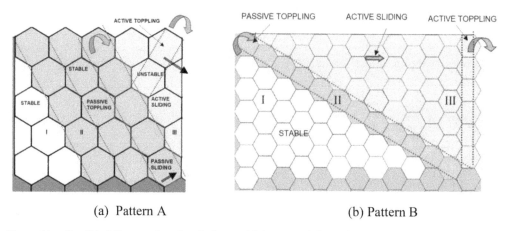

(a) Pattern A   (b) Pattern B

Figure 11.   Possible failure modes of rock slopes with hexagonal discontinuity patterns.

The failure regions within the slopes with Discontinuity Pattern B can be denoted as shown in Figure 11(b). Region III is fundamentally stable under gravity and it would become unstable first when the slope is subjected to shaking. The region denoted III will fail in the active sliding mode and/or active toppling mode under shaking (e.g. Aydan 2018). The active toppling mode will be governed by the highest column of hexagonal blocks. If the slope angle such that the region III does not exist, the region denoted as II, can only fail in the form of passive toppling and/or passive sliding mode when the slope is subjected shaking. Although passive sliding mode is possible, the passive columnar toppling would be the prevailing failure mode. The stable denoted by Region I will remain stable as expected.

6   CONCLUSIONS

The hexagonal discontinuity patterns are often observed in nature, particularly in extrusive volcanic rocks. Nevertheless, studies in rock mechanics and rock engineering is quite rare. First, a brief summary of rock masses with hexagonal discontinuity patterns is given and the possible mechanism of such discontinuity patterns is presented. A brief review of engineering problems is described. The results of these experiments are presented and their implications are discussed. And then the possible failure modes depending upon the discontinuity patterns are discussed. On the bases of theoretical formulations based on limiting equilibrium method (e.g., Aydan 1989, 2021; Aydan 2017) and results of frictional properties and geometry of model slopes, the critical acceleration levels may be estimated (Aydan et al., 2021). Despite some limitations of the limiting equilibrium method, the results of estimations can be quite close to those of experimental results. Besides the limiting equilibrium method, it would be desirable to carry out some numerical analyses based on Discrete Finite Element Method (DFEM) and compare both experiments and those from limiting equilibrium method.

ACKNOWLEDGEMENTS

The financial fund for the experimental studies by the Chuden Technological Consultants Company is gratefully acknowledged.

REFERENCES

Aydan, Ö., 1989, The stabilization of rock engineering structures by rockbolts, *Doctorate Thesis*, Nagoya University, 1989.

Aydan, Ö., 2016, Large Rock Slope Failures Induced by Recent Earthquakes, *Rock Mechanics and Rock Engineering*, 2503-2524, 2016.

Aydan, Ö., 2017, Rock Dynamics, CRC Press, ISRM Book Series 3, pp.147–186, 2017.

Aydan, Ö., 2019, *Rock Mechanics and Rock Engineering: Vol. I, Fundamentals*, CRC Press.

Aydan, Ö., Shimizu, Y. and Ichikawa, Y., 1989, The effective failure modes and stability of slopes in rock mass with two discontinuity sets, *Rock Mechanics and Rock Engineering*, Vol. 22, pp. 163–188.

Aydan, Ö., Tokashiki, N., Ito, T. and Murayama, Y., 2021, Dynamic stability of rock slopes with hexagonal discontinuity pattern (in Japanese with English abstract). 15th Japan Rock Mechanics Symposium, 453–458.

Rock Mechanics and Engineering Geology in Volcanic Fields – Ohta, Ito & Osada (eds)
© 2023 copyright the Author(s), ISBN 978-1-032-27657-1

Monitoring volcanic activity of Mount Agung, Indonesia by SBAS-DInSAR using Sentinel-1 data from 2014 to 2020

I Nyoman S. Parwata*
Udayana University, Denpasar, Indonesia

Shinichiro Nakashima & Norikazu Shimizu
Yamaguchi University, Ube, Japan

ABSTRACT: Differential Interferometric Synthetic Aperture Radar (DInSAR) is an effective method for measuring ground surface displacements over extensive areas without the necessity for contact with any objects on the ground surface. DInSAR has the potential to measure the displacements in dangerous areas, such as those with active volcanos. In this study, DInSAR is applied to measure the ground surface displacements induced by the activity of Mount Agung. The Small Baseline Subset DInSAR (SBAS-DInSAR) is employed to obtain the time transition of the surface displacements around Mount Agung. The satellite images observed by Sentinel-1A and -1B SAR are used. The results of SBAS-DInSAR reveal that the displacement increased once between May and November 21, 2017 and then increased a second time after May 2019 when the seismic activity began again. Finally, the volcano erupted two times, on November 25&29, 2017 and on June 13, 2019. This means that SBAS-DInSAR was able to detect the signs of the upcoming eruption, that is, increases in displacements in the northern part of Mount Agung. Thus, SBAS-DInSAR can be applied to monitor volcanic activity and will be useful for predicting future potential crisis periods.

Keywords: Mount Agung, Volcanic activity, Displacement monitoring, SBAS-DInSAR, Sentinel-1 SAR satellite

1 INTRODUCTION

Mount Agung is the largest volcano on the island of Bali in Indonesia. It is located in the eastern part of Bali and has an elevation of about 3,018 m above sea level, as measured by ALOS World 3D-30m resolution of the digital elevation model. The location and topography of Mount Agung are shown in Figure 1. Mount Agung had been dormant since 1963. However, it began to show signs of activity with intense seismicity, measurable ground surface displacements, and thermal anomalies in the summit crater (Syahbana et al., 2019). Although the seismic activity peaked from late September to early October 2017, the eruption of Mount Agung did not occur until November 21, 2017 (Syahbana et al., 2019). The most massive eruptions with rapid lava effusion occurred between November 25 and 29, 2017, and 140,000 people had to evacuate the area (Syahbana et al., 2019). Smaller and more infrequent explosions continued until June 2019. Monitoring the activity of Mount Agung is important for predicting the risks of eruptions and for providing announcements for evacuation.

Differential Interferometric Synthetic Aperture Radar (DInSAR) is an effective method for monitoring ground surface displacements over extensive areas without the necessity for direct

*Corresponding author: sudi_jbc@yahoo.com

DOI: 10.1201/9781003293590-7

contact with any objects on the ground surface. Due to its capability, DInSAR has the potential to measure the surface displacements in dangerous areas, such as those with active volcanos.

DInSAR has been successfully applied to monitor the volcanic activities of several volcanos around the world (Brunori et al., 2013; Chen et al., 2017; Kobayashi, 2018). For Mount Agung, Lingyun et al (2013) applied a Small Baseline Subset DInSAR (SBAS-DInSAR) time-series analysis using an ALOS-1 SAR dataset in the period of 2007 to 2009. Their work revealed that the Agung volcano inflated during this period at a nearly constant rate. However, no remarkable seismic activity was recorded.

In this study, DInSAR is applied to monitor the ground surface displacements induced by the activity of Mount Agung. The Small Baseline Subset DInSAR (SBAS-DInSAR) is employed to obtain the time transition of the surface displacements around Mount Agung. The satellite images observed by Sentinel-1A and -1B SAR are used. The results of SBAS-DInSAR reveal that the displacement increased once between May and November 2017 and then increased a second time after May 2019 when the seismic activity began again. The SBAS-DInSAR results were able to detect the signs of the upcoming eruption, that is, increases in displacements in the northern part of Mount Agung. Thus, SBAS-DInSAR can be applied to monitor volcanic activity and will be useful for predicting future potential crisis periods.

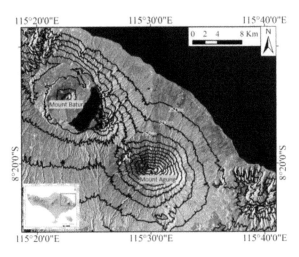

Figure 1. Location of Mount Agung and its topography presented by contour lines at vertical intervals of 250 meters.

2 OUTLINE OF DINSAR AND SBAS TIME-SERIES ANALYSIS

2.1 *Basic concept of SAR*

Synthetic Aperture Radar (SAR) is a radar system mounted on an artificial satellite or aircraft. It is a kind of active remote sensing system, whereby SAR transmits electromagnetic waves and receives the reflected electromagnetic waves from the Earth's surface. The reflected electromagnetic waves are stored in digital form in a complex number data format (Pepe and Calo, 2017). It contains the amplitude and phase of the reflected electromagnetic waves (Pepe and Calo, 2017). The most important advantages of SAR are that it can be used for day-night time observations, it is independent of the weather conditions, and it does not require the use of any devices in the area being monitored.

One of the principle applications of SAR is differential SAR (DInSAR). DInSAR is often used for measuring the displacements of the Earth's surface over vast areas in the direction of the satellite's line of sight (LOS). To conduct a DInSAR analysis, at least two sets of SAR data are required. Those SAR data are obtained from the same or an identical SAR satellite which observes the same area on the Earth's surface at different time acquisitions and satellite positions. A schematic view of DInSAR is presented in Figure 2.

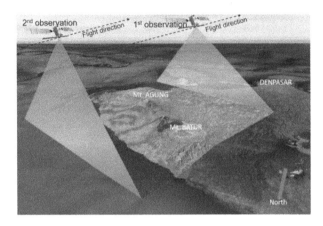

Figure 2.   Schematic view of SAR satellites observing Mount Agung.

To obtain the surface displacements, DInSAR analysis requires topographical (Digital Elevation Model/DEM) information on the monitored area. DEM is used to remove the topographic phase component from the interferometric phase (interferogram). To obtain reliable displacement results by DInSAR, several errors should be removed, i.e., orbit inaccuracies, atmospheric delays, and noises (Hanssen, 2002).

2.2   *SAR dataset and SBAS analysis*

In this research, the SAR dataset from the Sentinel-1A and Sentinel-1B satellites (operated by the European Space Agency: ESA) is employed to monitor the displacements of Mount Agung. The SAR images used in this research comprise 224 scenes. These images were taken from October 7, 2014 to October 23, 2020 in the descending orbit direction. Detailed information on the SAR dataset is presented in Table 1.

Table 1.   Sentinel-1A and -1B SAR dataset.

Satellite name	Sentinel-1A and -1B
Number of SAR data images	224 scenes
From (date)	October 7, 2014
To (date)	October 23, 2020
Orbit direction	Descending
SAR sensor direction	Right looking

The Small Baseline Subset (SBAS) DInSAR time-series analysis (Berardino et al., 2002) is employed to obtain the spatial distribution and temporal transition of the surface displacements. SBAS-DInSAR utilizes a series of SAR images and conducts multiple DInSAR processing to produce several interferograms. The number of possible combinations of SAR pairs for multiple DInSAR processing is based on the threshold of the spatial and temporal baselines. In this research, the temporal baseline threshold is 36 days and the spatial baseline threshold is 250 meters. The thresholds generate the sets of combinations shown in Figure 3. The total number of pairs to obtain the time-series displacements is 602.

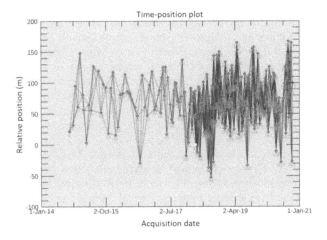

Figure 3. Combinations of SAR pairs for SBAS-DInSAR analysis. The green diamonds are the number of SAR images and the yellow diamond is taken as the super master.

3 RESULTS AND DISCUSSION

The yearly spatial distributions of the LOS displacements are shown in Figure 4 for the period of November 12, 2014 to October 23, 2020. The colors from green to red indicate that the LOS displacements are positive or compression, and the colors from green to blue indicate that the LOS displacements are negative or extension.

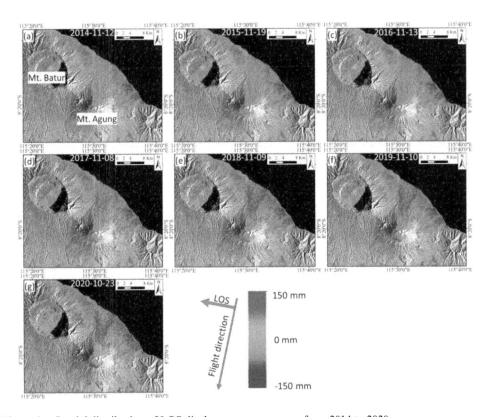

Figure 4. Spatial distribution of LOS displacements every year from 2014 to 2020.

It can be seen that no significant displacements occurred from November 12, 2014 to November 13, 2016 (Figures 4a-c). However, larger displacements were found from November 13, 2016 to November 8, 2017 in the northern part of Mount Agung.

Figure 5 shows the time transition of the LOS displacements at selected points. The locations of the selected points are shown in Figure 5a. The elevations of the selected points are given by Table 2. Points AGUNG-1 and AGUNG-14 are located around the top of Mount Agung, while points AGUNG-13 and AGUNG-26 are located in areas of low elevation. Unfortunately, there are no results (displacements) in this analysis for point AGUNG-1 because of low coherency.

It can be clearly seen that the displacements at points AGUNG-2 to AGUNG-7 and AGUNG-14 to AGUNG-18 have periodical behavior. This was possibly caused by tropospheric delays of the microwaves transmitted from SAR through the atmosphere. The tendency for such behavior is related to the elevation of each point; lower elevations are less affected by tropospheric delays (Figures 5b-c).

Table 2.   Elevations of selected points.

Point name	Elevation (m)	Point name	Elevation (m)	Point name	Elevation(m)
AGUNG-1	2874.18	AGUNG-10	786.52	AGUNG-19	1492.74
AGUNG-2	2738.53	AGUNG-11	503.73	AGUNG-20	1221.17
AGUNG-3	2447.01	AGUNG-12	256.96	AGUNG-21	986.26
AGUNG-4	2209.61	AGUNG-13	59.92	AGUNG-22	771.07
AGUNG-5	1974.36	AGUNG-14	2703.02	AGUNG-23	514.96
AGUNG-6	1762.73	AGUNG-15	2495.15	AGUNG-24	257.76
AGUNG-7	1520.42	AGUNG-16	2227.10	AGUNG-25	175.94
AGUNG-8	1249.26	AGUNG-17	1971.56	AGUNG-26	75.39
AGUNG-9	987.01	AGUNG-18	1724.46		

There were no significant displacements from October 7, 2014 to early March 2017 (Figures 5b-c). This is called the dormant period of Mount Agung. The pre-eruption period of Mount Agung lasted from the end of March 2017 to early August 2017. The co-eruption period lasted from the end of August 2017 to the end of November 2017. This is also called the crisis period. Significant displacements were also detected a few months prior to the main eruptions on November 25 and 29, 2017 (Figures 5b-c). The post-eruption period lasted from early December 2017 to May 2019. This is probably the first cycle of seismic activities for Mount Agung since November 2014.

After the seismic activities had been decreasing for about one and a half years, the second cycle of seismic activities for Mount Agung began on May 28, 2019. Increasing displacements were detected a few months prior to the eruption on June 13, 2019. However, this eruption was much smaller than the eruptions on November 25 and 29, 2017. At least SBAS-DInSAR detected increasing displacements prior to the eruption. This becomes important information for use in detecting the next possible seismic cycle of Mount Agung.

In addition, the spatial distribution of the displacements obtained in this study is compared to that in the previously published study, as seen in Figure 6 (Albino et al., 2019). Figure 6a presents the LOS displacements from April to August 2017 which coincide with the results of this study, as presented in Figures 4a-c. Both studies indicate that there were no significant displacements prior to August 2017. Figure 6b presents the detection of significant displacements in the northern area of Mount Agung from August to November 2017. These results are similar to those of this study, as presented in Figures 4d-g.

The displacements recently started to increase for the third time, in September 2020; and thus, careful monitoring should be continued.

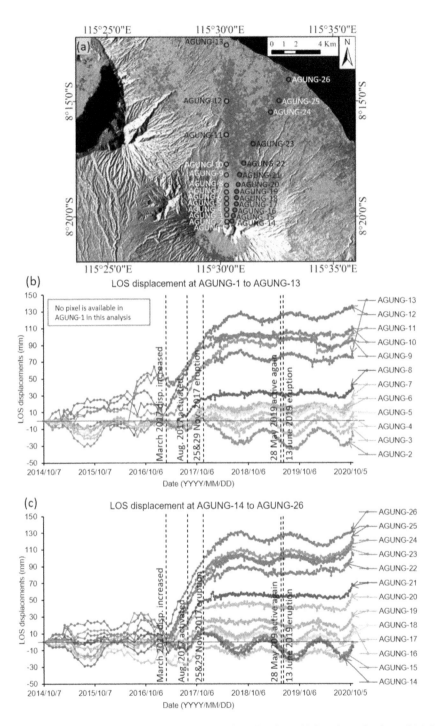

Figure 5. Time transition of LOS displacements at selected points; (a) location of points, (b) LOS displacement transition at points AGUNG-2 to AGUNG-13, and (c) LOS displacement transition at points AGUNG-14 to AGUNG-26.

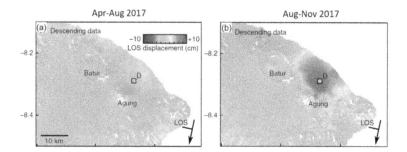

Figure 6. LOS displacement maps provided by Albino et al. (2019): (a) LOS distribution during period of April to August 2017 and (b) LOS distribution during period of April to August 2017.

4 CONCLUSION

This study has presented the monitoring results of ground displacements induced by the volcanic activity of Mount Agung by means of SBAS-DInSAR. The results indicated that there had been no remarkable displacements before May 2017. However, significant displacements were detected by SBAS-DInSAR from May to November 2017, and they increased again from May to July 2019 after a decrease in activity for one and a half years.

Furthermore, SBAS-DInSAR provided the surface displacement results from the pre-eruption, co-eruption, and post-eruption periods. These results will be very important for understanding the active cycle of Mount Agung and for predicting future potential crisis periods. Although tropospheric delays remain in areas of high elevation, SBAS-DInSAR presents reliable measurement results for the region surrounding Mount Agung, namely, areas of lower elevation.

Finally, it is concluded that SBAS-DInSAR will be useful for monitoring volcanic activity and for predicting future potential crisis periods.

ACKNOWLEDGEMENTS

The authors express their gratitude to the European Space Agency (ESA) and the Japan Aerospace Exploration Agency (JAXA) for providing the Sentinel-1 SAR data and the Digital Elevation Model (DEM), respectively, free of charge. This research was partially supported by JSPS KAKENHI Grant Number 16H03153.

The authors also extend their appreciation to Ms. H. Griswold for proofreading the manuscript.

REFERENCES

Albino, F., Biggs, J., Syahbana, D. K., 2019, Dyke intrusion between neighbouring arc volcanoes responsible for 2017 pre-eruptive seismic swarm at Agung, *Nature Communications, 10*. https://doi.org/10.1038/s41467-019-08564-9

Berardino, P., Fornaro, G., Lanari, R., Sansosti, E., 2001, A new algorithm for surface deformation monitoring based on small baseline differential SAR interferograms, *IEEE Transactions on Geoscience and Remote Sensing*, 40(11),2375–2383. doi: 10.1109/TGRS.2002.803792.

Brunori, C. A., Bignami, C., Stramondo, S., Bustos, E., 2013, 20 years of active deformation on volcano caldera: Joint analysis of InSAR and AInSAR techniques, *International Journal of Applied Earth Observation and Geoinformation*, 23, 279–287.

Chen, Y., Remy, D., Froger, J-L., Peltier, A., Villeneuve, N., Darrozes, J., Perfettini, H., Bonvalot, S., 2017, Long-term ground displacement observations using InSAR and GNSS at Piton de la Fournaise volcano between 2009 and 2014. *Remote Sensing of Environment*, 194, 230–247.

Hanssen, R. F., 2002, Radar Interferometry: data interpretation and error analysis. The Netherlands: *Kluwer Academic Publisher*.

Kobayashi, T., 2018, Locally distributed ground deformation in an area of potential phreatic eruption, Midagahara volcano, Japan, detected by single-look-based InSAR time series analysis. *Journal of Volcanology and Geothermal Research*, 357, 213–223.

Lingyun, J., Qingliang, W., Shanlan, Q., 2013, Present-day deformation of Agung volcano, Indonesia, as determined using SBAS-InSAR. *Geodesy and Geodynamics*, 4, 65–70.

Pepe, A., Calò, F., 2017, A Review of Interferometric Synthetic Aperture RADAR (InSAR) Multi-Track Approaches for the Retrieval of Earth's Surface Displacements. *Applied Sciences*, 7, 1264.

Syahbana, D. K., Kasbani, K., Suantika G., Prambada, O., et al., 2019, The 2017–19 activity at Mount Agung in Bali (Indonesia): Intense unrest, monitoring, crisis response, evacuation, and eruption. *Science Report*, 9, https://doi.org/10.1038/s41598-019-45295-9

Rock Mechanics and Engineering Geology in Volcanic Fields – Ohta, Ito & Osada (eds)
© 2023 copyright the Author(s), ISBN 978-1-032-27657-1

Resurgent magma characteristics of a super-volcano, the Youngest Toba Tuff, northern Sumatra, inferred from zircon U-Pb geochronology

Hisatoshi Ito*

Central Research Institute of Electric Power Industry, Chiba, Japan

ABSTRACT: Resurgent magma of the climactic Youngest Toba Tuff (YTT) super-eruption at ~74,000 years ago or ~0.074 million years ago (Ma), was investigated using zircons from a post-caldera lava dome. Here I reconfirmed that magmatic activity of the Toba Caldera Complex (TCC) started at ~1.3 Ma and that zircon-forming magmatic activity of YTT flared up at ~0.3 Ma and significantly declined at ~0.2 Ma, well before the ~0.074 Ma YTT eruption. Therefore, the climatic YTT eruption may not have been triggered by intensive/rapid magma supply at the time of eruption. The trigger may have been a small magmatic input or some other mechanisms. The remarkable accordance of zircon U-Pb age distribution between YTT and a post YTT lava dome indicates that post-caldera magmatic activity/resurgence occurred using essentially the same magma with YTT. A significant decline of magmatic inputs since ~0.2 Ma may indicate that another super-eruption is unlikely for at least hundreds of thousands of years to come in the TCC.

Keywords: Super-volcano, U-Pb dating, Zircon, Toba, Magmatic history

1 INTRODUCTION

The Toba Caldera Complex (TCC), northern Sumatra (Figure 1), is thought to originate from subduction of the fluid-rich Investigator Fracture Zone directly beneath the continental crust of Sumatra, leading to formation of a deep crustal hot zone that feeds the TCC (Koulakov et al., 2016). The TCC is well described as the site of super-eruption that severely affected the Earth's climate and human evolution (Petraglia et al., 2012). In fact, some researchers believe that the ~0.074 Ma Youngest Toba Tuff (YTT) super-eruption from TCC brought about "volcanic winter" (Rampino and Self, 1992), which nearly caused the demise of humankind (Ambrose, 1998). Therefore, understanding how the YTT super-eruption occurred is vital for our human society to recognize and hopefully prepare for our fate in the future.

Zircon geochronology and trace element chemistry can provide powerful windows into the duration and continuity of magma assembly in the build up to a super-eruption like the YTT (Reid and Vazquez, 2017). Reid and Vazquez (2017) revealed that the YTT zircons have nucleated episodically and persisted over 0.5 million years in the magma, eventually led to the ~0.074 Ma super-eruption.

The TCC is a 100×30 km topographic depression of several overlapping calderas resulting from four major eruptions during the Quaternary (Knight et al., 1986; Chesner et al., 1991) (Figure 1). It also is the largest resurgent Quaternary caldera on Earth (Smith and Bailey, 1968) and is elongated in a NW-SE direction parallel to the active volcanic front of Sumatra.

*Corresponding author: ito_hisa@criepi.denken.or.jp

DOI: 10.1201/9781003293590-8

TCC has been the locus of silicic volcanic activity for at least 1.3 million years. The cataclysmic ~0.074 Ma 2,800 km^3 (minimum volume) of non-welded to densely welded ignimbrite and coignimbrite ashfall associated with the YTT exceeds the combined volume of ignimbrite and ash deposits from the three preceding ignimbrites, the ~1.2 Ma ~35 km^3 Haranggaol Dacite Tuff (HDT), the ~0.8 Ma ~800–2,300 km^3 Oldest Toba Tuff (OTT), and the ~0.5 Ma ≥60 km^3 Middle Toba Tuff (MTT).

Here I analyzed zircons from a post-caldera (i.e., post-YTT) lava dome. As a result, I could further constrain the entire magmatic history of the TCC.

2 METHODOLOGY

2.1 *Sampling*

Figure 1.   Map of the Toba Caldera Complex (TCC) and the sampling locality (yellow star). Caldera locations are indicated by dashed lines where HDT = green, OTT = yellow, MTT = blue, and YTT = red. Post-caldera lava domes are outlined by orange circles (modified after Chesner (2012)). Towns are indicated in black dot. Inset: Map of Sumatra and its vicinity with TCC marked as a white star.

A sample from a post caldera lava dome (Tuk Tuk lava dome) was obtained at the northeastern part of Samosir Island (Figure 1). The Tuk Tuk lava dome is a crystal-rich rhyodacite with quartz, plagioclase, sanidine, biotite, and amphibole.

2.2 *Zircon U-Pb dating*

Zircons were extracted by conventional magnetic and heavy liquid separations. No acid (HF nor HCl) treatments were performed because zircons are mostly clean without glass coating. Hundreds of zircon grains were obtained from a 1 kg sample. Zircons were embedded in a PFA Teflon sheet and unpolished surface was used for the analyses.

Zircon U-Pb dating was performed at the Central Research Institute of Electric Power Industry, using laser ablation inductively coupled plasma mass spectrometry (LA-ICP-MS: a Thermo Fisher Scientific ELEMENT XR magnetic sector-field ICP-MS coupled to a New

Wave Research UP-213 Nd-YAG laser) with experimental conditions followed by Ito (2020). A 30 μm laser beam with ~7 J/cm^2 energy density for 30 s was employed for laser ablation after 30 s background measurement. U-Pb (^{238}U-^{206}Pb) ages were obtained using 10–20 s laser ablation data for shallow section and using 20–30 s data for deep section. As for the initial Th disequilibrium correction, (Th/U)$_{magma}$ of 6.0 was used according to the literature (Reid and Vazquez, 2017; Mucek et al., 2017). The common Pb (^{207}Pb/^{206}Pb) was assumed to be 0.832, which is essentially the same with 0.834 adopted by Reid and Vazquez (2017).

3 RESULTS

Fifty zircons from the Tuk Tuk lava dome were analyzed for U-Pb dating by LA-ICP-MS. Data with high (>75%) common Pb contamination or high (>50%) uncertainty were excluded for further analyses to keep data quality. As a result, the Tuk Tuk lava dome zircons younger than 3 Ma yielded 0.38 ± 0.08 Ma (MSWD = 3.3; n = 11) for shallow section and 0.30 ± 0.04 Ma (MSWD = 6.2; n = 21) for deep section, respectively. In Figure 2, both shallow and deep section ages were grouped together and shown as a probability density plot with a histogram.

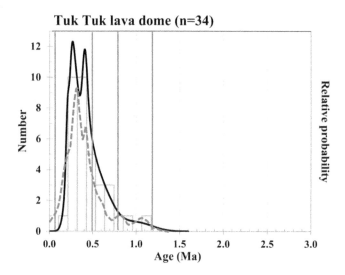

Figure 2. U-Pb (^{238}U-^{206}Pb) age distributions (a histogram and a probability density plot) for zircons younger than 3 Ma obtained from the Tuk Tuk lava dome. Vertical red lines indicate eruption ages of either ~1.2 Ma HDT, or ~0.8 Ma OTT, or ~0.5 Ma MTT, or ~0.074 Ma YTT. U-Pb ages of both the shallow and deep sections of zircon are grouped and shown in black. The YTT U-Pb age distribution from Reid and Vazquez (2017) is shown as a dashed red curve. n = number of ages that meet quality criteria (see text for further information).

4 DISCUSSION

The Tuk Tuk lava dome U-Pb data (Figure 2) show that most zircons grew ~0.1–0.4 Ma before eruption at ~0.074 Ma (Mucek et al., 2017). Its U-Pb age distribution overlaps closely with that of YTT (Reid and Vazquez, 2017). Therefore, it is reasonable to mention that post caldera lava domes erupted using essentially the same magma with YTT. In other words, post-caldera magmatic activity/resurgence occurred using essentially the same magma with YTT, or remnant YTT magma "intruded" into shallow levels of crust causing resurgence (de Silva et al., 2015). As with YTT zircons, post-caldera lava dome zircons nucleated at various times before but mostly after the MTT eruption at ~0.5 Ma.

The Tuk Tuk lava dome U-Pb data (Figure 2) also indicate that zircon-forming magmatic activity of TCC started at ~1.2–1.3 Ma. This is in agreement with the literature: the abundance of volcanic particles since 1.4 Ma in the South China Sea indicates that TCC activity started at ~1.4 Ma (Lee et al., 2004) and the HDT was dated at ~1.2 Ma (Nishimura et al., 1984). It is clear that the zircon-forming magmatic activity of YTT significantly declined at ~0.2 Ma, well before the ~0.074 Ma YTT eruption. Therefore, the climatic YTT eruption may not have been triggered by intensive/rapid magma supply at the time of eruption. The trigger may have been a small magmatic input (Matthews et al., 2012) or some other mechanisms (such as mechanical roof failure and roof subsidence into the magma reservoir) (Budd et al., 2017; Troll et al., 2002; Gregg et al., 2012). The YTT and post-YTT magma was stored incrementally over 1 million years and it vented >2,800 km^3 monumental tephras paroxysmally at ~0.074 Ma. Finally, a significant decline of magmatic inputs since ~0.2 Ma may indicate that magmatic build up has been weak since then and therefore another super-eruption is unlikely for at least hundreds of thousands of years to come in the TCC.

5 CONCLUSIONS

Zircon U-Pb ages from a post-caldera lava dome at Toba indicated that the post caldera magma was essentially the same with the caldera-forming magma and the magma at Toba significantly declined its activity since ~0.2 Ma well before the climatic ~0.074 Ma YTT eruption, which may indicate that another super-eruption at Toba is unlikely for at least hundreds of thousands of years in the future. Contrary to Toba, zircon crystallization continued immediately before major eruptions in the Taupo Volcanic Zone, New Zealand (e.g., Wilson and Charlier, 2009), which indicates continued active magmatism and hence, humans may have to pay particular attention to another supereruption there.

ACKNOWLEDGEMENTS

I thank S. de Silva for organizing the 7th International Workshop on Collapse Calderas, held in Toba Caldera in 2018, which motivated this study. I also thank M. Yukawa for her help with sample preparation and LA-ICP-MS data collection, and Y. Adachi for her technical assistance on LA-ICP-MS.

REFERENCES

Ambrose, S.H., 1998, Late Pleistocene human population bottlenecks, volcanic winter, and differentiation of modern humans, *J. Hum. Evol.*, 34, 623–651.

Budd, D. A., Troll, V.R., Deegan, F.M., Jolis, E.M., Smith, V.C., Whitehouse, M.J., Harris, C., Freda, C., Hilton, D.R., Halldórsson, S.A. and Bindeman, I.N., 2017, Magma reservoir dynamics at Toba caldera, Indonesia, recorded by oxygen isotope zoning in quartz, *Sci. Rep.*, 7, 40624.

Chesner, C.A., Rose, W.I., Deino, A., Drake, R. and Westgate, J.A., 1991, Eruptive history of Earth's largest Quaternary caldera (Toba, Indonesia) clarified, *Geology*, 19, 200–203.

Chesner, C.A., 2012, The Toba caldera complex, *Quatern. Int.*, 258, 5–18.

de Silva, S.L., Mucek, A., Gregg, P. and Pratomo, I., 2015, Resurgent Toba–field, chronologic, and model constraints on time scales and mechanisms of resurgence at large calderas, *Front. Earth Sci.*, 3, Article 25.

Gregg, P.M., de Silva, S.L., Grosfils, E.B. and Parmigiani, J.P., 2012, Catastrophic caldera-forming eruptions: Thermomechanics and implications for eruption triggering and maximum caldera dimensions on Earth, *J. Volcanol. Geotherm. Res.*, 241–242, 1–12.

Ito, H., 2020, Magmatic history of the Oldest Toba Tuff inferred from zircon U–Pb geochronology, *Sci. Rep.*, 10, 17506.

Knight, M.D., Walker, G.P.L., Ellwood, B.B. and Diehl, J.F., 1986, Stratigraphy, paleomagnetism, and magnetic fabric of the Toba tuffs: Constraints on the sources and eruptive styles, *J. Geophys. Res.*, 91, 10, 355–10, 382.

Koulakov, I., Kasatkina, E., Shapiro, N.M., Jaupart, C., Vasilevsky, A., Khrepy, S., Ak-Arifi, N. and Smirnov, S., 2016. The feeder system of the Toba supervolcano from the slab to the shallow reservoir. *Nat. Commun.*, 7, 12228.

Lee, M.-Y., Chen, C.-H., Wei, K.-Y., Iizuka, Y. and Carey, S., 2004, First Toba supereruption revival, *Geology*, 32, 61–64.

Matthews, N.E., Huber, C., Pyle, D.M. and Smith, V.C., 2012, Timescales of magma recharge and reactivation of large silicic systems from Ti diffusion in quartz, *J. Petrol.*, 53, 1385–1416.

Mucek, A.E., Danišík, M., de Silva, S.L., Schmitt, A.K., Pratomo, I. and Coble, M.A., 2017, Post-supereruption recovery at Toba Caldera, *Nat. Commun.*, 8, 15248.

Nishimura, S., Abe, E., Nishida, J., Yokoyama, T., Dharma, A., Hehanussa, P. and Hehuwat, F., 1984, A gravity and volcanostratigraphic interpretation of the Lake Toba region, North Sumatra, Indonesia, *Tectonophysics*, 109, 253–272.

Petraglia, M.D., Korisettar, R. and Pal, J.N., 2012, The Toba volcanic super-eruption of 74,000 years ago: Climate change, environments, and evolving humans, *Quat. Int.*, 258, 1–4.

Rampino, M.R. and Self, S., 1992, Volcanic winter and accelerated glaciation following the Toba super-eruption, *Nature*, 359, 50–52.

Reid, M.R. and Vazquez, J.A., 2017, Fitful and protracted magma assembly leading to a giant eruption, Youngest Toba Tuff, Indonesia, *Geochem. Geophys. Geosyst.*, 18, 156–177.

Smith, R.L. and Bailey, R.A., 1968, Resurgent cauldrons, *Geol. Soc. Am. Mem.*, 116, 613–662.

Troll, V.R., Walter, T.R. and Schmincke, H.-U., 2002, Cyclic caldera collapse: Piston or piecemeal subsidence? Field and experimental evidence, *Geology*, 30, 135–138.

Wilson, C.J.N. and Charlier, B.L.A., 2009, Rapid rates of magma generation at contemporaneous magma systems, Taupo Volcano, New Zealand: Insights from U–Th model-age spectra in zircons, *J. Petrol.*, 50, 875–907.

Rock Mechanics and Engineering Geology in Volcanic Fields – Ohta, Ito & Osada (eds)
© 2023 copyright the Author(s), ISBN 978-1-032-27657-1

Geological investigation of the excavation-induced landslide in a geothermal area in Sumatra, Indonesia

Keiji Orihara, Tomohiro Yasuda*, Koji Nishida & Koki Kimura
Kiso-Jiban Consultants Co., Ltd., Tokyo, Japan

Achmad Sri Fadli & Reza Ardiansyah Suyono
Pertamina Geothermal Energy, Jakarta, Indonesia

ABSTRACT: A geothermal power plant was planned to be constructed in mountainous and geothermal area in South Sumatra, Indonesia. The site is underlain by volcanic rocks and their decomposed soils in the Quaternary period. High cut slopes with a maximum height of about 80 m were required for the site formation works on a mountain ridge. The cut slopes were designed based on limited soil investigation data in the thick tropical rainforest. A small scale landslide took place immediately after the excavation started on one of the slopes. To find out the cause of the landslide, an additional soil investigation was carried out. The removal of the vegetation from the slope made an examination of rock conditions by a geologist possible. The additional investigation and examination revealed that there exist thick residual soils, expansive clay minerals related to hydrothermal alteration, discontinuities/faults with slickenside planes irregularly on the slopes, typically formed by geothermal activities under tropical conditions. A few potential landslide masses were also recognized on the slopes.

A remedy slope design was carried out using newly obtained information for the entire slopes including rectification of landslides. The rectification work includes soil removal and installation of rock anchors at the upper part and foot of landslides respectively based on results of stability analysis. Because of complex geological features, we decided to resume the slope excavation after examination of the slopes and guidance by an on-site geologist and monitoring of ground using geotechnical devices. The slope design including slope angle and protection was altered in a timely manner once the adverse geology was encountered. After downtime for 9 months, the remaining slope excavation was safely completed. This paper presents the landslide feature related to geothermal geology and how it was rectified.

Keywords: Geothermal, landslide, hydrothermal alteration, cut slope, slope design

1 INTRODUCTION

In the land formation works for the construction of geothermal power plant in South Sumatra, a small scale landslide took place at the very beginning of excavation works. The cause of landslide was investigated and it was found to have been related to the typical geology in geothermal areas. However, the geological conditions of the slope were revealed in detail only after the landslide took place. It was because the thick tropical rainforest covered the entire site and it made the geological and geotechnical investigation difficult prior to the design and construction.

*Corresponding author: yasuda.tomohiro@kiso.co.jp

DOI: 10.1201/9781003293590-9

2 GEOLOGICAL SETTING

2.1 *General geology*

The site is situated in the mountainous area and the elevation thereof is 1,400 m or higher above sea level. The nearest volcano is located about 4 km southwest of the site. According to Geological Map of Indonesia (U.S. Geological Survey, 1965), the project site coincides with the Quaternary volcanic rocks, namely Banatan Volcanics formation, which consists of andesite-basalt lava, volcanic breccia, and bedded tuffaceous sandstone flow deposits.

2.2 *Geological and geotechnical features*

The bedrock is comprised of highly fractured andesite and highly to moderately fractured andesite breccia.

The overlying soils predominantly comprise cohesive residual soils with some interbedded layers of silty sand. The upper cohesive soils are generally firm becoming stiff at approximately 4 m depth and have a relatively high moisture content of about 50% typically of volcanic soils. The thickness of residual soils ranges from 8 to 14 m.

Groundwater level is at approximately 8 to 10 m deep below the existing ground surface. The water table appears to be shallower towards the foot of the slope.

3 LANDSLIDE AND POSSIBLE CAUSE

3.1 *Landslides*

The first landslide took place on the southern slope of the site. It has a dimension of 60 m wide and about 16 m high. Initially, it was assumed this landslide had a slip circle occurred in residual soil. However, after the second landslide assessment, the failure mode was revised to block failure with a near horizontal sliding plane.

A second landslide covers almost the whole of the first landslide. The width and length of the landslide were approximately 60 m and 80 m respectively as shown in Figure 1. The crown of landslide was depressed by about 3 m and landslide scarp (tension crack) was formed. Figure 2 shows an overview of the second landslide from the north.

Figure 1. Landslide mass in the southern slope.

Figure 2. Overview of the second landslide.

3.2 *Possible cause of landslide*

After the second landslide, an extensive investigation was carried out to reveal the cause of landslide, which is required for remedy measures. The investigation consisted of a number of boreholes, geological reconnaissance and laboratory testing. The investigation revealed the type of landslide and typical geological features owing to the geothermal conditions, which might be the causes of landslide.

1. Type of landslides and sliding planes

Figure 3 shows a presumed cross section of landslides based on the additional boreholes. The type of landslide is a block failure with a flat sliding plain. The sliding plain is formed in the residual soil. Layers of clay minerals (smectite and kaolinite) and slickenside form a shear zone of a sliding plane. Figure 4 shows slickenside appeared on a main scarp face, which is a part of sliding plane observed at the top of landslide mass. A potential sliding plane consisting of slickenside was also observed in the borehole No A4 at a depth of about 9.5 m. Figure 5 presents results of X-ray diffraction analysis of clay minerals obtained from the sliding plane.

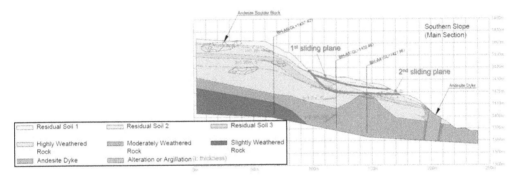

Figure 3.   Presumed cross section of landslides in the southern slope.

Figure 4.   Slickenside at the top of landslide.

Tuff Breccia has been intruded by numerous hydrothermal alteration. Then, alteration veins (thin layers) of secondary clay minerals of smectite and kaolinite are formed. The thickness of alteration veins ranges from a few millimeters to 50 mm or more in general. They are typically yellowish in color. Smectite is known as "expansive clay", which expands its volume and reduces its strength drastically once it is wet. Kaolinite has low

65

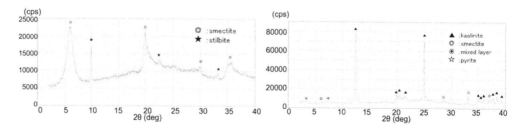

Figure 5.   X-ray diffraction analysis of samples from a sliding plane of landslide in the southern slope.

strength. As shown in a stereo net in Figure 6, strikes and dips of alteration veins have no predominant trend and the alteration veins were found randomly over the site during the slope excavation.

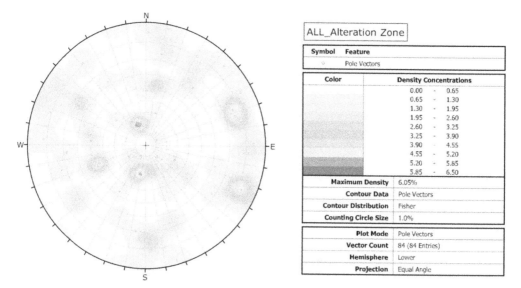

Figure 6.   Stereo net of strikes and dips of alteration veins.

A slickenside has a smooth polished surface caused by previous frictional movements in faults and landslides. The contact surface of slickenside has very low strength. Several faults were found on the exposure and in the boreholes. The faults accompanied slickenside, breccia and/or clay as fault gouge. Similar to alteration veins, strikes and dips of slickensides have no predominant trend.

The slaking feature was also recognized in the tuff breccia. Slaking tests were carried out on the rock core samples in accordance with JGS 2124 (Japanese Geotechnical Society, 2009). The test results indicated that the tuff breccia is categorized in Classes 2 to 4, which are likely to slake.

(2) Groundwater

Water seepage was observed from the edge of the landslide and elsewhere on the slope. The groundwater was trapped in the landslide mass at the shallow depth. In the boreholes located on the second landslide mass, the groundwater level was first encountered at the shallow depth near the ground surface. With progress of drilling, the groundwater level dropped to a depth of about 10 m below the sliding plane. This means there are two levels of groundwater in the landslide area. Accordingly, it is presumed that the groundwater was trapped on the impervious clay of hydrothermal alteration vein in the landslide mass.

(3) Back analysis

Figure 7 shows a back calculation of stability analysis employing Janbu's simplified method for landslide mass based on a critical factor of safety of 1.0. The effective shear strength was back-calculated to be c'=0 and an angle of internal friction of 20 degrees for the presumed sliding plane. The back-calculated strengths of the sliding plane are much smaller than the material strengths of residual soils surrounding the sliding plane. Deep iron staining of material on slickenside contact is an evidence of a relict geological feature, i.e. the sliding plane was pre-existed. Clay mineral and contact of slickensides have very low strength. Therefore, it is reasonable to assume that the sliding plane results from the pre-existing discontinuity with slickenside and veins of clay minerals or the combination of both.

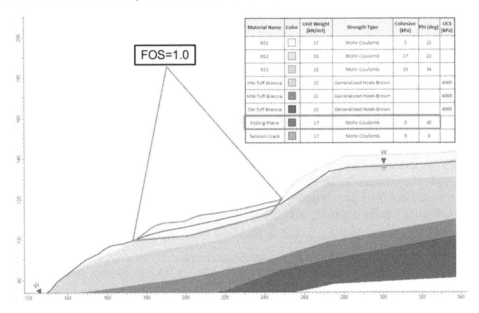

Figure 7.   Back analysis of landslide in the southern slope.

(4) Landslide mechanism and contributing factors

Based on the foregoing findings, we concluded the mechanism and contributing factors as follows:

In the soil and rock mass of slopes, there exist multiple potential sliding planes formed along the discontinuities with slickenside and alteration veins with clay minerals. A rise of groundwater level in the landslide mass is a primary contributing factor and made the landslide mass unstable. The rainfall and surface runoff water infiltrated through discontinuities in soil and rock mass. The groundwater was trapped on the impervious clay and slickenside sliding plane in the landslide mass. With the water level rise, pore water pressure was built up along the sliding plane and eventually reduced the effective stress, i.e. a resisting force. A groundwater rise increased the weight of landslide mass, i.e. a driving force. Then, the landslide took place along the most critical sliding plane.

4   SLOPE DESIGN

4.1   *Original design*

The slope excavation was designed originally based on the results of soil investigation of boreholes, seismic survey and site reconnaissance, which were carried out before the project started. The information obtained by the soil investigation was limited due to the thick tropical rainforest. A geologist was unable to obtain the typical geological features in the geothermal area due to few outcrops because the ground surface was covered almost by vegetation. The information from exploratory boreholes was also limited in both quantity and quality because of difficult access and un-updated local drilling skill. The thin alteration veins of clay minerals and slickenside would have been easily missed even by skilled drillers if their operation was not careful enough.

Therefore, the original design was basically carried out based on the material strength of mass of residual soils and weathered rocks and stability analysis of slip circles. The material strengths were determined basically by laboratory tests on soil samples for soil mass and by Hoek-Brown criterion (Hoek and Brown, 1997) for rock mass. Alteration veins and discontinuities of slickenside were not considered in the design.

4.2 *Remedy design*

1. Remedy of landslide

   The design of the southern slope where the landslide took place was totally changed from the original design. The landslide mass was removed from the slope by changing the config-uration of the slope.

2. General slope design

   It is important for the remedy design to take into consideration the mechanism of landslide and its contributing factors. The complication of existence of clay layers and trapped groundwater make the effective drainage from the slope difficult. It is there-fore concluded that the stability of the slope will not be achieved by drainage of groundwater but by adopting a gentle slope and/or other protective measures.

The slope design was reviewed and changed to the following methods:

   i. Identify potential landslides from topographical features.
   ii. Determine material strengths of soil and rock mass by laboratory tests and Hoek-Brown criterion respectively as shown in Table 1.

Table 1. Material parameters in slope design.

Material name	Depth (m)	Unit weight (kN/m^3)	Undrained shear strength, s_u (kPa)	Drained shear strength		Unconfined compres-sive strength (kPa)
				c' (kPa)	phi' (degree)	
RS1	0 to 4	17	40	5	22	-
RS2	4 to 7.5	18	80	17	22	-
RS3	7.5 to 10	18	130	10	34	-
HW (tuff breccia)*[1]	10 or deeper	22	-	-	-	4,000

* 1: Geological Strength Index (GSI)=40, Intact rock constant=19, Disturbance factor=0.5

   iii. Reduce material strengths of soil mass forming the slopes where alteration veins are concentrated. The ratio of reduction depends on the degree of concentration of the veins. In the reduction, an idea of GSI of Hoek-Brown criterion was referred to as GSI is a reduction factor from degrees of discontinuities. Table 2 lists reduced strengths.
   iv. Assume the design groundwater level at 2 to 3 m higher than the observed groundwater level and passing through the toe of the cut slope.

Table 2. Reduced strengths of soil mass with alteration vein concentration.

Material name	Unit weight (kN/m^3)	Undrained shear strength		Drained shear strength	
		c_{ud} (kPa)	phi$_{ud}$ (degree)	c' (kPa)	phi' (degree)
RS1	17	16	7	3	15
RS2	18	18	7	9	15

v. Conduct slope stability analysis using the effective shear strengths for a minimum factor of safety of 1.3 in the normal condition. The undrained shear strengths and a factor of safety of 1.0 are used in the earthquake condition.

vi. Adopt a horizontal pseudo static force of k=0.19 in the earthquake condition. The force is derived from half of the peak ground acceleration of 0.345g multiplied by a site coefficient of 1.1. The peak ground acceleration is determined locally based on Standard National Indonesia SNI 1726 (2012).

vii. Adopt the following general gradients of slopes and berms without reinforced measure where possible.
- 1(V) in 2(H) for Residual Soil 1 (RS1) and Residual Soil 2 (RS2)
- 1(V) in 1.5(H) for Residual Soil 3 (RS3)
- 1(V) in 1(H) for Highly Weathered (HW) to Slightly Weathered (SW) Rock
- A 4 m-wide berm for each 10 m or less in height

viii. Provide surface drainage system of U-ditches and culverts based on the assumed rainfall of 225 mm/hour for a return period of 100 years.

ix. Provide surface protection against water runoff and slaking. The protection includes vegetation, lean concrete, concrete frame depending on the slope gradients and soil/rock types forming slopes.

x. Provide soil nails, ground anchors, gravity wall, etc. in the area with site constraints and potential landslides.

5 SLOPE EXCAVATION

There are hazardous geological conditions of alteration veins and clay minerals, faults associated with slickenside, and presence of slaking-prone materials in a complex manner at the site. They can be a potential risk of forming a sliding plane of landslide and slopes to be unstable. It is practically impossible to foresee all the risks arising from these geological features in the design stage. The actual soil and rock conditions of slopes would be different from the conditions estimated in the design.

To enhance safe slope excavation, the following observational methods were adopted.

i. Monitor regularly slope movement during excavation using inclinometers and displacement pegs.

ii. Monitor regularly groundwater levels by water standpipes.

iii. Provide on-site geologist full time to identify geological risks during slope excavation.

iv. Change the slope gradient or provide additional slope protection measures where necessary in a timely manner.

The slope excavation was safely completed after the downtime of about 9 months. The downtime was required for additional investigation and remedy designs of slopes.

ACKNOWLEDGEMENTS

The authors are grateful to Pertamina Geothermal Energy, a project owner who allows us to present this paper and also Toshiba Corporation Energy Systems & Solutions Company, a main contractor of the project who conducted all the works related to the slope excavation in collaboration with PGE Construction Team. We specially thank all members on site who have accomplished this challenging task as a team.

REFERENCES

Hoek, E., Brown, E.T., 1997, Practical estimates of rock mass strength, *Int. J. Rock Mech. Min. Sci.*, 34 (8),1165–1186.
Japanese Geotechnical Society, 2009, Method for rock slaking test, Japanese Geotechnical Society Standard JGS 2124.
Standard National Indonesia, 2012, SNI 1726 Design procedure for earthquake resistance for building and non-building structure.
U.S. Geological Survey, 1965, Geologic map of Indonesia.

Rock Mechanics and Engineering Geology in Volcanic Fields – Ohta, Ito & Osada (eds)
© 2023 copyright the Author(s), ISBN 978-1-032-27657-1

Characteristic topography of large scale landslides occurred in the 2016 Kumamoto earthquakes in Kumamoto Prefecture, Japan

Katsuhisa Nagakawa*, Kiyoyuki Tsumita, Kenzo Fukui & Hiroshi Kochihira
Kiso-Jiban Consultants Co., Ltd., Tokyo, Japan

Yoshito Kitazono
Chuodoboku Consultant Co., Ltd., Kumamoto, Japan

Takashi Jitousono
Kagoshima University, Kagoshima, Japan

ABSTRACT: The 2016 Kumamoto earthquakes caused large-scale landslides in Minamiaso, Aso District, Kumamoto Prefecture in Japan, which is situated on a caldera wall of the Aso volcano. The length and width of a landslide studied are approximately 700 m and 200 m respectively, and its volume reached 500,000 m^3.

We conducted geological reconnaissance surveys, exploratory drilling and topographical analysis based on laser profiler (LP) data immediately after the disaster in order to grasp topographic and geological features of the site for disaster recovery.

From the investigations, we found the followings:

- The site is underlain by alternating layers of andesite and pyroclastic rocks. Andosols and debris of volcanic origin, which cover the base rocks remained on the ground surface above the landslide.
- The landslide occurred at the location with typical topographic features of seismic slope failure, i.e. convex break line, slope with an angle of 50 degrees or more, and at high elevation.
- Unstable soil/rock mass lies on such a topographically unstable slope. The steeper slope below the convex break line is formed by andesite with cavities in loose conditions. This condition is resistive against erosion but not in earthquake.
- A seismic survey detected a 20 to 25 m-thick layer with a P-wave velocity of 1 km/sec or so on the unstable slope.
- The landslide is located about 1 km away from an active fault. A peak acceleration of 1,316 gal was recorded at a meteorological station of Kawayou of Minamiaso, which is situated 3.7 km away from the landslide.
- The landslide mass containing fluidal volcanic ash flowed long distance and reached Kurokawa valley, about 1 km away from the top of the landslide.

Conclusively, a large-scale landslide took place due to highly intense earthquake (moment magnitude of 7.0) on the unique geology and typical topography.

We propose to improve an accuracy of risk assessment by a study of topography and geology sensitive to landslides, which includes historic ground movement analysis by using remote sensing and damage scales of existing roads, railways and houses.

Keywords: Topographical Analysis, Laser Profiler Data, Volcanic Rock, Caldera Wall, Seismic Slope Failure

*Corresponding author: nagakawa.katsuhisa@kiso.co.jp

DOI: 10.1201/9781003293590-10

1 INTRODUCTION

The landslide is located in Minamiaso, Aso District, Kumamoto Prefecture, a central part of Kyushu Island of Japan. It is situated on the western part of a caldera wall of the Aso volcano. As shown in Figure1, the landslide is located about 1 km close to the active fault, namely Futagawa Fault, which appeared on the ground surface during the 2016 Kumamoto earthquakes.

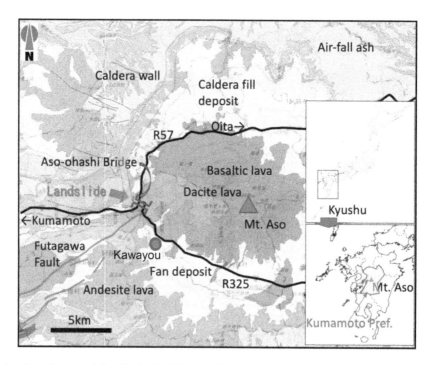

Figure 1.   Location map (after Geological Survey of Japan website: Seamless Geo Map).

The 2016 Kumamoto earthquakes occurred on April 16, 2016. It destroyed National Route 57, Aso-ohashi Bridge along National Route 325 and Hohi Main Line of Japan Railway. Their operation was suspended for a long period.

A further disaster was anticipated due to the loose humus volcanic ash soils, namely Andosols and debris overhanging on a scarp of the landslide. Therefore, urgent remedy measures were required. This paper presents results of investigations carried out immediately after the earthquake and landslide, reviews topographic and geological features causing the landslide under intense earthquakes, and proposes mitigative and preventive measures for the future.

2 OUTLINE OF DISASTER, TOPOGRAPHY AND GEOLOGY

2.1 Outline of disaster

The dimension of the landslide is about 700 m long and about 200 m wide. The landslide mass has a volume of about 500,000 m^3 and flowed over roads and railway. It destroyed Aso-ohashi Bridge as shown in Figure2 (Yonamine and Nagakawa, 2018). Cracks appeared and unstable rock/soil mass remained behind the landslide scarp as shown in Figure3 (Yonamine and Nagakawa, 2018). It was anticipated that this remained soil/rock mass might drop due to the aftershock and rainfall, and might cause further damages. Therefore, urgent investigation was required for the remedy measure design and works.

Figure 2.   Overview of the landslide.

Figure 3.   Cracks appeared behind the landslide scarp.

2.2   *Topography and geology*

The landslide is located at the western part of the outer rim of a caldera of the Aso volcano. The landslide occurred on the caldera wall inside the outer rim. The base rock is the pre-Aso volcanic rocks, formed about 300,000 years ago. Andosols composed of a typical volcanic clay of the Aso volcano cover the ground surface. The pre-Aso volcanic rocks are formed by alternating layers of andesite and pyroclastic rocks. The layers lie horizontally or slightly dip against the slope. Debris deposits, which comprise unsorted angular gravels of andesite, are present between the Andosols and the base rocks.

3   RESULTS OF INVESTIGATION

3.1   *Topographical survey*

(1) Method of study

Based on topographical data obtained by airborne laser profiler before and after the landslide, the following 3 types of maps are created.

• CS 3D map (Toda, 2012) created by Kagoshima University:

Colored and transparent GIS layers of elevation, slope and topographic curvature data

• Slope angle map: Difference in slope angle before and after the landslide
• Differential elevation map: Difference in elevation before and after the landslide

The following laser profiler (LP) data are used:

• Before the landslide: Digital Elevation Model (DEM) obtained by aerial photograph and laser profiler in the central part of Kyushu in April to June 2010

- After the landslide: DEM obtained by airborne laser profiler by Kokusai Kogyo Co., Ltd. on April 17, 2016

DEM data were disturbed by ground movement due to the earthquake. Therefore, DEM data were adjusted when creating the differential elevation map (Kobayashi, 2011).

(2) Findings

1) CS 3D map in Figure 4 (1)

Before the landslide, the convex terrain is found linearly at the elevation between 600 and 650 m, i.e. between convex and concave break lines at the middle of slope. The landslide area has a clear scar, which indicates the past landslide. A terrain near a forest road above the upper break line is slightly convex. The presence of trough above this convex terrain may be a sign of toppling failure. Japan Society of Erosion Control Engineering (2016) reported possible rock creep taken place at this point before the landslide.

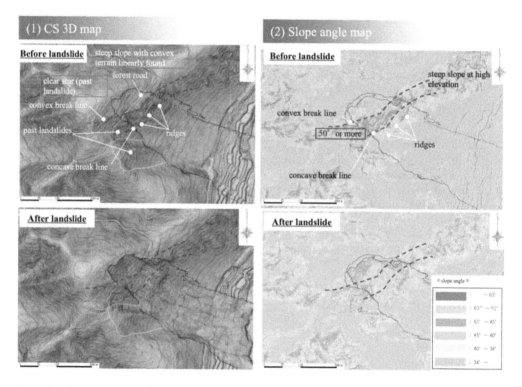

Figure 4.   CS 3D map and slope angle map (Kyushu Regional Development Bureau of MLIT, 2016).

2) Slope angle map in Figure 4 (2)

Before the landslide, the slope below the upper break line has an angle of 50 degrees or greater. The concave break line is clear below this steep slope. The steep slope clearly shows a convex terrain and has changed its angle to as gentle as 30 degrees after the landslide.

3) Differential elevation map in Figure 5

The maximum thickness of the landslide mass was about 25 m. The ground above the landslide sank by 1 to 2 m. The slope surrounding the landslide has swelled out, and the swell is significant on the slope in the northwest.

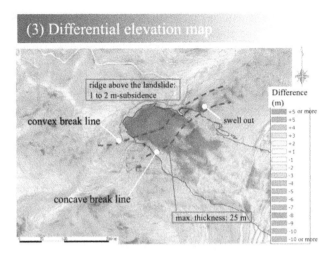

Figure 5. Differential elevation map (Kyushu Regional Development Bureau of MLIT, 2016).

3.2 Results of geological investigation

(1) Reconnaissance survey

Cracks appeared on the ground surrounding the landslide were surveyed and their drop, opening, extent and distance from the landslide were recorded. Figure 6 (a) (Yonamine and Nagakawa, 2018) shows magnitude of drop and opening of cracks. The magnitude of drop is proportional to the opening. Based on the relationships among the data, 6 blocks are categorized as shown in Figure 6 (b). The most critical block is identified as Block 1, which should be protected urgently from the further failure.

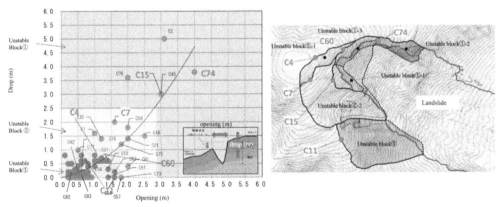

(a) Relationship between drop and opening of cracks.   (b) Zoning map of unstable blocks.

Figure 6. Reconnaissance survey result.

(2) Exploratory drilling and seismic survey in Figure 7 (Yonamine and Nagakawa, 2018)

Exploratory drilling and high-resolution seismic survey using a stacking method were carried out to reveal the landslide mechanism for remedy measure design.

1) Geological feature

Alternating layers of strong andesite and weak pyroclastic rock of autobrecciated lava, tuff breccia, etc. deposit almost horizontally.

75

Strong andesite is fractured and has local cavities. As vertical and horizontal cracks are stained with brownish color, the cracks would have been created as cooling joints before the landslide. No slickenside and clay gouge have been observed along the cracks. Pyroclastic rocks can be decomposed into soils with weathering.

2) High-resolution seismic survey

The seismic survey revealed that the landslide took place in a 20 to 25 m-thick layer near the ground surface with a seismic P-wave velocity of around 1 km/sec.

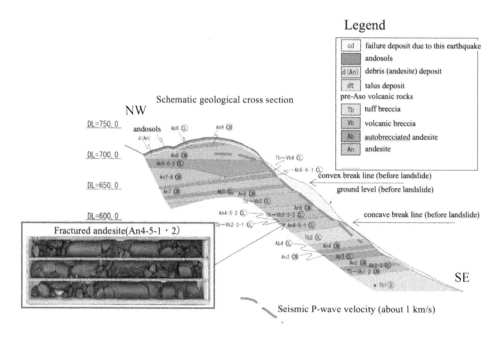

Figure 7. Longitudinal geological section along the landslide (vertical exaggeration = 1).

4 CONCLUSION

The mechanism of the landslide is inferred as follows:
Contributing factors

- Topographic features of the past landslides
- Poor rock condition of andesite with cracks susceptible to loosening while the intact rock is strong enough to resist erosion
- Vertical and horizontal cracks in the ground due to cooling process of lava
- Soft volcanic ash soil of Andosols with 2 to 5 m thick and debris deposits on the slope
- Steep slope of inside wall of the outer rim, i.e. caldera wall

Trigger

- Multiple Mw6 shocks including Mw6.2 pre-earthquake on April 14 repeated loosening of ground
- Mw7.0 main shock with a peak acceleration of 1,316 gal caused further loosening and landslide

(A peak acceleration of 1,316 gal was recorded at a meteorological station of Kawayou of Minamiaso, which is situated 3.7 km away from the landslide.)

It is known that the ground motion is amplified on the convex slope near the mountain top, where the rock and soil conditions are poor (Abe and Hayashi, 2011). We concluded that the debris on the slope above the convex break line caused the large-scale landslide and the intense disaster.

5 RISK MITIGATION

We have conducted various topographical studies using LP data before and after the land-slide. The studies revealed that the landslide took place at a slope with topographic features of convex terrain, break line, steep slope and high elevation, which are all well known on seismic slope failure. In addition, a large volume of unstable debris on the slope and underlying poor rock conditions (resistive against erosion but weak in earthquake) are contributed to the landslide.

Accordingly, the topographical and geological studies using LP data can be a potential approach for evaluation of seismic slope failure.

In addition to the topographic and geological factors, the landslide is located very close, about 1 km away from the active Futagawa fault, which is a source fault of the 2016 Kumamoto earthquakes and the intense shake caused the large ground movement of the slope at the site. The landslide mass contains pyroclastic rocks with a large amount of volcanic soils with high fluidity (equivalent friction coefficient of 0.2 to 0.3) may be a reason why the mass flows over a long distance.

Further to a lesson from the present studies, disaster prevention and mitigation measures can be evaluated more practically by employing a risk matrix based on likelihood and severity, i.e. the topographic feature and historic movements monitored using remote sensing techniques and damage scales of existing roads, railways, houses, etc. as shown in Figure 8.

				Likelihood		
				Judged from topography and ground movement		
				Low [1]	Midium [2]	High [3]
	grade	existing facilities	road conditions after disaster	no sign of landslide	sign of landslide	clear sign of landslide
Severity / Estimated damage scale	Low [1]	no facilities	no damage or early recovery	C	B	B
	Midium [2]	roads, railways, houses, etc.	one-side or short-term closure	B	A	A
	High [3]	roads, railways, houses, etc. and important routes	total or long-term closure	B	A	AA

| | AA | A | B | c |
| Risk level | High | | | Low |

Figure 8. Geological risk matrix for disaster prevention and mitigation.

ACKNOWLEDGEMENTS

This paper is based on the initial results of the "Engineering Committee for Restoration in Aso-ohashi Bridge Area" organized by the Kyushu Regional Development Bureau of the

Ministry of Land, Infrastructure, Transport and Tourism. Authors are grateful to various advice, guidance and suggestions from many academic researchers, related authorities and local people. We also thank to Kumagai Gumi Co., Ltd., a main contractor for remedy works, Kokusai Kogyo Co., Ltd. for topographical survey and analysis, Hiruzen Institute for Geology and Chronology Co., Ltd. for reconnaissance survey, Kuusatu Japan, and others.

All the remedy measures of Aso-ohashi Bridge, roads and railway have been completed and their operation has started. We look forward to early recovery and further development of the local people.

REFERENCES

Abe, S. and Hayashi, K., 2011, Morphologic features of landslides induced by recent large-scale earthquakes and their geomorphologic and geologic features. *Journal of the Japan Landslide Society*, 48, No.1, 52–61.

Geological Survey of Japan website, 2021, Geomap Navi - Seamless Geo Map, *https://gbank.gsj.jp/geonavi/geonavi.php#12,32.90091,131.07248.*

Japan Society of Erosion Control Engineering, 2016, Emergency survey report on sediment disaster caused by the 2016 Kumamoto earthquakes (in Japanese).

Kobayashi, Y., 2011, DEM horizontal position correction and difference target range to improve accuracy of laser differential analysis. *Proceedings of 33rd Survey Technology Presentation organized by Association of Precise Survey and Applied Technology.*

Kyushu Regional Development Bureau of the Ministry of Land, Infrastructure, Transport and Tourism (MLIT), 2016, Summary of 3rd engineering committee for restoration in Aso-ohashi bridge area (in Japanese).

Toda, K., 2012, Development of the microtopographical map by using airborne LIDAR DEM. *Journal of the Japan Society of Erosion Control Engineering*, 65, No.2, 51–55.

Yonamine, A. and Nagakawa, K., 2018, Study on remedy measures for landslide in Aso-ohashi bridge area, *Kisokou Journal (in Japanese)*, 46, No.3, 48–51.

Rock Mechanics and Engineering Geology in Volcanic Fields – Ohta, Ito & Osada (eds)
© 2023 copyright the Author(s), ISBN 978-1-032-27657-1

Aso volcano: Estimating the probabilistic likelihood of a future Aso4-scale eruption from stochastic uncertainty analysis of volcanological evidence using importance sampling

Willy P. Aspinall*
University of Bristol, Bristol, UK
Aspinall & Associates, Tisbury UK

R. Stephen J. Sparks
University of Bristol, Bristol, UK

Charles B. Connor
University of South Florida, Tampa, Florida USA

Brittain E. Hill
University of South Florida, Tampa, Florida USA
Independent Consultant, Jefferson, Maryland, USA

Antonio Costa
Istituto Nazionale di Geofisica e Vulcanologia, Bologna, Italy

Jonathan C. Rougier
University of Bristol, Bristol, UK
Rougier Consulting Ltd., Bristol, UK

Hirohito Inakura
West Japan Engineering Consultants, Inc., Fukuoka, Japan

Sue H. Mahony
University of Bristol, Bristol, UK
Independent Consultant, Woolacombe, UK

ABSTRACT: The Aso4 explosive eruption on Kyushu, Japan, 89,500 years ago was one of the biggest global eruptions in the last one hundred millennia, with a magnitude of approximately M8. Its effects were widespread throughout Japan; a similar scale event now would have huge societal impact. Today, 6.5 million persons live within 100 km of Aso. The key question for disaster preparedness and mitigation is: what is the likelihood of another M8 eruption from Aso-san? We estimate the probability of such an event within the next 100 years so that the scenario and its threats can be compared to other potential natural disasters. To evaluate this probability, we performed a comprehensive stochastic uncertainty analysis using advanced computational Bayes Net (BN) software. Our BN eruption process model is informed by multiple strands of evidence from volcanology, petrology, geochemistry and geophysics, together with estimates of epistemic (knowledge) uncertainty, adduced from reviews of published data, modelling and from expert judgement elicitation. Several lines of scientific evidence characterise the likely structure, magma composition and eruptibility state of the present-day volcano,

*Corresponding author: willy.aspinall@bristol.ac.uk

DOI: 10.1201/9781003293590-11

which has had numerous smaller eruptions since Aso4. An initial analysis indicated that another Aso4-scale event has an extremely low likelihood, being less than 1 - in - 10 million in the next 100 years (i.e. $< 10^{-7}$ probability). To further constrain this probability, we implemented probabilistic 'importance sampling' in our BN to allow even smaller probabilities to be enumerated. We find that the chance of an Aso4-scale eruption (characterised by mean volume 500 km^3 DRE, 90% credible interval [370 .. 685] km^3 DRE) is less than $1 - in - 1$ billion in the next 100 years (i.e. $<10^{-9}$ probability). We doubt that this conclusion, based on current understanding and evidence, could be different by more than an order of magnitude.

Keywords: Volcanic Eruption, Volcanic Hazards, Stochastic Uncertainty Analysis, Importance Sampling, Bayes Net (BN)

1 INTRODUCTION

We describe a Bayes Net (BN) model designed to estimate the probability that a future explosive eruption of Aso volcano (Figure 1), occurring within the next 100 years, could produce a total volume of erupted magmatic products that could equal or exceed the volume erupted in the Aso4 event, 89.5 kyr BP. Almost all the volcanological, geological and geophysical aspects of the problem involve substantial scientific uncertainties; these are represented in our BN model to the extent they are informed by observational data, theory or expert judgment.

In this study, we conducted an exhaustive literature collection in 2019-2020, built an initial BN model, and completed the BN model by considering the most recent and important litera-

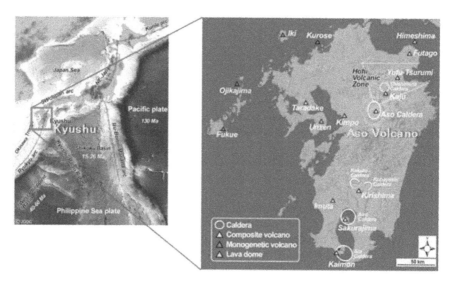

Figure 1.   Location of Aso volcano in Kyushu, Japan (from Chapman et al., 2009).

ture published by March 2021. Published research is the main evidential basis for our BN model, supplemented by a series of separate, commissioned specialist studies.

We discuss the way these various appraisals of the Aso4 eruption are combined with a fundamental advance in stochastic uncertainty modelling to refine the basis for re-calculating the probability of a near future eruption on the scale of Aso4. In this study, we argue – on compelling geological, geophysical and volcanological grounds – that the likelihood of such an eruption within the next 100 years is extraordinarily low, so low that it could not be calculated previously using conventional methods for estimating event potential occurrence probability.

2 APPROACH

In the light of our new work and with the publication of an important paper discussing Aso4 eruption deposit volumes (Takarada and Hoshizumi, 2020), a key part of this study has been to re-determine the magnitude of the Aso4 eruption and, critically, to estimate the uncertainty associated with that magnitude estimate. While this partly relied on information in the literature, we also took advantage of significant improvements in numerical analysis techniques applied to field data. Our approach involved a detailed review and comparative study of previous Aso4 erupted volume estimates and methods, and the application of an advanced statistical modelling technique for integrating contoured isopach (thickness) data to estimate total deposit volumes. The key point here is that the Aso4 eruption volume/magnitude – with formally assessed uncertainty – is the critical scaling parameter against which we test for the probability that a near-future explosive eruption of Aso could equal (or exceed) the Aso4 volume.

Recently published investigations on the composition and origins of Aso4 magma (Ushioda et al., 2020) and post-Aso4 magma (Kawaguchi et al., 2021) help constrain our understanding of the petrogenesis of Aso volcano and provide an enhanced framework to interpret evolutionary trends of post-Aso4 silicic and mafic magmatism. This, and other published petrological information, is used to develop a conceptual model of the current and near-future Aso volcano magma system. Thus, our new analyses of relophysical data, together with improved petrological insights, are synthesized through expert elicitation to inform parameter revisions for nodes in our BN model that represent reservoir sizes, compositions, and eruptible volumes.

Benefit is also taken from new numerical advances in performing importance sampling analysis (Rubin, 1987), within the BN framework, enabling stochastic modelling calculations to be undertaken which relate to events and situations with hitherto unquantifiable extremely low probabilities of occurrence; coupled with these stochastic modelling developments, advantage is taken of the increased capacity to process massive spreadsheet datasets afforded by the Data Model and Power Query capabilities in the current version of Excel software (Microsoft 365 Apps for Enterprise version 2104).

3 ASO BAYES NET DESIGN

The essential theme of the Aso BN is a magma-volume accounting model with two main elements: the total volume of magma that was erupted 89.5 kyr BP, and the estimated total volume of eruptible magma that may currently reside within the volcano (or become present within the next 100 years). The lines of evidence informing these aspects are many, varied, and all entail uncertain parameters; thus, these are treated as random variables (RV) in a BN framework. Magma composition is another critical element: the Aso3-4 large magnitude explosive eruptions were silicic in composition. There is no evidence from the geological records that silicic volcanoes are capable of large explosive eruptions of the kind that create large calderas, such as that formed by Aso4. Thus, our Aso BN includes an assessment of the probability that the present stored magma is mafic or silicic.

The numerical analysis uses stochastic sampling of the multiple RV uncertainty distributions (and other variables and factors which feed into them) to ascertain the probability that the available eruptible magma volume could equal or exceed the estimated Aso4 erupted volume. This is implemented using the UNINET software package, developed originally by TU Delft in The Netherlands (now maintained by LightTwist Software https://lighttwist-software.com/). UNINET is an advanced analytical graphical program for high-dimension stochastic uncertainty modelling, multivariate data mining and machine learning, using the Bayes Net paradigm, probability vines and dependence trees.

The framework of our complete Aso BN comprises three parts:

[a] a network of the nodes to model contributory lines of evidence for numerical estimation of the composite total volume of the Aso4 deposits, shown schematically in Figure 2;

[b] a second network for the nodes that represent lines of evidence informing the numerical estimation of the total volumes of eruptible magma available in reservoirs in the present-day volcano, which includes the next 100 years (BN model not shown here);

[c] a sub-net of nodes which contain numerical information about partial statistical cumulative distribution functions (CDFs) for total Aso4 volume and for eruptible reservoir magma volume, used for importance sampling and for estimating the probability of an Aso4-scale eruption in the next 100 years, described below. These partial CDFs are drawn from the range of sample volumes that jointly overlap in both the Aso4 total volume samples probability density function (PDF - obtained from enumeration of the nodes and branches comprising Figure 2) and in the total eruptible volume samples PDFs from the second part BN (not shown).

Here, we present only the first part of our BN: i.e. the framework for estimating the total volume of the Aso4 eruption from deposit data – see Figure2.

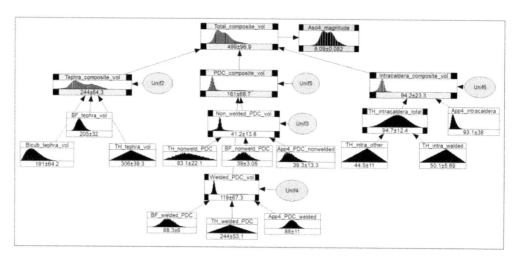

Figure 2.   Aso4 deposits volume synthesis: BN model nodes and resulting PDFs in UNINET graphical form. Numbers in each panel indicate mean value and standard deviation; full PDF statistics can be extracted for further analysis. Deposit volumes are expressed as Dense Rock Equivalent (DRE), for direct comparison with magma volumes.

*Key: App4_* relates to geophysical data or values from a study appendix report; *BF_* basis function values (determined in a separate study); *Bicub_* indicates values obtained from bicubic spline fitting modelling (previous report); *Comb* means "combination"; *Nonweld* is for non-welded PDC/ignimbrite data; *PDC* "pyroclastic density current"; *TH_* denotes data or values from Takarada and Hoshizumi (2020); *Unif_*is uniform distribution on interval [0..1]; *_vol* or *Vol_* denote "volume".

In Figure 2, nodes representing input data, variables or parameter distributions are shown as plain rectangles (or ellipses); calculational (functional) nodes are shown as with black corner/ blocks. The latter contain equations or dependence conditions for operating on variable samples from data nodes to which they are linked by arrows ('arcs' in BN terminology). Top level nodes, representing the key outputs from the BN networks have coloured borders for identification purposes. Green nodes, labelled "*Unif*" etc., are simple uniform PDFs on the interval [0 .. 1], sampled randomly to activate relative weights where needed for alternative data options selection in linked calculational nodes.

For input nodes derived for the BN, our various independent volume estimates and associated uncertainties are expressed as Normal distributions with means and standard deviations to match analysis results. For parameters which can take only positive real values (e.g. physical volumes), we adopt lognormal PDFs to avoid having negative value samples in such distributions.

For Takarada & Hoshizumi (2020) parameter spreads (i.e. Figure 2 nodes labelled: *TH_xxxx*), these are represented by triangular distributions to accord with the way those authors reported their findings. In our model, the triangular distributions are extended below and above the TH minimum and maximum values allowing the reported endpoint values to be actively sampled as actual values. In effect, this is tantamount to ascribing arbitrary – but realistic – uncertainties to minimum and maximum values for stochastic sampling purposes. To implement this intrinsic range adjustment to their variables, the distributions are here extended beyond Takarada and Hoshizumi's bounding values by 10% below the relevant minima to 10% greater than their maxima. Simple tests suggest these extension values are not critical and appear reasonable for the present application and circumstances.

Hints of bimodality in the top-level nodes arise from combining different contributing subtype deposit PDFs (second row nodes); they are not considered to imply any substantive physical meaning.

The results of computing the various contributory Aso4 volumes can be read off the BN chart Figure 2; the key composite (summed/combination) results are summarised on Table 1:

Table 1.   Aso4 deposit volumes: summary means and standard deviations.

	Mean	Stdev	
Vol_tephra	209	60	km^3 DRE
Vol_PDC	161	69	km^3 DRE
Vol_intracaldera	94	23	km^3 DRE
Aso4_total_vol	499	97	km^3 DRE

The absolute smallest sample volume is about 212 km^3 DRE, which is less than Takarada and Hoshizumi (2020) minimum estimate of 465 km^3 DRE. The greatest BN sample volume, 1202 km^3 DRE, also exceeds Takarada and Hoshizumi (2020) maximum bound of 962 km^3 DRE.

When converted to equivalent eruption magnitudes, the BN determined DRE volumes equate to:

$$\text{Mean magnitude M8.1} \pm 0.08 \text{ magnitude units}$$

M7.7 (minimum); M8.0 (5th % ile); M8.1(median); M8.2 (95th % ile); M8.5 (maximum).   (1)

The LaMeve database (Crosweller et al, 2012) reports Aso4 eruption magnitude as M7.7; Takarada and Hoshizumi (2020) suggest M8.1 – 8.4. The magnitude uncertainty distribution from our study, enumerated by the BN analysis, spans these published magnitude values; our mean magnitude M8.1 accords with the lower of Takarada and Hoshizumi's values.

This correspondence engenders confidence that our BN-derived PDF for the volume/magnitude of Aso4 is an appropriate basis for quantifying the probability of an eruption on this scale in the next 100 years, discussed in the next section.

4   COMPUTATIONAL ENUMERATION OF ASO4-SCALE ERUPTION EXCEEDANCE PROBABILITY

To estimate the probability of an Aso4-scale eruption in the next 100 years, we introduce a novel 'importance sampling' technique into our BN analysis. Attempts to resolve a very low event probabilities (e.g. well below 10^{-6} probability) by 'brute force' Monte Carlo simulation requires upward of a billion samples and is simply not feasible with ordinary software.

However, 'importance sampling' inference methods (e.g. Rubin, 1987) can be used to estimate posterior densities or expectations in state- or parameter estimation probabilistic models that are too hard to treat analytically, for example in Bayes nets.

In the present study, a separate sub-net is constructed that can selectively use just those fractional parts of parameter statistical distributions in the main BN model that are relevant to the uncertainty space of interest. That is, we analyse only those samples with large-scale potential eruption volumes, large enough to be close to or exceed corresponding 'low-end' stochastic samples from the estimated Aso4 volume distribution. Importance sampling is, thus, a numerically feasible replacement for an unattainable full-scale sampling approach.

Importance sampling inference is applied here, as follows. With the dual branched BN model, outlined in Section 3 above – i.e. comprising the Aso4 volume estimator in Figure 2 and the available eruptible volume estimator net (not shown) – the upper tail of the PDF for available eruptible volume just overlaps the lower tail of the PDF for node <Total_composite_vol> (i.e. the Aso4 eruption volume estimate, per Figure 2); the overlap covers the range $213 - 350$ km^3 DRE. When the dual BNs are run jointly with 20×10^6 samples, just 73 samples are found in the overlap range for <Total_eruptible_vol>, while 23,138 samples are obtained for <Total_composite_vol>. The latter $\approx$23k samples represent only about 0.1% of the original 20 million samples!

UNINET's conditionalized sampling option is used to export those samples, drawn from each BN top-level node <Total_eruptible_vol> and <Total_composite_vol>, which fall inside the overlap range. These samples are processed with Excel to define cumulative density functions (CDF) to characterise the tail properties of the two variable distributions. The resulting CDFs are input back into the importance sampling sub-net of the BN, for the target exceedance probability calculations. This probability is determined by multiplying the importance samples' exceedance probability test distribution <Eruptible_ge_Aso4> jointly with the two sample size ratios; thus, the equation for enumerating the probability of an Aso4-scale eruption in 100 years is shown (2):

$$\text{Pr[Eruptible\_ge\_Aso4 | importance sampling]}^{*}(73/\text{Number\_Samples})^{*}(23138/\text{Number\_Samples})$$

(2)

where Number_Samples from the main BN is 20×10^6.

One million iterations of the importance sampling BN sub-net are sufficient for calculating the required exceedance probability to an appropriate precision.

Without importance sampling, the main BN model – with maximum 20 million samples allowed by the software – fails to find a single instance where the potential future eruption volume exceeds the smallest volume quantified in the Aso4 eruption distribution. This demonstrates that the probability of an Aso4-scale eruption in the next 100 years is definitively lower than 5×10^{-8}, but fails to enumerate exactly how low it is.

Thus, to resolve this issue, recourse to the advanced and innovative importance sampling technique, just outlined, is indispensable. When the BN model top-level volume nodes overlap samples are jointly pooled in importance sampling mode, as just described, the mean probability of an Aso eruption in the next 100 years is enumerated about 5.6×10^{-10}. The associated standard deviation on this mean – derived from all the parameter uncertainties included in the BN – is: $\pm 1.4 \times 10^{-9}$. Numerical uncertainty analysis indicates the corresponding 99% confidence level eruption probability is not greater than 4.2×10^{-9}.

These results are determined numerically from computations in our BN framework, which is based on all the geological, geophysical and volcanological evidence that contributes to model parameterisation.

For most purposes, and to avoid spurious precision, the probability values given here should be rounded to the nearest single significant digit, i.e. 6×10^{-10} (mean probability) and 4×10^{-9} (99% confidence).

5 DISCUSSION AND CONCLUSIONS

To provide some context for our new assessment of the likelihood of an Aso4-scale eruption in the next 100 years, previous calculations with an earlier version of the Aso initial BN model using conventional numerical techniques, could show only that the probability of such an eruption was likely to be less than 1×10^{-7} (i.e. 1-in-10 million chance); in a following study, with some advances in stochastic modelling techniques, the Aso4-scale eruption probability in the next 100 years estimate was refined and evaluated at about 2×10^{-9}.

With the current BN model run in UNINET with importance sampling and relying on our re-appraised lines of volcanological evidence, **the mean probability of an Aso eruption in the next 100 years is enumerated at about 6×10^{-10}**. The standard deviation on this mean – derived from all the parameter uncertainties now included in our model – is: $\pm 1.4 \times 10^{-9}$. This numerical uncertainty analysis indicates the corresponding 99% confidence level eruption probability is not greater than 4×10^{-9} in the next 100 years. It is noteworthy that this, our 99% confidence probability, is radically smaller than 3×10^{-4} probability in 100 years, which would be a naive interpretation of the fact that there has been one Aso4-scale eruption over 300 kyr BP.

Assuming the latter, simplistic basis for inferring a mean century rate for Aso4-scale eruptions of Aso is invalidated by the geological evolution of the volcano, especially since the Aso4 event. Our conceptual model for the present Aso magma system is represented in the BN model by shallow, intermediate, and deep magma reservoirs, which are evaluated for ranges of eruptible volumes and compositions. With our latest revisions to BN node parameters, the minimum plausible value for the Aso4 total eruption volume is 212 km^3 DRE. In order for a present-day (i.e., including next 100 years) eruption to reach this lowest bound volume – without involving the deep magma reservoir – would require 86% evacuation of the maximum sample volume from the shallow magma reservoir (i.e. 95 km^3 DRE) and of the maximum sample volume from the intermediate magma reservoir volume (i.e. 150 km^3 DRE). Alternatively, 100% evacuation of the shallow reservoir maximum volume plus 78% of the intermediate reservoir maximum volume would achieve the required total volume. Both these scenarios, however, are dependent on the shallow and intermediate magma reservoirs being wholly silicic in composition – a very improbable scenario.

Clearly, if the 'target' volume for a future Aso4-scale eruption is greater than the sum of these two reservoir maxima, i.e. 95 + 150 = 245 km^3 DRE, then silicic magma from the deep reservoir must become involved in a prospective future eruption on the scale of Aso4. The critical assumption in this line of argument is that the shallow and intermediate reservoir volumes are both currently charged entirely with silicic magma to their (barely credible) maximum volumes. If either, or both, has substantially smaller capacity – and only low fractions of available reservoir volumes can be evacuated in eruption (as would be expected from volcanological considerations) – then a very substantial volume of deep reservoir silicic magma would need to be present, and eruptible, to plausibly match the scale of an event approaching our estimate of the mean volume of the Aso4 eruption.

From all petrological and geophysical considerations, however, the existence of such massive volumes of silicic magma at intermediate-to-lower crustal depths below Aso volcano, at the present time, is not deemed credible. Moreover, the prevailing compositions of the current shallow and intermediate magma systems are judged to be predominantly mafic, with only small likelihoods that the magmas which exist in these reservoirs are predominantly silicic.

ACKNOWLEDGEMENTS

We are grateful to Japanese experts, Prof. Toshiaki Hasenaka, Prof. Hiroshi Shimizu, Prof. Masaya Miyoshi, Assoc. Prof. Koji Kiyosugi, and Asst. Prof. Tomohiro Tsuji, and to Prof Roger M. Cooke, Dr Dan Ababei, Dr Hideki Kawamura and Dr Samantha Engwell for their advice and assistance with this study.

REFERENCES

Chapman, N., et al. (2009), Development of Methodologies for the Identification of Volcanic and Tectonic Hazards to Potential HLW Repository Sites in Japan: The Kyushu Case Study, NUMO-TR-09-04.

Crosweller, H.S., Arora, B., Brown, S.K. et al. Global database on large magnitude explosive volcanic eruptions (LaMEVE), 2012, *J Appl. Volcanol.* 1, 4. https://doi.org/10.1186/2191-5040-1-4

Kawaguchi, M. and eight others, 2021, Persistent gas emission originating from a deep basaltic magma reservoir of an active volcano: the case of Aso volcano, Japan. *Contributions to Mineralogy and Petrology*, 176:6, doi.10.1007/s00410-020-01761-6

Rubin, D.B., 1987, *Multiple Imputation for Nonresponse in Surveys*. John Wiley & Sons Inc., New York. doi.10.1002/9780470316696

Takarada, S. and Hoshizumi, H., 2020, Distribution and Eruptive Volume of Aso-4 Pyroclastic Density Current and Tephra Fall Deposits, Japan: A M8 Super-Eruption. *Frontiers* 8: 170. doi.10.3389/feart.2020.00170

Ushioda, M., Miyagi, I., Suzuki, T., Takahashi, E. and Hoshizumi, H., 2020, Preeruptive P-T conditions and H2O concentration of the Aso-4 silicic end-member magma based on high-pressure experiments. *Journal of Geophysical Research: Solid Earth*, 125: e2019JB018481. 10.1029/2019JB018481.

Rock Mechanics and Engineering Geology in Volcanic Fields – Ohta, Ito & Osada (eds)
© 2023 copyright the Author(s), ISBN 978-1-032-27657-1

Distribution recognition of lava flows at the southern foot of Nakadake and Takadake volcanoes in Aso caldera, SW Japan

Mitsuru Okuno*
Fukuoka University, Fukuoka, Japan

Toshihiko Koyanagi
Fukuoka University, Fukuoka, Japan
CTI Ground Planning Co., Ltd., Fukuoka, Japan

Masaya Miyoshi
Fukuoka University, Fukuoka, Japan

ABSTRACT: The Nangodani Valley is part of the caldera floor at the southern foot of Nakadake, a post-caldera volcano of Aso Volcano, SW Japan. In this valley, lava flows from the Nakadake and Takadake volcanoes are widely distributed. However, their exact distribution was not known because they were covered by fan deposits and mantle-betting tephra layers. In this study, we used various topographic maps and aerial photographs provided by the Geospatial Information Authority of Japan (GSI). Although outcrops of the lava itself are scarce, we have succeeded in estimating the distribution of lava flows by carefully observing the topography, especially the flow paths of rivers. The marginal cliffs of lavas are not easily eroded and the natural flow paths from the volcanic slopes are blocked. In fact, we found outcrops of lava were outside the previous distribution area of the lava. This fact indicates that the lava is likely to be distributed in a wider area than the previously indicated distribution area. As for the outflow area of the lava, it was confirmed from the microscope observation that the lava flowed at the point which was not considered as the outflow area of Takadake lava. The distribution of the Nakadake old-edifice lava is partly hidden by the Takadake lava, which flows around it to the vicinity of the Takamori lava. Therefore, the Nakadake and Takadake lava flowed almost at the same time, but the Takadake lava is considered to have flowed slightly after the Nakadake. There are many natural springs in Nangodani Valley, which is consistent with the distribution of the lava revealed in this study.

Keywords: Nagodani valley, Nakadake volcano, Takadake volcano, lava flow, volcanic geomorphology

1 INTRODUCTION

Also volcano, central Kyushu, SW Japan, is a large caldera volcano, 25 km from north to south and 18 km from east to west (Figure 1), and Nakadake, a post-caldera volcano, erupts frequently and is one of the most active volcanoes in Japan (Ono and Watanabe, 1985; Watanabe, 2001). This volcano had four major pyroclastic-flow (ignimbrite) eruptions between 300 ka and 87 ka, which are named Aso-1, Aso-2, Aso-3, and Aso-4 from the oldest to the youngest. The total eruption volume of the latest huge eruption, Aso-4 pyroclastic flow (Watanabe, 1978), is estimated to

*Corresponding author: okuno326@gmail.com

DOI: 10.1201/9781003293590-12

be 930 - 1860 km^3 (1.2 - 2.4 × 10^{15} kg) by Takarada and Hoshizumi (2020). The post-caldera volcanoes located in the center of the caldera divides caldera floor into the Asodani Valley to the north and the Nangodani Valley to the south (Figure 1B). The eruptive history of the post-caldera volcanoes after the Aso-4 ignimbrite eruption has been comprehensively described by Watanabe (2001), the framework of tephrostratigraphy has been established in detail by Miyabuchi (2009), and the K-Ar ages of the lava flows have been reported by Matsumoto et al. (1991) and Miyoshi et al. (2012).

Lava flows from Nakadake and Takadake volcanoes (Ono and Watanabe, 1985; Sakai et al., 1994; Baba, 1999; Baba et al., 1999) are widely distributed at the southern foot of these volcanoes in the Nangodani Valley (Figure 1B). The distribution of the lava flows has not been clarified because they are covered by fan deposits and mantle-betting fallout tephra layers. We examined the distribution of these lava flows using topographic information, occurrence in the field and petrographic character to distinguish them. In this paper, we propose revised distribution of these lava flows, and discuss on relationship with location of natural springs in this region.

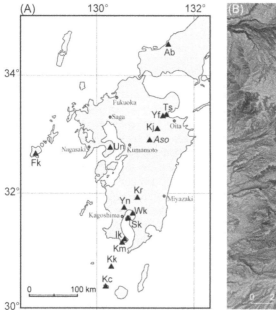
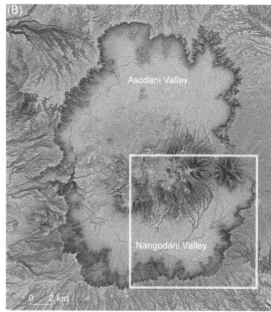

Figure 1.   Index maps of Aso volcano. (A) Map showing active volcanoes in and around Kyushu Island (Japan Meteorological Agency, 2013). The solid triangles and open circles indicate active volcanoes and prefectural capitals, respectively. Ab: Abu volcanoes, Fk: Fukue volcanoes, Ik: Ikeda caldera and Yamakawa maar, Kc: Kuchinoerabu volcano, Kj: Kuju volcanoes, Kk: Kikai caldera, Km: Kaimondake volcano, Kr: Kirishima volcanoes, Sk: Sakurajima volcano, Ts: Tsurumidake and Garandake volcanoes, Un: Unzendake volcano, Wk: Wakamiko in Aira caldera, Yf: Yufudake volcano, Yn: Yonemaru and Sumiyoshi-ike maars. (B) Topographic map in and around Aso volcano. The rectangle indicates the studied area shown in Figure 2. Red relief image map (Chiba et al., 2008) using elevation data provided by the Geospatial Information Authority of Japan (GSI) was used as the base map. A map is oriented with north at the top.

2   OUTLINE OF GEOLOGY OF NAKADAKE AND TAKADAKE VOLCANOES IN ASO CALDERA

2.1   Teprochronological framework

During the last 30,000 years, the Kusasenri-ga-hama tephra (Kpfa), Aira Tn tephra (AT), and Kikai-Akahoya tephra (K-Ah) in ascending order are useful as maker tephra in this region. The Kpfa is largest pumice fall during the post-caldera volcanism of Aso volcano. The AT and K-Ah are widespread tephra originated from Aira and Kikai calderas (Machida and Arai, 1983) in southern Kyushu (Figure 1A). These tephras have been recognized in the varve sediments of Lake Suigetsu, and their eruption ages are estimated to be ca. 32.5 cal ka BP (MacLean et al., 2020), ca. 30 cal ka BP and ca. 7.3 cal ka BP (Smith et al., 2013).

Nakadake volcano is composed of the Old volcanic edifice, Young volcanic edifice and Youngest pyroclastic cone (Ono and Watanabe, 1985; Koyama and Taniguchi, 2004). The basaltic lava flows of Old volcanic edifice (Baba, 1999) and the Yamasaki scoria 20 to 15 (YmS20–YmS15; Miyabuchi, 2009) erupted successively at 22-21 cal ka BP. Subsequently, the basaltic Izumikawa pyroclastic flow flowed down to the northeastern foot of Nakadake at ca. 19 cal ka BP (Miyabuchi et al., 2006). After the deposition of K-Ah tephra in the Holocene, the Young volcanic edifice and the youngest pyroclastic cone were also formed, which are dealt with together in this paper.

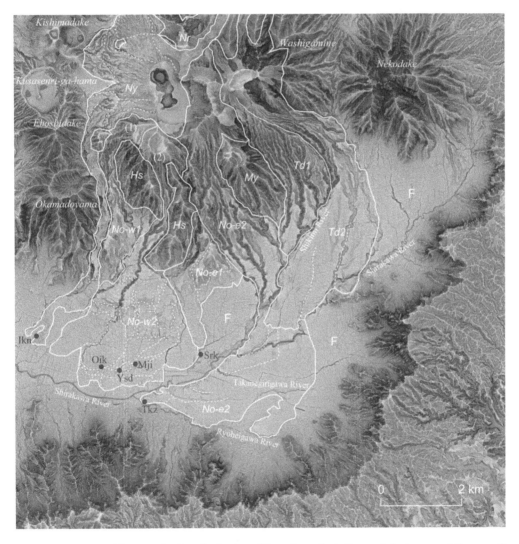

Figure 2. Topographic map showing distribution of lava flows (including agglutinate around the summit craters) from Nakadake, Takadake and Naraodake volcanoes, southeast part of Aso caldera. Red relief image map was used as the base map. F: volcanic fan deposit, No: Nakadake Old Volcanic Edifice, Ny: Nakadake Young Volcanic Edifice, Nr: Naraodake volcano, Td: Takadake volcano, Hs: Hakusui volcano, My: Maruyama volcano. The No can be subdivided into four units (-w1, -w2, -e1 and e2). The Td can also be subdivided into Td1 and Td2. Blue circles indicate locality of representative natural springs. Ikn: Ikenokawa Suigen, Oik: Oike Suigen, Ysd: Yoshidajo-gokenjo Kumiba, Mji: Myoujin-ike, Tkz: Takezaki Suigen, Srk: Shirakawa Suigen.

2.2 K-Ar ages of lava flows

The Nakadake and Takadake volcanoes were formed by covering previous volcanoes such as Hakusui (Hs), Maruyama (My), and Washigamine (Figure 2). Among them, K-Ar ages of 36 ± 4 ka, 37 ± 2 ka (Miyoshi et al., 2012), and 30 ± 6 ka (Matsumoto et al., 1991) for Hakusui and 53 ± 28 ka and 43 ± 8 ka (Miyoshi et al., 2012) for Maruyama were obtained using the sensitivity method. Naraodake volcano (Nr), which is adjacent to the north of Nakadake and Takadake volcanoes (Figure 2), also has a K-Ar age of 22 ± 3 ka (Miyoshi et al., 2012), which overlaps with the age of Nakadake Old volcanic edifice revealed from tephrostratigraphy (Baba, 1999; Baba et al., 1999).

3 METHODOLOGY

We observed volcanic topography using various topographic maps and aerial photographs provided by the Geospatial Information Authority of Japan (GSI), noting on development of river channels. The natural flow paths from volcanic slopes are blocked because the lateral and terminal cliffs of lava flows are not easily eroded. Surficial gullies formed by water erosion reflect the geological characteristics of the constituent landforms. These pattern changes allow us to classify the volcanic geological units. The Minamiaso Village Spring Group (Kagabu et al., 2011), such as the Shirakawa Suigen in the Asodani Valley, are also plotted on Figure 2.

Although there are few outcrops of lava flows in this area due to the cover of fan deposits and thick fallout tephras, we confirmed the occurrence of lava flows mainly along the river channel and collected rock samples from representative outcrops to observe their petrological characteristics.

4 RESULTS AND DISCUSSION

4.1 Nakadake old volcanic edifice

The Nakadake Old volcanic edifice (Ono and Watanabe, 1985; Baba et al., 1999) can be sub-divided into four units (No-w1, -w2, -e1, -e2) at least. The No-w1 and -w2 may be successive eruptive products, but the eruptive center shifted slightly from (1) to (2), and the lava flowed down in separated lobes along the erosional valley of the western part of Hakusui volcano and merged at the southern foot. The relatively later flow, No-w2, cut through the preceding -w1 and spreads downstream.

As for the No-e1 and -e2, the No-e1 flowed along the eastern margin of Hakusui volcano in the early stage of the eruption, but the No-e2 flowed straightly to south due to the increase of outflow rate. It is widely distributed on Asodani Valley including Takamori lava (Sakai et al., 1994). Since the No-e2 covers the tip of the eastern lobe of No-w2, it is considered to have erupted afterwards.

The biggest spring in the Minamiaso Village Spring Group, Shirakawa Suigen, is located on boundary among the No-w2, No-e1 and fan deposit. The Oike Suigen, Yoshi-dajo-gokenjo Kumiba and Myoujin-ike are located on the flow unit boundaries in the No-w2. The Takezaki Suigen on the tip of the No-e2. Ikenokawa Suigen is located on the tip of the Nakadake Young volcanic edifice. These locations are consistent with these lava distribution (Figure 2).

4.2 Takadake and Naraodake volcanoes

Takadake volcano can also be divided into the Td1 and Td2, which are considered to be continuous eruptions. First, the Td1 flowed along the eastern margin of Maruyama volcano, but its terminal part stagnated in contact with the No-e2 and gradually changed its flow path to the Td2. In the Td2 phase, the Td2 also flowed to the north and east, and the southern terminal of Td2 covered part of No-e2, and the same relationship is confirmed in other contact boundaries. Therefore, Takadake erupted slightly later than the Nakadake Old volcanic edifice. The channel formed on the surface of Td1 and No-e2 joins the Shikimigawa River formed along the levee of the Td2 margin.

Naraodake volcano (Nr) has a K-Ar age similar to the stratigraphic age of Nakadake Old volcanic edifice and Takadake. However, it is limited to the summit area, and most of the northern main part is considered to be the Hakusui volcano. The Izumikawa pyroclastic flow has been considered to originate from Nakadake (Miyabuchi et al., 2006), but pyroclastic eruptions with agglutinate formation around the summit of Takadake and Naraodake also occurred at similar period. Therefore, it is possible that they originated from either of these volcanoes.

5 CONCLUSIONS

In this paper, we report the results of a study on the distribution of lava in the southern foot of Nakadake and Takadake volcanoes based on detailed observations of volcanic landforms. The Nakadake Old volcanic edifice can be divided into at least four units (No-w1, -w2, -e1, -e2), and the Takadake volcano can also be divided into Td1 and Td2. Based on the topographic relationship, at least three eruptions (No-w, No-e, and Td) are considered to have occurred sequentially. Naraodake may have erupted at a similar period, but the detail relations are unknown because the boundary with Takadake is hidden by the Nakadake Young volcanic edifice. The Izumikawa pyroclastic flow may have originated from either Takadake or Naraodake volcanoes. The location of natural springs on the southern foot of Nakadake is consistent with these lava distribution by this study.

ACKNOWLEDGEMENTS

This paper is based on the 2020 graduation thesis submitted by co-author TK to the Faculty of Science, Fukuoka University, and has been revised with additional research. We would like to express our gratitude to the Aso-Kuju National Park Administration Office and the Forest Bureau, Department of Agriculture, Forestry and Fisheries, Kumamoto Prefecture, for their assistance in applying permissions to conduct this research.

REFERENCES

Baba, M., 1999, Lava flows in the southern part of Aso caldera, *J. Kumamoto Geosci. Assoc.*, 120, 2–8. (in Japanese)

Baba, M., Watanabe, K. and Miyabuchi, Y., 1999, Stratigraphy of eruption products from Nakadake in the southern part of the central cones of Aso volcano, Japan, *Mem. Fac. Educ. Kumamoto Univ., Nat. Sci.*, 48, 133–146.(in Japanese with English abstract)

Chiba, T., Kaneta, S. and Suzuki, Y., 2008, Red relief image map: New visualization method for three dimensional data. *Int. Arch. Photogramm., Remote Sens. Spat. Inf. Sci.*, 37, 1071–1076.

Japan Meteorological Agency, 2013, National Catalogue of the Active Volcanoes in Japan (The Fourth Edition, English Version). http://www.data.jma.go.jp/svd/vois/data/tokyo/STOCK/souran_eng/menu. htm

Kagabu, M., Shimada, J., Shimano, Y., Higuchi, S. and Noda, S., 2011, Groundwater flow system in Aso caldera, *J. Jpn. Assoc. Hydrolog. Sci.*, 41, 1–17. (in Japanese with English abstract)

Koyama, M. and Taniguchi, H., 2004, Volcanic geology of Nakadake youngest pyroclastic cone in Aso caldera, *Northeast Asian studies*, 9, 221–242. (in Japanese with English abstract)

Machida, H., and Arai, F., 1983, Extensive ash falls in and around the sea of Japan from large late Quaternary eruptions, *J. Volcanol. Geotherm. Res.*, 18, 151–164.

Matsumoto, A., Uto, K., Ono, K. and Watanabe, K., 1991, K-Ar age determinations for Aso volcanic rocks: concordance with volcanostratigraphy and application to pyroclastic flows, *Program Abstr. Volcanol. Soc. Japan*, 1991-2, 73. (in Japanese)

McLean, D., Albert, P.G., Suzuki, T., Nakagawa, T., Kimura, J.-I., Chang, Q., Miyabuchi, Y., Manning, C. J., MacLeod, A., Blockley, S.P.E., Sftaf, R.A., Yamada, K., Kitaba, I., Yamasaki, A., Haraguchi, T., Kitagawa, J., SG14 Project Member and Smith, V.C., 2020, Constraints on the timing of explosive volcanism at Aso and Aira calderas (Japan) between 50 and 30 ka: New insights from the Lake Suigetsu sedimentary record (SG14 Core), *Geochem., Geophys., Geosyst.*, 21 (8) 10.1029/2019GC008874

Miyabuchi, Y., 2009, A 90,000-year tephrostratigraphic framework of Aso Volcano, Japan, *Sediment. Geol.*, 220,169–189.

Miyabuchi, Y., Watanabe, K. and Egawa, Y., 2006, Bombrich basaltic pyroclastic flow deposit from Nakadake, Aso volcano, southwestern Japan, *J. Volcanol. Geotherm. Res.*, 155, 90–103.

Miyoshi, M., Sumino, H., Miyabuchi, Y., Shinmura, T., Mori, Y., Hasenaka, T., Furukawa, K., Uno, K. and Nagao, K., 2012, K–Ar ages determined for post-caldera volcanic products from Aso volcano, central Kyushu, Japan, *J. Volcanol. Geotherm. Res.*, 229–230, 64–73.

Ono, K. and Watanabe, K., 1985, *Geological Map of Aso Volcano, 1:50,000, Geological Map of Volcanoes 4*, Geological Survey of Japan, AIST.

Sakai, H., Nakano, Y. and Watanabe, K., 1994, "Takamori lava flow", A newly discovered lava flow, In post calderas stage of Aso volcano, *Abstr. Japan Earth Planet. Sci. Joint Meet. 1994*, 389. (in Japanese)

Smith, V.C., Staff, R., Blockley, S.P.E., Bronk Ramsey, C., Nakagawa, T., Mark, D. F., Takemura, K., Danhara, T. and Suigetsu 2006 Project Members, 2013, Identification and correlation of visible tephras in the Lake Suigetsu SG06 sedimentary archive, Japan: Chronostratigraphic markers for synchronising of east Asian/West Pacific palaeoclimatic records across the last 150 ka, *Quat. Sci. Rev.*, 61, 121–137.

Takarada, S. and Hoshizumi, H., 2020, Distribution and eruptive volume of Aso-4 Pyroclastic Density Current and Tephra Fall Deposits, Japan: A M8 super-eruption. *Front. Earth Sci.*, **8**,170, 10.3389/feart.2020.00170

Watanabe, K., 1978, Studies on the Aso pyroclastic flow deposits in the region to the west of Aso Caldera, Southwest Japan, I: Geology. *Mem. Fac. Educ. Kumamoto Univ. Nat. Sci.*, **27**, 97–120.

Watanabe, K., 2001, *Geology of Aso Volcano, Ichinomiya Chosi (History of Ichinomiya Town, Kumamoto Prefecture, Japan)*, 7, 238p. (in Japanese)

Rock Mechanics and Engineering Geology in Volcanic Fields – Ohta, Ito & Osada (eds)
© 2023 copyright the Author(s), ISBN 978-1-032-27657-1

Local tephra as an age-determination tool: Example of 2.3 ka Yakedake volcano tephra in Nagano Prefecture, central Japan

Satoru Kojima*
Department of Civil Engineering, Gifu University, Gifu, Japan

Saya Kagami & Tatsunori Yokoyama
Tono Geoscience Center, Japan Atomic Energy Agency, Toki, Japan

Yoshihiko Kariya
Department of Geography, Senshu Univeristy, Kawasaki, Japan

Yoshikazu Katayama & Gaku Nishio
Department of Civil Engineering, Gifu University, Gifu, Japan

ABSTRACT: A local tephra embedded in a hand-auger boring core drilled at the near-shore of the Kinugasanoike Pond approximately 4.6 km northeast of Mt. Yakedake, which is one of the most active volcanoes in central Japan, was found to be composed mainly of several kinds of volcanic glass shards (microlite-bearing, blocky, fluted, and micro-vesicular types) and minor amounts of crystal minerals, including quartz, plagioclase, hornblende, biotite, and pyroxene. The plant remains recovered from the horizon 10 cm below the tephra layer revealed ^{14}C ages of 2,331 – 2,295 cal years BP and 2,270– 2,155 cal years BP (19.2% and 76.2 % probability distributions, respectively). In this study, we measured the major element compositions of 241 individual glass shards using an electron probe microanalyzer, and found that they plotted on SiO_2–K_2O, SiO_2–Na_2O+K_2O, and FeO^*–K_2O diagrams in different regions than those of the major regional tephras distributed in central Japan. Moreover, the observed clast and chemical compositions coincide with those of a tephra embedded in the Nakao pyroclastic flow deposit distributed approximately 2 km north-northwest of Mt. Yakedake, which is dated at approximately 2,300 cal years BP. Thus, this tephra could be used as a local marker of 2,300 cal years BP for the southern part of the Northern Japan Alps.

Keywords: Tephrochronology, Yakedake volcano, Local tephra, Northern Japan Alps, Kamikochi

1 INTRODUCTION

Tephrochronology is an important method for determining the ages of Cenozoic sediments, especially in countries located along subduction zones. For example, in Japan, the Kikai submarine caldera, located 50 km south of Kyushu Island, erupted in 7307 – 7196 cal years BP, dispersing approximately 100 km^3 volcanic ash, which was named the Kikai Akahoya tephra (K-Ah), and dispersed the characteristic volcanic glass shards over much of Japan and the surrounding oceans (Machida and Arai, 2003; McLean et al., 2018). The K-Ah regional tephra is used as a key age marker in Quaternary geology, engineering geology, and geomorphology.

*Corresponding author: skojima@gifu-u.ac.jp

DOI: 10.1201/9781003293590-13

However, near the active volcanoes, important signals of the regional tephra are obscured by local volcano-clastic materials of unknown ages. While conducting landslide research near Mt. Yakedake, one of the most active volcanoes in central Japan, we found a distinctive tephra in a hand-auger boring core drilled at the near-shore of the Kinugasanoike Pond, approximately 4.6 km northeast of Mt. Yakedake (Figure 1). In this study, we describe the characteristics of the obtained tephra, including the clast composition, morphological features, and chemical composition of its glass shards. We also examine the characteristics of tephra included in the most probable source, a Nakao pyroclastic flow deposit (PFD), distributed approximately 2 km north-northwest of the Mt. Yakedake volcano (Figure 1).

2 GEOLOGIC SETTING

Mt. Yakedake (2,455 m), located in the southern part of the Northern Japan Alps, is one of the most active volcanoes in central Japan (Figure 1). Its volcanic activities are subdivided into older (120 – 70 ka) and younger (26 ka to present) stages (Oikawa, 2002). Oikawa et al.

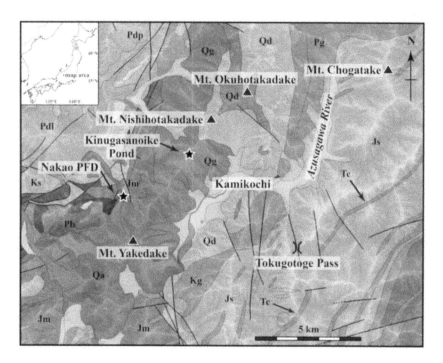

Figure 1. Geologic map of the study area (modified from GeomapNavi, Geological Survey of Japan). Inset map shows the location of the map area. Qa: Quaternary andesite, Qd: Quaternary dacite, Qg: Quaternary granodiorite, Pdl: Paleogene dacite lave, Pdp: Paleogene dacite pyroclastics, Kg: Cretaceous granite, Ks: Cretaceous sandstone, Jm: Jurassic mélange, Js: Jurassic sandstone, Tc: Triassic chert, Pb: Paleozoic basalt, and PFD: pyroclastic flow deposit.

(2002) studied the last 3,000 years of volcanic activity and found that the last magmatic eruption produced the Nakao PFD approximately 2,300 cal years BP. The volcanic rocks of Mt. Yakedake are mostly composed of andesite and dacite.

Meanwhile, the basement of the Yakedake volcanic consists of Paleozoic to Mesozoic sedimentary, metamorphic, and igneous complexes of the Hida Gaien belt, Jurassic accretionary complexes of the Mino terrane, and Cenozoic (mostly Quaternary) plutonic and volcanic rocks (Figure 1, Harayama, 1990). The rocks of the Mino terrane are composed of Jurassic sandstone and mudstone, with Triassic chert interbeds, and are distributed mainly on the east side of the Azusagawa River flowing in the Kamikochi valley (Figure 1). These rocks make wide ridges that have been gravitationally deformed to form double and multiple ridges at its top (Figure 1).

The Kinugasanoike Pond, approximately 4.6 km northeast of Mt. Yakedake (Figure 1), is a depression that was formed by deep-seated gravitational slope deformation. Kariya and Takaoka (2019) drilled a 280 cm-long core (KNG-2017) to obtain sediments accumulated in this depression (Figure 2). They found that the upper part of the core sediment is rich in

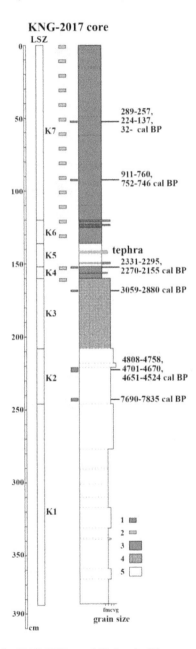

Figure 2. Columnar section of the KNG-2017 core drilled at the Kinugasanoike Pond (Modified from Kariya and Takaoka, 2019). K1 to K7: Local sedimentary facies zones (LSZ). 1 and 2: samples for ^{14}C age determination, 1: measured, 2: not measured, 3: organic sediment, 4: slightly organic sediment, 5: inorganic sediment. f: fine-grained sand, m: medium-grained sand, c: coarse-grained sand, v: very coarse-grained sand, and g: granule.

organic materials, such as leaves, twigs, and wood fragments, and the lower part is inorganic. A glassy tephra bed was found at 143 – 141 cm intervals approximately 10 cm above the carbonaceous sample horizon, yielding ^{14}C ages of 2,331 – 2,295 (19.2%) and 2,270 – 2,155 (76.2%) cal years BP (Figure 2; Kariya and Takaoka, 2019). We also collected tephra samples embedded within the Nakao PFD (Figure 1) that are approximately 2,300 cal years BP in order to check the origin of the tephra recovered from the Kinugasanoike Pond.

3 METHODOLOGY

The tephra samples from the Kinugasanoike Pond and Nakao PFD were washed with water, and then sieved using #250 (62 μm) and #100 (200 μm) sieving cloths. Then, the 62 – 200 μm fractions of tephra were utilized for the analyses.

The clast compositions were counted under a polarizing microscope for more than 400 grains. The grains were subdivided into glass shards, minerals, and others. The glass shards were then separated into blocky glass, glass with microlites, pumiceous glass, fluted glass, and microvesicular glass groups (Figure 3; McLean et al., 2018). Although some glass shards had several characteristics of these categories and could be classified as an intermediate type (Figure 3), we placed them in an end-member category according to their predominant feature. The minerals observed include quartz, plagioclase, biotite, hornblende, and pyroxene. Note that the clastic grains of non-volcanic origin (e.g., plant remains) were not counted.

The chemical compositions of the glass shards were measured using an electron probe microanalyzer (EPMA) installed at the Tono Geoscience Center, Japan Atomic Energy Agency. The specific EPMA used was a JXA-8530F (EOL Ltd.) and its measurement conditions were as follows: 15 kV accelerating voltage, 6×10^{-9} A current, 5 μm beam. The analytical results were corrected by using the conventional ZAF method (Z: atomic number factor, A: absorption factor, F: characteristic fluorescence correction). We selected the blocky glass shards for the chemical analyses because they had homogeneous inner structures and were free from microlites (Figure 3).

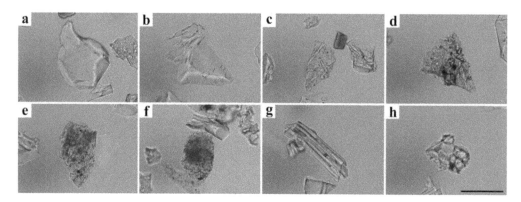

Figure 3. Representative glass shards included in the tephra of the Kinugasanoike Pond core. a: blocky glass, b: blocky glass with few microlites, c: glass with microlites, d: glass with several kinds of microlites, e: partly pumiceous glass with microlites, f: pumiceous glass, g: fluted glass, and h: micro-vesicular glass. Scale bar, shown in h, is 0.1 mm for all the photographs.

4 RESULTS

Both tephra samples from the Kinugasanoike Pond and Nakao PFD had similar clast compositions, in which approximately 70% of the clasts were volcanic glasses and the microlite and blocky types were the main constituents (Figure 4). The chemical compositions of the blocky glass shards were also similar, as summarized in Table 1. The age of the Nakao PFD was estimated to be ca. 2300 cal years BP based on the ^{14}C ages of the humic soil immediately below the glassy tephra distributed 1 km north-northeast of Mt. Yakedake (Oikawa et al., 2002). The Nakao PFD tephra sample treated in this study was collected from an exposure approximately 2 km north-northwest of the volcano (Figure 1). Meanwhile, the age of the tephra recovered from the core of Kinugasanoike Pond was 2,331 – 2,155 cal years BP on the basis of the ^{14}C age of the carbonaceous sediments below the tephra. These results strongly indicate that the Kinugasanoike Pond tephra was derived from the eruption of the Yakedake volcano during the activity of the Nakao PFD.

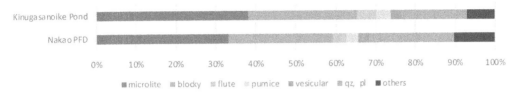

Figure 4.   Clast composition of tephras included in the core of the Kinugasanoike Pond and Nakao PFD. Total number of counts are 511 and 431, respectively. PFD: pyroclastic flow deposit, qz: quartz, and pl: plagioclase.

Table 1.   Average and standard deviation values of major elements (weight%) of glass shards included in the tephra of Kinugasanoike Pond and Nakao pyroclastic flow deposit (PFD). In addition to the elements listed, we measured F, S and Cr. Each value is normalized to achieve a total of 100%.

		SiO_2	TiO_2	Al_2O_3	FeO*	MnO	MgO	CaO	Na_2O	K_2O	P_2O_5	Cl
K	Av	77.48	0.17	12.13	0.80	0.03	0.11	0.96	3.32	4.82	0.02	0.10
	SD	0.70	0.06	0.44	0.13	0.03	0.04	0.28	0.26	0.29	0.02	0.02
N	Av	77.83	0.17	12.01	0.85	0.03	0.11	0.89	3.03	4.89	0.02	0.11
	SD	0.64	0.06	0.39	0.13	0.03	0.03	0.20	0.27	0.26	0.02	0.03

K: Kinugasanoike Pond (n=241), N: Nakao PFD (n=126), Av: average, SD: standard deviation, FeO*: total iron as FeO.

The chemical composition of the glass shards in the Kinugasanoike Pond and Nakao PFD were quite different from those of the Quaternary regional tephra found in Japan. McLean et al. (2018) examined Holocene tephras in cores recovered from Lake Suigetsu in the Fukui Prefecture of central Japan. The chemical compositions of the glasses in the tephra treated in this study varied from those of the regional tephra, and were especially discriminated in the SiO_2–K_2O (Figure 5), SiO_2–K_2O+Na_2O, and FeO*–K_2O plots. Based on these chemical characteristics and the observed clast compositions, this tephra could be used as a Holocene age marker for the southern part of the Northern Japan Alps.

5   DISCUSSION

In this study, we showed that the eruption of the Yakedake volcano approximately 2,300 years ago dispersed a large amount of tephra with a characteristic clast composition (rich in glasses, especially microlite-bearing and blocky glass shards, Figure 4) and diagnostic chemical composition (high silica and alkali contents, Figure 5). Although the Nakao PFD was the most recent magmatic eruption of the Yakedake volcano, older eruptions might have produced volcaniclastic materials similar to the tephra treated in this study. Therefore, it might be difficult to differentiate the 2,300 year old tephra from other older events, and additional information is needed regarding older tephras.

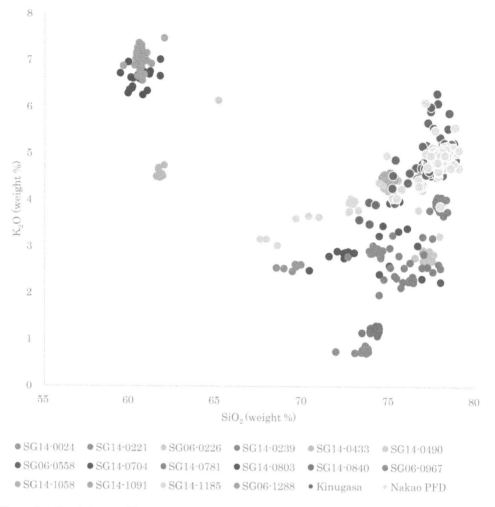

Figure 5. Chemical composition of glass shards included in the Quaternary regional tephras found in the Suiget-suko Lake cores in central Japan (plots with SG-prefix; after McLean et al., 2018) and that for the tephras found in the Kinugasanoike Pond core (Kinugasa) and embedded with the Nakao pyroclastic flow deposit (PFD).

6 CONCLUSIONS

The tephra recovered from the hand-auger boring core at Kinugasanoike Pond had characteris-tic clast composition (rich in glass shards of microlite and blocky types) and chemical compos-ition (high silica and alkali contents) that differed from the Holocene major regional tephras found in Lake Suigetsu in central Japan (McLean et al., 2018). This tephra could potentially be used as a 2,300 cal years BP time marker for researches on Quaternary geology, applied geol-ogy, and geomorphology in the southern part of the Northern Japan Alps, although additional information concerning older tephras and the volcanic history of the Yakedake volcano group are still necessary.

ACKNOWLEDGEMENTS

We thank H. Nagata for his help in the field, and M. Niwa and Y. Ogita for their supports in the lab. Part of this study was financially supported by JSPS KAKENHI Grant Numbers 26400487, 17H02033, and 19K03990. Another part of this study was carried out under a contract with the

Ministry of Economy, Trade and Industry (METI) as part of its R&D supporting program titled "Development of Evaluation Technology for Long-term Geosphere Stability (2013-2018 Fy)".

REFERENCES

Harayama, S., 1990, *Geology of the Kamikochi District*. With Geological Sheet Map at 1:50,000, Geol. Surv. Japan, 175p. [JE]

Kariya, Y. and Takaoka, S., 2019, Geological description of a recovered core from fill deposits in a linear depression on the northern Japanese Alps: An example from Kinugasano-ike pond in the Kamikochi District, Nagano Prefecture. *Bull. Inst. Natural Sci. Senshu Univ.*, 50, 1–11. [JE]

Machida, H. and Arai, F., 2003, Atlas of Tephra in and around Japan. University of Tokyo Press, 336p. [J]

McLean, D., Albert, P.G., Nakagawa, T., Suzuki, T., Staff, R.A., Yamada, K., Kitaba, I., Haraguchi, T., Kitagawa, J., SG14 Project Members, and Smith, V.C., 2018, Integrating the Holocene tephro-stratigraphy for East Asia using a high-resolution cryptotephra study from Lake Suigetsu (SG14 core), central Japan, *Quat. Sci. Rev.*, 183, 36–58.

Oikawa, T., 2002, Geology, volcanic history and eruptive style of the Yakedake Volcano Group, Central Japan. *Jour. Geol. Soc. Japan*, 108, 615–632. [JE]

Oikawa, T., Okuno, M. and Nakamura, T., 2002, The past 3000 years' eruption history of Yakedake volcano in the Northern Japan Alps. *Jour. Geol. Soc. Japan*, 108, 88–102. [JE]

Rock Mechanics and Engineering Geology in Volcanic Fields – Ohta, Ito & Osada (eds)
© 2023 copyright the Author(s), ISBN 978-1-032-27657-1

Study on the factors determining vegetation around the summit of the Kuju Volcano Group

Yuya Noguchi* & Takehiro Ohta
YamaguchiUniversity, Yamaguchi, Japan

ABSTRACT: The Kuju volcano Group is a volcanic field whose highest peak is Nakadake (1,791m above sea level), the highest peak on Kyushu Island in Japan. The summits of the Kuju Volcano Group have a unique vegetation which consists of shrubby forests of *Rhododendron kiusianum* and other trees, which are similar to the vegetation of the alpine zone. On the other hand, the summit of Sobo Mountain (1,756m above sea level), which is almost same altitude as the Kuju volcano and is located about 50km southeast of that volcano, is covered with tall forest.

In this study, the distribution of vegetation was examined from a geoecological approach, which is a methodology that tries to explain each environmental factor that defines the distribution of vegetation based on geomorphology, geology, climatology and so on. The purpose of this study is to clarify the factors that led to the formation of the unique vegetation around the summit of the Kuju Volcano Group.

As a result of multivariate analysis for geomorphological and geotechnical features, it was found that the elevation and the pH of soil involve the distribution of vegetation. The elevation is correlated with temperature, and the pH of soil indicates the degree of decomposition of organic matter in the soil. These results can be concluded as follows; 1) The tall forests are distributed in areas in which the temperature is high and the decomposition of organic matter in the soil is active. 2) The shrubby forests are distributed in areas in which the temperature is low and the decomposition of organic matter in the soil is relatively active. 3) The temperature and soil conditions at grassland are similar to shrubland.4) The temperature at the marshland is high and the decomposition of organic matter in the soil is inactive.

Keywords: Geoecology, Geomorphology, Vegetation distribution, Environmental factor, GIS

1 INTRODUCTION

The Kuju volcano Group is a volcanic field whose highest peak is Nakadake (1,791m above sea level), the highest peak on Kyushu Island in Japan. At the summits of the Kuju Volcano Group, shrubby forests of *Rhododendron kiusianum* and *Vaccinium vitis-idaea L*, which are similar to alpine vegetation, form a unique vegetation (Umezu and Suzuki, 1970). And, deciduous tall trees such as *Clethra barbinervis, Quercus crispula community, Fagus crenata* are distributed near the foot of the mountain, forming a forest. However, the summit of Sobo Mountain (1,756 m above sea level), located 50 km southeast of the Kuju Volcano Group, is covered with tall forest.

Koizumi (1974) studied the distribution of alpine plants like as the Kuju Volcano Group, that showed the effect of snow on plants due to differences in topography, such as relief and

*Corresponding author: b029vcv@yamaguchi-u.ac.jp

DOI: 10.1201/9781003293590-14

incline orientation, as a factor regulating the distribution of vegetation. The relationship between vegetation and soil in the Kuju Volcanic Group (Umezu and Suzuki, 1970) and the relationship between vegetation and soil temperature (Suzuki and Arakane, 1972) are some of the studies on the factors that caused the unique vegetation in the Kuju Volcanic Group. However, the relationship among vegetation, topography, and soil has not been studied in the entire Kuju Volcano Group. In this study, we examined the distribution of vegetation from a geoecological approach (Koizumi, 2018), which is a method that explain each environmental factor that defines the distribution of vegetation based on geomorphology, geology, and climatology and so on.

In order to compare the Kuju Volcano Group and Sobo Mountain, which are almost at the same altitude but have different vegetation, the purpose of this study is to first clarify the factors that led to the formation of the unique vegetation in the Kuju Volcano Group from a geoecological approach.

2   METHODOLOGY

2.1   GIS analysis

The digital elevation models (DEMs) of the study area were downloaded from the Technical Report of the Geospatial Information Authority of Japan. The elevation, incline, and aspect were analyzed from the DEMs using QGIS. The relationship between the distribution of vegetation and these calculated factors were examined.

2.2   Field survey

Soil samples were collected at 83 sites (Show Figure 1). Each sampling point was representative site on each vegetation, which are provided by Ministry of the Environment in Japan (1999-2012, 2013-). When collecting soil samples, the organic matter layer (O layer) which is top layer, was removed from the ground surface, and the soil samples were obtained at a depth of about 10 cm from top of the A layer.

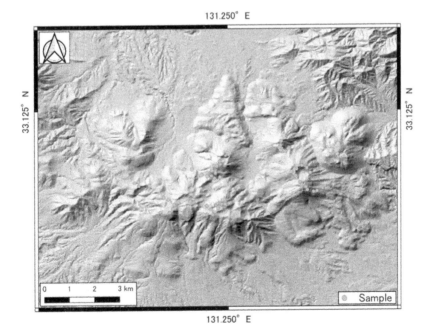

Figure 1.   Sites of soil samples.

2.3 pH and EC measurement

In order to identify the features of pore water in soil among different vegetation types, the pH and EC of soil samples were measured. In this study, the pH and EC of pore water are defined as those of soil suspension. Preparing of soil suspensions and measurement of pH and EC were followed the Laboratory Testing Standards of Geomaterials (Japanese Geotechnical Society Standard, 2000).

2.4 Multivariate analysis

The discriminant analysis, a type of multivariate analysis, was used to determine the extent to which environmental factors affect vegetation growth. The discriminant analysis used the results of GIS analysis for the sampling sites and the data of pH and EC measurements of the soil suspension.

3 RESULTS

This study included 38 vegetation species above 550 m elevation out of 69 vegetation species which are shown in the vegetation map of the 6th-7th vegetation survey (1999-2012/2013-) of this study area provided by Ministry of the Environment in Japan (2020). Figure 2 shows the areas where each type of vegetation is distributed.

The vegetations in the study area are classified into the following 6 types based on field survey and the description of the vegetation published by Ministry of the Environment in Japan (2020).

 i. The tall forests: trees with a height of 5 m or more.
 ii. The shrubby forests: trees with a height of less than 0-5 m.
iii. The marshlands: herbaceous plants distributed where periods of time exist when the water table is higher than the ground surface.
iv. The grassland: herbaceous plants distributed outside of the marshlands.
 v. The bare area: areas where plants cannot grow under natural conditions.
vi. The plantation and pasture: areas that have been artificially planted or cut.

Table 1 shows the results of the vegetation classification.

3.1 pH and EC measurement

Figure 3 Shows the results of pH measurement for each vegetation classification. There is not significant difference among each vegetation classification in the mean value of pH. On the other hand, the marshlands have a larger mean value of pH than other vegetation classifications.

The EC of soil for each vegetation classification is illustrated in Figure 4. The mean values of EC in each vegetation classification decreases in the following order: the tall forests, the shrubby forests, the marshlands, the grassland, and the bare area.

3.2 Multivariate analysis (discriminant analysis)

In this study, the environmental factors of each vegetation classification, i.e., the tall forest, the shrubby forest, the marshland, and grassland, were abstracted using the discriminant analysis. The bare area was not analyzed because of the small number of samples for pH and EC measurements. Plantations and pastures were ineligible because those are managed artificially. The parameters using in the discriminant analysis were the elevation and incline obtained from GIS analysis, and measurements of pH and EC of soil suspensions.

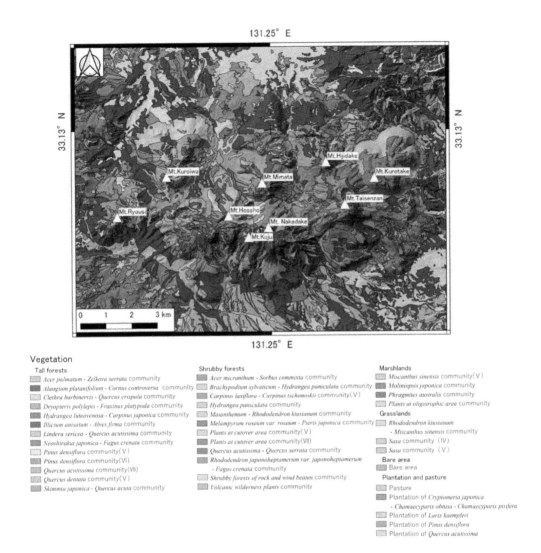

Figure 2. Vegetation map.

Figure 5 shows the results of the discriminant analysis. Comparing the mean values of each vegetation classifications on axis-1, shrubby forests and grasslands are plotted closer together, while tall forests and marshlands are plotted farther apart. On axis-2, shrubby forests and grasslands are plotted near 0, with marshlands showing minimum values and tall forests showing maximum values.

The values of each sample shown on the vertical and horizontal axes in Figure 5 are determined by the canonical coefficients, calculated from the discriminant analysis. The calculated canonical coefficients are listed in Table 2. The larger the absolute value of canonical coefficients, the more it affects the value of the axis. The absolute value of elevation is largest on axis-1, and the absolute value of pH is largest on axis-2.

4 DISCUSSION

The most dominant parameter affecting the canonical coefficient at axis-1 is elevation, followed by EC. Therefore, axis-1 is affected by the characteristics of elevation and soil

Table 1. Vegetation classification.

classification	representative vegetation
i. Tall forests	*Acer palmatum - Zelkova serrata* community, *Alangium platanifolium - Cornus controversa* community, *Clethra barbinervis - Quercus crispula* community, *Dryopteris polylepis - Fraxinus platypoda* community, *Hydrangea luteovenosa - Carpinus japonica* community, *Illicium anisatum - Abies firma* community, *Lindera sericea - Quercus acutissima* community, *Neoshirakia japonica - Fagus crenata* community, *Pinus densiflora* community(V), *Pinus densiflora* community(VI), *Quercus acutissima* community(VII), *Quercus dentata* community(V), *Skimmia japonica - Quercus acuta* community
ii. Shrubby forests	*Acer micranthum - Sorbus commixta* community, *Brachypodium sylvaticum - Hydrangea paniculata* community, *Carpinus laxiflora - Carpinus tschonoskii* community(V), *Hydrangea paniculata* community, *Maianthemum - Rhododendron kiusianum* community, *Melampyrum roseum var. roseum - Pieris japonica* community, *Plants at cutover area* community(V), *Plants at cutover area* community(VII), *Quercus acutissima - Quercus serrata* community, *Rhododendron japonoheptamerum var. japonoheptamerum - Fagus crenata* community, *Shrubby forests of rock and wind beaten* community, *Volcanic wilderness plants* community
iii. Marshlands	*Miscanthus sinensis* community(V), *Moliniopsis japonica* community, *Phragmites australis* community, *Plants at oligotrophic area* community
iv. Grassland	*Rhododendron kiusianum - Miscanthus sinensis* community, *Sasa* community(IV), *Sasa* community(V)
v. Bare	Bare area
vi. Plantation and pasture	Pasture, Plantation of *Cryptomeria japonica - Chamaecyparis obtusa - Chamaecyparis pisifera*, Plantation of *Larix kaempferi*, Plantation of *Pinus densiflora*, Plantation of *Quercus acutissima*

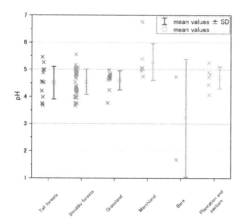

Figure 3. pH for each vegetation classification.

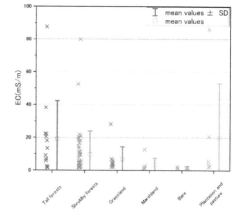

Figure 4. EC for each vegetation classification.

EC for each vegetation classifications. In other words, the interpretation of the discriminant analysis shows that as the value of axis-1 increases, the elevation increases and the value of EC of the soil decreases. Comparing the mean values of each vegetation classification in axis-1 of Figure 5, shrubby forests and grasslands have similar values, and the mean values decrease in the order of marshlands and tall forests. This suggests that shrublands and grasslands are distributed in areas of similar elevation and soil EC, whereas marshlands are distributed in areas of lower elevation and higher soil EC than shrublands and grasslands, and tall forests are distributed in areas of lowest elevation and highest soil EC.

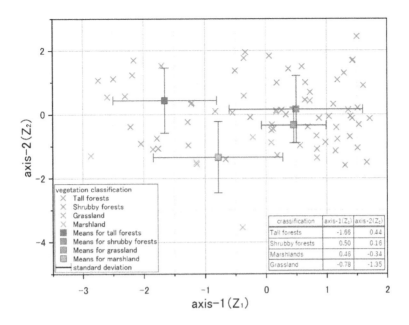

Figure 5. Results of discriminant analysis.

Table. 2. Canonical coefficients.

classification	axis-1(Z_1)	axis-2(Z_2)
Tall forests	-1.66	0.44
Shrubby forests	0.5	0.16
Marshlands	0.46	-0.34
Grassland	-0.78	-1.35

The difference in elevation shown in axis-1 of Figure 5 is thought to indicate the difference in temperature due to the rate of temperature lapse rate (Uchida et al., 1968). Therefore, it is considered that shrublands and grasslands distributed in the high elevation region are distributed in the cold temperature region, while tall forests distributed in the low elevation region are distributed in the warm temperature region. This idea matches the results of GIS analysis.

The most dominant parameter affecting the canonical coefficient of axis-2 is soil pH, followed by incline. Therefore, axis-2 shows the characteristics of soil pH and incline for each vegetation classification. In other words, the interpretation of the discriminant analysis shows that as the value of axis-2 increases, the value of pH of the soil acidifies and the incline decreases. Comparing the mean values of each vegetation classification in axis-2 of Figure 5, marshlands, grassland, shrubland, and tall forest are larger in this order. This suggests that they are distributed in areas with acidic soil pH and a large amount of incline, in the order of tall forests, shrubland, grassland, and marshland.

The difference in soil pH for each vegetation classification shown in axis 2 of Figure 5 may represent the difference in the degree of decomposition of organic matter in the soil (Shibata, 2018). Shibata (2018) concluded that, the more active the decomposition of organic matter in the soil by nitrification bacteria, the more acidic the pH of the soil becomes. In the discriminant analysis, tall forests distributed in the most acidic region of the soil are distributed in the region where the decomposition of organic matter in the soil is most active. And the marshes distributed in the most alkaline region are considered to be distributed in the region where the decomposition of organic matter in the soil is inactive.

5 CONCLUSIONS

The summit of the Kuju Volcanic Group has a unique vegetation distribution. In this study, GIS analysis and measurements of pH and EC of soil suspension were conducted to clarify the factors that define vegetation from a geoecological perspective. Discriminant analysis was performed on each result to show the extent to which each factor affected the distribution of vegetation. Discriminant analysis showed that among elevation, incline, and soil pH and EC, vegetation distribution was mainly influenced by elevation and soil pH.

It is concluded to estimate the environmental factors that influence the distribution of each vegetation category in the study area based on GIS analysis, pH and EC measurements of soil suspensions, and discriminant analysis as follows.

i. The tall forests
Distributed in areas of low elevation and acidic soil pH, i.e., areas of warmth and most active decomposition of soil organic matter.

ii. The shrubby forests
Distributed in areas of high elevation and relatively acidic soil pH, i.e., areas of cold and relatively active decomposition of soil organic matter.

iii. The marshlands
The elevation is relatively high and the soil pH is distributed in the most alkaline region, that is, in the region that is relatively cold and where the decomposition of soil organic matter is most inactive. The location also has the smallest amount of incline.

iv. The grasslands
Distributed at similar elevations to the shrubby forests, and in areas of soil pH and incline between the shrubby forests and the marshlands.

v. The bare area
Located in the area with the lowest soil EC, although this is unreliable due to the small number of soil samples.

vi. The plantation and pasture
Distributed in areas that are artificially controlled and where soil pH values are relatively alkaline.

REFERENCES

Japanese Geotechnical Society Standard., 2000, *Laboratory Testing Standards of Geomaterials*, 159–169.
Koizumi, T., 2018, *Vegetation in Japan from the perspective of geoecology*, Bun-ichi Co.,Ltd., 444p.
Ministry of the Environment in Japan, 2020,6th-7th vegetation survey (1999-2012/2013-) http://gis.biodic.go.jp/webgis/index.html (April 9th, 2020).
Shibata, H., 2018, *Series in Forest* Science 7 *Forest and Soil*, Kyoritsu Shuppan Co., Ltd., 240p
Suzuki, T., and Arakane, M., 1972, Plant societies and geothermal temperatures in the Kuju Volcanic Group, *Japanese Journal of Ecology.*, 22, 180–189.
Technical Report of the Geospatial Information Authority of Japan, Fundamental Geospatial Data Download Service,2020, https://fgd.gsi.go.jp/download/mapGis.php?tab=dem (March 18th, 2020).
Umezu, Y., and Suzuki, K., 1970, Vegetation and soils at the top of the Kuju Mountain Group, *Japanese Journal of Ecology.*, 20, 188–197.
Uchida, T., and Yoshida, H., 1968, On the distribution and morphological characters of Dymecodon pilirostris true from Kyusyu, *The Journal of the Mammalogical Society of Japan.*, 8, 17–26.

Rock Mechanics and Engineering Geology in Volcanic Fields – Ohta, Ito & Osada (eds)
© 2023 copyright the Author(s), ISBN 978-1-032-27657-1

Volcanic ash impacts on railway signal facilities and utilizing of the volcanic ash fall forecast

Yuichiro Nishikane*, Natsuki Terada, Takuya Urakoshi & Shoichi Kawamura
Railway Technical Research Institute, Tokyo, Japan

ABSTRACT: Ash fall affects various railway operations in a wide range. One of the most critical and likely impact of ash fall on railway is an obstacle to the train detection using track circuit.

The track circuit, which is generally used in Japanese railway and controls the signal and level crossing security devices, detects the existence of train on rails when the rails are electrically shunted by the axles of the train. When volcanic ash is deposited on rails, the track circuit is failed because the volcanic ash interferes with the energization between the wheels and the rails. However, the thickness of the volcanic ash which prevents train detection is unknown so far.

In this study, we examined the impact of volcanic ash on the shunting of track by running trains on rails which are artificially covered with volcanic ash. The experimental track used in this study simulates the track generally used in Japan. As a result, it is clarified that 0.025 mm thick volcanic ash can make the shunting unstable, and that over 0.05 mm thick volcanic ash causes the malfunction of shunting. Additionally, it is also clarified that the electronic train detector, which controls the operation of the level crossing, fails to detect the existence of trains when volcanic ash is deposited over 0.2mm thick on rails.

Based on the results, we propose prevention actions at normal time and at the time of ash fall for railway companies to mitigate the impact, using public information on eruption. The Volcanic Ash Fall Forecast, for example, can be utilized for narrowing the range where inspection or cleaning of rails should be carried out according to the predicted thickness of volcanic ash.

Keywords: Volcanic ash, Railway, Track circuit, Volcanic Ash Fall Forecast

1 INTRODUCTION

There are currently 111 volcanoes in Japan designated as active volcanoes. It is noted that the active volcanoes are defined as volcanoes which have erupted within the last 10,000 years and are currently actively fuming. 50 active volcanoes of them have been designated as particularly active volcanoes which should be monitored continuously based on the possibility of eruption and the impact on society. Phenomena associated with the volcanic activities, such as, ash fall, pyroclastic flow, lahar, and ground movement possibly cause damage to social infrastructure including railway. Among the phenomena, especially, ash fall affects various railway operations in a wide range. One of the most critical and likely impacts of the ash fall on railway is an obstacle to the train detection using track circuit.

The track circuit, which is generally used in Japanese railway and controls the signal and level crossing security devices, detects the existence of train on rails when the rails are electrically shunted by the axles of the train (Figure 1). When volcanic ash is deposited on rails, the track circuit is failed because the volcanic ash interferes with the energization between the wheels and the rails. As an example, malfunctions of railway signal and crossing alarm due to the failure of shunting, which

*Corresponding author: nishikane.yuichiro.48@rtri.or.jp

DOI: 10.1201/9781003293590-15

were caused by the volcanic ash overlying rail surface, occurred during the 2011 eruption of Mt. Shinmoedake (Iguchi, 2011). However, it is unknown how thick the volcanic ash causes the malfunction of the track circuit.

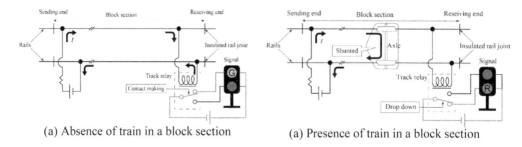

(a) Absence of train in a block section   (a) Presence of train in a block section

Figure 1. Schematic figure of structure of a normally closed track circuit.

In this study, we conduct track circuit tests to evaluate the impact of volcanic ash on the shunting characteristics of the track circuit by running trains on rails which are artificially covered with volcanic ash. Based on results of the tests, we propose prevention actions at normal time and at the time of ash fall for railway companies to mitigate the impact, using public information on eruption, such as the volcanic ash fall forecast.

2   METHOD

2.1   Track circuit test

We carried out track circuit tests on a test track in RTRI using a test train. The type of track circuit system constructed in the test track is the normally closed track circuit, most popular way for detecting trains (Figure 1). The shunting sensitivity of the track circuit was 0.12 Ω. Under normal time, when a train enters a blocked section, which is electrically separated from the other blocked sections with insulated rail joints, the rails are electrically shunted by axles of the train, and the track relay drops down. While the train is running or stopping in the block section, this track relay is kept as down state, which is the situation that the presence of the train is detected. As shown in Figure 2, a test segment was configured of which the length was 5-8 m from a receiving end of a track circuit. Testing devices were mainly composed of the track circuit, an axle detector and a recorder. The axle detector was independent of the track circuit and installed near the receiving end of the track circuit to catch the timing when the train enters the test segment. Using these devices, we measured the residual voltage of the track relay, the state of the track relay (contact making or drop down) and the output signal from the axle sensor while the train passed through the test track.

Before the start of each test, we scattered volcanic ash originated in Sakurajima Island on the rail head in the test segment, while controlling the thickness of volcanic ash. The methods to control the thickness of volcanic ash on rails are as follows: When the thickness is needed to be reduced to 0.5 to 1.0 mm a comb-shaped device with a gap of a prescribed height at the bottom is moved to even out the volcanic ash spread on the rail. When the thickness is needed to be reduced to 0.2mm or less, the required volume of volcanic ash is sprinkled evenly using a perforated apparatus. In the latter case, since the surface of rails is not perfectly covered by the volcanic ash, the thickness is the mean value. When volcanic ash of thickness of d mm was deposited, the required volume of the volcanic ash deposited on a rail of length of 1m was calculated by Eq. (1), since the width of the rail head is 5 cm.

$$V = 5 \times 100 \times \frac{d}{10} = 50d \qquad (1)$$

where V (cm³/m) is the required volume deposited on a rail of length of 1m.

Additionally, in order to reproduce the ash fall under a fine and rainy day, we conducted test on volcanic ash in dry and wet conditions. In order to wet the volcanic ash, water was sprayed on the rails after the volcanic ash was scattered. The water content of the volcanic ash in the dry and wet condition was less than 1 % and between 5 to 26 %, respectively. The electrical resistivity of volcanic ash strongly depends on the water content according to the measurements reported on another paper (Urakoshi et al., in press).

The test train enters the test segment from the receiving end and keep coasting at a speed of 10 km/h while running through the test segment.

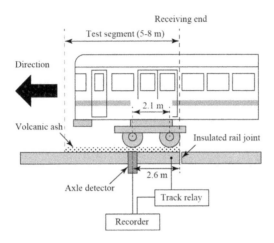

Figure 2.   Simplified schematic figure of the track circuit test.

2.2   *Electronic train detector test*

An electronic train detector is a track circuit with short length used for controlling level crossings. It makes possible to detect the presence of trains even if the short-circuit resistance between both rails through the wheels is relatively high. We carried out shunting tests of a crossing control in the presence of the volcanic ash. The type of electronic train detector used in the test is the normally opened track circuit, which works to stop a level crossing warning after detecting a train passage through level crossing.

The method of the tests is almost same as that of the track circuit test. An electronic train detector used in the test was constructed near the receiving end of the track circuit used in the track circuit test. The test segment of the electronic train detector test was common to that of the track circuit test. When the train enters the test section at the normal time, the rails are electrically shunted by axles of the train and the track relay of the electronic train detector is lifted. We measured the voltage between rails in the electronic train detector, the state of the relay of the electronic train detector and the output signal from the axle detector while the test train was passing through the test segment. The shunting sensitivity of the electronic train detector was 1.4 Ω.

3   RESULTS

3.1   *Track circuit test*

As examples of track circuit tests, the results for (a) no volcanic ash, (b) 0.05 mm thick volcanic ash in dry condition, (c) 0.025 mm thick volcanic ash in dry condition, and (d) 0.1 mm thick volcanic ash in wet condition are shown in Figure 3. In the case (a), the track relay dropped down in response to a sudden reduction of its residual voltage before the axle detector detected the first axle. On the other hand, in the case (b), the track relay dropped down just after the axle detector detected the second axle. When the second axle was above the axle detector, the first axle was at a position 0.3 m before the end point of the test segment. This is because that the length of the test segment is 5 m in this case whereas the length between the starting point of the test segment and the first axle is 4.7 m. (c.f. Figure 2). Therefore, it is thought that the timing when the track relay drops down corresponds to the timing when the first axle passes out the test segment in the case (b). The result of (b) shows that the rails in the test segment are not electrically shunted by the first axle because the volcanic ash interferes with the electric conduction between the wheels and the rails.

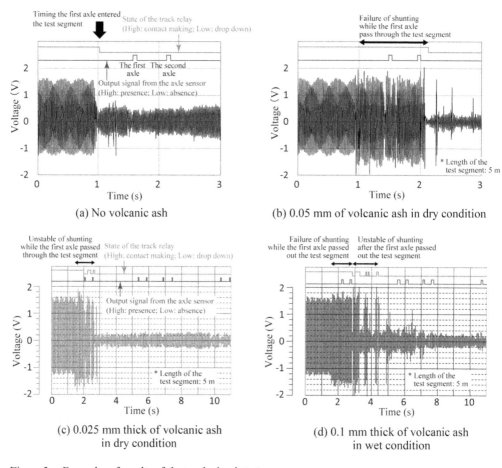

Figure 3.   Examples of results of the track circuit tests.

In the case (c), since the residual voltage of the track relay fluctuated significantly, the state of the track relay is unstable while the first axle passes through the test segment.

Table 1 shows the results of the track circuit tests. It is clarified that volcanic ash of 0.025 mm thickness possibly makes the shunting of the track circuit unstable, and that volcanic ash of 0.05 mm thickness possibly causes the malfunction of shunting. Additionally, in the case of the thickness of volcanic ash of 0.1 mm or more in wet condition, the shunting of track circuit was failed or unstable even if the test train passed the test segment (Figure 3(d)). The reason for this is that wet volcanic ash adhered to wheels and interfered with the electric conduction between the wheels and the rails even outside of the test segment.

Table 1. Thickness of the volcanic ash against the shunting of track circuit.

Thickness	Train detection	
	Dry condition	Wet condition
1 mm	Failed	Failed (failed continuously in outside of the test segment)
0.5 mm	Failed	Failed (unstable in outside of the test segment)
0.1 mm	Failed	Failed (unstable in outside of the test segment)
0.05 mm	Failed	Failed
0.025 mm	Unstable	Unstable
0.01 mm	Succeeded	Succeeded

3.2 *Electronic train detector test*

Examples of the results of the electronic train detector tests are shown in Figure 4. In the case of no volcanic ash, the relay of the electronic train detector is lifted before the axle detector detects the first axle (Figure 4(a)). When there is 0.05 mm of dry volcanic ash on rails in the control section of the electronic train detector, the state of the relay of the electronic train detector is unstable while the first axle passes through the test section (Figure 4(b)).

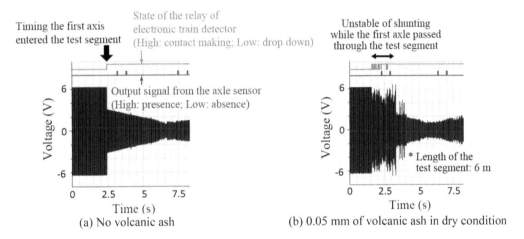

(a) No volcanic ash   (b) 0.05 mm of volcanic ash in dry condition

Figure 4. Examples of results of the electronic train detector tests.

Table 2 shows the results of the electronic train detector tests. It is clarified that volcanic ash of 0.025 mm thickness makes the shunting of the electronic train detector unstable, and that volcanic ash of 0.2 mm thickness causes the malfunction of shunting.

Table 2. Relation between thickness of the volcanic ash and detection by the electronic train detector.

Thickness	Train detection	
	Dry condition	Wet condition
0.2 mm	Failed	Not conducted
0.1 mm	Unstable	Unstable
0.05 mm	Unstable	Unstable
0.025 mm	Unstable	Succeeded
0.01 mm	Succeeded	Succeeded

4 CONDITION OF VOLCANIC ASH AFFECTING TRACK CIRCUITS

From the results of our tests, we find that 0.025 mm thick volcanic ash makes the shunting of track circuit unstable. The electrical resistivity of the volcanic ash used in our tests is about 100 Ωm when its water content is 20 % according to the measurements reported on another paper (Urakoshi et al., in press). Considering a two-axle (four-wheel) bogie on rails with volcanic ash of $d = 0.025$ mm thickness, the contact area between a rail and a wheel $S = 1 \times 10^{-4}$ m^2 and the electrical resistivity of volcanic ash $\rho = 100$ Ωm, the electric resistance between the rails and wheels (R) is estimated by Eq. (2). The value of S is a rough value calculated based on the Hertzian theory of elastic contact.

$$R = \rho \times 1000d/4S = 6.25\Omega \qquad (2)$$

This electric resistance is higher than the shunting sensitivity of the track circuit (0.12 Ω) and that of electronic train detector (1.4 Ω). This is consistent with the results of our tests in which shunting is failed or unstable. From the above discussion, it is thought that there exists a risk of malfunction of track circuit detecting system due to ash fall of 0.025 mm thickness. However, it is noted that the electrical resistivity of the volcanic ash used in our tests is relatively high (Urakoshi et al., in press) and then this volcanic ash would be easy to cause the failure of the shunting of track circuit.

5 MITIGATION OF IMPACTS OF ASH FALL ON RAILWAY USING PUBLIC INFORMATION

Public information on ash fall includes ash fall hazard maps issued by Commission on Mitigation of Volcanic Disasters and the Volcanic Ash Fall Forecast (VAFF) issued by the Japan Meteorological Agency (Hasegawa et al., 2015). Note that the VAFF has three types: Regular Information, Preliminary Forecast and Full Forecast. Among the types of the VAFF, the Preliminary Forecast is issued within 5–10 minutes of the onset of an eruption, and the Full Forecast is issued within 20–30 minutes of the onset of an eruption. The assumed area of ash fall in the Preliminary Forecast and the Full Forecast is classified into thickness levels: Little (0.0001 to 0.1 mm), Moderate (0.1 to 1 mm) and Much (over 1 mm) (Figure 5).

Based on the results of our tests, we propose some prevention actions at normal time and at the time of ash fall for railway companies to mitigate its impacts on railway, using public information on ash fall (Table 3).

During normal time, ash fall hazard maps are useful for railway companies to recognize areas that are affected by volcanic ash. This assumption is usable for determining deployment place of apparatus to clean volcanic ash. Additionally, it is possible to check the kind of track circuit, the presence or absence of crossing back-up device in assumed areas of ash fall described in the ash fall hazard maps. Most ash fall hazard maps describe the assumed areas of ash falls of 0.1–1 m thickness, whereas there is no ash fall hazard map that describes assumed areas of ash falls of thinner than 1 mm thickness. Therefore, in the assumed areas of ash falls described in ash fall hazard maps, regardless of the thickness of ash fall, there is a risk that the track circuit becomes incomplete.

When a volcanic warning is issued, railway companies can strengthen their emergency contact system and confirm their actions at the time of ash fall. However, it should be noted that an actual eruption might occur before a volcanic warning is issued.

When Preliminary Forecast or Full Forecast are issued after an eruption, railway companies can utilize them for narrowing assumed areas of ash fall and then confirm actual condition of ash fall by visual observation. Considering the thickness of volcanic ash, they can suppose possible crisis and examine the handling of that such as inspection or cleaning of facilities. Weather forecast is also important information to determine a timing of cleaning because dry volcanic ash after rain tends to adhere to rails.

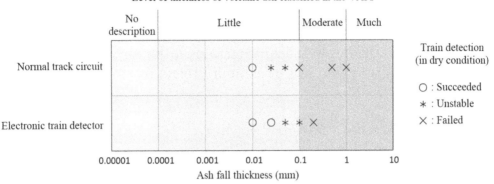

Figure 5.   Results of our test with the classification of thickness of volcanic ash in the VAFF.

Table 3.   Examples of possible prevention action of railway companies using public information on ash fall.

Timing	Public information	Prevention actions of railway companies
Normal time	Ash fall hazard map	Recognize areas affected ash fall impacts Determine deployment place of apparatus to clean volcanic ash Check the type of track circuit, the presence or absence of crossing back-up device
Volcanic warning	Volcanic alert level VAFF (Regular Information)	Establish headquarters and strength emergency contact system Arrange workers and apparatus for cleaning volcanic ash
After eruption	VAFF (Preliminary Forecast, Full Forecast)	Inspect facilities
After ash fall	VAFF (Full Forecast) Weather forecast (precipitation)	Determine timing and range of cleaning volcanic ash

6   CONCLUSIONS

We examined the impact of volcanic ash on the shunting of track circuit by running trains on rails which are artificially covered with volcanic ash. The results of these examinations are described below.

1) Volcanic ash thickness of 0.025 mm makes the shunting of normal track circuit unstable, and that of 0.05 mm thickness makes it failure.
2) Volcanic ash thickness of 0.025 mm makes the shunting of electronic train detector unstable, and that of 0.2 mm makes it failure.

Based on the obtained results, we propose some methods to activate public information on ash fall such as VAFF to mitigate of ash fall impacts on railway. We think that this study would be useful for railway companies to decide a disaster prevention plan.

REFERENCES

Hasegawa, Y., Sugai, A., Hayashi, Y., Saito, S. and Shimbori, T., 2015, Improvements of volcanic ash fall forecasts issued by the Japan Meteorological Agency, *Journal of Applied Volcanology*, 4, Article 2.
Iguchi, T., 2011, Impact of eruption of Mt. Shinmoedake on railway electrical facilities and correspondence on the eruption, Railway and Electrical Technology, 22, 33–37 (in Japanese).
Urakoshi, T., Nishikane, Y. and Kawamura, S., in press, Resistivity of Volcanic Ash And Its Relation to Chemical Composition, *The 5th International Workshop on Rock Mechanics and Engineering Geology in Volcanic Fields*, Fukuoka, Japan.

Rock Mechanics and Engineering Geology in Volcanic Fields – Ohta, Ito & Osada (eds)
© 2023 copyright the Author(s), ISBN 978-1-032-27657-1

Spread-type landslides in volcanic terrains in Japan: A case study on the Edamatsu spread in Fukushima Prefecture

Yasuo Katoh*
Department of Civil Engineering, Gifu University, Gifu, Japan
Kawasaki Geological Engineering Co., Ltd., Aichi, Japan

Satoru Kojima
Department of Civil Engineering, Gifu University, Gifu, Japan

ABSTRACT: Records of spread-type landslides (STLs) in Japan, with their geomorphic and geologic characteristics, such as locations (coordinates of the center of the main scarp), heights of the top of the scarp and bottom of the slide, length and width of the landslide body, and lithology and age of the landslide, are summarized in a database. The sizes of the STLs range from 70–3200 m in length and 200–3000 m in width, and they occur mostly in the Oligocene to Pleistocene sedimentary and volcano-sedimentary formations. Some of the STLs are closely related to cap-rock geological structures developed in the volcanic terrains. The ages of their activities are mostly unknown, except for the STL in the Chiba Prefecture, which was active during the 1923 Great Kanto earthquake. The Edamatsu STL in the Fukushima Prefecture, Northeast Japan, developed in the Quaternary Tonohetsuri caldera, and it has a diameter of approximately 20 km. The sliding block, 1000 m long and 2500 m wide, is composed of massive andesitic lava overlying the lacustrine sediments accumulated in the caldera lake. In most cases, the boundary layer, i.e., siltstone, plays the role of a sliding surface.

Keywords: Landslide, Spread-type landslide, Caldera, Caldera deposits, Cap rock, Database

1 INTRODUCTION

A type of landslide called lateral spread (Varnes, 1978), spread (Cruden and Varnes, 1996), or (lateral) spreading (Hungr et al., 2014), is characterized by the lateral extension or movement of the landslide mass. Varnes (1978) stated that the lateral spread is characterized by fracturing and extension of the coherent material, owing to liquefaction or plastic flow of the subjacent material. Cruden and Varnes (1996) defined spread as "an extension of a cohesive soil or rock mass, combined with a general subsidence of the fractured mass of cohesive material, into softer underlying material. The surface of the rupture is not a surface of intense shear." They introduced two types of spreading: block spreads and liquefaction spreads. They noticed that the soft underlying material was squeezed into the cracks of the fractured and separated stronger upper blocks. Hungr et al. (2014) revised the classification of landslides by Varnes (1978) and divided them into three types: rock slope spread, sand/silt liquefaction, and sensitive clay spread. The rock slope spread and sand/silt liquefaction are equivalent to the block spreads and liquefaction spreads (Cruden and Varnes 1996), respectively. In this paper, we present a detailed description of the geologic and geomorphologic characteristics of the Edamatsu spread, which has the characteristics of a rock slope spread (Hungr et al. 2014) or block spreads (Cruden and Varnes 1996).

*Corresponding author: katohy@kge.co.jp

DOI: 10.1201/9781003293590-16

Although uncommon, spread-type landslides (STLs) are found in Japan. Oyagi (2003a) conducted a pioneering study of STLs in Japan, showing several examples of STLs and described their topographic features. In this paper, we summarize the cases presented by Oyagi (2003a) along with our own research. Some of these cases are topographically similar to STL, but it is difficult to determine whether their dynamic and geological characteristics are similar. We have included these cases in our database, tentatively.

Caldera-forming volcanic activities provide suitable environments for STL development; massive, coherent, and competent lava and pyroclastic rocks erupted from the central volcanic cone cover the thinly laminated, incoherent, and incompetent sedimentary rocks deposited in the caldera lake. Oyagi (2003b) showed that landslides are common in caldera volcanoes, some of which are classified as STLs. In this paper, we present a database of probable STLs in Japan and describe the geomorphic and geologic characteristics of a typical example: the Edamatsu spread, which developed in a Pleistocene caldera in the Tohoku region of Northeast Japan.

2 DATABASE OF STLS IN JAPAN

A database was made to determine the overall characteristics of STLs in Japan (Table 1, Figure 1) from Oyagi (2003a), with each landslide named after the geographic name of the STL site. In this database, the following geomorphological and geological features are summarized: location (coordinates of the center of the main scarp), height difference between the top of the main scarp and the toe of the landslide body, length and width of the landslide body, age and lithology of the landslide mass, and the underlying sheared materials. STL sizes are recognized in a wide range, 70–3200 m in length and 200–3000 m in width.

Table 1. List of spread-type landslides (STLs) in Japan.

No	Name of Spread	Location (Prefecture)	latitude (N) longitude (E)	highest elevation lowest elevation (m)	difference in elevation (m)	width length (m)	hight of scarp (m)	inclination (degree)	geological age lithology (deformation layer/ landslide body)
1	Edamatsu	Fukushima	37° 17' 21.1" 139° 56' 57.9"	830 450	380	2500 1000	120 ∼ 200	15.5 ∼ 17.2	Ps Ms, Ss/Ad
1	Omatsugawa	Fukushima	37° 13' 17.5" 139° 54' 29.1"	840 650	190	900 1050	80	8.7	Ps Ms, Ss/Ad
1	Kan-nonyama	Fukushima	37° 11' 32.2" 139° 55' 52.2"	1450 820	630	1100 2450	480	14.9	Ps Df
2	Ikenouchi	Chiba	35° 1' 25.9" 139° 54' 28.0"	20 20	0	370 350	1 ∼ 2	-	H Al
3	Nishihata	Hyogo	34° 50' 07.1" 135° 9' 49.0"	197 169	28	200 70	6	22.0	Pg Ms, Ss, Tf
4	Tamugi	Niigata	37° 10' 29.1" 138° 33' 06.7"	390 160	230	1500 2300	10	5.6	Pg Alt. Ss & Ms

(Continued)

Table 1. (*Continued*)

No	Name of Spread	Location (Prefecture)	latitude (N) longitude (E)	highest elevation lowest elevation (m)	difference in elevation (m)	width length (m)	hight of scarp (m)	inclination (degree)	geological age lithology (deformation layer/ landslide body)
5	Gozodake	Nagasaki	33° 14' 47.3" 129° 42' 43.5"	440 40	400	3000 3200	150 ~ 200	6.7	Pg Ms, Ss/Bs
5	Terakami	Nagasaki	33° 19' 42.4" 129° 46' 23.5"	270 40	230	1400 1700	40 ~ 110	6.9	Pg Ms, Ss, Cg, Cl/Bs
5	Kobune	Nagasaki	33° 12' 01.9" 129° 45' 05.5"	320 100	220	1200 1150	110	9.7	Pg Ms, Ss/Bs
5	Sasebo	Nagasaki	33° 10' 13.7" 129° 44' 21.0"	150 5	145	1400 1360	-	6.5	Pg Ms, Ss, Cl
6	Aratozawa	Miyagi	38° 54' 16.8" 140° 50' 45.6"	465 260	205	870 1480	150	7.2	N Sls, Ss, Tf/Ltf, Tf
7	Osawa	Yamagagta	38° 54' 41.1" 140° 12' 52.4"	190 100	90	2050 1550	20	17.2	N Sls

Age abbreviation (H: Holocene, Ps: Pleistocene, N: Neogene, Pg: Paleogene)
Lithology abbreviation (Ad: andesite, Al: Allvium, Alt. Ss & Ms: alternating sandstone and mudstone, Bs: basalt, Cg: conglomerate, Cl: coal, Df: debris flow deposit, Ltf: lapilli tuff, Ms: mudstone, Sls: siltstone, Ss: sandstone, Tf: tuff)

STLs occur mainly in Quaternary volcano-clastic complexes in the Tohoku region of Northeast Japan (Edamatsu (this study), Omatsugawa, Kan-nonyama, Aratozawa, and Osawa spreads) and in the Kyushu region of southeast Japan (Gozodake, Terakami, Kobune, and Sasebo spreads). The Nishihata spread in the Hyogo Prefecture and the Tamugi spread in the Niigata Prefecture occur in sedimentary rocks of the Eocene and Miocene ages, respectively. The Ikenouchi spread in the Chiba Prefecture is recognized as a Holocene unconsolidated alluvial deposit.

The ages of these STLs are unknown, except for the Ikenouchi spread that was active during the Great Kanto Earthquake of 1923 and the Aratozawa spread that was active during the Iwate-Miyagi Nairiku Earthquake in 2008.

3 THE EDAMATSU SPREAD

The Edamatsu spread occurred inside the Tonohetsuri caldera in the central Fukushima Prefecture (No. 1 in Figure 1). The landslide mass is a thick andesitic lava that erupted onto the lacustrine sediments accumulated in the caldera lake during the post-caldera stage.

Figure 1. STL locations in Japan. The numbers indicate the spread number shown in Table 1.

3.1 *Geologic setting*

The Tonohetsuri caldera, formed during the Pleistocene, consists of pyroclastic flow and debris avalanche deposits that accumulated during the caldera-forming stage along with lacustrine sediments and volcanic rocks, including lava and pyroclastic flow deposits of the post-caldera stage (Figure 2; Yamamoto, 1999). The Early Pleistocene lacustrine sediments are mainly composed of a few centimeters to a few meters-thick formations of sandstone and conglomerate, with minor amounts of parallelly laminated

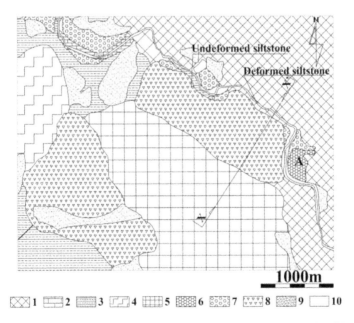

Figure 2. Geological map of the northeastern part of Tonohetsuri caldera (simplified from Yama-moto, 1999).

1: basement rocks, 2: pyloclastic flow deposit, 3: lacustrine deposits, 4: dacite lava, 5: andesite lava, 6: Lower III terrace deposits, 7: Lower IV terrace deposits, 8: landslide deposits, 9: slope sediments, 10: Alluvium, A: Edamatsu village, X-X':geological profile(Figure 4).

siltstone, rich in plant fossils. These formations are often accompanied by several meters-thick massive tuff breccia and dacitic pyroclastic rocks. The maximum thickness of the lacustrine sediment layer is approximately 350 m. The Futamatayama volcano, situated in the northwestern part of the Tonohetsuri caldera, erupted between late Middle Pleistocene to early Late Pleistocene as post-caldera volcanism, and unconformably covered the lacustrine sediments (Figure 2). The Futamatayama volcanics are divided into three layers: lower lava, pyroclastic flow deposits, and upper lava. The uppermost lava constitutes the dome-shaped summit of Mt. Futamatayama.

Figure 3. Map showing topographic characteristics of the Edamatsu spread. a: scarp, b: linear depression, c: small linear depression, d: knick line, e: slide-type landslide, f: alluvial cone, g: landslide mass of the Edamatsu spread, A: Edamatsu village, B: large scarp with alluvial cone (for the explanation see text).

3.2   *Geomorphic features of the Edamatsu spread*

The Edamatsu spread occurs on the left bank of the Tsurunuma River, downstream of the Edamatsu village (Figure 3) and on the northwestern end of the Futamatayama andesites. The andesite lava ridge, approximately 12 km long and 1.5 km wide, extends northwest from the top of Mt. Futamatayama. At elevations below approximately 1000 m above sea level, the upper surface is nearly flat with a gentle slope (~4°). Recurrent northeast-southwest oriented depressions with small scarps on their northwest-southeast sides are observed on this ridge. (Figure 3). These systematic geomorphic features are considered to be extensional grabens caused by tensile stress in the northwest-southeast direction. From the geologic cross section (Figure 4), it is observed that the andesitic landslide mass sinks approximately 200 m into the underlying lacustrine sediments, decreasing the thickness of the sediments by several tens of meters.

A relatively large scarp (Point B in Figure 3) is recognized, approximately 750–800 m above sea level, on the lava ridge. An alluvial cone was formed at the northeastern foot of the scarp, suggesting sediment transport by surface water. The scarp extends up to the gap that bisects the geomorphic features inside the Edamatsu spread, suggesting its formation prior to the landslide activity.

The Edamatsu spread formed on the northeast slope of the lava ridge. The direction of landslide movement is northeastward, and the main scarp of the slide strikes northwest-southeast. Several northwest-southeast oriented linear depressions were observed within the landslide body (Figure 3). The southwestern slope of the ridge is also bounded by a scarp, with a sharp slip surface, probably formed due to a landslide

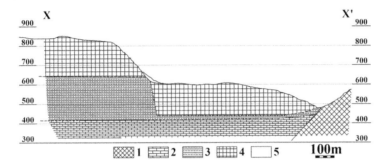

Figure 4. Geological profile of the Edamatsu spread. Line of cross section is shown in Figures 2 and 3. Symbols are same as for Figure 2.

3.3 *Occurrence of siltstone underlying the landslide body*

An outcrop of undeformed siltstone (Figure 5) is observed on a slope ~50 m above the left bank of Tsurunuma river, which comprises millimeter-scale alternating layers of siltstone and very fine-grained sandstone, dipping at 4°–5° westward. It is thought that the landslide occurred at least above this elevation.

AOn the other hand, severely deformed siltstone was observed at two locations (~20 m apart) on the left bank of the Tsurunuma River (Figures 3, 5–7). The outcrop toward downstream stretches 2.2 m along the river direction (width, 1.3 m), and covers the modern riverbed gravel. The upstream outcrop stretches 3.0 m along the river (width, 1.5 m), includes up to 17 cm of riverbed gravel (Figure 8), and is in contact with riverbed boulders (diameter, ~1 m) (Figures 6, 7).

Figure 5. Undeformed lacustrine siltstone.

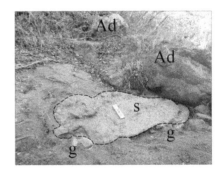

Figure 6. Outcrop of deformed lacustrine siltstone(s). Ad: Andesite (fragments of landslide body), g: gravels.

A Figure 7 shows a sketch of the downstream outcrop. The siltstone preserves very few primary sedimentary structures and is broken into centimeter-sized or smaller rounded fragments (Figure 8). The gray area in Figure 7 indicates a fine-grained siltstone with no rock fragments. The oval-shaped rock fragment in the center of Figure 8 is deformed and shows a right-lateral shear sense. In the lower half of the outcrop, the rock fragments become rounded, and the planar structures (considered to be shear surfaces) dip gently to the southwest (Figures 6 and 7). The bottom surface of the deformed siltstone is generally smooth and dips ~30° toward southwest (Figure 6).

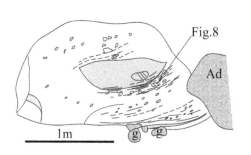

Figure 7. Sketch map of the deformed lacustrine siltstone. Lines show foliations. Ad: Andesite (fragments of landslide body), g: gravels.

Figure 8. Deformation of lacustrine siltstone. Scale bar is 5 cm.

4 CONCLUSIONS

The STLs in Japan vary in size (length, 70–3200 m; width, 200–3000 m), but are mostly confined to the Cenozoic, especially Quaternary volcanic terrains. One example is the Edamatsu spread, which developed in the Quaternary Tonohetsuri caldera in the central Fukushima Prefecture, Northeast Japan. Middle to Late Pleistocene thick andesite lava caps Early Pleistocene lacustrine sediments; the former is massive and hard, while the latter is laminated and soft. We observed the presence of a deformed siltstone member of lacustrine sediments below the landslide mass. The siltstone has lost its original sedimentary structures and is broken into small sheared clasts, covering the modern riverbed boulders. On the northeast of the landslide scarp, the landslide mass has a "horst-graben" topography, indicating lateral extension of the mass. The geological cross section shows the andesitic landslide mass to have sunk approximately 200 m into the underlying lacustrine sediments. These geomorphic and geologic evidence clearly indicate that the landslide can be classified as a spread.

ACKNOWLEDGEMENTS

We thank T. Yokoyama for his guidance of early stage of the YK's research on STL in the Nishihata area, Hyogo Prefecture, and the late N. Oyagi for his invaluable suggestions on our STL studies.

REFERENCES

Cruden, D. M. and Varnes, D. J., 1996, Landslide types and processes, *Landslides Investigation and mitigaition (Turner and Schuster eds)*, *Transportation Research Board, National Research Council*, Spec. Rep. 247, 36–75.

Hungr, O., Leroueil, S. and Picarelli, L., 2014, The Varnes classification of landslide types, an update, *Landslides*, 11, 167–194.

Oyagi, N., 2003a, Examples of spread type landslides in Japan, *Annual Report, Fukada Geological Institute*, 4, 133–153. [in Japanese]

Oyagi N., 2003b, Genesis of landslides in the dissection process of caldera volcanos -An overview in northern Honshu, Japan-. *Jour. Japan Landslide Soc.*, 40, 10–21. [in Japanese with English abstract]

Varnes, D. J., 1978, Slope movement types and processes, *Landslides Analysis and Control (Schuster and Krizek eds)*, *Transportation Research Board, National Academy of Sciences*, Spec. Rep. 176, 11–33.

Rock Mechanics and Engineering Geology in Volcanic Fields – Ohta, Ito & Osada (eds)
© 2023 copyright the Author(s), ISBN 978-1-032-27657-1

Reconsideration of caldera classification at the summit of Olympus Mons, Mars

Mayuko Nakamura* & Tomohiro Tsuji
Graduate School of Science and Technology for Innovation, Yamaguchi University, Yamaguchi, Japan

ABSTRACT: The Olympus Mons, located in the northwestern part of the Tharsis Highlands on Mars, is the largest volcano in the solar system. The summit caldera was classified into six complex calderas (Zeus, Hera, Hermes, Dionysus, Apollo and Athena Paterae). There are numerical grabens on the caldera floors (Paterae) and around the rim of the calderas. Mouginis-Mark and Robinson (1992) interprets this terrain as a circumferential graben. However, the classification and formation of graben has not been fully investigated. In order to study the detailed understanding of the graben and Paterae, this study using that data taken by High Resolution Imaging Science Experiment (HiRISE) and Mars Orbiter Laser Altimeter (MOLA) DEM data. As a result, we show some notable remarks below: 1) The grabens around the summit caldera are classified into four Graben unit A to D.

Two patterns of Graben unit A can be seen, inside the Zeus Patera and the other in parallel of the rim around the caldera. Graben unit B is restricted to the outer of the caldera rim along the Athena patera and intersects intricately. Graben unit C is distributed on Hera Patera and is cut by Hermes Patera. Graben unit D intersects the boundary between the Zeus Patera and the Hermes Patera. These Graben units have the different curvatures and the center coordinates. 2) A north-south arc-shaped cliff topography with concentrated contour lines was observed in the central part of Hermes Patera. The patera on the east side of the cliff is lower than that on the west side.

These results suggest that Hermes Patera could be divided into two: the west to be Hermes-w Patera and east to be Hermes-e Patera. The grabens could be associated with the formation of the caldera. The center coordinates of Graben unit C and D suggest that Graben unit C is associated with the formation of the eastern part of Hermes Patera and Graben unit D with the formation of the western part of Hermes Patera. Therefore, the history of formation can be re-examined by subdividing the graben around the caldera.

Keywords: Caldera, Martian Volcano, Volcanic field, Topography

1 INTRODUCTION

The Olympus Mons, located in the northwestern part of the Tharsis Highlands on Mars. It is the highest and most prominent shield volcano in the Solar System (Mouginis-Mark. 2018), 840 × 640 km in diameter (Plescia. 2004) and the elevation is 21150 m (MOLA DEM). The Olympus Mons has Caldera (Olympus Paterae) that is located about 30 km apart of the summit. It has a nested caldera 60×80 km in

*Corresponding author: b028vcv@yamaguchi-u.ac.jp

DOI: 10.1201/9781003293590-17

diameter that is strikingly similar to those of certain terrestrial volcanoes (Mouginis-Mark et al., 2007). Olympus Patera formed by six large collapse events (Mouginis-Mark. 2018) during the latest phase of the active history of Olympus Mons. Although the caldera is general formed in a circular or ellipse, Olympus Patera has a unique and complicated. Especially, the Hermes Patera form is disagree with the caldera rim the rim shape is complex and ridge topography is developed on the floor. Therefore, the topography of the caldera floor needs to be re-examined.

Numerous tectonic features that have been interpreted to be formed by compression (wrinkle ridges) and extension (circumferential graben) were found within the caldera (Greeley and Spudis, 1981; Zuber and Mouginis-Mark, 1992). The relationship of the grabens and the caldera collapse has not been fully understood. Thus, it is necessary that detail analysis of the morphology and distribution of the circumferential grabens. Therefore, in older to consider the relationship between the caldera and the grabens, we analyze the topography of the Hermes Patera and the circumferential grabens of Olympus Patera, around the Hermes Patera.

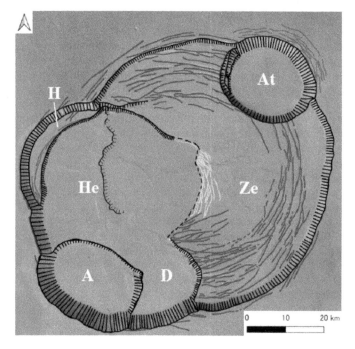

Figure 1.   Classification of grabens around Olympus Patera. Patera name are A:Apollo, At:Athena,D: Dionysus, H:Hera, He:Hermes, Z:Zeus. Patera's name follows Mouginis-Mark.2018.

2   METHODOLOGY

2.1   *Analysis of Hermes Patera*

Mouginis-Mark (2018) reported a ridge of north-south trend in the central Hermes Patera. MOLA DEM data was used to create a 10 m-intervals topographic map for Hermes Patella (Figure 2 [b]). In addition, the topographic profile(A-A') is created in an east-west direction through this ridge (Figure 2 [c]).

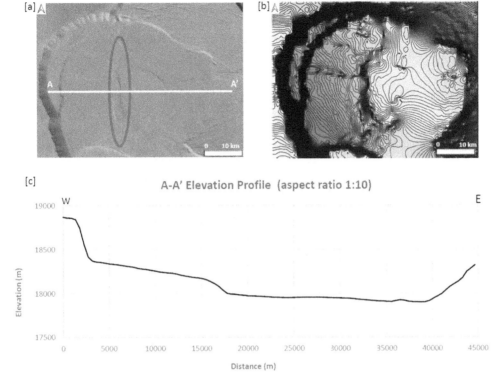

Figure 2. Analysis of the Hermes Patera. [a] The caldera floor of Hermes Patera. Red ellipse shows north-south trend ridge. [b] 10 m interval elevation map. [c] Profile derived from MOLA DEM data. See [a] for location.

2.2 Curvature analysis

Mouginis-Mark (2018) has recoginized the circumferential grabens distribution, detailed morphological classification has not been carried out. The circumferentlal grabens (Figure 1) can be divided into three categories in terms of distribution: those on Zeus Patera (Graben unit A), those on the outer edge of Athena Caldera (Graben unit B), and those forming an arc toward to Hermes Patera. In addition, those arcs distribution to Hermes Patera can be divided into two graben groups by eastern - western, Graben unit C and D. The two Graben units has different from factors and are related to the topography of Hermes Patera.

To classify the Graben units around Hermes Patera, we defined the grabens on the west and east sides of the Hermes Patera as Graben unit C and D (Figure 1). The methods ate described below.

1. Using Google Earth Pro, drop pins on the graben to obtain coordinate data.
2. Create to ellipses and circles that approximate the center coordinate by using Origin Pro 2021, the curvatures are calculated by radius for the circle, miner and major diameter for the ellipses.

We calculated the curvature of the ellipses averaging the maximum and minimum curvatures. Therefore, we use the miner diameter, which is the maximum curvature, and the major diameter, which is the minimum curvature adds the reciprocals of each and averages them to get the curvature value of the ellipse. Similarly, we calculated the curvatures of Hermes-w Patera and Hermes-e Patera. we used the western rim of Hermes Patera for the calculation of

Hermes-w Patera and eastern rim of Hermes Patera for the calculation of cliff to Hermes-e Pateraastern rim.

As a result of the elliptical and circular studies, the circular results are used only for Graben unit D, the elliptical results are used for the rest.

3   RESULT

3.1   *Topographic analysis of the Hermes Patera*

Figure 3 [b] indicates topographic map with 10m intervals for Hermes Patera. On the topographic map, north-south trend ridge shows the cliff that slope direction of east. The topographic profile across the floor of Hera-Hermes-Zeus Patera, shows that the specific height of this steep cliff is about 160 m. The eastern part of the cliff is lower than western and two different topographic floor (Figure3 [c]). It is decided to eastern floor of Hermes-e Patera and western floor of Hermes-w Patera. The cliff length of about 20 km, north edge incured to northeast and south edge to southeast that is surrounding east of Hermes Patera.

3.2   *Curvature*

The approximate circle was created by latitude-longitude coordinate systems Graben unit C, D, Hermes-e Pateraast rim and from the cliff to Hermes-w Pateraest rim (Figure 3). Table 1. shows result of all data. The approximate ellipse of Graben unit C has a minor diameter of 14 km and a major diameter of 15.5 km. Hermes-e Pateraast rim has minor diameter of 12.7 km and major diameter of 15 km and from the cliff to Hermes-w Pateraest rim has 12 km,12.6 km. The curvature of each approximate ellipse is 0.06 for Graben unit C, 0.06 for the east rim of the Hermes, and 0.07 from the cliff to Hermes-w Pateraest rim. The radius of the Graben unit D is 12.1 km. From the radius, the circular curvature of the Graben unit D is 0.08. For the two graben unit curvature values, Graben unit D was larger than C. This indicates curvature is larger in Graben unit D which curve is sharp. Figure 3 [a], [c] indicate, Graben unit C and Hermes-e Pateraast rim approximate ellipse center coordinate point is nearby location. In addition, Figure 3 [c], [d] shows Graben unit D and from cliff to Hermes-w Patera east rim center coordinate point like location.

Table 1.   Data of approximate ellipse and circle.

	Graben unit C	Graben unit D	Hermes-w Patera	Hermes-e Patera
Ellipse center coordinate point	133°32′2.87”W, 18° 20′3.06”N	133°5′6.27”W, 18° 17′7.32”N	133°29′9.37”W, 18° 22′3.33”N	133°17′6.09”W, 18° 20′5.49”N
Minor diameter (km)	14	1.84	12.7	12
Major diameter (km)	15.5	9.73	15	12.6
Ellipse Curvature	0.06	0.03	0.07	0.08
Circle Center coordinate poin	133°32′1.92”W, 18° 20′3.17”N	133°16′4.48”W, 18° 19′4.72”N	133°31′5.18”W, 18° 21′8.85”N	133°17′6.09”W, 18° 20′5.49”N
Diameter (m)	15.5	12.1	14.3	12
Circle curvature	0.06	0.08	0.07	0.08

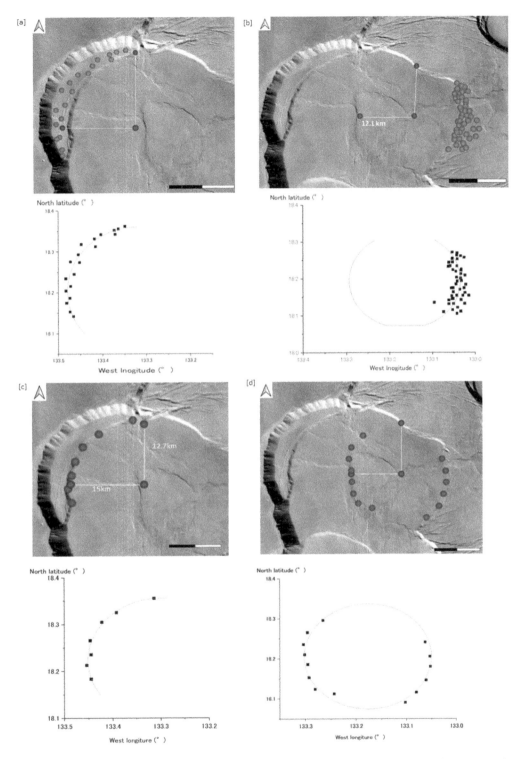

Figure 3. Analysis of the grabens and rims. [a], [c] and [d] are approximate ellipse. [b] is Graben unit C approximate circle center coordinate point. [a] Graben unit C. [b] Graben unit D. [c] Hermes-e Patera. [d] Hermes-w Patera. Scale bar is 10 km.

125

3.3 *Topographic analysis of the Hermes Patera*

Figure 3 [b] indicates topographic map with 10 m intervals for Hermes Patera. On the topographic map, north-south trend ridge shows the cliff that slope direction of east. The topographic profile across the floor of Hera-Hermes-Zeus Patera, shows that the specific height of this steep cliff is about 160 m. The eastern part of the cliff is lower than western and two different topographic surfaces (Figure 3 [c]).

The cliff length of about 20 km, north edge incurved to northeast and south edge to southeast that is surrounding east of Hermes Patera

4 DISCUSSION

Using data of Graben unit C and D obtained in the result, approximate circles were created on Google Earth Pro (Figure 4 [a]). The center coordinate of the approximate ellipse of Graben unit C is located west of the cliff, and Graben unit D is east side. Therefore, the approximate ellipses of Graben unit C and D were different. The two-center coordinate are about 15 km apart and different curvature, Graben unit C and D were not formed by the same single factor. One of the factors that may cause the formation of graben is the formation of cracks as an extension of the rim at the outer edge during caldera formation. These grabens aren't extended into the Hermes Patera, but there are distributed outside of the rims. We think so that these related to the formation of the caldera. The center coordinates of approximate ellipse are different between Hermes rim and cliff that we can infer that two calderas have been formed in the Hermes Patera (Figure 4 [b]). The east surface of the cliff was lower in elevation than the west. We decided to divide different two calderas that the western floor of Hermes-w and the eastern floor of Hermes-e.In other words, these different topographic surfaces may represent another caldera, and the cliffs is considered to be the eastern caldera rim.

Therefore, Graben unit C is considered to be associated with the Hermes-e caldera subsidenced, and D with Hermes-w caldera subsidenced. The Olympus Patera, which currently consists of six calderas, can be divided into seven from this study, and the order to caldera formation can be re-examined. At this point, we can't determine which caldera formed first on the east and west sides of the Hermes Patera, but we will analyze this issue in the future.

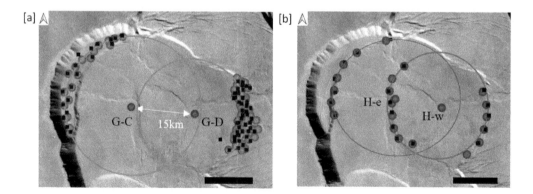

Figure 4.   Approximate circle. [a]Approximate ellipse of Graben unit C and circle of Graben unit D. [b] Approximate ellipse of Hermes-e Patera and Hermes-w Patera. Scale bar is 10 km.

5 CONCLUSION

- Analysis of topographic and curvature of Hermes Patera and the grabens were carried out in order to re-examine the topography of the Hermes Patera and to classify the grabens (Graben unit C and D).
- The topographic ridge was seen Hermes Patera that revealed to be east sloping cliff which divide to two caldera floor Hermes west and east Patera.
- The grabens around of Hermes Patera are classified to Graben unit C outside of the Hermes-w Patera and Graben unit D outside of the Hermes-w Patera, based on curvature and the location of the center coordinates of the approximate circle.
- The grabens are considered to haven't been formed in same process. It is considered that Graben unit C was related to the subsidence Hermes-w Patera and Graben unit D was related to the subsidence Hermes-e Patera.

REFERENCES

Greeley R, Spudis PD., 1981, Volcanism on Mars. *Reviews of Geophysics.*, **19**, 13–41.

Zuber, M. T. and Mouginis-Mark, P. J., 1992, Caldera subsidence and magma depth of the Olympus Mons volcano, Mars. *Jonal of Geophysical Research.*, **97**, 18295–18307.

Mouginis-Mark, P. J. and Robinson, M.S., 1992, Evolution of the Olympus Mous Caldera, Mars, *Bullten of Volcanology.*, **54**, 347–360.

Mouginis-Mark P. J., Andrew J. L., Harris and Scott K. Rowland., 2007, Terrestial analogs to the caldera of the Tharsis volcanoes on Mars, In: Chapman, M. (Ed.), *The Geology of Mars: Evidence from Earth-based Analogs.*, Cambridge University Press, 71–94.

Mouginis-Mark, P.j., 2018, Olympus Mons volcano, Mars: A photogeologic view and new insights, *Chemie der Erde.*, **78**, 397–431.

Plescia, J.B., 2004, Morphometries of Martian volcanoes, *Jornal of Geophysical research.*, **109**, 1–26.

Rock Mechanics and Engineering Geology in Volcanic Fields – Ohta, Ito & Osada (eds)
© 2023 copyright the Author(s), ISBN 978-1-032-27657-1

Variation of petrological features caused by style of eruption in the western volcanic chain of the Aonoyama Volcano Group, Southwestern Japan

Daichi Okazawa* & Takehiro Ohta
Yamaguchi University, Yamaguchi, Japan

ABSTRACT: Abu Monogenetic Volcano Group and Aonoyama Volcano Group, which are active in the Quaternary, are distributed in the northern part of Yamaguchi Prefecture. The Abu Monogenetic Volcano Group are composed of more than 50 volcanoes with diverse topography, such as pyroclastic cones, lava plateaus, and lava dome with flat tops. This magma was produced by melting the lithosphere due to the rise of the mantle asthenosphere. On the other hand, the Aonoyama Volcano Group is a Quaternary island arc volcano related to the subduction of the Philippine Sea Plate and consists of more than 20 lava domes. Both the Abu Monogenetic Volcano Group and the Aonoyama Volcano Group are aggregates of small volcanic bodies. There are significant differences between these groups in topographical features and magma origin in spite of neighboring position of them. Recent studies for Aonoyama Volcano Group have reported that Nabeyama shows a different topography from the lava dome, and that Chojagahara represents a lava plateau. Because the Aonoyama Volcano Group has not only lava domes but also various topography, it is necessary to review the definition of the Aonoyama Volcano Group.

This study aims to clarify the style of eruption of the western volcanic chain of the Aonoyama Volcano Group, which remains volcanic topography well, based on geology and petrography and to consider the variation of petrological features caused by these styles.

Keywords: Volcanology, Volcanic geology, Petorology, Southwest Japan, Island arc

1 INTRODUCTION

In southwestern Japan, the Quaternary volcanic front has been formed due to the subduction of the Philippine Sea Plate. The volcanic front is characterized by adakitic volcanic rocks, which are distributed from the Chugoku region to northern Kyushu, including Daisen, Sanbe, Oe Takayama, Aonoyama Volcano Group, Himejima Volcano Group, Futago Volcano Group, and Yufu Tsurumi Volcano. (Kakubuchi and Nagao, 1994; Morris, 1995; Kimura et al, 2003; Shibata et al, 2005; Sugimoto et al, 2006; Horikawa and Nagao, 2009). The Aonoyama Volcano Group, the subject of this study, consists of more than 20 hornblende andesite-dacite lava domes distributed over 50 km from north to south from Tsuwano, Shimane Prefecture to Yamaguchi Prefecture. (Kamata et al., 1988; Kakubuchi et al., 1995; Moriya, 1983).

In addition to the Aonoyama Volcano Group, the Abu Monogenetic Volcano Group is also distributed as Quaternary volcanos in northern Yamaguchi Prefecture. The Abu

*Corresponding author: okzwdic@gmail.com

DOI: 10.1201/9781003293590-18

Monogenetic Volcano Group are composed of more than 50 volcanoes with diverse topography, such as pyroclastic cones, lava plateaus, and lava domes with flat tops (Moriya, 1983). This magma was produced by melting the lithosphere due to the rise of the mantle asthenosphere (Kimura et al., 2003).

The Aonoyama Volcanic Group and the Abu Monogenetic Volcano Group are both collections of small volcanic bodies, and although they are distributed near to each other, there are significant differences in their topographic characteristics and styles of eruption.

Previous studies have reported that the volcanoes of Aonoyama Volcanic Group is composed of lava domes. However, recent studies have suggested that Nabeyama presents a different topography from the lava dome (Marumoto and Nagao, 2009), and Chojagahara presents a lava plateau (Takahashi and Furuyama, 2006; Kakubuchi et al., 1995). Hence, the Aonoyama Volcanic Group has not only lava domes but also various topography. Therefore, it is necessary to review the definition of the Aonoyama Volcanic Group.

The purpose of this study is to clarify the volcanic geological and petrological features of the western volcanic chain of the Aonoyama Volcanic Group, which has well-preserved topography, and to consider the petrological features due to the style of eruption and its differences.

2 METHODOLOGY

2.1 Topographical survey

For Aono-yama and its surrounding volcanoes, referring to Moriya, (1983) and Suzuki (1997), we conducted aerial photographic interpretation of the volcano topography, and topographic interpretation from the cross-sectional shape of the volcano slope.

2.2 Field survey

Geological surveys were conducted to characterize volcanic ejecta and clarify the style of eruption.

2.3 Microscopic observation

The petrographic texture and the feature of the rock forming minerals were revealed by microscopic observation.

2.4 Petrological studies

The chemical composition of rocks and minerals were analyzed to reveal the petrological characteristics of the ejecta.

2.4.1 Bulk rock chemistry

Major and trace (Ba,Cr,Nb,Ni,Rb,Sr,V,Y,Zn and Zr) elements were measured by X-ray fluorescence (Rigaku ZSX Primus-II;Rh) at Center of Instrumental Analysis (CIA) of Yamaguchi University. Analytical procedure is described by Eshima and Owada (2018).

2.4.2 Mineral chemistry

The rock forming minerals were analyzed by the electron-probe microanalyzer (EPMA:JAX-8230) at CIA in Yamaguchi University. The measurement conditions are as follows: acceleration voltage: 15 kV, irradiation current: 20 nA, and irradiation time: 30 s.

3 RESULTS

3.1 *Topographical survey*

Two types of topography were identified, lava dome and pyroclastic cone (Figure 1), and each is described below.

(1) Lava dome

At lava dome type topography, the cross-sectional shape of the slope near the summit is convex ridge-slope type (Suzuki, 1997) and from the foot to the middle of the mountain is concave ridge-slope type (Suzuki, 1997), and the summit also shows dome-shaped topography without crater-like depressions. Aono-yama, Koaono-yama, Chikura-yama, Benten-yama, Kareki-yama, Tatuboushi-yama, 412m peak, Taiin-yama, Nosaka-yama, Danbara-yama, Mihara-yama, Kannon-yama, Egusa-yama, and Funahira-yama are classified in this type.

(2) Pyroclastic cone

Pyroclastic cone shows a uniform ridge-slope type (Suzuki, 1997) from the summit to the base of the mountain and has crater-like depressions near the summit. Kumoi-mine is classified as a topographic feature of a pyroclastic cone.

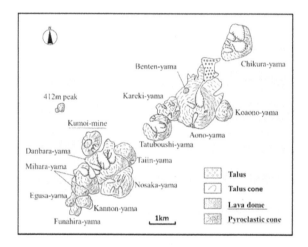

Figure 1.   Topographic classification diagram.

3.2 *Field survey*

The route maps are shown in Figure 2. The results of the field survey are as follows.

Only massive lava was distributed as ejecta in each volcanic body classified as lava dome according to the results of topographic interpretation.

At Kumoi-mine, which is classified as pyroclastic cone according to the topographic interpretation, pyroclastic rocks, sintered rocks, and agglutinates were distributed as ejecta.

3.3 *Microscopic observation*

All volcanic rocks in the Aono-yama volcanic group have phenocrysts of hornblende and plagioclase, and the groundmass is composed of plagioclase, hornblende, and opaque minerals (Figure 3). Some of phenocrysts of hornblende and plagioclase have dusty zoning and others have zoning.

All volcanoes classified as lava domes, except 412m peak, have intersertal texture. Kumoi-mine, classified as a pyroclastic cone, and 412m Peak, classified as a lava dome, are composed of short columnar plagioclase, which is a unique groundmass structure.

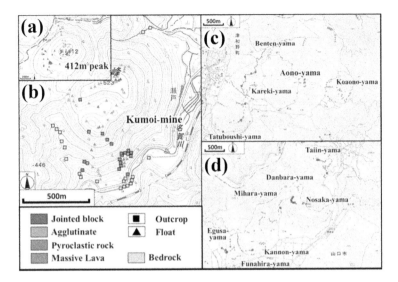

Figure 2. (a) Route Map of 412m Peak area. (b) Route Map of Kumoi-mine area. (c) Route map of Aono-yama area. (d) Route map of Nosaka-yama area.

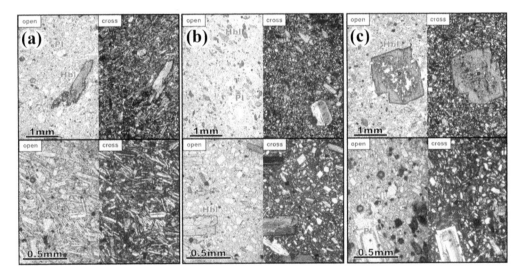

Figure 3. Micrograph of thin section. (a) Aono-yama. (b) Kumoi-mine. (c) 412m Peak. Pl: plagioclase, Hbl: hornblende.

3.4 Bulk rock chemistry

Of the volcanic rocks from the western part of the Aonoyama Volcanic Group, bulk rock chemistry was measured for a total of nine samples: Aono-yama, Kumoi-mine, 412m peak, Benten-yama, Kareki-yama, Nosaka-yama, Taiin-yama, Mihara-yama, and Egusa-yama.

The analysis results are shown in Table1, and the TiO_2, Al_2O_3, Fe_2O_3, MnO, MgO, CaO, Na_2O, K_2O and P_2O_5 contents relative to SiO_2 content are shown in Figure 4. Figure 5 shows the Geochemical diagrams.

The SiO_2 content of the measured sample ranges from 61.30 to 65.86wt.%. (Na_2O + K_2O) versus SiO_2 (wt.%) diagram (Le Bas et al., 1986) shows the composition of andesite - dacite. In the FeO*/ MgO versus SiO_2 (wt.%) diagram (Miyashiro, 1974), all samples show calc-alkaline rock type composition. The Sr/Y versus Y(ppm) diagram (Defant and Drummond, 1990) shows that all the samples have composition of adakite, or a similar composition.

131

Table. 1.   Results of Bulk rock chemistry analysis. LOI: loss on ignition.

	Si	Ti	Al	Fe	Mn	Mg	Ca	Na	K	P	Lol	Total	Ba	Cr	Nb	Ni	Rb	Sr	V	Y	Zn	Zr
	(wt.%)											mass%	(ppm)									
Aono	64.02	0.58	18.19	3.99	0.08	1.94	4.85	4.43	1.49	0.19	1.54	99.75	387.20	10.20	6.80	8.40	16.50	1065.40	70.10	16.70	49.40	123.90
Kumoi	64.17	0.53	16.68	4.42	0.09	2.38	5.11	4.09	1.86	0.17	1.05	99.49	728.90	11.90	7.70	1.90	44.10	1005.20	91.80	13.70	61.20	94.10
412m	62.58	0.59	17.48	4.80	0.10	2.76	5.10	3.88	1.78	0.17	2.58	99.23	768.30	13.20	7.80	4.60	41.40	1061.50	109.90	13.20	61.40	96.60
Benten	64.65	0.48	19.17	3.27	0.07	1.59	4.32	4.33	1.52	0.08	2.97	99.47	514.90	11.40	8.50	10.70	31.50	1078.80	56.80	12.90	49.10	124.50
Kareki	62.78	0.61	17.91	4.25	0.10	2.10	4.82	4.74	1.74	0.27	0.72	99.32	881.40	12.40	8.00	17.40	27.10	1406.20	62.70	17.10	66.30	156.70
Nosaka	62.36	0.53	18.27	4.01	0.10	1.78	4.31	4.21	1.97	0.24	1.35	97.78	853.30	17.60	7.40	17.80	37.30	1378.10	36.90	19.30	60.90	143.50
Taiin	65.86	0.53	17.41	3.48	0.07	1.79	4.55	4.43	1.68	0.17	0.75	99.96	507.60	10.90	7.50	13.30	29.30	1034.90	66.20	15.20	47.30	114.90
Mihara	61.30	0.63	19.14	4.72	0.11	2.71	5.32	4.40	1.75	0.30	1.52	100.38	1018.00	15.40	6.60	26.50	19.30	1431.50	37.60	22.40	67.80	159.10
Egusa	61.58	0.65	18.97	4.95	0.10	2.83	4.24	4.00	2.03	0.28	3.05	99.63	902.80	17.50	7.70	31.90	39.30	1052.40	57.40	21.30	67.70	164.50

3.5 *Mineral chemistry*

The analysis results are shown in Figures 6 and 7. Figure 6 illustrates the results of phenocrysts of hornblende from Aono-yama, Kumoi-mine and 412m peak on the discriminant diagram by Leake et al. (1997). Figure 7 shows the results of phenocrysts of feldspar from Aono-yama, Kumoi-mine and 412m peak on the discriminant diagram by Smith (1997).

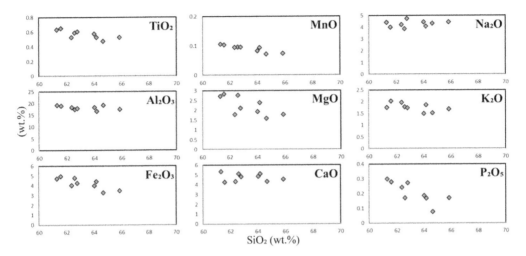

Figure 4.   TiO₂, Al₂O₃, Fe₂O₃, MnO, MgO, CaO, Na₂O, K₂O and P₂O₅ contents relative to SiO₂.

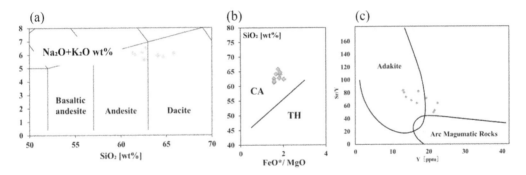

Figure 5.   Geochemical diagrams (a) (Na₂O + K₂O)-SiO₂ (wt.%), (b) FeO*/ MgO-SiO2 (wt.%) diagram, (c) Sr/Y-Y (ppm) diagram.

Almost all of the phenocrysts of hornblende in Aono-yama, Kumoi-mine and 412m peak are classified as magnesiohornblende-tschermakite.

The majority of the phenocrysts of feldspar in Kumoi-mine, 412m peak, are classified as labradorite - andesine, with some anorthosite and sanidine compositions in the dusty zoning.

4 DISCUSSION

Kumoi-mine and 412m Peak show a different groundmass structure from other volcanoes in Aonoyama Volcano Group. Kumoi-mine and 412m Peak were erupted in the mountain sides, despite of other lava domes was erupted near the lowlands. This may be one of the reasons for the formation of the unique groundmass structure in Kumoi-mine and 42 m peak. However, these facts, along with their relevance to eruption style, are issues which must be clear in the future.

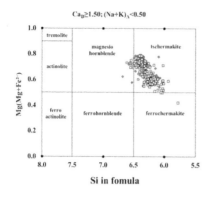

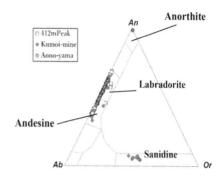

Figure 6.   Diagram to classify hornblende.   Figure 7.   Diagram to classify plagioclase.

5   CONCLUSIONS

i. Definition of Aonoyama Volcano Group
Aonoyama Volcanic Group is a member of the Quaternary Volcanic Front, which is formed by the subduction of the Philippine Sea Plate and is characterized by adakitic volcanic rocks.

ii. Style of eruption of Aonoyama volcano group
The Aonoyama volcanic group have been said to consist only of lava domes, howwever it was revealed that it has a variety of landforms such as lava plateaus, explosion craters, and pyroclastic cones.

iii. Petrology of Aonoyama Volcano Group
The rocks of the Aonoyama volcano group are adakite or similar composition, showing calc-alkaline rock type composition. Kumoi-mine and 412m peaks show unique groundmass structures such as short columnar plagioclase richness.

REFERENCES

Defant, M. J. and Drummond, M. J., 1990, Derivation of Some Modern Ark Magmas by Melting of Young Subducted Lithosphere, *Nature, 347,* 662-665.
Eshima, K. and Owada, M., 2018, Whore-rock geochemistry of diorite and dikes from Mt.Shaku-dake area, Fukuoka, Kyushu, *Jour. Geol. Soc. Japan, Vol. No. 10,* 857-862. (*in Japanese*)
Horikawa, Y. and Nagao, T., 2009, Petrology of Futago Volcano Group in Kunisaki Peninsula, Kyushu The early activity of adakite in Quaternary volcanic front on northern Kyushu, *Programme and abstracts the Volcanological Society of Japan,* 29. (*in Japanese*)
Kakubuchi, S. and Nagao, T., 1994, Are the dacites from Aono volcanic group Adakites?, *Programme and abstracts the Volcanological Society of Japan,* 137. (*in Japanese*)
Kakubuchi, S., Nagao, T. and Shiraki, K., 1995, Cenozoic volcanic rocks in Yamaguchi Prefecture, Japan, *Programme and abstracts Geological Society of Japan,* 102. (*in Japanese with English*)
Kamata, K., Uto, K. and Utumi, S., 1988, Geochronology and evolution of the post-Shishimuta caldera activity around the Waitasan area in the Hohi volcanic zone,Kyushu,Japan, *Bulletin of Volcanological Society.*
Kimura, J., Kunikiyo, T., Isaku, O., Magano, T., Yamaguchi, S., Kakubuchi, S., Osaka, S., Fujibayashi, N. and Okada, R., 2003, Late Cenozoic volcanic activity in the Chugoku area,southwest Japan arc during back-arc basin opening and reinitiation of subduction, *The Island Arc, 12,* 22-45.
Le Bas, M. J., Lemaitre, R. W., Streckisen, A. and Zanettin, B., 1986, A Chemical Classification of Volcanic Rocks Based on the Total Alkali-Silica Diagram, *Jour. Petrol., 27,* 745-750.
Miyashiro, A., 1974, Volcanic rock series in island arc and active continental margins. Am, *Jour. Sci., 274,* 321-355.

Marumoto, K. and Nagao, T., 2008, Petrology of Nabeyama Volcano in Aonoyama Volcano Group, SW Japan, *Programme and abstracts the Volcanological Society of Japan,* 31. (*in Japanese*)

Moriss, P. A., 1995, Slab melting as an explanation of Quaternary volcanism and aseismicity in southwest Kyushu, Japan, *Geology, 23,* 395-398.

Moriya, I., 1983, Volcanic topography of Japan, *University of Tokyo press.* (*in Japanese with English*)

Shibata, T., Itho, J., Ujike, O. and Takemura, K., 2005, Southwest edge of slab melting in the southwest Japan arc. Am, *Programme and abstracts the Volcanological Society of Japan,* 45.

Sugimoto, T., Shibata, T., Yoshikawa, M. and Takemura, K., 2006, Magma genesis of Yufu-Tsurumi volcanoes, Kyushu, Japan based on Sr, Nd and Pb isotopic and trace element compositions, *Annual Meeting of the Geochemical Society of Japan, 53,* 261. (*in Japanese*)

Sumith, J, V., 1974, Feldsper Minerals. 1 Crystal Srtucture and Physical Properties, *Springer-Verlag, pp,* 627.

Suzuki, T., 1997, Intoroduction to Map Reading for Civil engineers, Volume1, *Kokon Syoin, 122-123,* 160-162. (*in Japanese*)

Takahashi, K. and Furuyama, K., 2006, Geology of Chyojagahara monogenetic volcano, Yamaguchi Pref, *Programme and abstracts the Volcanological Society of Japan,* 45. (*in Japanese*)

Rock Mechanics and Engineering Geology in Volcanic Fields – Ohta, Ito & Osada (eds)
© 2023 copyright the Author(s), ISBN 978-1-032-27657-1

The relationship between the unidentified ash fall layer just under Kuju D ash layer and 54 ka eruption at Kuju volcano in Japan

Masashi Fukuoka*, Takehiro Ohta & Tomohiro Tsuji
Yamaguchi University, Yamaguchi, Japan

ABSTRACT: The largest eruption at Kuju volcano occurred at 54 ka. The previous studies have already reported the distribution and properties of the ejecta at this eruption, i.e. Kuju-D Ash Fall (Kj-D) layer, Handa Pyroclastic Flow (Kj-Hd) layer, and Kj-P1 Pumice Fall (Kj-P1) layer in ascending order. Moreover, Kj-D which is lowest layer of 54 ka ejecta is divided into Lower (L) part and Upper (U) part. The undefined ash fall layer was identified just under Kj-D L, and was named Yukouji Ash Fall (Kj-Y) layer. Because Kj-Y is situated just under Kj-D L without sandwiching soil layer, Kj-Y is considered as a series of ejecta at 54 ka eruption. This study aims to make the relationship between Kj-Y and 54 ka eruption clear based on field survey and microscopic observation.

Kj-Y is composed of silt to clay sized particles, which is a characteristic of phreatic eruption ejecta. On the other hand, Kj-Y contains colorless and clear volcanic glasses. Therefore, we consider that Kj-Y eruption was phreato-magmatic eruption.

Because Kj-Y is situated just under Kj-D L without sandwiching soil layer, it is considered that there is no time gap between these layer's deposition. As Kj-Y and Kj-D L have similar mineral assemblages, it is concluded that these tephra layers were derived from same magma. From these evidences, we conclude that Kj-Y eruption was precursory activity of 54 ka eruption.

The isopach map of Kj-Y illustrates that the eruption center of Kj-Y was in the eastern region of current Kuju volcano. On the other hand, it has assumed that Kj-D erupted at the central and eastern region, and that the Kj-Hd and Kj-P1 erupted at the central region. Therefore, it is suggested that the eruption center of 54 ka eruption migrated from eastern region to central region.

Keywords: Kuju volcano, 54 ka eruption, phreato-magmatic eruption, mineral assemblage, isopach map

1 INTRODUCTION

The largest eruption at Kuju volcano occurred at 54 ka (Okuno et al., 2017; Nagaoka and Okuno, 2015). And the volcano's eruption rate changed significantly before and after 54 ka eruption (Figure 1; Nagaoka and Okuno, 2015). At that time, the active eruption region moved to center region from western region in Kuju volcano (Kawanabe et al., 2015). Therefore, it is important to clarify 54 ka eruptive process and understand changes in magma of 54 ka eruption.

The previous studies have already reported the distribution and properties of the ejecta at this eruption, i.e. Kuju-D Ash Fall layer (Kj-D), Handa Pyroclastic Flow deposit (Kj-Hd), and Kj-P1 Pumice Fall layer (Kj-P1) in ascending order (Kawanabe et al., 2015). Tsuji et al (2017) divided Kj-D into Lower (L) part and Upper (U) part.

*Corresponding author: b030vcv@yamaguchi-u.ac.jp

DOI: 10.1201/9781003293590-19

On the other hand, the undefined ash fall layer was situated just under Kj-D L and named Yukouji Ash Fall layer (Kj-Y). Because this layer is situated just under Kj-D L without sandwiching soil layer, that is considered as a series of ejecta at 54 ka eruption. thus, this study aims to make the relationship between Kj-Y and 54 ka eruption clear based on field survey and microscopic observation.

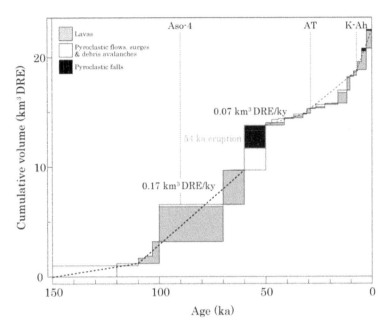

Figure 1.   Cumulative volume of erupted magma during the past 150 kyr. (modified from Nagaoka and Okuno, 2015)

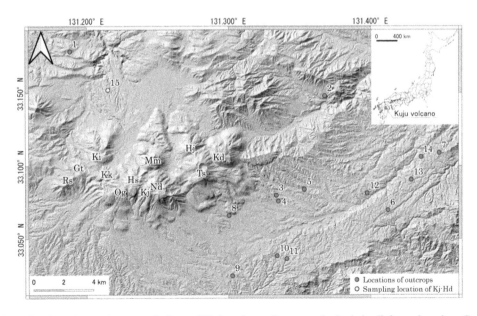

Figure 2.   Location and geomorphology of Kuju volcano. Base map is shaded-relief map based on Geospatial Information Authority of Japan. Rs: Ryoshiyama, Gt: Gotosan, Ki: Kuroiwayama, Kk: Kutsukakeyama, Og: Ogigahana, Hs: Hosshozan, Kj: Kujusan, Mm: Mimatayama, Nd: Nakadake, Hj: Hiijidake, Ts: Taisenzan, Kd: Kurodake.

2 METHODS AND RESULTS

2.1 *Field survey*

This method was used to discuss the style of Kj-Y eruption and make a comparison between Kj-Y and 54 ka eruption ejecta. Figure 2 shows outcrops, where Kj-D and Kj-P1 can be recognized. At those locations, stratigraphy of volcanic ejecta was observed, such as Kj-Y (some locations of those), Kj-D, Kj-P1 from bottom to top, and samples of ejecta were collected. Also, at loc. 15 Kj-Hd was observed and obtained samples (Figure 2).

2.1.1 *Yukouji Ash Fall layer (Kj-Y)*
Kj-Y's type locality is loc. 4 (Figure 2). The layer is yellowish white to light brown colored and com-posed mainly of silt to clay. Moreover, the layer contains particles like accretionary pellet at loc. 4 (Figure 4). The layer is situated just under Kj-D L and the soil layer is not between the layer and Kj-D (Figure 3).

2.1.2 *Kuju-D Ash Fall layer (Kj-D) (Ono, 1963)*
Tsuji et al (2017) divided the Kj-D into Lower (L) part and Upper (U) part (Figure 3).

Kj-D L consists of brown a lapilli layer, a pinkish gray ash layer, and a reddish brown ash layer in ascending order. On the other hand, the brown lapilli layer is similar to the characteristics of the Ariuji Lithic Fall layer (Kj-AL, Nagaoka and Okuno, 2014; 2015). Nagaoka and Okuno (2014) recognized soil just upper Kj-AL, but since soil has not been confirmed in this study, this lapilli layer is included in Kj-D L at present.

The brown lapilli layer includes volcanic rock fragments with mainly 4–6 mm and the maximum size is 60 mm. The layer also contains a small amount of pumice grains. The pinkish gray ash layer is composed mainly of silt to clay size grains and black grains with a diameter of about 4 mm and orange particles with a diameter of about 1–4 mm. The reddish brown ash layer contains mainly of coarse to medium sand sized particles and a small amount of pumice grains.

Kj-D U is composed of characteristic bluish gray ash layers and inserted some brown ash layers (Figure 3). The two brown ash layers contains pumice grains.

2.1.3 *Handa Pyroclastic Flow deposit (Kj-Hd) (Kamata and Mimura, 1981)*
This deposit is poorly sorted and is composed of grayish white pumice fragments and surrounding ash matrix (loc. 15). The pumice in Kj-Hd contains a large amount of hornblende.

2.1.4 *Kj-P1 pumice fall layer (Kj-P1) (Ono, 1963)*
The layer consists of yellowish white pumice fragments and andesitic lapillus. The pumice in Kj-P1 contains a large amount of hornblende and the lapillus in that layer concentration zone exists at the bottom of Kj-P1.

2.2 *Microscopic observation*

To discuss the style of Kj-Y eruption and make a comparison between Kj-Y and 54 ka eruption ejecta, we observed for Kj-Y, Kj-D L (matrix of brown lapilli layer), Kj-D U (matrix of bluish gray layer), Kj-Hd (matrix), and Kj-P1 (matrix and pumice). Samples of Kj-Y, Kj-D L, Kj-D U, and Kj-P1 were collected at loc. 4 and that of Kj-Hd was collected at loc. 15.

2.2.1 *Methods*
First, samples were washed by ultrasonic washing machine and panning. Afterwards, those were dried in an oven at 110°C and brought to room temperature. The matrix samples (excluding pumice of Kj-P1) were sieved at 1φ (0.5mm), 2φ (0.25mm), 3φ (0.125mm), and 4φ (0.063mm) intervals. The pumice of Kj-P1 was also crushed finely and sieved. Finally, we observed all samples for grain sizes of $4-3\varphi$, $3-2\varphi$, and $2-1\varphi$ using a stereomicroscope and a polarized microscope.

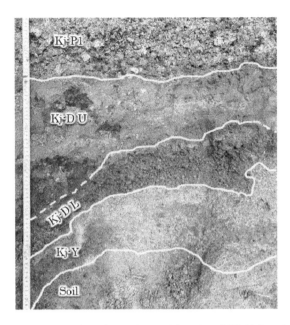

Figure 3.　The outcrop at loc. 4 showing Kj-Y, Kj-D L, Kj-D U and Kj-P1 from bottom to top.

0　　　　　　2 mm

Figure 4.　The particles like accretionary pellet (a dashed circle) in Kj-Y at loc. 4.

2.2.1 *Results*

All the observed samples contain a large amount of plagioclase and hornblende and a small amount of quartz at any observed grain sizes (Table 1).

Kj-Y contains moderate amount of orthopyroxene and clinopyroxene at any grain sizes (Table 1). A small amount of biotite is contained in grain sizes of 3-2φ and 2-1φ (Table 1). On the other hand, this layer contains a small amount of colorless and clear volcanic glasses (Figure 5).

Kj-D L contains moderate amount of clinopyroxene in grain sizes of 4-3φ and 3-2φ and a small amount of it in grain size of 2-1φ (Table 1). This layer contains moderate amount of orthopyroxene in grain size of 4-3φ and a small amount of it in the grain sizes of 3-2φ and 2-1φ (Table 1). A small amount of biotite is contained at any grain sizes (Table 1). Although this layer contains slightly less clinopyroxene and orthopyroxene than Kj-Y, that has similar mineral assemblages to Kj-Y (Table 1).

Kj-D U contains a small amount of orthopyroxene and clinopyroxene in grain sizes of 4-3φ and 3-2φ (Table 1). A small amount of Biotite is contained in grain sizes of 3-2φ and 2-1φ (Table 1). This layer contains less orthopyroxene and clinopyroxene than Kj-Y (Table 1).

Kj-Hd contains a small amount of orthopyroxene at any grain sizes and clinopyroxene in only grain size of 3-2φ (Table 1). A small amount of biotite is contained in grain sizes of 3-2φ and 2-1φ (Table 1). This layer contains less clinopyroxene than Kj-Y (Table 1).

The matrix of Kj-P1 contains moderate amount of clinopyroxene and orthopyroxene in grain size of 4-3φ and a small amount of the two in grain sizes of 3-2φ and 2-1φ (Table 1). Moderate amount of biotite is contained in grain sizes of 3-2φ and 2-1φ and a small amount of it is contained in grain size of 4-3φ (Table 1). This tephra has a higher biotite content and lower pyroxene content than Kj-Y (Table 1).

The pumice in Kj-P1 contains moderate amount of biotite in grain size of 2-1φ and a small amount of it in grain sizes of 4-3φ and 2-1φ. But clinopyroxene and orthopyroxene are not identified (Table 1).

Table 1. Mineral assemblages of Kj-Y and 54 ka eruption ejecta.

| Unit name | Grain size (φ) | Mineral | | | | | |
		Hbl	Cpx	Opx	Bt	Pl	Qz
Kj-P1 Pm	2-1	++	−	−	+	+++	+
	3-2	++	−	−	tr	+++	tr
	4-3	++	−	−	tr	+++	tr
Kj-P1 Mtx	2-1	++	tr	tr	+	+++	+
	3-2	++	tr	tr	+	+++	+
	4-3	++	+	+	tr	+++	tr
Kj-Hd Mtx	2-1	++	−	tr	tr	+++	tr
	3-2	++	tr	tr	tr	+++	tr
	4-3	++	−	tr	−	+++	tr
Kj-D U Mtx	2-1	++	−	−	tr	+++	tr
	3-2	++	tr	tr	tr	+++	tr
	4-3	++	tr	tr	−	+++	tr
Kj-D L Mtx	2-1	++	tr	tr	tr	+++	tr
	3-2	++	+	tr	tr	+++	tr
	4-3	++	+	+	tr	+++	tr
Kj-Y Mtx	2-1	++	+	+	tr	+++	+
	3-2	++	+	+	tr	+++	tr
	4-3	++	+	+	−	+++	tr

Hbl, Hornblende. Cpx, Clinopyroxene. Opx, Orthopyroxene. Bt, Biotite. Pl, Plagioclase. Qz, Quartz. Pm, Pumice. Mtx, Matrix. Mineral relative volumes: +++ > ++ > + > tr, trace. −, not recognized.

Figure 5. Minerals and volcanic glasses of Kj-Y. Hbl, Hornblende. Cpx, Clinopyroxene. Opx, Ortho-pyroxene. Bt, Biotite. Pl, Plagioclase. Qz, Quartz. Pm, Pumice type volcanic glass. Bw, Bubble wall type volcanic glass.

3 DISCUSSION

3.1 *The style of Kj-Y eruption*

Kj-Y is composed of silt to clay sized particles and contains particles like accretionary pellet at loc. 4 (Figure 4). These characteristics are the same as those of phreatic eruption (Oikawa et al., 2018). Kj-Y contains a small amount of magmatic particles such as colorless and clear volcanic glasses (Figure 5) and has similar mineral assemblages to Kj-D L (Table 1). Therefore, the style of Kj-Y eruption is considered to be "phreato-magmatic eruption".

3.2 *The relationship between Kj-Y and 54 ka eruption*

As Kj-Y and Kj-D L have similar mineral assemblages (Table 1), these tephras were possibly derived from same magma. Therefore, it is considered possibly that Kj-Y was one of the ejecta in 54 ka eruption from the evidence of similarity of the mineral compositions of Kj-Y and Kj-D L. Also, Kj-Y is situated just under Kj-D L without sandwiching soil layer (Figure 3). From the above, it is considered possibly that Kj-Y was precursory activity of 54 ka eruption. Since it is difficult to make definitive comparisons of those magmas based on mineral compositions alone, it is necessary to compare magmas based on the chemical composition of volcanic glasses and minerals.

3.3 *Eruption centers of Kj-Y and 54 ka eruption*

The isopach map of Kj-Y was drawn based on the thickness (Figure 6). The map illustrates possibly that the eruption center of Kj-Y was in the eastern region of current Kuju volcano (Figure 6). On the other hand, it has assumed that Kj-D erupted at the central and eastern region (Figure 6, Nagaoka and Okuno, 2014), and that the Kj-Hd and Kj-P1 erupted at the central region (Figure 6, Kamata and Mimura, 1981; Nagaoka and Okuno, 2014). Therefore, it is suggested possibly that the eruption center of 54 ka eruption migrated from eastern region to central region.

4 CONCLUSIONS

It is important to clarify 54 ka eruptive process for considering the history of magma at Kuju volcano. The previous studies considered 54 ka precursory activity to be Kj-D eruption, however Kj-Y is situated just under Kj-D L. Thus, this study aims to make the relationship between Kj-Y and 54 ka eruption clear. The results of the study are summarized below.

Kj-Y is composed of silt to clay sized particles and contains particles like accretionary pellet at loc. 4. These characteristics are considered to be the ejecta of phreatic eruption. Kj-Y also contains a small amount of magmatic particles such as colorless and clear volcanic glasses and has similar mineral assemblages to Kj-D L. Therefore, the style of Kj-Y eruption is considered to be "phreato-magmatic eruption".

Kj-Y has similar mineral components to Kj-D L, the two layers have possibly same magma. From this, Kj-Y is considered possibly to be one of the ejecta in 54 ka eruption. Also, Kj-Y is situated just under Kj-D L without sandwiching soil layer. Thus, it is considered possibly that Kj-Y was precursory activity of 54 ka eruption.

The isopach map of Kj-Y illustrates that the eruption center of Kj-Y was possibly in the eastern region of current Kuju volcano. On the other hand, it has assumed that Kj-D erupted at the central and eastern region, and that the Kj-Hd and Kj-P1 erupted at the central region. Therefore, it is suggested possibly that the eruption center of 54 ka eruption migrated from eastern region to central region.

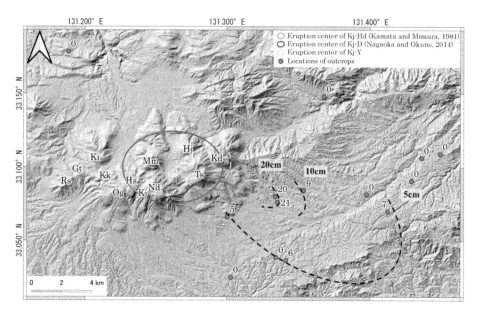

Figure 6.   Isopach map and estimated the eruption center of Kj-Y.

ACKNOWLEDGEMENTS

We are grateful to Nakagawa, M. for guidance in fieldwork. We also appreciate the material and emotional support of the persons at the inn "Kujira". This work was supported by the Ministry of Education, Culture, Sports, Science and Technology (MEXT) of Japan, under the Integrated Pro-gram for Next Generation Volcano Research and Human Resource Development. We thank anonymous reviewers for constructive comments that improved the manuscript.

REFERENCES

Kamata, H. and Mimura, K., 1981, Imbrications and the Flow Directions of Handa Pyroclastic Flow Deposits from Kuju Volcano, Southwestern Japan, *Bul. Volcanol. Soc. Japan*, 26, 281–292.

Kawanabe, Y., Hoshizumi, H., Itoh, J. and Yamasaki, S., 2015, Geological map of Kuju Volcano, *Geological Map of Volcanoes*, no. 19, Geological Survey of Japan, AIST.

Nagaoka, S. and Okuno, M., 2014, Tephra-stratigraphy of Kuju volcano in southwestern Japan, *Earth Monthly (Gekkan Chikyu)*, 36(8), 281–296.

Nagaoka, S. and Okuno, M., 2015, Eruptive history of Kuju volcanic group, SW Japan, *Transaction of Japanese Geomorphological Union*, 36(3), 141–158.

Oikawa, T., Oba, T., Fujinawa, A. and Sasaki, H., 2018, Geological study of phreatic eruptions, *J. Geol. Soc. Japan*, 124(4), 231–250.

Okuno, M., Nagaoka, S., Saito-Kokubu. Y., Nakamura, T. and Kobayashi, T., 2017, AMS radiocarbon dates of pyroclastic-flow deposits on the southern slope of the Kuju volcanic group, Kyushu, Japan, *Radiocarbon*, 59, 483–488.

Ono, K., 1963, Geology of the Kuju District. Quadrangle Series 1: 50,000 Map, *Geol. Surv. Japan*, 106p.

Tsuji, T., Kishimoto, H., Fujita, K., Miura, D., Ikeda, M., Nishizaka, N. and Onishi, K., 2017, Stratigra phy and estimation of eruptive parameter of very large eruption at Kuju Volcano at 54 ka, *Programme and Abstracts the Volcanol. Soc. Japan, 2017*, p71.

Rock Mechanics and Engineering Geology in Volcanic Fields – Ohta, Ito & Osada (eds)
© 2023 copyright the Author(s), ISBN 978-1-032-27657-1

Deposit temperature and sedimentation process of Aso-4 pyroclastic flow deposit

Kent Osawa* & Tomohiro Tsuji
Graduate School of Science and Technology for Innovation, Yamaguchi University, Japan

ABSTRACT: Large-scale pyroclastic flows sometimes travel long distances. There are various debates about the long-distance movement mechanism of pyroclastic flows. In particular, the influence of the sea is still under debate. The steam generated by the interaction between sea surface and the pyroclastic flow promotes the current of the pyroclastic flow. However, as the propulsion is not fixed to one direction on a flat sea surface, the propulsion force is considered to be decreased. The Aso-4 pyroclastic flow erupted about 90,000 years ago and the deposits are observed in Yamaguchi Prefecture that has a distance of more than 120 km from the source. In the process of travelling, it should have overcome obstacles such as Seto Inland Sea and topographical heights. It is a demanding research target for elucidating the movement mechanism of pyroclastic flows. To discuss the effects of the sea on the Aso-4 pyroclastic flow, we measured the deposition temperature using charcoal in Aso-4 pyroclastic flow deposits (PFD) in Yamaguchi Prefecture. Charcoal are often found in Aso-4 PFD. The temperatures of the Aso-4 PFD were measured from the reflectances and the Raman spectroscopy thermometers using them. The samples were taken from the Aso-4 PFD at three localities; Setobara, Hanaka, and Matsubara, (126.7 km, 136.7 km and 145.3 km from Nakadake crater, Aso volcano, respectively). The reflectance (Rm) measurement result Rm = 0.83-1.29 %. Based on the results, the heated temperatures were estimated to be 330-522 °C in the Setobara, and Hanaka and 288-429 °C in the Matsubara. The results of Raman spectroscopy showed that the D and G bands were measured. Based on these results, the heated temperature was estimated to be 192-324 °C in Matsubara and 151-319 °C for Hanaka. Compared to the estimated temperature of 337 °C for Aso-4 pyroclastic flow charcoal reported in Saga Prefecture, Kyushu (80.3 km from Nakadake crater, Aso volcano), no significant decrease of the temperature beyond the Seto Inland Sea was observed.

Keywords: **Aso-4** pyroclastic flow, Temperature, Charcoal, Raman spectroscopy, Reflectance thermometer

1 INTRODUCTION

Pyroclastic flow generally occurs due to the collapse of the eruption column or of the lava dome. The pyroclastic materials flow down the mountain body at high temperature and high speed. Large pyroclastic flows sometimes travel more than 100 km. The factors that control the flow consist of external factors such as topography and internal factors such as the initial velocity, temperature, erupted volume and the height of the eruption column (Hayakawa, 1991). However, there are various debates on the mechanism of long-distance movement of pyroclastic flows, one of which is the effect of sea. Because of the flat sea surface, there are

*Corresponding author: a037@yamagchi-u.ac.jp

DOI: 10.1201/9781003293590-20

few obstacles to the currents of pyroclastic flows. However, the flat sea surface is considered to disperse the propulsive force of pyroclastic flows and lose unidirectional flowability compared to valleys.

The Aso-4 pyroclastic flow erupted about 90,000 years ago and crossed mountains and oceans, leaving deposits as far away as Yamaguchi Prefecture, 160 km away. The Aso-4 pyroclastic flow deposit (PFD) has been named Ube volcanic ash layer in Yamaguchi Prefecture. They are divided into upper (PYU) and lower (PYL) parts (Geology of Japan "Chugoku Region" Editorial Committee, 1987), and the PYU and PYL are correlated with the Tosu orange pumice flow deposits and the Yame white pumice flow deposits, respectively (Watanabe, 1978). Matsuo (2009) newly defined the Yubeda unit, and the Yame clay, which was previously considered to be the Yame white pumice flow, was designated as the Yubeda unit. All the charcoal used in this study were collected from the Yubeta unit.

In this study, to discuss the effect of the sea on the Aso-4 pyroclastic flow, we attempt to measure the heated temperature of charcoal in the Aso-4 PFD.

2 METHODOLOGY

2.1 *Reflectance thermometer*

Charcoal is found in pyroclastic flow deposits. It is known that the reflectance of such charcoal changes sensitively under a reflecting microscope depending on the temperature at which it was heated. A method to estimate the heated temperature from the reflectance is called a reflectance thermometer. In addition, in recent years, there has been a lot of research on the use of spectra obtained by Raman spectroscopy to estimate the degree of ripening and temperature (Delano et al, 2019). In this study, these two methods are used to estimate the heating temperature of charcoal.

The collected charcoal was washed and dried. The samples were put into a resin (unsaturated polyester resin) to prepare a chip for measure reflectance. When the samples were small, the samples were mounted on a slide glass coated with a resin (ultra-low viscosity epoxy resin), stretched thinly, and cured. Created chips and glass slides were polished with # 500, # 1000, and # 2000 water resistant paper until charcoal was exposed on the surface. Then, the surface was polished with an alumina turbid liquid.

For reflectance measurement, a calibration curve was prepared using reflectance standard samples. The standard samples used for calibration were Spinel (R = 0.45 %), Sapphire (R = 0.55 %), Yttrium-Aluminum. The standard samples used for liberation were Spinel (R = 0.45 %), Sapphire (R = 0.55 %), Yttrium-Aluminum-Garnet (R = 0.96 %), Gallium-Gadolinium-Garnet (R = 1.73 %), Diamond (R = 5.228%) and Silicon Carbide (R = 7.57%). Each sample measured the random reflectance of up to 100 particles. The average was taken as the random reflectance (Rm). In addition, since Rm < 3.0 % can be regarded as average rotational reflectance (Ro) = average random reflectance (Rm) (Houseknecht and Weesner, 1997), Rm < 3.0 % of the Rm obtained in this study is Ro. It was assumed that it was the same value as.

2.2 *Raman spectroscopy*

Raman spectroscopy may be used to determine vitrinite reflectance equivalent organic matter maturity values for petroleum exploration, to provide temperature data for metamorphic studies, and to determine the maximum temperatures reached in fault zones (Delano et al, 2019). Raman spectroscopy will be applied to the temperature measurement of pyroclastic flows. The problem is that carbonized charcoal have different formation processes. Especially for pyroclastic flow deposits, it is necessary to consider the effects of weathering. For this reason, chemical treatment was used to remove corrosive substances. Since the effect of polishing on Raman spectroscopy has been suggested (kouketsu et al, 2019), we prepared an unpolished samples. In addition, we also measured low-grade metamorphosed (LGM) sandstone samples to confirm the validity of the measurements (Table 1).

The chemical treatment was performed according to the following procedure (Ujiie et al. 2011). It should be washed for 1 hour in an ultrasonic cleaner by adding 30 ml of NaOH (0.5 mol/L). Remove the supernatant liquid and repeat until the color of the supernatant liquid disappears. After leaving overnight, discard the supernatant liquid and wash. Next, 10 ml (1.0 mol/L) of hydrochloric acid is added and washed in an ultrasonic cleaner for 1 hour. Leave overnight and discard the supernatant liquid.

A Raman spectrometer (HORIBA JOBIN YVON T64000) at Kochi Core Center was used. The light source is an Argon laser (wavelength: 514.5nm). The obtained Raman spectroscopy data were subjected to fitting correction and then various Raman spectroscopies were separated. 30 points were measured in the Hanaka-Medium sample and 5 points in the other samples.

Table. 1. Sampling localities, experimental methods and conditions.

Sample	Latitude	Longitude	Reflection	Raman	
				Chemical treatment	Polishing
Setobara	34° 00′ 32.59″	131° 17′ 08.63″	○	○	×
Hanaka Top1			×	○	×
Hanaka Top2	34° 06′ 04.24″	131° 18′ 50.37″	○	×	○
Hanaka Medium			○	○	×
Hanaka Bottom			○	○	×
Matsubara	34° 10′ 05.47″	131° 18′ 03.6″	○	○	×
sandstone			○	×	○

3 RESULT

3.1 *Reflectance thermometer*

Several large tree-derived charcoals were identified in Aso-4 PDF in Yamaguchi Prefecture (Table 1, Figure 1), 127-145 km from Mt Aso. The observed biggest charcoals in each locality were 10 cm long and 7 cm wide in Setobara, 15 cm long and 10 cm wide in Matsubara, and 150 cm long and 40 cm wide, the biggest charcoal in all the locality, in Hanaka area. (Figure 1 C-2). The area in contact with the PDF was on powder with a lot of water. The middle portion retained its wood structure.

The maximum reflectance of the obtained charcoal was Rm 1.29 % in Hanaka, and the minimum was Rm 0.82 % in Matsubara (Figure 1 D).

3.2 *Raman spectroscopy*

As a result of Raman spectroscopy, two major peaks were identified. The first peak was found in the range of 1334 cm^{-1} to 1458 cm^{-1} and the second peak was found in the range of 1557 cm^{-1} to 1608 cm^{-1} (Figure 2). The two major peaks in charcoal are called G-band (located around 1340-1360 cm^{-1}) and D-band (located around1580 cm^{-1}), respectively. The peaks obtained in this study do not deviate significantly from this range, and it is considered that the G- and D-bands have been measured.

No clear peak was observed in the Hanaka-Bottom sample (Figure 2). However, except for the Hanaka-Bottom, D and G bands were observed in all the samples of the Hanaka-Top1, Hanaka-Medium, and LGM sandstone. The obscure observations tend to be high in intensity. Peak is sharper in LGM sandstone samples than in charcoal. The Raman spectrum of the peak is then larger than 1600 cm^{-1}.

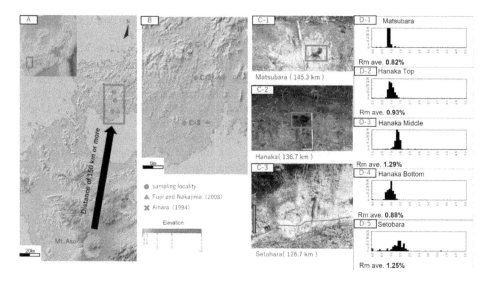

Figure 1. Study area (A), sampling area (B), occurrence of charcoal, the locality names and distances from the crater of Nakadake (C), and reflectance results (D). The figure above is from The Geospatial Information Authority of Japan (GSI).

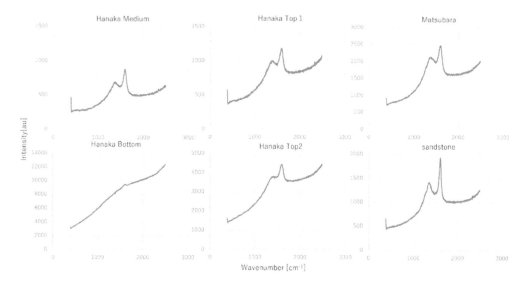

Figure 2. The Raman spectra of charcoal and LGM sandstone samples.

4 DISCUSSION

The largest charcoal in the Aso-4 PFD are significantly larger than the pumice in the PFD in Yamaguchi Prefecture, suggesting that the charcoal were not transported for a long distance with pumice fragments but were eroded by the pyroclastic flow near the localities.

Scott and Glasspool (2005) reported that the reflectance of charcoal stabilizes after 24 hours or more. It is known that pyroclastic flows retain heat for several days after deposition. Therefore, the reflectance of charcoal included in a pyroclastic flow is considered to be stable. However, since the heat during deposition decreases with time, several references were used to estimate the reflectance and temperature in this study. The heat-catalyzed temperatures of the

Table.2.　Conversion equations in the literature that have been used for reflectance thermometer and Raman spectroscopy.

Methodology	Reference	conversion equation.T (°C)	Temperature range (°C) [error]
Reflectance	Scott and Glasspool (2005)	Use approximate line	-
	Ujiie et al. (2008)	$201 \times (Ro) + 264$	-
Raman spectroscopy	Aoya et al. (2010)	$221 \times R2^2 - 637.1 \times R2 + 672.3$	$340 - 655$ [$\pm$ 30]
	Aoya et al. (2010)	$91.4 \times R2^2 - 556.3 \times R2 + 676.3$	
	Beyssac et al. (2002)	$-445 \times R2 + 641$	$330 - 640$
	Rahl et al. (2005)	$737.3 + 320.9 \times R1 - 1067 \times R2 - 80.638 \times R1^2$	$100 - 700$ [$\pm$ 50]

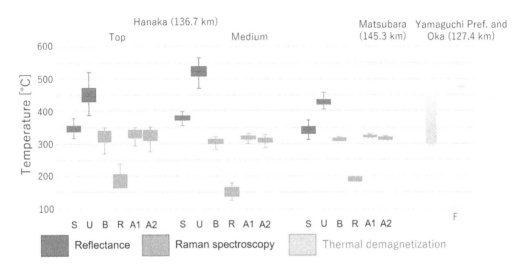

Figure 3.　Estimated temperatures obtained from reflectance thermometer and Raman spectroscopy. Comparison with thermal demagnetization (Fujii and Nakajima, 2008). S: Scott and Glasspool. (2005), U: Ujiie et al. (2011), B: Beyssac et al. (2002), R: Rahl et al. (2005), A1, A2: Aoya et al. (2010) F: Fujii and Nakajima (2008). Maximum temperature 475 °C (Oka 127.4 km), and other areas abave 300 °C.

charcoal, determined from the reflectance, were 288-429°C for Matsubara, 334-522°C for Hanaka, and 330-515°C for Setobara (Scott and Glasspool. 2005, Ujiie et al. 2011).

　Raman spectra of LGM sandstone samples and charcoal samples are similar (Figure 2). This can be considered as a confirmation of the validity of the measurements. The peak position of the LGM sandstone sample is shifted. The reason for the shift of the peak position is thought to be that the pressure and maceral species are different from those of pyroclastic flow samples. In addition, the peaks are known to shift due to factors such as heat from laser irradiation, so it is preferable not to use the wavenumbers of these peaks but use the full widths at half maximum and area ratios. The Raman intensity ratio R1 (D/G) and the area ratio R2 (D1/(G + D1 + D2)) of various Raman spectra are widely used as general parameters of temperature change in Raman spectroscopy (Delano et al, 2019). The D2-band was not identified in the measured samples. D2-band is not separable from low-grade charcoal and is included in the G-band (Beyssaac et al, 2005, Delano et al, 2019).

　In the present study the D2-band is assumed to be included in the G-band and is R2 (D1/(G + D1)). In this study, R1 and R2 were calculated and the temperatures were specified according to these parameters (Figure 3). The Raman parameters of the charcoal obtained in this study were applied to conversion equations (Beyssac et al. 2002, Rahl et al. 2005, Aoya et al. 2010), and the values were Matsubara 192-324 °C, Hanaka-Top 176-325 °C and Hanaka-Medium 151-319 °C. The estimated temperature derived by the conversion equation of Rahl et al. (2005) is lower than others. This is explained by the fact that the value of R2 does not change much below 330°C in the experiment of Rahl et al. (2005) and the sensitivity is not good.

The estimated temperatures based on the reflectivity thermometer and the Raman spectroscopy are consistent. The reported maximum deposit temperature of Aso-4 PFD in Yamaguchi Prefecture, obtained by magnetic susceptibility thermometry in Fujii and Nakashima (2008), is reported to be 475°C in the Oka area, Ube City (Figure 1b), Yamaguchi Prefecture. As a result of this study, it is estimated that the temperature of the Aso-4 pyroclastic flow that reached Yamaguchi Prefecture was higher than about 300°C. Further studies for the deposit temperature of Aso-4 PFD in Kyushu could contribute to discuss the influence of the sea on the Aso-4 pyroclastic flow migration.

In the following, we make a preliminary discussion citing the former literature on the temperature of the Aso-4 PFD in Kyushu Island. Buried charcoal was found from the Yame unit, the same unit that we studied, at Kamimine Town, Saga Prefecture (80.3 km from the crater of Nakadake, Figure 1c; Aihara, 1995). The average reflectance of the charcoal is 0.83 (Aihara, 1995). Applying the thermometer given by Scott and Glasspool (2005) to the data, the temperature is estimated to be 337°C that is similar to those of Aso-4 PFD in Yamaguchi prefecture. Thus, no significant decrease is found in the temperature of Aso-4 PFD beyond the Seto Inland Sea. Also, it is reported that the sea level just before the Aso-4 eruption was about 20 m lower than the present level (Sugitani, 1983). According to this, the area of the Seto Inland Sea, where the water depth is shallower than 20 m, might have been dried up at that time. That area coincides with the assumed path area of the Aso-4 pyroclastic flow from Kyushu Island to Yamaguchi Prefecture (the black arrow in Figure 1a). Although this is only a hypothesis, the Aso-4 pyroclastic flow might have maintained its temperature because it passed through the area that had become land due to the coastline depression. Further studies of the temperature of Aso-4 PFD in Kyushu Island are needed.

REFERENCES

Aihara, Y. 1995, Pyroclastic flow temperature estimated from charred wood fragments, *The Aso-4 pyroclastic flow with buried forest in the Saga plain*, **109**, 47–48.

Aoya, M., Kouketsu, Y., Endo, S., Shimizu, H., Mizukami, T., Nakamura, D., Wallis, S., 2010, Extending the applicability of the Raman carbonaceous material geothermometer using data from contact metamorphic rocks, *J. Metamorph. Geol.*, **28**, 895–914.

Beyssac, O., Goffe, B., Chopin, C., Rouzaud, J.N., 2002. Raman spectra of carbonaceous material in metasediments: a new geothermometer. *J. Metamorph. Geol.*, **20**, 859–871

Fujii, J., Nakajima, M., 2008, paleomagnetic Estimate of the Emplacement Temperature of the Aso-4 Distal Ahs-flow Deposits Distriduted in Yamaguchi Prefecture, Japan, Bulletin of the Center for Regional Environmental Research and Education, *University of Fukui "Nature and Environment in the Sea of Japan Region".*, **15**, 45–52

Hayamawa, Y., 1991, Volcaniclastic Flows and Their Deposits: Pyroclastic Flow, Surge, Lahar and Debris Avalanche, *Kazan.*, **36**, 357–370

Henry. D. G., Jarvis, I., Gillmore, G., Stephenson, M., 2019 Raman spectroscopy as a tool to determine the thermal maturity of organic matter: Application to sedimentary, metamorphic and structural geology, *Earth-Science Reviews.*, 198.

Houseknecht, D. W. and Weesner, C. M. B., 1997, Rotational reflectance of dispersed vitrinite from the Arkoma basin, *Ore. Geochem.*, **26**, 233–241.

Kouketsu, Y., Miyake, A., Igami, Y., Taguchi, T., Kagi, H. and Enami, M., 2019, Drastic effect of shearing on graphite microtexture: attention and application to Earth science, *Prog Earth Planet Sci.*, **6**, 23.

Matsuo, S., Horikawa, Y., Hoshizumi H., 2007, Flow units and distribution of the Aso-4 pyroclstic flow deposits in Yamaguchi Prefecture, *Abstracts, the 113th Annual Meeting of the Geological Society of Japan*, 280–280.

Rahl, J. M., Anderson, K. M., Brandon, M., Fassoulas, C., 2005, Raman spectroscopic carbonaceous material thermometry of low-grade metamorphic rocks: Calibration and application to tectonic exhumation in Crete, Greece, *Earth Planet. Sci. Lett.*, **240**, 339–354.

Scott A. C, Glasspool I. J, 2005, Charcoal reflectance as a proxy for the emplacement temperature of pyroclastic flow deposits, *Geology*., **33**(7): 589–592.

Sugitani, T, Geomorphological development of the north Ariake bay lowland, Kyushu, since the last – interglacial epoch: a quantitative study, *Geographical Review of Japan*, **56**, 403–419.

Ujiie, Y. Ito, A. Oyama, G. 2011, Estimation of emplacement of pyroclastic flows using carbonized wood in the case of Pleistocene Towada-Ofudo and Towada-Hachinohe Pyroclastic Floe Depsits, *Earth Science (Chikyu Kagaku)*., **65**, 111–124.

Watanabe, K., 1978, Studies on the Aso Pyroclastic Flow Deposits in the Region to the West of Aso Caldera, Southwest Japan, I: Geology, Mem. Fac. Educ. Kumamoto Univ, **27**, 97–120.

Geology of Japan "Chugoku Region" Editorial Committee, 1987, Geology of Japan 7 Chugoku Region, 290p, KYORITSU SHUPPAN CO., LTD.

Rock Mechanics and Engineering Geology in Volcanic Fields – Ohta, Ito & Osada (eds)
© 2023 copyright the Author(s), ISBN 978-1-032-27657-1

Issues of prolonged volcanic eruption disaster for revival

Hideki Kosaka*

Kankyo Chishitsu Co., Ltd

ABSTRACT: Volcanic disasters have been protracted after an eruption. Responding to prolonged disasters after an eruption is an important for coexistence with volcanoes. The purpose of this study is to investigate the relationship between the duration of the disaster after the eruption and the revival, and to clarify the issues for the revival of the volcanic disaster. Histogram of eruption duration based on the global eruption database for the last 100 years shows that 90% are within 2 years. On the other hand, the disaster duration after the eruption of Volcanic Explosivity Index (VEI) 4 or higher was 50 to 100 years based on satellite images. The reason for the continued disaster is mainly the flooding of rivers due to deforestation and rising riverbeds. While there are many volcanoes that continue to erupt in Japan, in the past 100 years, tourism, geoparks, and the agriculture and livestock industry at the foot of the mountain have developed, and in the case of VEI 3 class, they coexist with volcanoes. In the case of VEI 4-5 class, there are cases where the disaster has continued for more than 100 years after the eruption, and there are cases where it does not return to before the eruption even after the revival. These results indicate that the important issue for revival is the response to the continuation of disasters after the eruption of VEI 4 or higher. In recent years, especially in Japan, there is little experience of large-scale eruptions of VEI 4 or higher, and it is the key to disaster mitigation to spread the understanding of infrequent large-scale eruptions to the public. In some cases, the duration of a disaster exceeds 100 years, so It is important to consider measures for medium- to long-term risk diversification and disaster mitigation in advance, including utilizing the benefits of volcanoes.

Keywords: Duration of eruption, Disaster duration after eruption, Reconstruction cases with eruption, Long-term disaster, Coexistence with volcanoes

1   INTRODUCTION

Volcanoes have many merits such as fertile soils, abundant groundwater, vast foothills, geothermal energy, hot spring, deposit resources, and tourism resources. Even if there are many volcanic disasters, Japan has continued and developed as a country and region by taking advantage of the merits of volcanoes (Watanabe, 1998). In the history of coexistence with volcanoes, the long-term continuation of volcanic eruption disasters is often a problem. The purpose of this study is to introduce the prolongation of volcanic eruption disasters and their reconstruction cases, and to discuss issues related to geological engineering.

2   DURATION OF GLOBAL VOLCANIC DISASTERS

2.1   *Duration of eruption*

Using the eruption database (Smithsonian institution, 2019), I collected data on eruption locations and eruption durations for the past 100 years (1919-2019) in the world

*Corresponding author: kosaka@kankyo-c.com

DOI: 10.1201/9781003293590-21

(Figure 1A). Based on the same data, it was shown that the peak eruption duration of Volcanic Explosivity Index (VEI) 1-4 is within 100 to 1000 days (Nishimura, 2019). I have shown a detailed histogram of 100-1000 days for eruption duration in VEI 1-5 eruptions (Figure 1B). The histogram of eruption duration shows a power law, although 90% of eruptions tend to end within 3 years. These trends have little to do with the VEI. These results indicate that the duration of the eruption is difficult to predictions and it is important to be prepared for various scenarios.

2.2 *Disaster duration after eruption*

To investigate the duration of the disaster after the eruption, I used the satellite image of Google Map in 2019 to trace the eruption disaster (pyroclastic flow, ash fall, lava flow, etc.) and the trace of the sediment disaster after the eruption (debris flow, riverbed up), vegetation change and land use were read. The subject of interpretation was an eruption of VEI 4 or higher.

The tendency of vegetation invasion or recovery is fast in the pyroclastic flow and ash fall site and slow in the lava site (Figure 2). In the pyroclastic flow and ash fall site, grassland appears approximately 5 years after the eruption, the entire grassland is covered after 15 years, and trees are covered after 35 years, in low to mid latitude areas. Vegetation invasion and recovery takes longer in highlands, high latitudes, dry and cold regions than in mid-latitude regions. Vegetation is less likely to invade the site of lava than the site of pyroclastic flow and ash fall, and it is not covered with vegetation even 50 years after the eruption. At the volcano of Izu Oshima, Japanese knotweed and moss grow in the place where scoria sand is deposited, but in 1950 lava about 50 years after the eruption, the vegetation coverage is 10 to 30% and soil is not formed (Inagaki, 1998).

As a result of investigating the scale of the pyroclastic flow and ash fall, the distance from the crater to the braided river due to the riverbed up is longer than that of the pyroclastic flow and ash fall (Figure 3). Even 50 to 100 years after the eruption, the riverbed continues to up.The plots of the pyroclastic flow and ash fall site and the distance from the village to the crater shows that the village can be formed within 5 years after the eruption at a distance from the pyroclastic flow and ash fall site. It is necessary to pay attention to the influence of volcanic gas and volcanic ash, but From 20 to 50 years after the eruption, there were cases where villages were formed in the pyroclastic flow and ash fall site.

3   RECONSTRUCTION CASE OF LONG-TERM ERUPTION DISASTER IN JAPAN

3.1 *Reconstruction cases with eruptions of VEI 3 or less*

Since the middle of the 20th century, many eruptions of VEI 3 or less have occurred in Japan. I investigated cases of coexistence with volcanoes in Japan, focusing on land use and tourism resources peculiar to volcanoes, and volcanic disaster prevention measures of each project (Table 1).

The results show that a life utilizing the merits of the volcano has been established while the eruption continues and intermittently. There are many land uses that utilize the gentle slopes at the foot of volcanoes such as golf courses, ski resorts, ranches, and SDF training grounds. Leisure development can be seen utilizing lakes and marshes formed by lava and debris avalanches. Mud volcanoes and archaeological sites related to volcanic activity are also tourism resources. In the plains near the foot of the volcano, agricultural land and urban areas develop, benefiting from the abundant groundwater of the volcano. Around the crater, the unique landscape of the volcanic devastation is a tourist resource. These tourism resources are also important as a place for volcanic disaster prevention education. Geoparks are abundant in volcanic areas. Mt. Usu and Mt. Aso are UNESCO Global Geoparks and attract tourists

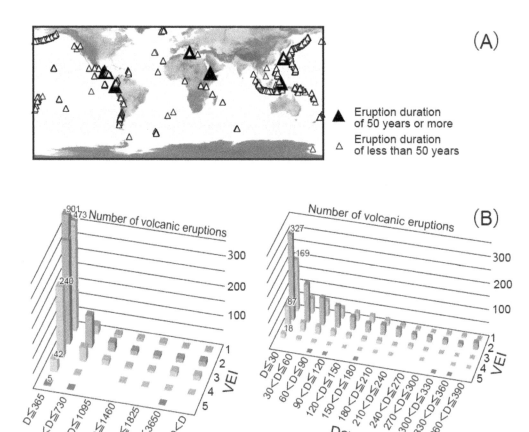

Figure 1.   Distribution of volcanic eruption event data (A) and histogram of eruption duration (B).

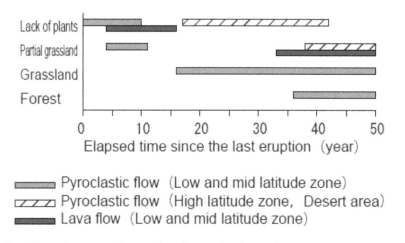

Figure 2.   Invasion and recovery of vegetation after a volcanic eruption.

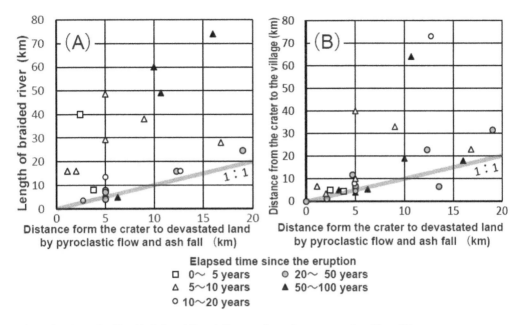

Figure 3.　Length of braided river (A) and distance from the crater to the village (B).

from all over the world. Volcano museums are effective places for disaster prevention education and outreach in tourism.

The Japan Meteorological Agency, Geospatial Information Authority of Japan, Ministry of Land, Infrastructure, Transport and Tourism, Ministry of Education, Culture, Sports, Science and Technology, National Research Institute for Earth Science and Disaster Prevention are working together to monitor and observe volcanoes and predict eruptions. In the sabo project, in addition to volcanic countermeasures centered on the response to mud and debris flows that cause prolonged disasters, disaster prevention education and outreach are being tackled as soft countermeasures. The Fire and Disaster Management Agency is promoting the development of evacuation and rescue facilities and supporting the transmission of information. The Ministry of Agriculture, Forestry and Fisheries is making key efforts for restoration and reconstruction, such as stockpiling seeds, promoting participation in the disaster compensation system, and maintenance of ash fall cleaning machines. Disaster prevention education, outreach, promotion of volcanic disaster risk research (financial insurance business), medium-term management plan (railway), and BCP (company), etc. are being promoted for private businesses.

3.2　*Reconstruction cases with an eruption of VEI 4 or higher*

In this paper, we have organized a chronological table of natural disasters and their responses with reference (Cabinet Office Central Disaster Prevention Council, 2006). Furthermore, statistical data on annual tribute and taxes is shown as quantitative information on reconstruction for a prolonged volcanic disaster.

Hoei eruption of Mt. Fuji volcano (Figure 4, left): Mainly caused damage by ash fall. The disaster was prolonged due to forest devastation and changes in slopes and river environment. According to the annual tribute data5) of Kanaishima in the alluvial fan of the Sakawa River, the annual tribute did not recover for 30 years after the eruption. When the effect of the reconstruction of the Sakawa River embankment appeared, the reconstruction accelerated, but the annual tribute rice leveled off. Now that civil engineering technology has improved, the development of agriculture and industry is recognized, responding well to disasters caused by heavy rain after the eruption.

Table 1.　List of cases of living with active volcanoes in Japan.

		Tokachi dake	Shikotsu (Tarumae)	Toya (Usu)	Hokkaido Komaga take	Izu-Oshima	Miyake jima	Asama yama	Unzenda ke	Asosan (Naka-dake)	Aira (Sakura jima)	Kikai (Iwo-dake)	Suwanos ejima
Eruption history (1990-2020)	Number of eruptions VEI	15 times VEI 1~3	24 times VEI 1~3	4 times VEI 1~3	13 times VEI 1~4	38 times VEI 1~2	13 times VEI 1~3	50 times VEI 1~3	2 times VEI 1~2	77 times VEI 1~3	Continue VIE 1~3	12 times VEI 1~2	20 times VEI 1~3
	Longest eruption duration	2 years	6 months	5 years	4 months	3 years	2 years	5 years	4 years	2 years	61 years	1 year	28 years
	Recent eruption VEI	2004 VEI 1	1996 VEI 1	2001 VEI 2	2000 VEI 1	2013 VEI 2	2010 VEI 1	2019 VEI 1	1996 VEI 2	Continue	Continue	Continue	Continue
Land use	Radius of devastated land	2.5km	1.5km	1km	2km	2km	3km	2.5km	2km	1km	2km	1km	1.5km
	Length of baided river	15km	5km	3km	7km	5km	4km	5km	5km	4km	5km	2km	3km
	Distance from the crater to the farmland	7km	3km	1km	3km	3km	3km	4km	3.5km	2.5km	3km	2km	2km
	Distance from the crater to the village	7km	11km	2km	5km	3km	3km	5km	3.5km	5km	4km	2km	3km
	Distance from the crater to the city	17km	12km	2.5km	7km	4km	4km	11km	6km	5.5km	8km	67km	172km
	Use at the foot of the volcano	Farm Forest Hot spring	Farm Forest Hot spring	Farm Forest Hot spring Village Golf	Farm Forest Hot spring Village Lake Training grounds	Farm Forest Hot spring Village Golf Airfield	Farm Forest Hot spring Village Lake Airfield	Farm Forest Hot spring Village Villa Golf Ski	Farm Forest Hot spring Village Villa, Golf Lake, Mud volcano Remains	Farm Forest Hot spring Village Golf	Farm Forest Hot spring Village	Farm Forest Hot spring Village	Farm Forest Hot spring Village
Tourist attractions	Sightseeing around the crater	○	○	○	×	○	×	○	○	○	×	×	×
	Sightseeing at the foot of the volcano	○	○	○	○	○	○	○	○	○	○	○	○
	Geopark	○	—	○	—	○	Promotion	○	○	○	○	○	—
	Volcano Museum	○	—	○	—	○	—	○	○	○	○	—	—
	Special product	Natural diversity Livestock products	Alpine plants Lava dome	Lava dome Ropeway	Livestock products	Camellia oil Ashitaba Deep sea salt Kusaya	Spiny lobster Gelidiaceae Ashitaba Kusaya Miyake Glass	Plateau vegetables Kyoho grapes Sake Sake brewing	Livestock products Sake brewing School excursion	Livestock products Akado-zuke Limonite rice Wood ear Volcanic ash pottery Lava processing	Mandarin orange Sakurajima radish Loquat Lava processing Volcanic ash dried fish	Daimyo bamboo shoots Shochu Camellia oil	Daimyo bamboo shoots Camellia oil
Volcano disaster prevention	Country, municipality, business	Act on Special Measures for Active Volcanoes, National resilience, Volcano expert dispatch (Cabinet Office). Volcano disaster prevention conference, Volcano disaster prevention map, Disaster Management Plan (Municipality). Monitoring of volcanic activity, Eruption forecast and warning, Volcanic Alert Level (Japan Meteorological Agency). Monitoring of crustal movements, Volcanic Land Condition Map (Geospatial Information Authority of Japan). Survey of submarine volcanoes (Japan Coast Guard), Volcano research, Volcano observation, Volcano education (Ministry of Education). Development of observation and prediction technology for volcanic activity (National Research Institute for Earth Science and Disaster Prevention). Measures against mud flow and debris flow, Monitoring system construction, Outreach (Sabo business). Evacuation pit / shelter, Maintenance of helicopter detachment plaza, Support for information transmission, Sophistication of mountain rescue activities, Expansion and strengthening of regular firefighting (Fire and Disaster Management Agency). Mountain disaster prevention measures, Disaster prevention farming measures business, Stockpile of seeds, etc., Promotion of participation in the disaster compensation system, Maintenance of mechanical facilities for cleaning ash fall (Agriculture, forestry and fisheries business). Disaster prevention education, Outreach, Volcanic disaster risk research promotion (Financial insurance business). Volcano disaster prevention plan・Medium-term management plan (Railway business). BCP formulation (Enterprise).											

Tenmei Asama eruption (Figure 4, right): Caused damage due to pyroclastic flow, lava flow, mud flow, etc. In particular, the Tenmei Asama mud flow changed the environment to the Agatsuma and Tone rivers, and the effects continued for more than 100 years. Looking at the cultivated areain Kamahara Village at the northern foot of Mt. Asama, only about 40% has recovered 70 years after the eruption. In the Tone rivers, the influence of the riverbed rises due to the mud flow continued for more than 100 years. Even with the subsequent Kathleen typhoon (1947), sediment runoff and floods in the basin became prominent, and the construction of the Yamba dam is currently required for the Tone rivers hydraulic control. The development of cultivated land had peaked, but now the development of agriculture centered on highland cultivation and the tourism industry utilizing the blessings of volcanoes is recognized.

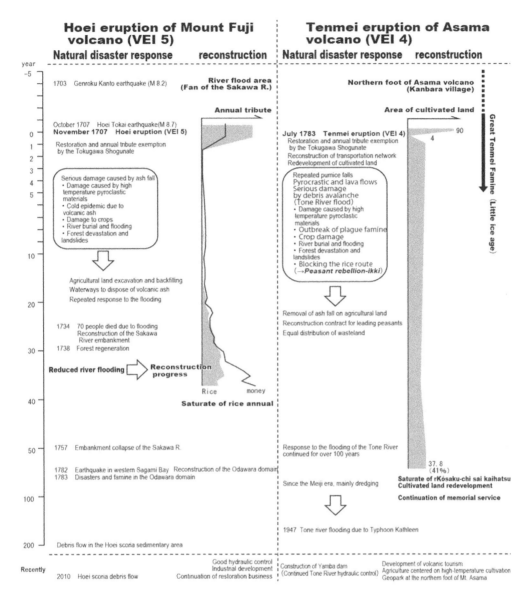

Figure 4.   Disaster chronology of Hoei and Tenmei eruptions.

4   APPLIED GEOLOGICAL ISSUES FOR RECONSTRUCTION OF LONG-TERM VOLCANIC DISASTERS

The applied issues related to geological engineering are as follows.

➢ The duration of a volcanic eruption is difficult to predict in advance. Research on the process from the start to the suspension of a volcanic eruption will be important in the future.

➢ The time required for vegetation and soil recovery is longer for lava than for pyroclastic flows. In addition, the main factors for the prolongation of many disasters are forest devastation, highlands, slopes, and changes in the river environment. We would like to clarify the factors that change the environment due to such an eruption and use it for recovery and reconstruction. There are many applied geological issues including slope problems, such as slope changes due to long-term crustal movements and slope failures due to changes in the

155

eruption environment, as well as the scale of the eruption. I would like to mention disaster prevention education and outreach rooted in the region, centering on various volcanic disasters including slope problems, as one of the applied geological problems.

➣ Looking at the reconstruction information, the recovery of agriculture and economy will reach a plateau, but I would like to pay attention to the cases of tourism utilizing volcanic merits such as volcanic landscapes and hot springs, and disaster succession by Geoparks. It is necessary to convey that wasteland has high value as disaster succession and that it is important for tourism resources and ecosystem diversity. In outreach activities, it is necessary to convey about coexistence with volcanoes as well as disaster inheritance.

➣ In recent years, especially in Japan, there is little experience of large-scale eruptions of VEI 4 or higher, and it is the key to disaster mitigation to spread the understanding of infrequent large-scale eruptions to the public. In some cases, the duration of a disaster exceeds 100 years, so it is important to consider medium- to long-term risk distribution in advance and recovery and reconstruction utilizing the benefits of volcanoes for disaster mitigation.

ACKNOWLEDGEMENTS

Problems in Volcanic Areas of the Japan Society of Engineering Geology for many discussions. I would like to thank the members of the Research Subcommittee on Applied Geological Applied geological issues for reconstruction of long-term volcanic disasters.

REFERENCES

CabinetOfficeCentralDisasterPreventionCouncil,2006, http://www.bousai.go.jp/kyoiku/kyokun/kyoukun nokeishou/index.html.

Inagaki, H. (1998): Volcano Engineering Society Research Subcommittee 1st Subcommittee "Oshima Field Study Group Report", Japan society of civil engineers, 15p. (in Japanese)

Nishimura, T., 2019, Characteristics of Volcanic Eruptive Activities: Analysesof Global Data Base, *Bulletin of the volcanological society of Japan*, 64, 2, 53–61. (in Japanese with English abstract)

Smithsonian institution, 2019, https://volcano.si.edu/search_eruption.cfm (July 2019).

Sumiya, H. Inoue, K., Koyama, M., and Yomita, Y., 2002, Distribution of sediment disasters after the 1707 Hoei eruption of Fuji Volcano in central Japan, based on historical documents, *Historical Earthquakes*, 18, 133–147. (in Japanese with English abstract)

Watanabe, N., 1998, History and culture library "Mt. Asama eruption", Yoshikawakoubunkan, 204p. (in Japanese)

Resources and energy in volcanic fields

Rock Mechanics and Engineering Geology in Volcanic Fields – Ohta, Ito & Osada (eds)
© 2023 copyright the Author(s), ISBN 978-1-032-27657-1

The geothermal energy resource developments and their hazards of the Indonesia Volcanic Areas

Asnawir Nasution*

Applied Geology Group, Geology Department, Faculty of Earth Science and Technology, Bandung Institute of Technology, Bandung, West Java, Indonesia

ABSTRACT: Geothermal energy in Indonesia is located along a volcanic arc, and is distributed and related to the tectonic setting of Indonesia. Most geothermal power plants are located along this volcanic arch at the flank of in-active volcanoes. The total install capacity of these geothermal power plants has reached 2318 MW, which is now the second largest geothermal capacity in the world.

A few exploration or production wells are located in un-stable volcanic areas, which has led to the blowouts of wells and the release of volcanic gases. The instability of the rock masses is caused by hydrothermal clay and the porous structure of the primary rock, which was intensively altered. The altered rock is affected by the hydrothermal processes, including the acidity of the volcanic gases. Acidity changes the petro physical properties. The physical alteration of the rock gradually causes a transformation of the hydrothermal system structure, which leads to changes in its hydrodynamic and temperature regimes.

Gases are dangerous and often overlooked hazards in volcanic and geothermal regions. The effects of gases from geothermal areas may cause asphyxiation, respiratory diseases, and skin burns. The hazards mostly stem from CO_2, H_2S, CO, SO_2, and other minor gases. New cases of gas hazards have involved the death of five people and more than 40 victims of anxiety and headaches. Volcanic gases may cause blowouts during geothermal drilling. Monitoring systems are being developed to mitigate such risks due to volcanic gases. Other geothermal hazards are landslides, phreatic eruptions, and earthquakes.

Keywords: Rock Mechanics, Engineering Geology, Volcanic field, Slope Stability, Permeability

1 INTRODUCTION

Intensive geothermal exploration and development in Indonesia began in the 1970s (Nasution and Supriyanto, 2011). They have been carried out in an attempt to solve the problem of electrical demands and to substitute renewable energy for traditional energy resources from oil, coal, and gas. Geothermal energy, which is related to volcanic areas, can be derived in abundance from the volcanic belt along the Indonesian Islands.

Volcanic geothermal prospects are most associated with high-temperature discharged fluids and with Quaternary volcanism. They are distributed along the volcanic arcs of Sumatra, Java, Nusa Tenggara, North Sulawesi, and Moluccas islands (Figure 1). The geothermal heat sources are derived from shallow cooling magma and igneous rock intrusions.

Geothermal Law No. 27/2003, then a re-new Law in 2014, regulates the upstream geothermal business and volcanic area development. Private investors are encouraged to develop and

*Corresponding author: a.nasution50@gmail.com; nasution@gl.itb.ac.id

DOI: 10.1201/9781003293590-22

run geothermal businesses as Independent Power Producers (IPPs). In accelerating geothermal exploration and production, Energy Sales Contracts (ESCs) or Joint Operation Contracts (JOCs) and Power Purchase Agreements (PPAs) are negotiated to PLN. These regulations have significantly accelerated the increase in the geothermal capacity from the 1980s, 1990s, 2000s, and up to the year 2021.

However, geothermal development in the volcanic areas of Indonesia involves many problems as well. Volcanic hazards, such as landslides, volcanic gases, phreatic eruptions, blowouts, scaling, toxic water running into rice fields, and probably earthquakes, are encountered.

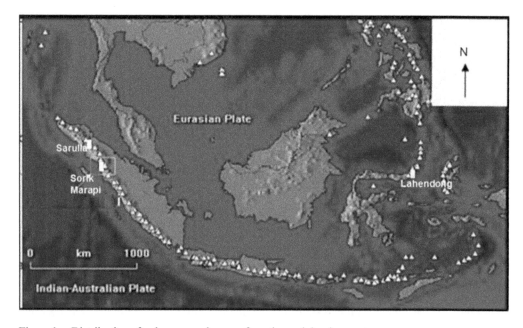

Figure 1.   Distribution of volcanoes and areas of geothermal development.

The objective of this paper is to report the current geothermal resources, install capacities, and geothermal hazards of the volcanic areas of Indonesia. This will lead to the knowledge of the number of units and the installed geothermal power generating capacity over the last 30 years and their possible hazards.

The objective of this paper is to inform the current geothermal resource, install capacity and geothermal hazards of volcanic area. This will know the number of units and installed geothermal power generating capacity over the last 30 years and their possible hazard.

2   VOLCANIC AREAS AND TECTONIC SETTING

In general, the geothermal areas of Indonesia are associated with volcanic arcs (Figure 2). These volcanic arcs are the result of interaction from the Indian – Australian, Eurasian, Pacific, and Philippine Plates. The arcs display micro continental arc volcanism related to the oceanic trench subduction zones (Cas and Wright, 1984). The volcanic arcs can be divided into four arcs, the Sunda arc in the east, the Banda arc in the east, and two small arcs situated to the north of the Banda arc, the Sangihe-north Sulawesi and Halmahera arcs (Figure 2).

The Sunda arc represents part of the collision zone between the Indian-Australian plate to the south and the Eurasian plate to the north. The Banda arc is characterized by an anomalous tectonic setting compared to the Sunda arc. It represents part of the collision between the Pacific plate to the east and the Eurasian plate to the north and the Indian-Australian plate to the south. The north Sulawesi-Sangihe and Halmahera arcs are a complex junction between the Eurasian, Indian-Australian, Pacific, and Philippine plates (Silver and Moore, 1981).

The geothermal areas, which are mostly located along these volcanic arcs (Figure 1), are found at about 276 locations (Geological Agency, 2011). Volcanic fractures or fault zones generally control the geothermal surface manifestations, such as hot springs, fumaroles, mud pools, steaming grounds, and geysers. They have high temperatures, high flow rates, and low to neutral pH levels (50 to 100 °C, pH 2 to 7), and provide early signals for geothermal investigation and exploration.

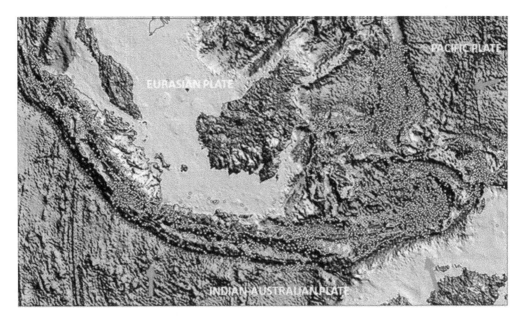

Figure 2.   Tectonic setting of volcanic island of Indonesia.

3   GEOTHERMAL REGULATIONS AND DEVELOPMENT

3.1   *Geothermal regulations*

A number of Indonesian laws and regulations apply to energy development activities. They include regulations on oil and gas, a law on mining development, and other laws. The geothermal industry needs a legal basis, which was declared in the 1980s. Presidential decree (PD) No. 22/1981 instructed the state-owned oil and gas company (PT. Pertamina) to explore Indonesia's geothermal potential, and ordered the national electricity state company (PT PLN) to buy its electricity production.

The first President Decree No. 22/1981 was renewed by Decree (PD) No. 45 Year 1991, which declared that there are two alternatives: firstly, JOC (Joint Operation Contract) and secondly, ESC (Energy Sales Contract). The JOC is Pertamina or contractors that can develop and operate the steam field, then sell the steam of electric generation to the National Electric State Company (PLN) or other parties. Secondly, Pertamina and its contractors are allowed to develop and operate the steam field, generate electricity, and then sell the electricity to PLN or other consumers, called as ESC (Energy Sales Contract) or a total project.

The New Geothermal Guidelines were declared by Presidential Decree No. 76/2000 which replaced the previous Decree. Under this PD 76/2000, the Government of Indonesia will handle all or part of the exploration. This means that the government will take the risk of exploratory drillings.

Geothermal Law No. 27/2003 was renewed by Law No. 21/2014. The ratified Geothermal Law represents geothermal activity and not mining, so this is a breakthrough for geothermal development activities.

The new law declares that the geothermal concession of permitted areas is controlled by the central government of Indonesia (The Department of Energy and Mineral Resources). By the new regulation, development of the accelerated geothermal areas for the geothermal business occurs faster than under the previous regulation.

The Independent Power Producers (IPPs) and private investors have an agreement to sell geothermal power under Energy Sales Contracts to PLN. Therefore, these results will accelerate exploration and production drilling and increase electricity production.

3.2 Geothermal prospects

The volcanic geothermal resources in Indonesia have shown abundant surface manifestations. They are located and distributed along volcanic belts, comprising about 276 surface indication areas. The surface manifestations have previously been identified by the Government of Indonesia, particularly The Geological Agency. The identified manifestations represent a total potential capacity estimated at approximately 29,000 MW (Geological Agency in Nasution and Supriyanto, 2011).

The Indonesian geothermal prospects are mostly related to in-active volcanoes, which are called C and B volcanic types. The C volcanic type is characterized by a fumarole field; there are no volcanic eruptions or magmatic gases (SO_2, HCl, or HF) occurring in this field. An example is the Kamojang volcanic geothermal complex. The B volcanic type is a volcano; it still represents a crater formed with an acid lake. Very small amounts of SO_2 and HCl gases are indicated, but no magmatic eruptions.

These volcanic geothermal systems are indicated by up-flow and out-flow geothermal systems. The up-flow zone or up-flow system is a geothermal manifestation (fumarole), mostly located in areas of high land with acid pH fluids (Eq. a fumarole field). The out-flow zone is a geothermal manifestation (a hot spring with neutral pH), which is located in areas of low land. They may have high temperatures or high enthalpy geothermal resources (>220°C). Therefore, prospects are mostly explored in up-flow zone areas.

3.3 Geothermal development areas

The potential resources are mostly explored and developed for electricity by PT. Pertamina, private investors, and PT.PLN (The Electric State Company). Right now, the high temperatures or high enthalpy geothermal resources are located in 23 areas for generating electricity (Figure 3). The areas mostly belong to the Pertamina concession (Table 1).

The potential exploration resources of Sungai Penuh, Iyang Argopuro, Bedugul, Kotamobagu, and Seulawah Agam have not yet been developed for the generation of electricity. In the near future, however, several geothermal fields will support the electrical capacity of Indonesia.

The other geothermal potential resources are also located in remote areas and on small islands, located in the eastern part of Indonesia, such as Flores, Lomblen, Ambon, Halmahera, and other small islands, which have high population densities (Figure 3).

A few small-scale geothermal power plants have been developed by PT in several districts. PLN and private companies, such as Ulumbu, Mataloko, Sokoria, Tulehu, and Jailolo. PT. ORKA have developed Sokoria (Ende District of Flores Island). Star Energy Company Jailolo of Halmahera island has also developed a plant. Their concessions are located outside of that of PERTAMINA. Therefore, in a near future, small-scale geothermal power plants will support additional geothermal electric capacity and will reduce the government subsidy of the electrical prices.

Geothermal fields which have not been developed commercially are still located in numerous areas, for example, in the western (Sumatra and Java islands) and eastern (Nusa Tenggara

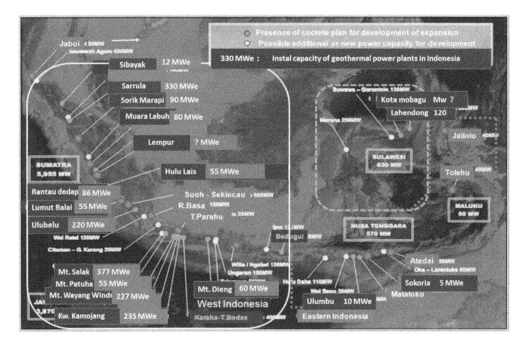

Figure 3. Map showing install capacities, and resource potential in promising geothermal fields (Modified from JICA, 2007).

and Maluku islands) parts of Indonesia, as shown in Figure 3. The Geological Agency (2011) identified speculative resources in the western Indonesia geothermal area of more than 7000 MW and to the east of approximately 2000 MW (Figure 3).

The commercially developed geothermal fields are Mt. Sibayak, Mt. Salak, Mt. Wayang-Windu, Mt. Kamojang, Mt. Darajat, Lahendong, the Mt. Dieng complex, the Muara Labuh volcanic field, the Rantau Dedap volcanic field, the Sarulla volcanic areas, the Mt. Sorik Marapi complex, the Ulubelu volcanic complex, the Hulu Lais volcanic complex, the Lumut Balai volcanic complex, the Kawah Cibuni volcanic complex, the Mt. Patuha complex, and the Mt. Karaha volcanic complex. These areas belong to the Pertamina Concession. The others outside of the Pertamina Concession areas, which have already been developed, are the Ulumbu volcanic complex, the Mataloko volcanic complex, and the Sokoria volcanic complex of Flores. Therefore, 20 developed areas have supported electricity of around 2318 MWe (Table 1).

4 VOLCANIC HAZARDS

As shown in Figure 3, all the geothermal fields are located in volcanic areas. They may be affected by the hazards of volcanic activities. The volcanic hazards which usually occur are volcanic eruptions (phreatic eruptions), gases, blowouts, toxic volcanic water, landslides, subsidence, and earthquakes. The hazards may composed of the following.

4.1 Volcanic gases

The small amounts of gases released from the solfatara on the low part of the crater are a function of the deep magma process, eq. vapour melt separation during the rise of the generated magma. The gases are mostly composed of CO_2, SO_2, HCl, HF, H_2S, H_2, O_2+Ar, CH_4, and NH_3. The existence SO_2, HCl, and HF in volcanic areas indicates a high temperature

Table 1. Power plants and install capacity of geothermal fields in Indonesia.

No	area	Drilling depth (m)	Capacity (MW) 2021	PPA (Cent US)	Plan & Capacity (MW) 2021-2025	Developer
1.	Kw.Kamojang, West Java	1200 to 1600	235	7.03	30	PT. PGE, JOC, ESC
2.	Kw.Darajat, West Java	1300 to 2500	271	6.95	Add 60	PT. Chevron, JOC, Esc
3.	G.Wayang Windu, West Java	1300 to 2400	227	8.39	Add. capacity 55	PT. Star Energy, Esc
4.	G. Salak, West Java	1300 to 3000	377	8.46	Add. capacity 55	PT. Chevron, JOC, Esc
5.	Kw.Patuha, West Java	700 to 1800	55	8.10	55	PT. Geodipa, JOC
6.	Kw.Krahabodas, West Java	1000 to 2000	30	8.46	55	PT. PGE, JOC,
7.	Kw.Cibuni, West Java	800 to 1400	-	6.90	10	PT. Yala Teknosa, Esc
8.	G.Dieng, Central Java	1200 - 2100	60	9.81	60	PT. Geodipa, JOC
9.	Kw.Bedugul, Bali	1200 - 2100	-	7.15	10	PT. Bali Energy, JOC
10.	G.Sibayak, North Sumatra	1400 to 2200	12	10. 7	30	PT. PGE, Esc
11.	Sarulla, North Sumatra	1300 to 2100	330	6.8	110	PLN & Consortium, Esc
12.	Sorik Marapi, North Sumatra	1000 to 2300	90	9.0	150	PT. SMGP(ORKA) ESC
13.	Muara Labuh, West Sumatra	1000 to 2400	80	13.0	55	PT. Supreme Energy
14.	Kw.Hulu Lais, Bengkulu	1000 to 1800	55	?	110	PT. PGE, Esc
15.	Lumut Balai, South Sumatra	1200 to 2000	55	?	110	PT. PGE, Esc
16.	Rantau Dedap, South Sumatra	1200 to 2400	86	9.4	55	PT. Supreme Energy
17.	Kw.Ulubelu, Lampung	1100 to 2000	220	7	55	PT. PGE, Esc
18.	Kw.Lahendong, North Sulawesi	1100 to 2200	120	4-4.7	40	PT. PGE, JOC
19.	Volc.Tompaso, North Sulawesi	1200 to 2100	-	6.00	60	PT. PGE, Esc
20.	Kotamobagu, North Sulawesi	1500 to 1900	-	6.00?	55	PT. PGE
21.	Kw.Ulumbu, West Flores	700 to 1800	10	7.00	20	PT. PLN
22.	Kw.Mataloko, Central Flores	250 to 750		7.30?	4 x 2.5	PT. PLN
23.	Sokoria, Ende, East Flores	1000 to 1500	5	9.0	3 x 20	PT. SMGP(ORKA) ESC
24.	Volc.Tolehu, Ambon - 9.60	900 to 1700	-	9.6	2 x 10	PT. PLN
	Sum		2318			

magmatic component, developing from the underlying magma. However, in the volcanic geo-thermal system, there is no SO_2 gas, but high levels of H_2S and CO_2. H_2S, assumed as a low temperature gas (less than 400°C), is present in the reservoir. The H_2, O_2+Ar, CH_4, and NH_3

gases probably indicate a secondary hydrothermal component slowly rising from a two-phase, saline brine vapor, and covering the magmatic system.

The gas hazards, which mostly occur during geothermal exploration and in the production wells are CO_2 and H_2S at high concentrations, particularly during and after drilling. The gas may cause suffocation in humans. The release of CO_2 from the Dieng Geothermal field in 1979, after an earthquake, killed 150 people. Another case of the release of CO_2 during the opening of a wellhead test occurred in the Sorik Marapi Geothermal field, killing five persons and leaving more than 50 unconscious. Therefore, monitoring such gases will be necessary for safety.

4.2 *Blowouts during drilling*

Blowouts may occur in a geothermal field during drilling, even if blowout preventive measures have already been taken. The blowouts may be caused by a high level of CO_2 gas or due to an increase in the pressure and temperature of the well from the subsurface geothermal field. These accidents will cause the stoppage of the drilling activity, rather than attempting to overcome them by intruding a heavy mud concentration to the deepest well. Blowout accidents have occurred in the Dieng geothermal field, the Mataloko geothermal field (2001), and Ijen Geothermal field (April 2020), as shown in Figure 4a.

In the Hulu lais geothermal field (Bengkulu Province), a blowout was followed by a landslide during a heavy rainfall event occurring in 2019 (Figure 4b).

Figure 4a. a) Geothermal drilling blowout of Ijen (2020)

Figure 4b. b) Blowout and landslide of Hululais geothermal field (2019).

Volcanic hydrothermally altered deposits contain a large number of zeolites and clay minerals that are broken by tectonic and sedimentary fractures, and are unstable to humidifying. They may be caused by hydrothermal processes, particularly when an increase in volcanic gas concentrations (H_2S, HCl, and HF) interact with the surrounding rock masses. The acid gases (low pH) will weaken the rock properties and lower the stability. The morphology factors, such as contrast relief, alteration zones, and rainfalls, will cause landslides, debris flows, and mudflows. Therefore, to increase the slope stability of geothermal development areas, civil engineers with knowledge of the rock properties of the geothermal development areas must be involved in order to reduce the hazards and to look for stabilized rock properties for drilling or power plant locations.

4.3 *Phreatic eruptions*

The hydrothermal or phreatic eruptions of geothermal areas may occur in several places in Indonesia. They are caused when the vapor pressure of the geothermal fluids exceeds the hydrostatic boiling pressure for a given temperature. These eruptions take place at a point where the convective rise in the geothermal fluids is impeded by a relatively impermeable layer, termed cap-rock (Facca and Tonani, 1967). The vapor pressure of a geothermal fluid receives significant partial contributions from CO_2 and H_2S, as well as H_2O. The Dieng Phreatic eruption, occurring in April 2021 (Figure 5a and Figure 5b), is an example of this type of eruption.

Figure 5a.   During phreatic eruption of geothermal field of Dieng Plateau (PVMBG, 2021)

Figure 5b.   After phreatic eruption of geothermal field of Dieng Plateau (PVMBG, 2021).

4.4   Earthquake hazard

The tectonic hazard of particularly active fault systems may occur in active geothermal fields (Figures 5a and 5b). The active fault zones will affect or damage the production wells. Earthquakes will increase the rock movement and release of gas from the subsurface. The high Vp/Vs of the Sarulla high fluid content, magneto-telluric, and structural geological data may indicate potential fields as well locations for geothermal exploration and exploitation, eq. the Sarulla Geothermal field (Muksin, U. et al. 2013).

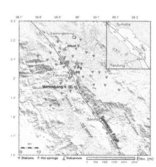

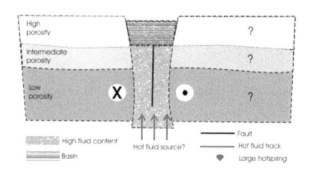

Figure 6a.   Tarutung area and presence of Tarutung pull-apart basin, Sarulla graben, Namora Langit field dome Martimbang, Imun volcano, and hot springs (Muksin, U. et. al. 2013)

Figure 6b.   Conceptual model for Sarulla graben based on seismicity distribution, V_p and V_p/V_s structure. The hot fluid from below the fault is transported directly to the surface along the weak fault (Muksin, U. et. al., 2013).

5   CONCLUSIONS

Recently, Indonesia has become one of the largest geothermal development countries in the world, after producing 2318 MWe in 2021. These renewable resources will become clean energy and may reduce the CO_2 being released into the atmosphere. The geothermal potential reserves and new development areas are expected to continue to increase on the prospected islands into the future.

By a new regulation, Geothermal Law No. 21/2014, the accelerated geothermal development areas are sharply and quickly increasing. The installed capacities have totaled more than 500 MWe for the last five years. Therefore, there will continue to be a gradual increase in the geothermal installed capacity for power over the next five years.

The geothermal and volcanic hazards composed of CO_2 and H_2S gases have occurred in the Dieng and Sorik Marapi geothermal fields. They have caused fatal accidents among the local people. Landslides, blowouts, phreatic eruptions, and possible earthquakes have occurred in the Hululais, Lempur, Mataloko, Dieng and Ijen geothermal areas. Therefore, monitoring systems are urgently needed for the geothermal development areas of Indonesia.

ACKNOWLEDGEMENTS

The author would like to offer his thanks and appreciation to the Organizing Committee of the 5th International Workshop on Rock Mechanics and Engineering Geology in Volcanic Fields (RMEGV2021) for accepting the present paper and for allowing him to participate in the workshop on September 9-11, 2021 in Fukuoka, Japan.

REFERENCES

Bixley and Browne, 1988. Hydrothermal eruption potential in geothermal area. *The 10th New Zealand Geothermal Workshop 1988.*

Facca, G and Tonani, F. 1967. The self-sealing geothermal field, *Bulletin Volcanologique*, 30, 271–273.

INAGA, 2018. Geothermal Resources of Indonesia, Geological Agency Survey 28.8 GWe. In *INAGA Conferences*, April 2018.

Nasution, A. and Supriyanto, E. 2011. Current Status and New Geothermal Development Areas in Indonesia. Proceedings of the 9th Asian Geothermal Symposium, 7-9 November 2011

PVMBG, 2021, The Phreatic eruption of Dieng Complex, Central Java, Indonesia, *Photo collection.*

Richter, A. 2020. Steam kick from geothermal exploration well at Blawan Ijen Geothermal Project. *Think Geo Energy Magazine 2020.*

Silver, E. A. and Moore, J.C. (1981) The Molucca sea collision zone, in The Geology and Tectonics of Eastern Indonesia, *Geological Research and Development Centre, Spec. Publ.*, 2.

Simkin, T. and Siebert, L. (1994) Volcanoes of the world, 2 nd ed. -A regional directory, gazetteer and chronology of volcanism during the last 1000 years-. *Smithonian Institution Global Volcanism Program*, Geoscience Press, Inc., Tuscan, Arizona, 349p.

Toni, A. Irawan, T. Prasetyo, A. Fakhri A. Andika, Prasetyo, I. Saputra,M. 2017. Success Story Handling the Blowout of Three Wells in Hululais Geothermal Field with Capping Method. *GRC Transactions*, 41, 2017.

Umar Muksin, Klaus Bauer and Christian Haberland, 2013. Earthquake tomography Seismic V_p and V_p/V_s structure of the geothermal area around Tarutung (North Sumatra, Indonesia) derived from local. *Journal of Volcanology and Geothermal Research*, 260, 27–42.

West -Jec., 2008. Pre-Feasibility Study for Geothermal Power Development Projects in Scattered Islands of East Indonesia. *Engineering and Consulting Firms Association*, Japan.

Rock Mechanics and Engineering Geology in Volcanic Fields – Ohta, Ito & Osada (eds)
© 2023 copyright the Author(s), ISBN 978-1-032-27657-1

Analysis of hot water flow in geothermal reservoirs by TOUGH2 considering vapor production and water reinjection

Akira Sato* & Eldan Arkin
Kumamoto University, Kumamoto, Japan

ABSTRACT: Many active volcanoes are located along the Japanese islands. Japan has abundant geothermal resources, which represent promising renewable energy sources. However, we still cannot fully harness these resources. Geothermal fluid power generation consumes large amounts of geothermal steam generated from production wells. Then, cold water can be inserted into the strata through injection wells. Poor reservoir management may cause a temperature drop in the reservoir or a halt in power generation. Therefore, to predict the distribution of the temperature change in the geothermal reservoir and to systematically maintain the geothermal source, appropriate simulation research is necessary. In this study, we investigated the flow of geothermal fluid and simulated the steady state of the geothermal reservoir using the standard geothermal reservoir simulator TOUGH2. Using the steady state model, the temperature change due to the vapor production and water reinjection process was estimated to simulate the operation of geothermal power plants. It was found that the temperature dropped in a short period when the heat supply was small. To conduct sustainable power generation, over 1000 kg/s of hot water supply was necessary. It was also found that the region where the temperature dropped downstream of the reinjection well was more than 400 m.

Keywords: Geothermal energy development, TOUGH2, reservoir parameters, hot water flow, reinjection of water

1 INTRODUCTION

The development of renewable energy is expected for measures against global warming and for the improvement of the energy self-sufficiency rate in Japan. Many active volcanoes are located along the Japanese islands. In fact, Japan has the world's third-largest geothermal resource. Therefore, geothermal energy is expected to become a stable power source at a low power generation cost that could play the role of a base-load power source (Ministry of Economy, Trade and Industry, 2018).

In many geothermal power plants in Japan, the power generator is driven by the steam produced from the production well. Then, the condensed water is reinjected underground through injection wells. In this system, the reinjected water may reach the production region and cause the temperature of the steam to drop. As a result, the geothermal reservoir could no longer be used for power generation. Therefore, an appropriate and precise evaluation of the water circulation process in a geothermal reservoir is necessary (Faust & Mercer, 1979; O'sullivan, 1985; Yano and Ishido, 1998; Eylem et al., 2011; Gudni, 2012), and numerical simulations are effective ways to understand the water circulation process. In this study, a reservoir model based on the geological structures of the target region and the steam quantity was constructed. By comparing the results obtained from the simulation and the results of measurements such

*Corresponding author: asato@kumamoto-u.ac.jp.ac

DOI: 10.1201/9781003293590-23

as temperature distributions in wells, the natural structure of the reservoir was reconstructed, and the state after development was simulated.

In this study, the hot water circulation process and temperature distributions in the geothermal reservoir were simulated using the simulator TOUGH2 (Pruess, 1999). TOUGH2 is a well-known and useful simulator for analyzing a multitude of problems, such as geothermal reservoir processes, including oil and gas reservoir engineering (Adrian and O'Sullivan. 2008; Tanaka and Itoi, 2010; Pruess, 2020). In this study, hot water at a temperature of 250 °C was supplied to the model, and from this, the initial state of the geothermal reservoir was reconstructed. By conducting simulations of vapor production and reinjection of condensed water, the flow of hot water and temperature change in the geothermal reservoir were investigated. Based on the results, the appropriate management process for the reservoir is discussed herein.

2   GEOLOGICAL CONDITION AND RESERVOIR MODEL

2.1   *Geological condition of the target site*

The geothermal site modeled in this study is the Oguni area in Kyushu, which is located at the bottom of Mt. Waita, and the area forms hilly terrain at an altitude of 700–800 m (Figure 1). This region is well known as one of the promising geothermal region in Japan (Nishijima et al., 1994; Tosha et al., 2018). The vertical cross section of the area is shown in Figure 2. The mudstone, which has a low permeability, consisted of Nogami mudstone formation at the shallow depth of the region, and plays the role of a caprock. It is assumed that the bedrock formation is 1500 m deep and that this formation also has low permeability. The layer between the caprock and bedrock consisted of various types of volcanic rocks, and its permeability is relatively high. Therefore, the layer at an altitude between 500 m and 1500 m forms a geothermal reservoir, and the hot water convects this layer.

2.2   *Geothermal reservoir model*

The grid model used in this study is illustrated in Figure 3. The size of the model is 5800 m in the x-axis direction, 4700 m in the y-axis direction, and 3400 m in the z-axis direction. The yellow cells in Figure 3 represent the geothermal reservoir region. The initial width of the reservoir (y-axis direction) was 500 m, and the region of the reservoir is also illustrated as yellow cells in Figure 4. By changing the reservoir width, the relationship between the reservoir width and the changes in temperature can be analyzed.

Each property was organized as shown in Figure 4, in accordance with the geological conditions of the target region. In this figure, the locations of the production and reinjection wells are also indicated. To discuss the interaction of reinjected water and the relative location between the production and reinjection wells, four production wells were positioned as shown in Figure 4. The red cells in the figure represent the wells from which the geothermal fluid was extracted. The blue cell represents the area where condensed water was injected into the reservoir. The mechanical parameters of each layer are listed in Table 1.

Initially, the model did not contain hot water. Then, hot water at a temperature of 250 °C was supplied from the bottom of the model (Figure 4) at a rate of 10 kg/s until the temperature in the reservoir became uniform. This state is defined as the initial state of the geothermal reservoir, and the simulations of heat production and reinjection of condensed water begin from this initial state.

3   RESULTS

3.1   *Evaluation of temperature drops of production well*

By continually producing geothermal fluid and reinjecting condensed water underground, the temperature and pressure in the reservoir will decrease over time. However, a higher temperature of the geothermal fluid is favorable for increasing the power generation capacity. For each well, sufficient number of days of operation were simulated so that the temperature would reach 100 °C. Here, the rate of supplied hot water, produced vapor, and reinjected condensed water was 10 kg/s. The temperatures of the supplied hot water and reinjection water were 250 °C and 50 °C, respectively. The vapor was produced in each well. The results of the

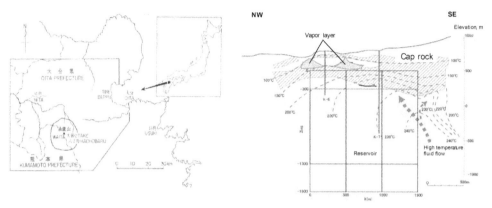

Figure 1. Modeled geothermal site, Oguni area, in Kushuku.

Figure 2. Vertical cross section of the target area.

Table 1. Mechanical properties of each layer.

	Density, kg/m^3	Porosity,%	Permeability, m^2			Heat conductivity, W/m°C	Specific heat, J/kg
			x direction	y direction	z direction		
Air	2.5×10^{-1}	10	1.0×10^{-16}	1.0×10^{-16}	1.0×10^{-16}	1.4	840000
Soil	2400	10	1.0×10^{-18}	1.0×10^{-18}	1.0×10^{-18}	2.95	1050
Hots pring	2400	10	1.0×10^{-15}	1.0×10^{-15}	1.0×10^{-15}	0.51	1050
Caprock	2400	1	1.0×10^{-22}	1.0×10^{-22}	1.0×10^{-22}	0.01	1050
Reservoir	2400	10	1.0×10^{-13}	1.0×10^{-13}	1.0×10^{-13}	2.9	1050
Bedrock	2400	10	1.0×10^{-18}	1.0×10^{-18}	1.0×10^{-18}	2.1	1050

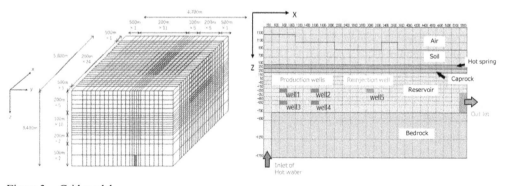

Figure 3. Grid model.

Figure 4. Assignment of geological formations.

temperature changes over time are shown in Figure 5. As this figure shows, wells 1 and 2 exhibited similar behavior, whereas wells 3 and 4 took longer to get colder. This is because hot water was supplied from the bottom of the model, as shown in Figure 4. In the case of wells 3 and 4, heat energy was supplied faster than the upper level, and the temperature drops were slower than those of wells 1 and 2. However, even in the case of wells 3 and 4, it took less than 1600 days (4.5 years) until the temperature dropped to 100 °C, and continuous power generation was impossible. This is because the amount of hot water supplied at 10 kg/s was exceedingly small compared with the size of the analytical region, and the stored hot water was easily consumed by the geothermal fluid production. To determine the location of geothermal power plants, the temperature and the amount of supplied hot water are important factors for sustainable power generation.

The necessary amount of hot water supplied from the heat source was estimated. The analysis was conducted under hot water supplies of 50 kg/s, 200 kg/s, 500 kg/s, and 1000 kg/s. The results are shown in Figure 6. When the supply rates were less than 200 kg/s, the reservoir temperature dropped to 100 °C in less than 2000 days. However, the temperature drops converged at some temperature. Moreover, even when the supply rate was 1000 kg/s, the temperature dropped 100 °C from the initial temperature. It was found that, to keep the reservoir temperature constant at various temperatures, sufficient amounts of hot water should be supplied continuously.

3.2 *Evaluation of temperature drops due to reinjection*

As aforementioned, condensed water is reinjected underground through reinjection wells in many geothermal power plants in Japan. In this case, the cool injected water might have migrated into the reservoir and influenced the temperature around the production well. Here, the temperature distribution around the reinjection well was simulated. As done previously, the rate of reinjection water was 10 kg/s, and the temperature of the reinjection water was 50 °C.

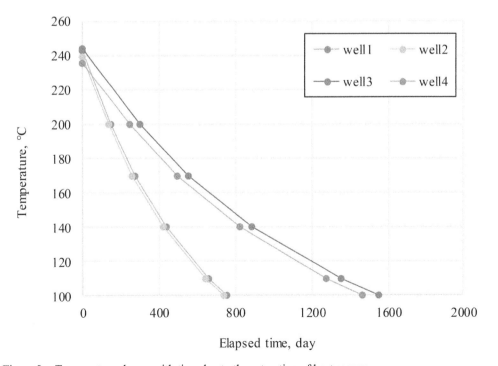

Figure 5.   Temperature change with time due to the extraction of heat energy.

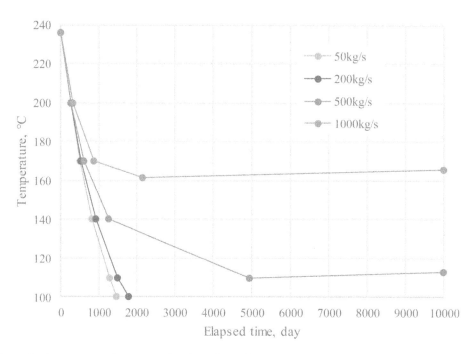

Figure 6.　Temperature change over time in well 4 under the different hot water supplies ratio.

The temperature distribution along the *x*-axis (flow direction) is shown in Figure 7, and that along the *y*-axis (perpendicular to the flow direction) is shown in Figure 8. In both cases, the results for each flow rate of supplied hot water 10 kg/s, 100 kg/s, 500 kg/s, and 1000 kg/s are plotted. The location of the reinjection well is shown in Figure 7. Figure 8 shows the *y-z* cross section at the location of the injection well. In both cases, the temperature decreased in the injection well. It can be seen that the temperature drops become small when the flow rate of the supplied hot water increases. This is because the larger hot water flowed into the region around the reinjection well, and thus the influence of cool reinjection water was trivial. However, as shown in Figure 7, the influenced region where the temperature becomes lower

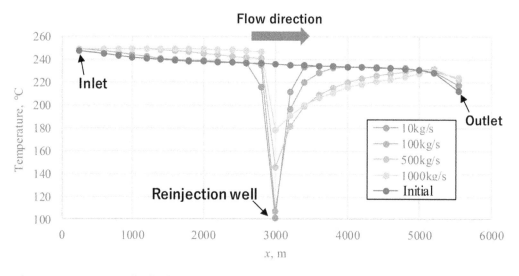

Figure 7.　Temperature distribution along the flow direction.

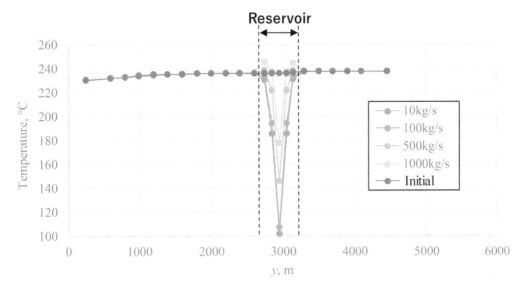

Figure 8. Temperature distribution perpendicular to the flow direction.

than the initial state increases. When the flow rate was 10 kg/s, the influenced area was limited to approximately 400 m downstream from the injection well. On the other hand, the influenced region spread more than 2000 m downstream from the injection well. Because the flow velocity in the reservoir was higher, the low-temperature region tended to spread wider.

4   CONCLUSIONS

In this study, the hot water circulation process and temperature distributions in the geothermal reservoir were simulated using the simulator TOUGH2. The geological condition of the Oguni area in Kyushu was modeled, and an analysis of the simulated temperature distribution was performed. It was found that a sufficient hot water supply into the geothermal reservoir is necessary to conduct continuous vapor production even when the initial temperature is sufficiently high. The influence of reinjected water was also simulated. When the amount of hot water supplied was small, the temperature drops became larger. However, even though the temperature decreased when the amount of supplied hot water increased, the influenced region of the temperature extended downstream from the injection well.

REFERENCES

Adrian E, C., O'Sullivan M., 2008, Application of the computer code TOUGH2 to the simulation of supercritical conditions in geothermal systems, Geothermics, 37, 622–634.
Eylem K., Sadiq J. Z., Michael J. O., 2011, Reinjection in geothermal fields: A review of worldwide experience, Renewable and Sustainable Energy Reviews 15, 47–68. https://www.sciencedirect.com/science/article/pii/S1364032110002121
Faust C, Mercer J., 1979, Geothermal reservoir simulation: 1. Mathematical models for liquid- and vapor-dominated hydrothermal systems. Water Resour Res, 15, 23–30. https://doi.org/10.1029/WR015i001p00023.
Gudni, A., 2012, Role and management of geothermal reinjection, Short Course on Geothermal Development and Geothermal Wells", organized by UNU-GTP and LaGeo, in Santa Tecla, El Salvador, March 11- 17.
Nishijima J, Ebara Y and Fujimitsu Y, 1994, Reservoir monitoring by observation of gravity changes at Oguni Geothermal Field, Kumamoto, Japan., Abstracts with Programs. Annual Meeting. Geothermal Research Society of Japan, 91. https://orkustofnun.is/gogn/unu-gtp-sc/UNU-GTP-SC-14-29.pdf

Ministry of Economy, Trade and Industry, Strategic energy plan, 2018. https://www.enecho.meti.go.jp/en/category/others/basic_plan/5th/pdf/strategic_energy_plan.pdf

O'Sullivan M., 1985, Geothermal reservoir simulation. Int J Energy Res, 9, 319–332. https://doi.org/10.1002/er.4440090309.

Pruess K, Oldenburg CM, Moridis GJ., 1999, TOUGH2 USER'S GUIDE. Technical Report. http://tough.lbl.gov/assets/docs/TOUGH2_V2_Users_Guide.pdf.

Pruess K., 2002, Numerical simulation of multiphase tracer transport in fractured geothermal reservoirs. Geothermics, 31:475–499. https://doi.org/10.1016/S0375-6505(02)00007-X.

Tanaka T and Itoi R, 2010, Development of Numerical Modeling Environment for TOUGH2 Simulator on the Basis of Graphical User Interface (GUI), Proceedings World Geothermal Congress 2010 Bali, Indonesia, 25–29.

Toshiyuki Tosha T, Kida, Y, Obara Y, Yamazaki T, Watanabe H, 2018, Geothermal Development in Oguni, Central Kyushu, Proceedings of 43rd Workshop on Geothermal Reservoir Engineering Stanford University, Stanford, California, February 12-14, 2018SGP-TR-213

Yano Y, Ishido T, 1998, Numerical investigation of production behavior of deep geothermal reservoirs at super-critical conditions, Geothermics, 27, 705–721.

Rock Mechanics and Engineering Geology in Volcanic Fields – Ohta, Ito & Osada (eds)
© 2023 copyright the Author(s), ISBN 978-1-032-27657-1

Volcanic caves at Azores Islands, Portugal. Conservation aspects

Luis Ribeiro e Sousa*
Construct, School of Engineering, University of Porto, Portugal

João Carlos Nunes
University of Azores, Ponta Delgada, Portugal

Ana Maria Malheiro & Filipe Marques
Regional Laboratory of Civil Engineering of Azores, Ponta Delgada, Portugal

ABSTRACT: The stability of volcanic caves particularly in urban areas and in touristic places presents high complex problems due to their nature, existing environment and mainly because the danger that represents for people and buildings, but also for the safeguard of those caves as natural monuments. There are risk sources which might cause large damages, deterioration and, in some cases, it is necessary to put forward conservation measures in order to prevent damage or destruction of the existing buildings above those caves. Volcanic caves are important parts of the geological heritage of the Azores Islands, Portugal. Therefore, it was considered relevant to establish a methodology for the conservation of these natural monuments, with particular emphasis of lava-tube caves. In this paper, cases are presented. Special emphasis was made to the Carvão cave because most of it is located beneath the more important city of the archipelago – Ponta Delgada, at São Miguel Island. Another situation that required special attention was the Torres cave at Pico Island since it is a touristic point of interest besides being an important natural heritage (it is the largest volcanic cave of Azores islands).

Keywords: Volcanic caves, Azores Islands, Stability, Conservation

1 INTRODUCTION

Volcanic caves, not being a very common geological phenomena, exist in various places in the world, such as in the Azores Islands, Canary Islands, Galapagos Islands, Japan, Iceland, Italy, Korea and USA (Ferreira 2000, Amigos dos Açores 2008, Costa et al. 2008, Pereira et al. 2015, Malheiro et al. 2015).

The Azores archipelago has a diversified speleological heritage. The Azores Archipelago is one of the two autonomous regions of Portugal, composed of nine volcanic islands situated in the North Atlantic Ocean, and is located about 1,600 km West of Portugal mainland. The island's volcanism is in close relation with the Azores Triple Junction, generally defined by the Mid-Atlantic Ridge and the Azores-Gibraltar Fracture Zone; the existing active faults and fractures controlled the volcanic activity and have produced many of the seismic events that occur in the islands. The islands have many examples of volcano-built geomorphology including volcanic caves and pits.

In the Azores, two main types of volcanic caves can be considered: lava-tube caves and volcanic pits. The lava-tube caves result from the cooling of the lava flow in contact with the air

*Corresponding author: Sousa-scu@hotmail.com

DOI: 10.1201/9781003293590-24

and the surrounding formations, forming a hardened crust, under which lava continues to drain. The volcanic pits are more often old volcanic conduits, more or less vertical, which emptied as a result of a new vent at a lower level, or a real reduction in emissions from the depth (Amigos dos Açores 2008; Costa et al. 2008; Pereira et al. 2015).

Some of the volcanic caves are important parts of the geological heritage of the Azores. Therefore, it was considered relevant to establish a methodology for the conservation of these natural monuments, with particular emphasis of lava-tube caves. Special emphasis was made to the Carvão cave because most of it is located beneath the more important city of the archipelago – Ponta Delgada, at São Miguel Island.

At Azores Islands, about 272 volcanic caves exist, and they are distributed on all the islands, except Corvo Island (Amigos dos Açores, 2008, Pereira et al. 2015). Pico Island is the one with the largest number of caves (129), followed by the Island of Terceira (69). On the Island of São Miguel, the cave system of the Gruta do Carvão is highlighted since it develops mostly underneath the city of Ponta Delgada.

2 AZORES GEOLOGY

All the Azorean islands are of volcanic origin and they correspond to the elevations of the ocean floor, due to the accumulation of volcanic products. The archipelago is located in a very active seismic and volcanic area that corresponds to an area where the American, Eurasian and African lithospheric plates meet (Gaspar et al., 2007), and this translates in the existence of important fault systems (Figure 1).

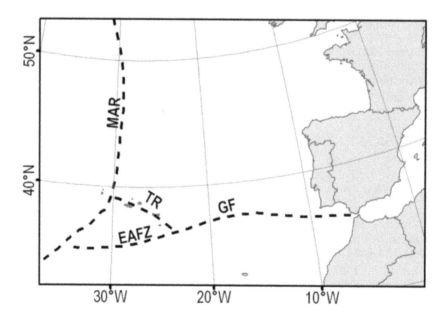

Figure 1. Main tectonic structures in Azores area (MAR – Mid Atlantic Ridge; EAFZ – East Azores Fracture Zone; TR – Terceira Rift; GF – Gloria Fault), (Gaspar et al., 2007).

Eruptions have been presented since its settling in the 15th century (Gaspar et al., 2007). Also, important landslides affected several islands associated to these events or associated with extreme meteorological conditions, as well as by coastal erosion processes. Volcanic activity includes several types of eruptive styles and products, with major relevance to extrusion of basaltic lava flows, especially on Pico Island, thus during effusive volcanic eruptions the formation of lava-tube caves is frequent at Azores islands (Costa et al., 2008).

3 CAVES AT SÃO MIGUEL ISLAND

The Azores region is composed by several islands. São Miguel, the largest one, is formed by three active, generally E-W trending, trachytic central volcanoes with caldera (Sete Cidades, Fogo and Furnas). An inactive trachytic central volcano (Povoação) and an old basaltic volcanic complex (Nordeste) comprise the easternmost part of the island (Valadão et al., 2002). The more recent area of the islands is the "Picos Volcanic Complex", a basaltic fissural area between the Sete Cidade and the Fogo central volcanoes. The Azores are an authentic natural laboratory for a great diversity of studies, in the area of volcanology, seismology, geothermal energy, and natural disasters, with particular incidence in landslides which gives these islands a status of very relevant interest (França et al. 2005).

Azores islands present different geological characteristics due to the volcanic nature of the rocks and the variability of eruptive styles and its stratigraphic sequences (Malheiro and Nunes 2007; Malheiro et al., 2018). Volcanic caves are important parts of the geological heritage of the Azores.

3.1 *Carvão cave system at Ponta Delgada*

This cave system is located in the western part of Ponta Delgada city. It has a total length of about 2,500m and is divided into three main separate sections (Figure 2). However, several old documents indicate a dimension far superior to currently mapped to this volcanic cave, which may have dimensions of around 5 km long (Sousa et al., 2014). Section I, designated as the Paim Street cave, has a total length of 880.2 m. The lava tube has an average height of 2-3 m, with maximum high values of 6.4 m in specific locations and the average width is 5.8 m. Section II, located downstream of section I and designated as Lisboa Street cave, has a total length of 701.8 m, 6.4 m maximum high (Figure 3) and reaches sometimes 13 m wide particularly where channels converged. Section III is located under the João do Rego Street, thus its name of João do Rego Street cave. This section has two branches almost parallel that joint at the end as it can be seen at Figure 4.

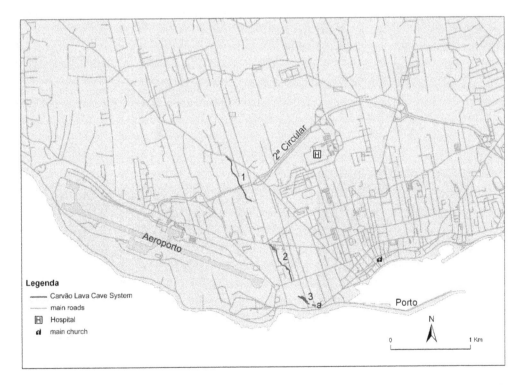

Figure 2. Location of the Carvão Cave system and its different sections at Ponta Delgada city area, S. Miguel Island.

This section of the lava tube has a total length of 278 m, a height of 3-6 m, and the width is variable with a maximum value of 13 m. This section of the cave was recently discovered during infrastructure works and has several buildings above the cave layout that were partially damaged by the collapses that occurred inside the cave. Further downstream is the José Bensaúde Street cave (Figure 5), a small lava tube with 43 m length, 7.5 m wide and 5 m maximum high. Images of the Carvão cave are illustrated in Figures 5 and 6.

Taking into account the scientific and touristic relevance and its importance in terms of natural heritage, the Carvão cave system was already classified as a Regional Natural Monument. Along the sections of the cave, it can be observed grooved walls, lateral benches at different levels, long sections of superimposed channels and branched galleries. The walls multi-colored show vestiges of gas bubbles, the ceiling presents several sets of lava stalactites, which confer the cave signs of extreme beauty (Costa et al., 2008).

As the cave develops, largely, in an urban area, during the years the roof of the cave was being damaged, presenting at several places collapsed areas, some with large volumes of material that blocked the passage. In order to protect the cave as well as to ensure the safety of people that lives in buildings that are constructed above the cave, several geological and geotechnical studies were performed and various recommendations were made.

Figure 3. Aspect of the lava tunnel.

Figure 4. João do Rego Street cave (section III). Collapsed blocks from the cave's roof/walls in red.

Figure 5. Aspect of the João do Rego Street cave.

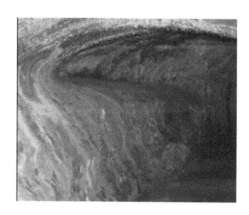

Figure 6. Aspect of the Lisboa Street cave.

Several interventions were performed at Section I at Paim Street cave, at Section II – Lisboa Street cave and Section III at João do Rego Street cave (Fragoso and Miranda 2003, Sousa and Oliveira 2004, Correia et al. 2010, Sousa et al. 2014, Malheiro et al. 2015, 2018).

3.2 *Extension of runway at Ponta Delgada airport*

An important study involved the analysis of the influence of natural cavities in the works to expand the runway at Ponta Delgada airport (Figure 2). The runway went from 1,700 m in length to 2,525 m. Laboratory tests were carried out on basaltic samples to determine the value of UCS (Uniaxial Compression Strength) and laboratory tests on fractures by using direct shear tests, with evaluation of the values of normal and tangential stiffness and values of cohesion and friction angle, peak and residual (Neves et al., 1987).

To assess safety, model calculations by boundary element method for various types of cavities were made, in an elongated diamond shape with a span of 10 m and in a circular shape with a 6 m diameter, in order to approximate the cavities that were evidenced during a visit made. They translate with some similarity the cavities detected in the Carvão cave. Figure 7 shows the natural cavities studied. In the elongated cavities, there are high compression stresses at the ends and tensile stresses at the top, which are limited to the upper part of the cavities. For circular cavities, there is a considerable reduction in the plasticized area.

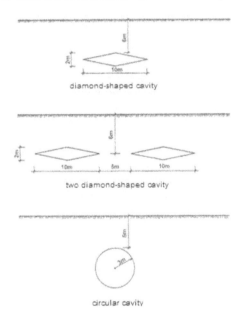

Figure 7. Natural cavities analyzed for the extension of the runway at Ponta Delgada airport.

Collapse analyzes of natural cavities were carried out using finite element models, considering various hypotheses of occurrence, in several possible locations. It was considered probable that the eventual collapse of the foundation would cause an effect of redistribution of tensions near the yield zone, so that the settlements at the level of the runway would be reduced (Neves et al., 1987).

4 FINAL CONSIDERATIONS

This paper deals with the stability and risk analysis for volcanic cavities with applications to caves, at Azores archipelago. The studies that have been done were mainly related to the stability problems that those caves presented. The main objectives of those studies were not only to ensure the safety of the visitors of those caves but also to preserve these natural monuments, already damaged in some sectors in the case of the Carvão cave system.

With this aim, recent legislation (cf. Regional Legislative Decree nr. 10/2019/A) advocates the preservation of such natural monuments and includes some measures, namely the prohibition of:

– New constructions or enlargement of the existing at a distance less than 10 m of the limits of the most important volcanic caves, except the ones for support for visitation or preservation of these caves.
– Execution of works of underground networks (water, sewer or telecommunications) above the delimitation of caves.
– Extraction of aggregates with the use of explosives in a distance less than 200 m of the limits of the caves.
– Extraction of aggregates in a distance less than 50 m of the limits of the caves.

This regional legislation also delimits a range of 5 m in which is prohibited the opening of new roads or widening the existing and the installation of underground electrical or telephone infrastructures. Finally, monitoring plans needs to be implemented for the mentioned caves, considering the monitoring activities as passive countmeasures for risk analysis.

ACKNOWLEDGEMENTS

This work was financially supported by: Base Funding - UIDB/04708/2020 and Programmatic Funding - UIDP/04708/2020 of the CONSTRUCT - Instituto de I&D em Estruturas e Construções - funded by national funds through the FCT/MCTES (PIDDAC). The authors would like to thank the experts of LREC and LNEC laboratories who worked in the studies related with the problems of the Carvão cave. Special thanks to Mr. M. Leal e Sousa for his help.

REFERENCES

Amigos dos Açores, 2008, Carvão cave. Geological heritage of São Miguel Island. Amigos dos Açores, Ponta Delgada, 43 (in Portuguese).

Correia, E., Oliveira, J.C.; Miranda, V., 2010, Monitoring system for Carvão cave. LREC Report 8/2010, Ponta Delgada, 24.

Costa, M.P., J.C. Nunes, J.P. Constância, P.A.V. Borges, P. Barcelos, F. Pereira, N. Farinha e J. Góis, 2008. Volcanic cavities in Azores. Amigos dos Açores/Os Montanheiros/GESPEA (Ed). 48 p. ISBN: 978-989-95432-2-5 (in Portuguese).

Ferreira, T., 2000, Characterization of the volcanic activity of the island of São Miguel (Azores): recent basaltic volcanism and degassing zones. Risk assessment. PhD thesis, University of Azores, Ponta Delgada, 233 (in Portuguese).

Pereira, F., J.C. Nunes, P.A.V. Borges, M.P. Costa, J.P. Constância, P.J.M. Barcelos, T. Braga, R. Gabriel, I.R. Amorim, E.A. Lima, P. Garcia e S. Medeiros, 2015. Catalog of Azores Volcanic Cavities (Lavic Caves, Ponds and Marine Erosion Caves). Os Montanheiros/GESPEA (Ed.). Edição online (in Portuguese).

Fragoso, M., Miranda, V., 2003, Structural safety assessment study at Santa Clara School EB/JI due to the static and seismic behavior of Algar do Carvão. LREC Report 101/2003, 44 (in Portuguese).

França, Z., Cruz, J.V., Nunes, J.C., Forjaz, V.H., 2005, Azores geology: a current perspective. Azores Vulcanological and Geothermal Observatory. Ponta Delgada, J. Açoreana, 11-140 (in Portuguese).

Gaspar, J.L., Queirós, G., Ferreira, T., 2007, Geological hazards at Azores region. Int. Workshop on Volcanic Rocks, Malheiro & Nunes Eds., Ponta Delgada, Azores, 11–18.

Malheiro, A., Nunes, J.C., 2007, Volcano stratigraphic for the Azores region: A contribution for the EC8 Regulations and the characterization volcanic rocks geomechanical behaviour. Int. Workshop on Volcanic Rocks, Malheiro & Nunes Eds., Ponta Delgada, Azores, 59–64.

Malheiro, A., Nunes, J.C., Sousa, L.R., Marques, F., 2015, Conservation of volcanic caves at Azores Islands, Portugal. Int. Symp. on Scientific Problems and Long-term Preservation of Large-scale Ancient Underground Engineering, Longyou, 1–8.

Malheiro, A., Amaral, P., Sousa, J.V., Miranda, V., Santos, A., 2018, Geomechanical characterization of volcanic rock materials from Azores and Madeira Archipelagos. 16th Portuguese Congress of Geotechnics, Ponta Delgada, 12 (in Portuguese).

Neves, E.M., Sousa, L.R., Rodrigues, L.F., Pinto, A.V., 1987, Landfill foundations for the extension of the runway at Ponta Delgada airport. 2nd Portuguese Geotechnical Congress, Lisbon, II.1–II.34 (in Portuguese).

Sousa, L.R., Malheiro, A. and Nunes, J. 2014, Methodology for the conservation of Carvão Volcanic Cave at São Miguel island, Azores. Symposium at Dunhuang, China, 81–87.

Sousa, L.R., Oliveira, M., 2004, Volcanic cave at João do Rego street. LNEC Report 142/04, Lisbon, 115 (in Portuguese).

Valadão, P., Gaspar J.L., Queiroz, G., Ferreira, T., 2002, Landslide's density map of S. Miguel Island, Azores archipelago. Natural Hazards and Earth System Sciences, 2, 51–56.

Mechanical behavior of volcanic rocks and soils

Rock Mechanics and Engineering Geology in Volcanic Fields – Ohta, Ito & Osada (eds)
© 2023 copyright the Author(s), ISBN 978-1-032-27657-1

Strength behaviour in monoaxial loading conditions in effusive rocks: The influence of porosity

Luca Verrucci* & Tatiana Rotonda
Sapienza Università di Roma, Italy,

Paolo Tommasi
Consiglio Nazionale delle Ricerche - IGAG, Roma, Italy

ABSTRACT: Relations between strength index properties from monoaxial tests (uniaxial compression, point load and indirect tensile tests) are extensively used in engineering practice and are object of suggestions by ISRM (1985). The results of an extensive test campaign, that investigated strength and static/dynamic deformability of effusive volcanic rocks under monoaxial loading conditions are presented. The tested lithotypes span a wide range of porosity (n = 1 - 50 %) and consist of an aphanitic groundmass with interspersed macropores and rare minute phenocrysts. The microstructure was observed through optical microscopy, scanning electron microscopy and 3D computed tomography images. The analysis of test results and literature data on similar lithotypes indicate that porosity controls quite strictly both the strength parameters and the ratios between them. The inadequacy of the ratios between point load and uniaxial/tensile strengths proposed in the suggested methods is apparent for $n >$ 10%. Some clues on the failure process were highlighted through theoretical and microstructural models in which the strengths obtained from different monoaxial tests are related to porosity.

Keywords: Porous rocks, Point load strength index, Uniaxial strength, Tensile strength, Effusive rocks

1 INTRODUCTION

The parameters measured through uniaxial compression test (UCT), point load test (PLT) and indirect brazilian tensile test (BTT) are used as index properties for rock characterization and classification in engineering practice. The uniaxial compressive strength and the point load strength index concur to the estimate of rock mass quality indexes (e.g. RMR, Q) widely used in tunneling. Tensile strength, even determined through indirect tests, is a basic parameter for predicting the TBM advance rate, as it is considered the main factor controlling rock breakage by cutters (Sapigni *et al.*, 2002; Yagiz, 2008).

Relations between these strength parameters have been widely recognized and it is common practice to estimate uniaxial compressive strength, UCS, and tensile strength, BTS, from the more easily measurable strength from PLT, $I_{S,50}$ (Broch & Franklin, 1972; Bieniawski, 1975; Wijk, 1980; Cargill & Shakoor, 1990). In this respect, ISRM still suggests to relate UCS and $I_{S,50}$ through a factor $K_{UCS} =$ UCS/$I_{S,50}$ varying between 20 and 25. It has been pointed out that a unique value of K_{UCS}, or variable within such a narrow range, is unrealistic and that K_{UCS} may vary widely depending on the rock

*Corresponding author: luca.verrucci@uniroma1.it

DOI: 10.1201/9781003293590-25

type (e.g. Read *et al.*, 1980; Forster, 1983; Abbs, 1985; Norbury, 1986). Moreover the K ratios (K_{UCS}, K_B = UCS / BTS and K_T = BTS / $I_{S,50}$) vary significantly even for the same lithotype. For instance Bowden *et al.* (1998) showed that for different rocks K_{UCS} increases with UCS. Textural factors, such as magnitude and type of porosity and/or size of crystals/clasts, influence the failure mechanism (see e.g. Liu *et al.*, 2005) and are likely to result in a variability of strength parameters, and in turn of the related strength ratios.

In this respect the analysis of experimental strength data of porous effusive volcanic rocks presented in this paper provides experimental evidence of this issue. All considered lithotypes are characterized by a common aphanitic groundmass with occasional small phenocrystals, in which macropores and microfissures are interspersed. Such a common microstructural features should diminish the strength variability caused by factors as crystal adhesion and crystal size, in order to highlight the influence of the porosity that is the most easy-to-determine quantitative parameter.

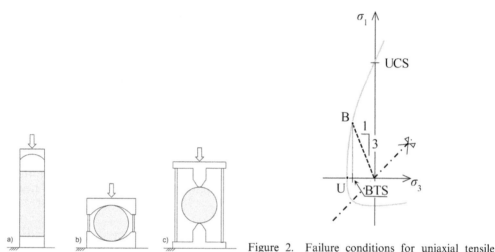

Figure 1. The monoaxial compressive tests. UCT (a), BT (b), PLT(c).

Figure 2. Failure conditions for uniaxial tensile (U) and indirect biaxial tensile (B) tests on a typical biaxial yield surface (grey line) for rocks.

2 THE MONOAXIAL COMPRESSIVE STRENGTH TESTS

The importance of UC, BT and PL tests is related to the possibility of acquiring fundamental properties of the rock through easy and fast test procedure. In this perspective the tests are described from the more strict and rigorous one to the faster and easier one.

2.1 *Uniaxial compression test*

The objective of the UC test (Figure 1a) is to apply in whole the specimen a state of homogeneous uniaxial compression represented by the principal stresses $\sigma_1 > 0$ and $\sigma_2 = \sigma_3 = 0$ (ISRM, 1979). However both the elasticity theory and the linear elastic fracture mechanics (LEFM) demonstrate that any geometric or physical dishomogeneity may produce tension cracks in brittle solids under such far-field stress state. The initial cracks are stable for perfectly uniaxial stress state and the compression load can grow up until a shear failure is attained (Horii & Nemat-Nasser, 1986).

Conversely, if the compression is carried out in presence of a far field radial tensile stress ($\sigma_2 = \sigma_3 < 0$), however small, after a certain length is reached the crack growth becomes unstable and failure occurs at a compression level significantly lower than the previous case, producing the so-called "axial splitting". It is worth noting that either significant irregularities in the material or a friction between the end faces and the load platens can generate a slight tensile confinement during the axial compression, thus inducing an axial splitting also in absence of an applied tensile confining stresses.

2.2 Indirect tensile (Brasilian) test

In the BT test the load is applied parallel to a diameter of a disc sample (ISRM, 1978, Part 2, Figure 1b). In this test the geometric and loading configuration is designed to generate an elastic stress field with a tensional stress on the axial plane passing through the compression line. This stress is independent on the Poisson ratio, v, when a plane stress condition is assured and its value, σ_{xx}, can be calculated through the following relation between compressive load, P, diameter, D, and thickness, t:

$$\sigma_{xx} = \frac{2P}{\pi D t} \tag{1}$$

The tensile strength, BTS, associated to the maximum measured load, P_{max}, is not reached in uniaxial conditions for the presence of a compressive stress in the orthogonal direction as high as three times the tensile stress in absolute value. However in such a biaxial stress field, rock failure is attained for a tensile strength near to the uniaxial tensile strength (Figure 2).

2.3 Point load test

The Point Load apparatus consists of a couple of spherical-truncated conical platens characterized by high rigidity and hardness (ISRM, 1985). The arrangement that gives the most repeatable results through this test is that with cylindrical specimens compressed along a diametral direction (Figure 1c). The maximum load at failure, P_{max}, is used to calculate the Point Load Index, $I_{s,50}$, corrected to taking into account the influence of the specimen dimension.

$$I_{s,50} = F \frac{P_{max}}{D^2} \tag{2}$$

where D is the specimen diameter and $F=(D/50)^{0.45}$ the correction factor (being D expressed in mm). Relationship (2) provides a conventional index that is proportional to the tensile strength through a costant factor. For specimens long enough ($L > D$) the tangential stress at failure, σ_θ, the longitudinal (axial) stress, σ_z, and radial stress (along the loaded diameter), σ_r, can be estimated through the solution of the elastic stress state (Serati et al., 2014):

$$\sigma_\theta = \sigma_z = \frac{\beta}{\pi} I_{s,50} \tag{3}$$

$$\sigma_r = -3.5\sigma_\theta \tag{4}$$

the coefficient β goes from 2.3 to 2.8 for Poisson coefficient ranging from 0.5 to 0.0 respectively.

3 TESTED MATERIALS

An extensive programme of laboratory mechanical tests was carried out on a series of volcanic rocks with the aim of including a range of porosity as wide as possible (Table 1). A minimum number of 20-30 specimens for each class of porosity, 5% wide, was tested at least in the range from 0 to 40% of porosity, that is the most common range for this lithotypes (Figure 3). Some further enrichment of the test campaign should be sought around $n = 25 - 30\%$. Excluding a small amount of microcracks and micro-

Table 1. Tested specimens and physical properties of tested lithotypes.

Origin	Lythotype	symb.	specimen number UCT	PLT	BT	ρ_{dry} Mg/m^3	ρ_s Mg/m^3	$\underline{n}$ %	n range %	V_P km/s	V_S km/s
Stromboli (IT)	basalt	SB	25	69	43	2.37	2.90	18.1	7.1 - 41.0	2.34	1.35
Vulcano (IT)	rhyolite	VR	19	6	5	2.40	2.51	4.40	3.2 – 7.0	5.04	2.91
Omo basin (ET)	massive basalt	EM	12	24	21	2.76	2.89	4.40	0.5 – 8.2	5.62	3.17
Jima (IT)	amygd. basalt	EA	3	6	7	2.52	2.79	9.80	3.4 – 13.1	3.75	1.94
Omo basin (ET)	rhyolite	ER	4	-	2	2.76	2.89	4.50	3.5 – 5.0	5.52	3.06
Bagnoregio (IT)	scoriaceous	BS	17	22	44	1.36	2.67	48.9	31.6 – 69.7	2.12	1.23
Bagnoregio (IT)	trachyte	BT	10	10	17	2.36	2.68	11.8	10.6 – 14.5	4.85	2.70
P. Borghese (IT)	leucitite	PL	9	10	12	2.37	2.91	18.8	15.0 – 27.4	4.91	2.82
Nemi (IT)	leucitite	NL	5	3	3	2.71	2.83	4.10	1.5 – 8.8	5.68	3.04
Nemi (IT)	scoriaceous	NS	11	8	9	1.67	2.85	41.5	34.2 – 47.0	3.07	1.49

Legend: ρ_{dry} = dry density, ρ_s = solid phase density, $\underline{n}$ = mean porosity, V_P, V_S = longitudinal and shear wave velocity

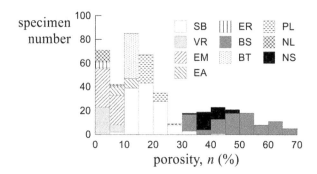

Figure 3. Porosity histograms of the tested specimens.

fissures, responsible for less than 5% of the total porosity, most of the voids of these rocks are macro-vesicles dispersed in the ground mass.

4 MICROSTRUCTURAL ANALYSES

The microstructure of the tested materials was observed in thin-section optic microscopy, scanning electron microscopy and optical incident-light stereo-microscopy. The Stromboli basalt (SB) was analyzed also through X-ray computer tomography (resolution 60 µm). A set of microphotographs taken from the SEM analyses are shown in Figure 4.

The less porous materials show quite regular surfaces of failure: they are rough without visible pores (Figure 4a). The solid mass is formed by small crystals of different minerals mixed with no (BT,PL,ER), modest (VR), and significant (SB,BS) amount of glass particles. When 5% of porosity is exceeded small pores can be observed (Figure 4b). This kind of porosity persists also in the mean porosity field (n about 20%) in association to large pores (vacuoles) that are interspersed in the ground mass (Figure 4c, 4d, 4e). Vacuoles can be spheroidal or ellipsoidal, as is expected in materials where dissolved gases nucleated in bubbles, but in other cases they are strongly flattened and distorted up to a nearly complete closure, thus evidencing that relevant flow deformations occurred before the complete solidification. At a detailed size analysis performed through the X-ray tomography, two different overlaid exponential distributions were evidenced for the small and the medium-large pore classes respectively.

At the highest porosity levels ($n > 30\%$) there is a change in the microstructure: the ground mass forms a lattice framework with thin septa (scoriaceous or pomiceous structure) (Figure 3f). However septa are thus far compounded largely by a crystalline solid.

Figure 4. SEM pictures of some tested lavas with different porosity. Ethiopia ryolite (a), Vulcano ryolite (b), Bagnoregio trachyte (c), Stromboli Basalt (d), Pantano leucitite (e), Bagnoregio scoriaceous (f).

5 LABORATORY CHARACTERIZATION

Cylindrical specimens 47 mm in diameter were cored in laboratory. End faces of the specimens for UC tests were ground up to a tolerance of 0.01 mm (ISRM, 1979). Geometry of the specimens was evaluated to determine both the exact volume and the length covered by the elastic wave paths (height of the specimens).

5.1 *Physical properties*

Solid matrix density, ρ_s, was determined through a helium pycnometer on the finely ground material (passing the No. 200 ASTM sieve). Measurements were performed on each block retrieved in situ and for at least two specimens for each lithotype. Total porosity was determined from solid matrix density and from dry density, ρ_{dry}, of cylindrical specimens oven-dried at 40°C. Longitudinal and shear wave velocities, V_P and V_S respectively, were measured in all the UCT specimens and most of the PLT specimens through a couple of ultrasonic pulse transducers (central frequency 1 MHz) subjected to a contact normal stress of about 2 MPa. In Table 1 the mean physical properties and the porosity ranges of each material are shown.

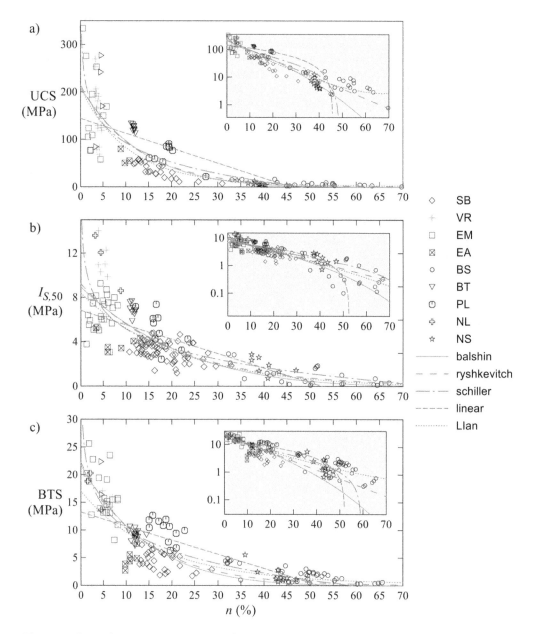

Figure 5. Strength parameters versus porosity of the tested specimens. Uniaxial compression strength (a), Point load strength (b), Indirect tensile strength (c). Log-log versions of the graphs are reported in the grey box for each plot.

5.2 Monoaxial strengths

The results of UC, PL and BT tests performed on the different lithotypes are reported in Figure 5. An inverse correlation of all the strength parameters with porosity over a wide range of values is apparent, although with a significant dispersion. The increasing dispersion of strengths for decreasing porosity is also apparent but it is somewhat illusory. If strength data are gathered in 10% wide porosity classes, the coefficient of variation of each class $CV=s/m$ (s and m being the standard deviation and the mean of the class respectively) does not diminish as porosity decreases.

Some of the most used empirical strength-porosity relations, derived from literature on rocks and other brittle materials, were used to interpolate the experimental data (Figure 5). In particular the power law relationship used the first time for metal-ceramic material by Balshin (1949), $\sigma = A(1\text{-}n)^B$, the exponential law used by Ryshkewitch (1953) for aluminum and zirconium oxides, $\sigma = Ae^{-B \cdot n}$, the logarithmic relation (Schiller, 1958), $\sigma = A \ln(B/n)$, a less usual composite relation obtained by Lian et al. (2011), $\sigma = (A(1\text{-}n)^B e^{-C \cdot n})^{0.5}$ and a linear regression are reported (being A, B, C generic parameters of regression). Regressions of the non-linear laws were obtained through the Levenberg-Marquardt iterative least squares algorithm.

A remarkable significance of all the regression model is shown in Figure 5 in the porosity range covered by the research. A simple analysis of variance shows that the mean square of each regression (MSR), representative of the strength variation accounted for by the postulated model, is always higher for at least two order of magnitude than the mean square error of data respect to the estimated values (MSE).

A comparison of the ratios between the MSE values of the five regressions through the F-test shows that relationships by Balshin (1949), Ryshkewitch (1953) and Schiller (1958) fit experimental data with similar efficacy. Among these three relationships the logarithmic regression by Schiller (1958) has slightly lower statistical robustness for the PL strength. The regression of Lian et al. (2011) is in general less suitable than the others. Finally, considering the three experimental data sets as a whole, the exponential regression (Ryshkewitch, 1953) exhibits the best adequacy in empirically reproducing the influence of porosity on the monoaxial strengths.

6 REGRESSIONS OF STRENGTH DATA AND STRENGTH RATIOS FROM LITERATURE AND EXPERIMENTAL RESULTS

Different authors observed and modelled the porosity influence on the strength of rocks. Relationships between uniaxial strength and porosity were studied by Kelsall et al. (1986) that analyzed the results on a vesicular basalt, highlighting the differences with respect to a fissured

Table 2. Classes of strength data from literature.

Reference	Lythotypes	symb.	number of classes			porosity range %
			UCT	PLT	BT	
Read et al., 1980	vesicular basalts	R80	3	3	-	9.0 – 25.0
Kahraman et al., 2005	Basalts	K05	1	1	-	8.0
Kurtulus et al., 2010	Andesites	K10	1	1	-	5.5
Guidicini et al., 1973	amygdaloid and low porosity basalts	G73	2	2	2	0.2 – 12.5
Karaman & Kesimal, 2015	basalts, andesites, dacites	K15	22	22	-	0.7 – 12.5
Tugrul & Gurpinar, 1997	low porosity basalts	T97	1	-	1	2.5
Kilic & Teymen, 2008	andesites, trachites	K08	3	2	3	4.4 – 12.4
Ersoy & Atici, 2007	andesites, dacites	E07	3	-	3	6.4 – 12.5
Hernández-Gutiérrez, 2014	Basalts	H14	4	4	4	3.0 – 16.8

basalt with a similar ground mass composition. Al Harthi *et al.* (1999) tested a vesicular basalt with a porosity varying quite continuously from 0 to 60% and obtaining correlations very similar to those described in the previous section. Palchik & Hatzor (2004) and Kahraman *et al.* (2005) already suggested that porosity controls not only the strength but also the mono-axial strength ratios for various rocks. In this context data collected from the experimental campaigns have been analyzed together with data of effusive rocks from the literature, with the exception of those retrieved from altered effusive rocks.

In order to perform a rational data analysis it is worth to observe that the strength of a single specimen is to be associated only to the porosity of that specimen. For this reason strength data were collected in classes with a restricted porosity range so that for each class the mean strength could be associated to the mean porosity and consistent strength ratios could be calculated between the mean strengths. Table 2 shows the literature data collected.

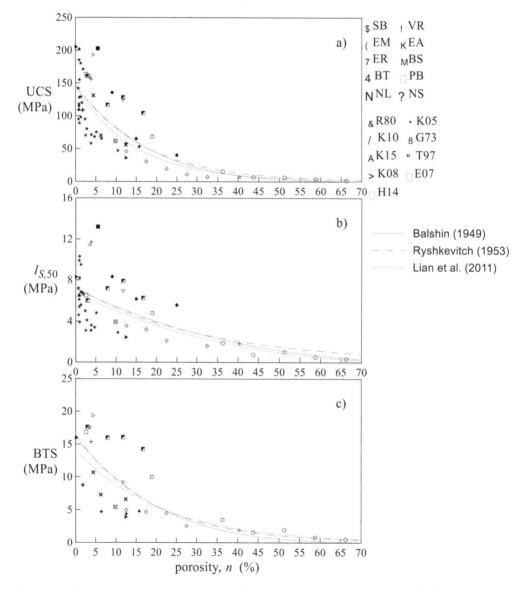

Figure 6. Strength parameters versus porosity from our experimentation (empty symbols) and literature (full and half full symbols) grouped in classes of restricted porosity. Uniaxial compression strength (a), Point load strength (b), Indirect tensile strength (c).

Among the data directly obtained in the experimental campaigns, Stromboli lava dataset was subdivided into 5 classes of porosity, 5% wide. A single class was identified for both the Vulcano rhyolite and each Ethiopian lava (massive basalt, amygdaloidal basalt and rhyolite, respectively). The Bagnoregio BL and BS lithotypes were grouped in one and six classes, respectively, 7.5% wide. Each of the Pantano and Nemi lavas were gathered in single classes.

The overall result in term of porosity classes confirm the observations made for the strengths of single specimens (Figure 6). Data from literature allow to cover quite uniformly the porosity range 1 - 40% and to strengthen the correlation of UCS also at very low porosities ($n < 2\%$). Nonetheless in the field of high porosity no further data are available in addition to the strengths experimentally measured on the BS lithotype.

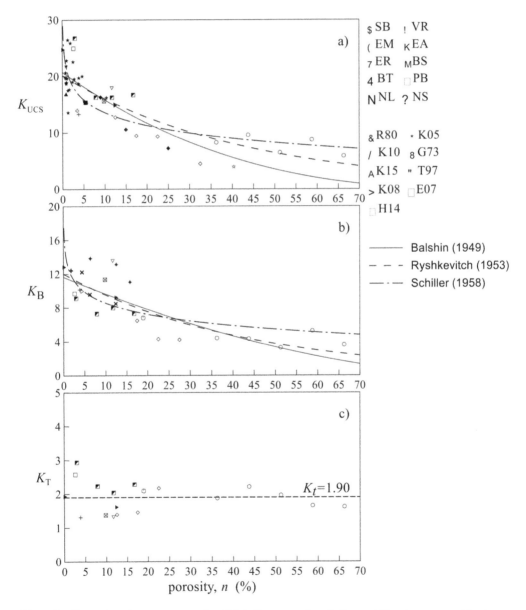

Figure 7.   Strength ratios versus porosity of data from experimentation and literature grouped in classes of porosity.

Only the power law, the exponential and the composite relation by Lian *et al.* (2011) were used to interpolate the new datasets obtained through the partition in porosity classes. The regression plots in Figure 6 show that, due to the augmented dispersion in strength data, regression including also literature data have lower significance. At the same time all the three mathematical expressions provide acceptable regressions with comparable mean square errors.

The strength ratios K_{UCS}, K_B and K_T were calculated for each class and plotted vs. porosity in Figure 7. The two most significant features of both K_{UCS} and K_B tendencies are the unequivocal decrease with porosity in the range 0 - 25% and a nearly constant trend for higher porosity. At this phase of the research the second aspect requires further investigation because of the reduced amount of data within the high porosity range. The power law and exponential function have similar adequacy in fitting both ratios; the latter, which provides higher ratios, has little more effectiveness. If the calculated ratios at high porosity should be confirmed, a different predictive law could be assumed accounting for of the constant trend of very porous rocks.

Conversely the K_T ratio is quite independent on porosity with a mean value of 1.90, as it is confirmed by the negligible MSR values with respect to the MSE of whichever regression. It is evident that porosity affects in the same way the two tests where failure occurs in tension (PLT and BTT) but influences differently the uniaxial compressive strength.

7 MODELS RELATING STRENGTHS AND STRENGTH RATIOS TO POROSITY

The porous and brittle materials are characterized by low deformability up to failure that is accompanied to an abrupt strength drop. Most rocks, concretes, ceramics, sintered metal powders, brittle biological tissues (e.g. bones) can be mentioned for example. In these cases the void volumetric ratio (i.e. the porosity) can be considered almost invariable during loading and porosity is a fundamental property controlling both stiffness and strength. A plenty of models relating strength to porosity have been developed in the last decades. Some of the most important ones are described in the following sections, highlighting the experimental agreement and the theoretical pursuits.

7.1 *The rule of mixture and the load bearing area models*

For heterogeneous materials the estimate of mechanical properties (as much as hydraulic and electromagnetic properties) can be approached through the choice of an appropriate mixture rule that in the most general case assumes the form:

$$\sigma^J = \sum_{i=1}^{N} \theta_i \, \sigma_i^{\,J} \tag{5}$$

being σ the generic equivalent property (the strength for example) of the material, σ_i and θ_i the properties and the volume fractions of the single component i (being $\Sigma\,\theta_i = 1$). In the simple case of a binary system with a "weak" component (i.e. having a negligible σ_i) this rule entails that the strength depends only on the strength of the solid phase, σ_0, and on its abundance in the mixture. The rule (5) becomes:

$$\sigma = \sigma_0 \theta_i^{\,1/J} \tag{6}$$

being $\theta = (1-n)$ the volumetric fraction of the solid phase.

In the most simple model $J = 1$ and the equivalent strength is determined as:

$$\sigma = \sigma_0(1 - n) \tag{7}$$

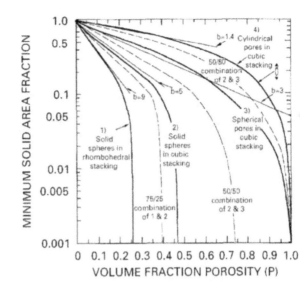

Figure 9. Models showing minimum solid area fraction (thick solid curves). Exponential approximations of the models for low porosity (thin solid lines) and respective exponent b values are shown. Models with combination of elementary assemblage modes are represented with dotted lines (from Rice, 1996).

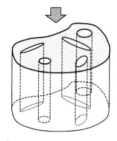

Figure 8. Condition of pores in cylindrical specimen parallel to the load direction (lotus-type porosity).

In general Equation (7) is not supported by the experimental observations, although for a particular combination of microstructure and loading conditions (anisotropic and translationally invariant) it can be theoretically justified and confirmed by experimental data: both the specimen and pores are to be cylinders parallel to the load direction (Figure 8). In these case the local stress σ_{loc} acting on the solid area is augmented with respect to the average stress just following the inverse of the solid area fraction, $(1-n)^{-1}$. The condition that failure occurs when solid material reaches its maximum strength ($\sigma_{loc} = \sigma_0$) entails equation (7).

For more generic microstructures the solid area fraction cannot be considered invariant with respect to the chosen cross section. Therefore researches on artificial porous materials (e.g. sintered metals and ceramic foams) developed models in which strength depends on the minimum value of the solid area over a generic load cross-section, the so-called minimum solid area fraction (Rice, 1996). This parameter can be easily calculated only for regular 3D assemblages of particles (or bubbles inside a solid). With these assumptions strength of a porous material should follow the same relationship between porosity and minimum solid area fraction of the specific modular framework to which the material conforms (Figure 8). Over the low porosity range all the geometrical models show a porosity dependence that can be approximate through an exponential function ($e^{-b \cdot n}$). In this way a theoretical justification for using the exponential relation by Ryshkewitch (1953) to interpolate data is strengthened.

The interpolation of the effusive rock strengths through the Ryshkewitch law provides an exponents b equal to about 6, 3, and 5 for UC, PL and BT strength, respectively (Figure 6), thus suggesting a structure to a cubic assembly of either spherical pores or solid spheres (Rice, 1996).

7.2 The stress concentration effects

In equation (6) the characteristic exponent J of the generalized mixture rule, the so-called microstructural coefficient (Ji et al., 2006), provides a scaling parameter for describing the effects of microstructure on the overall physical property. It is controlled mainly by the shape, size distribution, continuity and connectivity of the phases, and may reflects the formation processes of the materials. Porous materials have usually a $J < 1$ thus giving the typical upward concavity to the strength-porosity diagram.

In addition to modifications of the reacting solid area, the other critical effect of the voids is the stress concentration around them. The elastic theory allows to calculate the local tangential stress at the void boundary in uniaxial condition that is always amplified with respect to the average stress value with a "stress concentration factor", $m > 1$. A relationship between the coefficient J and the stress concentration factor m was experimentally evidenced in artificial materials with unequivocal microstructure (Ji et al., 2006). For lotus-type porous materials with long cylindrical pores oriented either parallel and perpendicular to the stress direction, $J = 1$ and $J = 1/3$ was obtained respectively. These results led several researcher to identify the exponent J of the mixture rule with the inverse of stress concentration factor for the respective void geometry, using the strength-porosity relationship:

$$\sigma = \sigma_0(1 - n)^m \tag{8}$$

These models were found to predict accurately the strength of porous ceramics containing dilute (i.e., non-interacting) pores, especially in uniaxial tensile test (Boccaccini, 1996). m factors for uniaxial load conditions applied on infinite elastic material with single (or dilute) cylindrical/spherical void are reported in Figure 10 (Timoshenko & Goodier, 1951; Obert & Duvall, 1967). Other estimates of m can be deduced from the elastic solution of the stress state around ellipsoidal cavities (Boccaccini, 1996; Sadowsky & Sternberg, 1947).

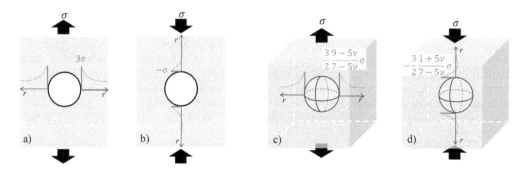

Figure 10.   Tangential stress along radial distance r (thin red lines) and maximum tensile stresses at the void boundary in elastic infinite medium subjected to uniaxial stress. (a) and (b): plane strain conditions with cylindrical void, (c) and (d) 3D condition with spherical void.

7.3   The linear elastic fracture mechanics of porous materials

In the field of LEFM the influence of porosity on the behaviour of brittle materials has been investigated first of all looking at the elastic modulus and fracture energy. From the results of these researches, Kendall et al. (1983) in a work on the characterization of porous cement pastes observed that, at least for low porosities, strength should depend mainly on the linear size of the defects rather than on their volume. In particular, following the Griffith (1921) energetic criterion for uniaxial tensile strength, crack-like pores some millimetres in length, though representing a small fraction of the total pore volume, influence the strength through their length $2c$:

$$\sigma_t = \sqrt{\frac{E \cdot R}{\pi c}} \tag{9}$$

where E and R are the Young elastic modulus and the fracture surface energy respectively. The hypothesis of the authors is that for increasing porosity the flaw size that is critical for the fracture process ($2c$) remains constant while the pore volume influence only E and R. The assumed relationships of these two parameters with porosity are a power law of the solid volume fraction, $E = E_0(1-n)^B$, and a negative exponential, $R = R_0 \, e^{-C \cdot n}$ respectively (Rice, 2000). Therefore Lian *et al.* (2011) derive the following equation to interpolate strength data of porous concrete:

$$\sigma_t = \sqrt{A \cdot (1-n)^B \cdot e^{-C \cdot n}} \tag{10}$$

Fracture mechanics suggests also some simple models for which the porosity influences the global failure mechanism, thus providing some insights on the evolution of both the compressive and tensile strength, and consequently of their ratio. For example the well-known model of Hoek and Bieniawski (1965), based on the basic hypotheses of Griffith (1921), assumes that in a rock with dispersed flat elliptical microcracks (as usual for hard brittle rocks) a fracture will initiate from the most critically oriented crack at the point of maximum tensile stress concentration (MTS criterion). In a biaxial plane stress condition this local criterion corresponds to a macroscopic one that can be expressed only as a function of the uniaxial tensile strength:

$$\sigma_3 = \sigma_t, \; \sigma_3 \leq -3\sigma_t \tag{11}$$

$$(\sigma_1 - \sigma_3)^2 = -8\sigma_t(\sigma_1 + \sigma_3), \; \sigma_3 \geq -3\sigma_t$$

The dependence on the micro-structural parameters (tensile strength of the solid and axis ratio of the elliptical crack) does not appear explicitly. The criterion is reported as a black solid line in the plane of the normalized principal stresses (Figure 11). The associated stress ratio, K_i, between uniaxial compressive and tensile stress at the onset of crack initiation is highlighted ($K_i = 8$). Further extensions of this theory in a complete triaxial condition, including a frictional coefficient on the faces of the closed cracks, do not change significantly the strength ratio at crack initiation, which can vary in the range 6 - 10. Furthermore if a similar procedure is repeated assuming a simpler micro-structural model like the spherical or cylindrical cavity, the K_i ratio can decrease to 3 or 4 respectively, as can be verified from the stress concentration factors of Figure 10.

Cai (2010) observed that these theories considerably underestimate the experimental K ratios (always higher than 10 for low porosity rocks) mainly because the crack initiation stress σ_{ci} corresponds to the stress at failure only in tensile conditions, when crack propagation is unstable and the failure is brittle and abrupt. On the contrary, crack initiation is stable in compression and the load has to grow to allow cracks to propagate and coalesce; therefore the macroscopic compressive strength UCS is always significantly higher than σ_{ci}. Cai (2010) proposed to express the failure strength ratio K between uniaxial compression strength and uniaxial tensile strength as the product of K_i to an amplifying factor:

$$K = \frac{UCS}{\sigma_t} = \frac{\sigma_{ci}}{\sigma_t} \cdot \frac{UCS}{\sigma_{ci}} = K_i \frac{UCS}{\sigma_{ci}} \tag{12}$$

To describe the σ_{ci}/UCS factor simple predictive models are missing, in spite the ratio was experimentally investigated in uniaxial compressive tests for several rocks. In particular σ_{ci} is usually identified as the stress level associated to the beginning of either a significant increase in the acoustic emission, or the decrease of elastic wave velocity. The σ_{ci}/UCS ratio is always in the range 0.4 - 1.0, where higher values are associated to more heterogeneous materials, larger grain/crystal size and higher porosity (Cai *et al.*, 2004). In particular, for rocks more and more porous the proliferation and propagation of microcracks yields to the macroscopic failure soon after the crack initiation and the σ_{ci}/UCS ratio is smaller and smaller. This

197

depends also on the reduced solid area that is to be damaged up to the formation of a global collapse mechanism, as confirmed by some models developed in the field of the damage mechanics (Cao *et al.*, 2010).

Therefore in the described framework the effect of increasing porosity can be associated to a decrease of K_i and a concomitant σ_{ci} UCS growth. The former is due to both the change of defect geometry and stress concentration effects whilst the latter is due to a shorter propagation path up to failure. Both effects drive towards a decreasing ratio between compressive and tensile strength (Figure 12).

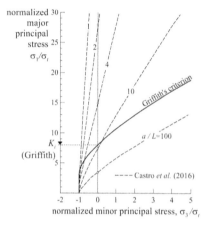 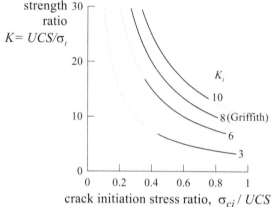

Figure 11. Biaxial normalized strength criterion at crack initiation as estimated by Hoek and Bieniawski (1965) through the Griffith's criterion and by Castro *et al.* (2016) (from Castro *et al.*, 2016, modified).

Figure 12. Relationship between strength ratio, crack initiation stress ratio for different values of the microstructural compression/tensile crack factor K_i (from Cai, 2010, modified).

Castro *et al.* (2016) developed also a model focused only on the evolution of the stress ratio at crack initiation, spreading K_i over a large interval. They used the theory of critical distance (Taylor, 2007). In this theory the material is characterized by a length parameter (the critical distance, L) and a strength parameter (the inherent strength, σ_0). The critical condition is assumed when at a distance $L/2$ from the defect (rather than just at the boundary as in the MTS criterion) the inherent strength is reached. In perfect analogy with the Hoek and Bieniawski (1965) method this local criterion can be transposed into a macroscopic one, using the elastic stress distribution impressed by a biaxial far field stress (σ_1, σ_3) around a sharp crack with length $2a$.

The biaxial criteria, normalized respect to the tensile strength, are shown as dotted curves in Figure 11 for a/L varying from 1 to 100. The variation of the strength ratio K_i is apparent. Castro *et al.* (2016) provide also a simplified expression for it:

$$K_i = \frac{49}{a/L} + 2.5 \tag{13}$$

For high a/L in this model the K_i ratio stabilizes at a not null value, similarly to the trend of the experimental K_{UCS} and K_B ratios this research, although for a direct comparison should be considered that the experimental ratios are calculated with respect to the indirect tensile strength rather than the uniaxial one. In particular for a K_B ratio variable from 12 to 4, as experimentally obtained, the associated micro-structural ratio a/L should vary from 5 to 33.

Using one of the regression relationships for the K_B ratio an equation linking the porosity to the micro-structural ratio a/L could be obtained. Although the critical distance L does not correspond to a precise microtextural character, studies on various materials and also specific studies on rocks (e.g. Cicero *et al.*, 2014) found a clear correlation with the magnitude of the grain size. Taylor (2007) observed that small values of L are simply related to the grain size while large values are associated to significant extension of damaged areas. In this respect an increase in porosity could contribute to augment the size, a, of the defects from which fractures begin (for the presence of bigger voids formed by coalesced pores). Also the critical distance L could decrease being related to the diminishing thickness of septa between pores, especially at high porosity.

8 CONCLUSIONS

The strengths in monoaxial compression are simple properties that are widely used to characterize rock materials. Literature offers a plenty of results from different monoaxial tests performed in parallel on twin specimens of the same lithotypes because the ratios between the strength properties are also diffuse parameters. The influence of the porosity on the mechanical behavior of rock materials can be usefully investigated inside the restricted field of monoaxial strengths also with reference to the significant but often underestimated variations of their ratios. The research focuses on the effusive rocks for their large porosity interval (2 - 70%) and to minimize the variability generated by other microstructural features.

Studied lavas have either a crystalline or partially glassy ground mass. Micropores feature rocks with a porosity less than 5%. Macropores, variable in shape and in size (up to several millimeters), feature the materials with 5% to 35% of porosity. The most porous rocks present a texture compounded by a continuous solid lattice which surrounds ubiquitous inter-connected voids.

The experimental campaign provided a relatively robust strength porosity relationship, at least in the more common porosity field ($n < 40\%$). The monoaxial strength parameters drop dramatically between 0 and 25% of porosity while further reduction develops with very low gradient. Data are effectively fit by the power law and the negative exponential law.

The complete data set obtained from both literature and experimental results was subdivided into narrow porosity classes, thus allowing to refer the mean strengths to the same mean porosity and to calculate realistic strength ratios. K ratios between compressive and tensile strength show a significant decrease (about the 50% of the maximum values) when porosity increases from 0 to 25% and stabilizes for higher porosity values. The ratio between the point load strength index and the indirect tensile strength is independent on porosity.

The presence of voids in rocks and other brittle materials affects first of all the solid area (which bears compression or tensile loads). Due to the complex three dimensional microstructure, this obvious effect does not accomplish in a simple manner, even for the elementary uniaxial stress state. Two factors in particular influence the strength: the casual disposition of both voids and solid areas, typical in natural materials, and the stress concentration around the void boundaries. The latter effect is quite sensitive respect to the shape and orientation of the pores and its complexity is augmented at high porosity when the interaction of near pores is significant.

The stress concentration effects, according to the Griffith theory, can give a simple explanation of the porosity effect on a strength criterion, thus providing indications about the compressive/tensile strength ratio. The most recent researches in this context suggest that when the porosity increases isometric macropores prevail microcrack pores and the stress of fracture initiation in compression diminishes with respect to the tensile strength. At the same time when porosity increases fracture can propagate throughout a reduced solid volume before a global failure is reached, thus reducing the ratio of the ultimate strength to the stress at fracture initiation.

ACKNOWLEDGEMENTS

Marco Albano (SEM laboratory, Department of Earth Science, Sapienza Università di Roma) is gratefully acknowledged for the SEM analyses. The help of dr. Remo Di Lorenzi, and dr. Francesca Maturani in carrying out mechanical tests is greatly appreciated.

REFERENCES

Abbs A.F., 1985. The use of the point load index in weak carbonate rocks. In *Strength Testing of Marine Sediments: Laboratory and In-Situ Measurements*, ed. R. Chaney and K. Demars (West Conshohocken, PA: ASTM International, 1985), 413–421, https://doi.org/10.1520/STP36349S

Al-Harthi A.A., Al-Amri R.M., Shehata W.M., 1999. The porosity and engineering properties of vesicular basalt in Saudi Arabia. *Eng. Geol.* 54, 313–320, https://doi.org/10.1016/S0013-7952(99)00050-2

Balshin M.Y., 1949. Relation of mechanical properties of powder metals and their porosity and the ultimate properties of porous metal-ceramic materials. *Doklady Akademii Nauk SSSR*,67, 5, 831–834 (in Russian).

Bieniawski Z.T. 1975. The point-load test in geotechnical practice. *Eng. Geol.*, 9, 1, 1–11, https://doi.org/10.1016/0013-7952(75)90024-1

Binal A., 2009. Prediction of mechanical properties of non-welded and moderately welded ignimbrite using physical properties, ultrasonic pulse velocity, and point load index tests. *Quarterly Journal of Engineering Geology and Hydrogeology*, 42, 107–122, https://doi.org/10.1144/1470-9326/07-067

Boccaccini A.R., Ondracek G., Mombello E., 1996. Determination of stress concentration factors in porous materials. *J. Mater. Sci. Lett.* 15, 534–536, https://doi.org/10.1007/BF00275423

Bowden A.J., Lamont-Black J., Ullyott S., 1998. Point load testing of weak rocks with particular reference to chalk. *Quarterly J. Engng. Geol.* 31, 95–103, https://doi.org/10.1144/GSL.QJEG.1998.031.P2.03

Broch E. & Franklin J.A., 1972. The point-load strength test. *Int. J. Rock Mech. Mining Sci.* 9(6): 669–676, https://doi.org/10.1016/0148-9062(72)90030-7

Cai M., (2010). Practical estimates of tensile strength and Hoek–Brown strength parameter m_i of brittle rocks. *Rock Mech Rock Eng*, 43, 167–184, https://doi.org/10.1007/s00603-009-0053-1

Cai M., Kaiser P.K., Tasaka Y., Maejima T., Morioka H., Minami M., 2004. Generalized crack initiation and crack damage stress thresholds of brittle rock masses near underground excavations. *Int J Rock Mech Min Sci* 41, 5, 833–847. https://doi.org/10.1016/j.ijrmms.2004.02.001

Cao, 2010. Statistical damage model with strain softening and hardening for rocks under the influence of voids and volume changes. Canadian Geotechnical Journal, 47 (8), 857–871, https://doi.org/10.1139/T09-148

Cargill J.S. & Shakoor A., 1990. Evaluation of empirical methods for measuring the uniaxial compressive strength of rock. *Int. J. Rock Mech. Mining Sci.* 27, 6, 495–503, https://doi.org/10.1016/0148-9062(90)91001-N

Castro J., Cicero S., Sagaseta C., 2016. A criterion for brittle failure of rocks using the theory of critical distances. *Rock Mech Rock Eng* 49, 63–77, https://doi.org/10.1007/s00603-015-0728-8

Cicero S., García T., Castro J., Madrazo V., Andrés D., 2014. Analysis of notch effect on the fracture behaviour of granite and limestone: An approach from the Theory of Critical Distances. *Engineering Geology*, 177, 1–9, https://doi.org/10.1016/j.enggeo.2014.05.004

Ersoy A., Atici U., 2007. Correlation of P and S-Waves with Cutting Specific Energy and Dominant Properties of Volcanic and Carbonate Rocks. *Rock Mech. Rock Engng.* 40, 5, 491–504, https://doi.org/10.1007/s00603-006-0111-x

Forster I.R., 1983. The influence of core sample geometry on the axial point-load test. *Int. J. Rock Mech. Min. Sci & Geomech. Abstr.* 20, 291–295, https://doi.org/10.1016/0148-9062(83)90599-5

Griffith A.A., (1921). The phenomena of rupture and flow in solids. *Philos. Trans. R. Soc. Lond. A* 221, 163–198. https://doi.org/10.1098/rsta.1921.0006

Guidicini C., Nieble C.M., de Cornides A.T., 1973. Analysis of point load test as a method for preliminary geotechnical classification of rocks. *Bull. Int. Ass. Eng. Geol.*, 7, 37–52, https://doi.org/10.1007/BF02635318

Hernández-Gutiérrez L.E., 2014. Caracterización geomecánica de las rocas volcánicas de las islas Canarias. *PhD Thesis.* Universidad de La Laguna, Tenerife

Hoek E., Bieniawski Z.T., (1965). Brittle fracture propagation in rock under compression. *Int J Fract* 1, 137–155, https://doi.org/10.1007/BF00186851

Horii H & Nemat-Nasser S., 1986, Brittle failure in compression: splitting, faulting and brittle-ductile transition. *Phil. Trans. R. Soc. Lond. A*, 319, 337–374, https://doi.org/10.1098/rsta.1986.0101

ISRM, 1978. Suggested methods for determining tensile strength of rock materials

ISRM, 1979. Suggested methods for determining the uniaxial compressive strength and deformability of rock materials

ISRM, 1985. Suggested method for determining Point Load Strength

Ji S., Gu Q., Xia B., 2006. Porosity dependence of mechanical properties of solid materials. *J Mater Sci* 41, 1757–1768, https://doi.org/10.1007/s10853-006-2871-9

Kahraman S., Gunaydin O., Fener M., 2005. The effect of porosity on the relation between uniaxial compressive strength and point load index. *Int. J. Rock Mech. Mining Sci.* 42, 4, 584–589, https://doi.org/10.1016/j.ijrmms.2005.02.004

Karaman K., Kesimal A., 2015. Evaluation of influence of porosity on the engineering properties of volcanic rocks from the eastern Black Sea region: NE Turckey. *Arab J Geosci* 8, 557–564. https://doi.org/10.1007/s12517-013-1217-6

Kelsall P.C., Watters R., Franzone J.G., 1986. Engineering characterization of fissured, weathered dolerite and vesicular basalt. *Proc. 27th U.S. Symposium on Rock Mechanics (USRMS)*, Tuscaloosa, Alabama, June, 77–84

Kendall K., Howard A.J., Birchall J.D., Pratt P.L., Proctor B.A., Jefferis S.A., 1983. The relation between porosity, microstructure and strength, and the approach to advanced cement-based materials. *Phil. Trans. R. Soc. Lond. A* 310, 139–153, https://doi.org/10.1098/rsta.1983.0073

Kılıç A., Teymen A., 2008. Determination of mechanical properties of rocks using simple methods. *Bull Eng Geol Environ* 67, 237, https://doi.org/10.1007/s10064-008-0128-3

Kurtulus C., Irmak T.S., Sertcelik I., 2010. Physical and mechanical properties of Gokceada Imbros (NE Aegean Sea) Island andesites. *Bull. Eng. Geol. Environ.* 69, 321–324, https://doi.org/10.1007/s10064-010-0270-6

Lian C., Zhuge Y., Beecham S., 2011. The relationship between porosity and strength for porous concrete. *Construction and Building Materials* 25, 4294–4298, https://doi.org/10.1016/j.conbuildmat.2011.05.005

Liu H., Kou S., Lindqvist P., Lindqvist J.E., Åkesson U., 2005. Microscope rock texture characterization and simulation of rock aggregate properties. SGU project 60-1362/2004 - Project Report 1, Geological Survey of Sweden, 89

Norbury D.R., 1986. The point load test. In: Site Investigation Practice. A. B. Hawkins (ed.) Assessing BS 5930, Geolog-ical Society, London, Engineering Geology Special Publications, 2, 325–329, ISBN 10: 0903317346

Obert L., Duvall W.I., 1967. Rock Mechanics and the Design of Structures in Rock. Wiley & Sons, New York NY. ISBN 13: 9780471652359

Pabst W., Gregorova E., 2015. Minimum solid area models for the effective properties of porous materials - A refutation. *Ceramics-Silikáty*, 59, 3, 244–249

Palchik V. & Hatzor Y.H., 2004. The influence of porosity on tensile and compressive strength of porous chalks. *Rock Mech. and Rock Eng.* 37, 4, 331–341, https://doi.org/10.1007/s00603-003-0020-1

Perucho A., 2016. Mechanical behavior of volcanic rocks. In *Volcanic Rocks and Soils*, Rotonda *et al.* (eds), Taylor & Francis Group, London, ISBN 978-1-138-02886-9

Read J.R.L., Thornton P.N., Regan W.M., 1980. A rational approach to the point load test. *Proc. Australian-New Zealand Geomech. Conf.*, Wellington, May 12-16, 1980, Wellington, N.Z., Institution of Professional Engineers New Zealand, 2, 35–39

Rice R.W., 1996. Evaluation and extension of physical property-porosity models based on minimum solid area. *Journal of Materials Science* 31, 102–118, https://doi.org/10.1007/BF00355133

Rice R.W., 2000. Mechanical properties of ceramics and composites. New York, CRC Press, Taylor & Francis Group. ISBN 9780367447373.

Ryshkewitch R., 1953. Compression strength of porous sintered alumina and zirconia - 9th Communication to Ceramography. *Journal of the American Ceramic Society*, 36, 2, 65–68. https://doi.org/10.1111/j.1151-2916.1953.tb12837.x

Sadowsky M.A. & Sternberg E., 1947. Stress concentration around an ellipsoidal cavity in an infinite body under arbitrary plane stress perpendicular to the axis of revolution of cavity. *J. Appl. Mech.*, 14, 3, A191–A201, https://doi.org/10.1115/1.4009702

Sapigni M., Berti M., Bethaz E., Busillo A., Cardone G., 2002. TBM performance estimation using rock mass clas-sifications. *Int. J. Rock Mech. Mining Sci.*, 39, 6, 771–788, https://doi.org/10.1016/S1365-1609(02)00069-2

Schiller K.K., 1958. Mechanical properties of non-metallic brittle materials. Ed. W.H. Walton (Butterworth)

Serati M., Alehossein H., Williams D.J., 2014. 3D elastic solutions for laterally loaded discs: generalised brazilian and point load tests. *Rock. Mech. Rock Eng.*, 47, 1087–1101, https://doi.org/10.1007/s00603-013-0449-9

Taylor D., 2007. The Theory of Critical Distances. Elsevier Science Ltd, ISBN 9780080444789

Timoshenko S., Goodier J.N., 1951. Theory of Elasticity. McGraw-Hill Book Company, Inc., New York

Tuğrul A., Gürpinar O., 1997. A proposed weathering classification for basalts and their engineering properties (Turckey). *Bulletin of the International Association of Engineering Geology*, 55, 139–149, https://doi.org/10.1007/BF02635416

Wijk, G. 1980. The point load test for the tensile strength of rock. *ASTM Geotech. Test. J.* 3, 2, 49–54, https://doi.org/10.1520/GTJ10902J

Yagiz S., 2008. Utilizing rock mass properties for predicting TBM performance in hard rock condition. *Tunneling Undergr. Space Tech.* 23, 3, 326–339, https://doi.org/10.1016/j.tust.2007.04.011

Rock Mechanics and Engineering Geology in Volcanic Fields – Ohta, Ito & Osada (eds)
© 2023 copyright the Author(s), ISBN 978-1-032-27657-1

Seismic behavior of Neapolitan pyroclastic soils

Anna d'Onofrio* & Giorgio Andrea Alleanza
Università degli Studi di Napoli Federico II, Napoli, Italy

ABSTRACT: Pyroclastic soil deposits, which cover a wide area of the Neapolitan metropolitan district (South Italy), have been often involved in catastrophic slope instability phenomena related to meteoric extreme events. For this reason, research attention on these soils was mainly devoted to the assessment of hydrogeological hazard of the Neapolitan area.

Nevertheless, the Neapolitan area is also characterised by a relevant seismic hazard, and extended coastal areas, where reworked pyroclastic soils are present at shallow depth in saturated conditions, are characterised by moderate to high susceptibility to liquefaction. Notwithstanding this, very few data are available on the cyclic behaviour of these soils.

In this study, the results of cyclic laboratory tests (torsional shear and cyclic triaxial tests) carried out on pyroclastic soil samples retrieved at different sites in the Neapolitan metropolitan district were collected and synthesised to analyse the main factors affecting their cyclic behaviour. The results indicate that due to the complex surface shape, the interlocking effect appears to be significant, resulting in higher linear threshold strain and initial damping ratio as well as higher liquefaction resistance when compared with hard-grained sands. Furthermore, the small and medium strain behaviour of pyroclastic soils seems to be ruled mainly by fine content. Samples exceeding 50% of fine content are plastic and 50% of fine content seems to be a threshold value beyond which soil behaviour is controlled by the fine matrix rather than by the granular skeleton. The peculiar cyclic behaviour of these soils significantly affects the seismic site response of the pyroclastic deposits, as shown by the results of site response analyses carried out on representative soil columns.

Keywords: Cyclic Behaviour, Seismic Site Response Analysis, Liquefaction

1 INTRODUCTION

Pyroclastic deposits are generated by the explosive activity of volcanoes. Their name derives from the Greek "πυρ" fire and "κλαστοσ" broken pieces. Accordingly, to the mode of transportation and deposition they can be classified as fall, flow, and surge deposits.

Fall deposits fall out from the eruption column (they are then characterised by aerial deposition); flow deposits are generated by fast moving currents of hot gases and volcanic matter; surge deposits are derived from flow, but they are characterised by a higher proportion of gas to rock.

The Neapolitan metropolitan area is mainly covered by the products of two main volcanic edifices: Campi Flegrei and Somma Vesuvio. The volcanic products of Campi Flegrei cover the urban area of Napoli and its neighbors, where the typical stratigraphic sequence includes the Neapolitan yellow tuff covered by the same formation in its unlithified condition, which in turn underlies the pyroclastic products of the volcanic

*Corresponding author: donofrio@unina.it

DOI: 10.1201/9781003293590-26

activity of Campi Flegrei younger than 15 ky (IPD- intra caldera deposits of Campi Flegrei (Picarelli et al., 2006).

The geological map of the Phlegraean fields (Figure 1a) shows that the coastal area of the city of Naples is mainly covered by reworked flow deposits (light green areas) and similar reworked flow deposits also settle the western part of the Phlegraean Island of Ischia, (extended green area reported in the map of Figure 1b). Pyroclastic fall deposits (dark green) instead cover a great part of the hillside of the city. In this case, they are generally in partially saturated condition and very often, during extreme meteoric events, they have been involved in catastrophic slope instability phenomena. For this reason, research on these pyroclastic soils has been mainly devoted to the assessment of the hydrogeological hazard of the Neapolitan area (Evangelista et al., 2002; Picarelli et al., 2006).

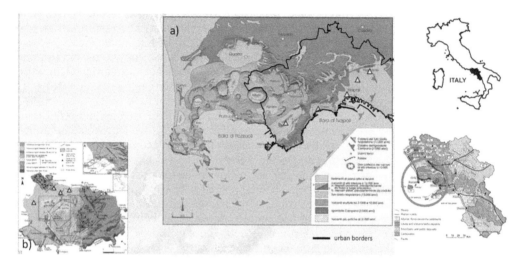

Figure 1. Geological map of the study area a) Phlegrean fields (modified after Orsi et al., 1996) and b) Ischia island (modified after de Vita et al., 2010).

These studies highlighted the significant role of grain size, void ratio, degree of saturation and structure on the mechanical behaviour of pyroclastic covers and their relationship with mechanisms of generation and formation of the deposits.

Figure 2 reports data about the main physical properties measured on more than 600 samples retrieved in different sites around Naples (Evangelista et al., 2002) showing that these deposits are characterised by quite a low particle weight ranging between 2.44 and 2.51 (in terms of specific gravity, Gs) and a porosity varying between 50% and 60% with a resulting low unit weight not exceeding 18.5 kN/m^3. The same soils are also characterised by narrow values of shear strength, with a friction angle ranging between 32 and 35 degrees (Figure 2b) but highly sensitive to the change in the degree of saturation (Picarelli et al., 2006). Nevertheless, the city of Naples is also featured by a relevant seismic hazard, as highlighted by the map reported in Figure 3 where the probabilistic hazard distribution of peak ground accelerations with a 10% of exceedance in 50 years is shown.

The expected peak ground acceleration values corresponding to a return period of 475 years vary between 0.15 and 0.175g, within the Neapolitan district. Furthermore, extended coastal areas of the urban territory are covered by reworked pyroclastic soils in saturated conditions, as shown in the simplified zonation map of Figure 4a, where red areas are classified as potentially susceptible to liquefaction, based on information about the grain size of the shallow covers and groundwater level. Santucci de Magistris

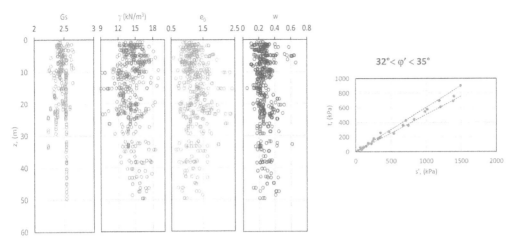

Figure 2. Main physical and mechanical properties of Neapolitan pyroclastic soils (modified after Evangelista et al., 2002).

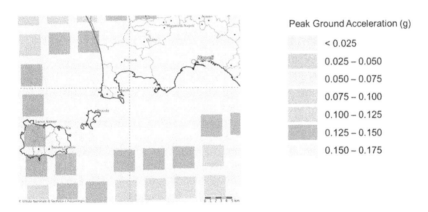

Figure 3. Hazard map of the study area: PGA distribution with 10% of exceeding in 50 years (modified after Stucchi et al., 2011).

and Evangelista (2007), Evangelista and Santucci de Magistris (2011) and Evangelista (2006; 2011) carried out a study at the urban scale, based on the results of more than 110 boreholes, as many static penetrometer tests and a small number of geophysical tests. The final product of this study is the map of potential damage expressed in terms of liquefaction potential, IL (Iwasaki et al., 1978) reported in Figure 4b, that confirms the moderate to high susceptibility to liquefaction of the urban coastal area. Notwithstanding these evidences, few studies have been carried out aimed at the characterization of the cyclic and dynamic behaviour of these soil deposits.

In this study laboratory experimental results of cyclic tests carried out on samples retrieved in various sites within the Neapolitan metropolitan area and in Ischia Island (where the volcanic edifice of Monte Epomeo produced similar pyroclastic deposits), have been collected, analysed and interpreted with the aim of detecting the main factors affecting the cyclic behaviour of these soils and their role in the seismic site response.

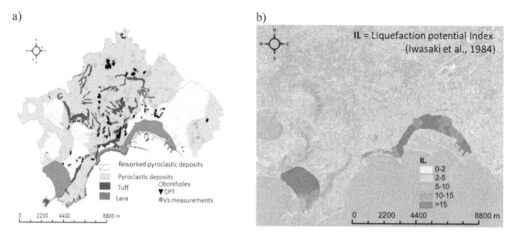

a)

b)

Figure 4. a) Simplified geological map of Naples (in red pyroclastic soil deposits susceptible to liquefaction) b) liquefaction susceptibility map expressed in terms of IL (modified after Evangelista, 2011).

2 TESTED MATERIALS AND EXPERIMENTAL PROGRAM

Data were collected from nine sampling sites where seismic site response or liquefaction susceptibility analyses have been carried out. Sampling site location is indicated on the map of Figure 1 with yellow triangles: all the samples (35) were retrieved in reworked fall deposits located within the Neapolitan urban area and on the west side of Ischia Island. A synthesis of the main physical properties of the tested samples is reported in Table 1 where the sampling depth and the confining stress applied in the laboratory tests are also indicated. The data confirm the peculiar physical properties of pyroclastic soils already observed in previous studies (Evangelista et al., 2002; Picarelli et al., 2006): a rather low particle weight (Gs ranges between 2.55 and 2.77), a mean value of porosity of about 50%, thus a low unit weight (16.5kN/m^3 on average).

The grain size distribution of tested samples is reported in Figure 5. The material ranges from clayey silt to well-graded silty sand. The coefficient of uniformity, $C_u = D_{60}/D_{10}$ is widely variable ranging between 2 and 50. The fine content varies between 20% and 100%, but only samples with a fine content exceeding 50% are plastic, these latter are characterized by a clay fraction higher than 10%. Particles are mainly constituted by pumices, lapilli and scoriae, whereas the fine fraction is made up of ash. This latter can be more or less argillified due to the suffered transport and leaching processes. Their structure, analysed through Scanning Electron Microscopy (SEM), is characterised on average by a non-homogeneous distribution of inter-particle porosity, an interlocked packaging with a high variability of grain shapes and an irregular grain surface (Licata et al, 2018).

The cyclic behaviour of the collected soil samples was analysed by resonant column and torsional shear tests most of which were carried out at the University of Naples by mean of a prototype torsional shear device (THOR) (d'Onofrio et al., 1999). Some undrained cyclic triaxial tests were also carried out on a limited number of samples to evaluate their liquefaction susceptibility. A summary of the experimental programme is reported in Table 1.

Table 1. Main physical properties of tested soil samples.

Site	BH	Sample	depth (m)	e_0	γ kN/m^3	w	Gs	σ' (kPa) RC	CTS	CTX
Arco Mirelli	S1	C3	14.30	0.970	17.04	0.35	2.43	100		
Arco Mirelli	S2	C6	20.30	1.340	15.79	0.49	2.43	150x		
Cardito	S6	1	3.25	0.851	16.00	0.29	2.69	50		
Cardito	S3	1	6.25	0.838	17.05	0.32	2.69	110		
Carmine	S1	F1	16.40	0.719	20.11	0.23	2.65	110		
Carmine	S1	B1	29.70	1.180	15.80	0.40	2.41	170		
Carmine	S1	D1	30.40	1.310	15.82	0.47	2.48	170		
Carmine	S1	H1	12.00	0.989	18.58	0.31	2.77	70		
Bagnoli	S2	C1	4.25	1.058	16.54	0.26	2.66	50		
Bagnoli	S2	C3	9.00	1.244	15.94	0.33	2.46	100		
Bagnoli	S1	C3	58.75	1.193	16.72	0.29	2.66	300		
Bagnoli	S1	C2	50.25	1.037	16.41	0.25	2.47	375		
Casamicciola	DH2	C2	10.75	0.883	16.43	0.30	2.38		186	
Casamicciola	DH2	C1	5.75	1.138	14.14	0.30	2.43		74	
Casamicciola	DH1	C2	14.75	1.255	13.92	0.24	2.39		210	
Poggioreale	S1	a	27.30	1.089	16.57	0.40	2.43		210	
Poggioreale	S2	a	27.30	1.086	16.65	0.40	2.42	100		
Poggioreale	S1	B2	24.30	1.041	16.95	0.42	2.43	100		
Poggioreale	S1	4	42.30	1.519	15.57	0.58	2.43	100		
Poggioreale	S1	5	51.25	1.984	14.06	0.70	2.37	100		
Poggioreale	S1	a	27.35	1.115	16.38	0.39	2.43	100		
Poggioreale	S2	a		1.041	16.79	0.43	2.43	250		
Poggioreale	S1	4	42.30	1.519	15.63	0.53	2.43	250		
Poggioreale	S1	5	51.25	1.984	14.28	0.70	2.37	250		
Poggioreale	S2	3	32.70	0.953	16.60	0.33	2.43	250		
Forio	S1	C2	14.50	0.535	19.30	0.21	2.46		137	
Forio	S1	C4	48.75	0.593	20.25	0.23	2.67		515	
Forio	S1	C3	34.25	1.130	16.48	0.44	2.42		282	
Lacco Ameno	S1	C1	4.50	1.247	16.29	0.51	2.43		89	
Lacco Ameno	S1	C1	4.50	1.204	16.29	0.51	2.43		139	
Lacco Ameno	S1	C4	31.75	0.877	17.62	0.32	2.55		270	
Lacco Ameno	S1	C5	47.40	0.729	17.84	0.30	2.43		360	
Lacco Ameno	S1	C3	23.50	1.246	15.94	0.52	2.35		181	
San Pasquale	S1	C1	18.00	1.224	16.25	0.25	2.46	89		100
San Pasquale	S1	C1	18.00	1.224	16.25	0.29	2.46	74		100

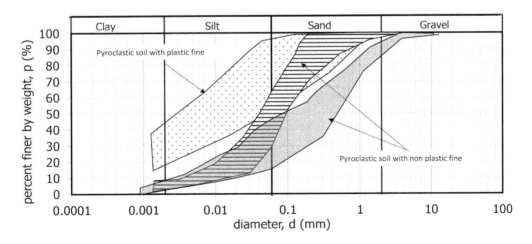

Figure 5. Grain size distribution of tested soil samples.

3 EXPERIMENTAL RESULTS

The results of the cyclic and dynamic torsional shear tests are shown in Figure 6, in terms of initial shear modulus (Figure 6a) and damping ratio (Figure 6b) as a function of the mean effective stress applied during the tests. The data are plotted with different symbols according to the plasticity of the fine. In the same figure, the initial shear modulus and damping data measured by Senetakis et al. (2012) on fine quartz and volcanic sands are reported for comparison.

The initial shear modulus measured on samples with nonplastic fine (open squares) is clearly higher than that measured on soil with plastic fine (full dots), but significantly lower than that measured on quartz sand (open dots). The non-plastic fine seems to slightly affect the initial shear modulus of the pyroclastic silty sand that behaves as the volcanic fine sand (open triangles), while it seems to strongly influence the initial damping ratio. As a matter of fact, both samples with plastic and nonplastic fine exhibit a higher damping ratio if compared to that measured on volcanic and quartz clean sands (Senetakis et al., 2012), and the plasticity of fine further enhance the dissipative properties of the pyroclastic sandy matrix.

The small strain shear modulus and damping ratio of pyroclastic soils with and without plastic fine content are correlated to the mean effective stress, p' through the general forms of Equations 1 and 2, respectively.

$$G_0 = S \cdot \left(\frac{p'}{p_r}\right)^n \tag{1}$$

$$D_0 = Z \left(\frac{p'}{p'_r}\right)^d \tag{2}$$

In the proposed relationships S and Z represent the initial shear modulus and damping ratio, respectively measured at a reference pressure equal to 1 kPa, whereas the exponents n and d express the sensitivity of the initial shear modulus and damping ratio to the current stress state.

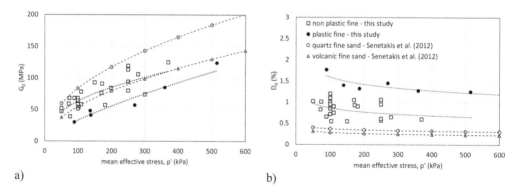

Figure 6. Experimental results: a) initial shear modulus and b) initial damping ratio vs. mean effective stress.

The values of the parameters obtained by the regression analysis are reported in Table 2 and compared with those measured on quartz sand and volcanic sand by Senetakis et al (2012).

Table 2.   Initial shear modulus and damping parameters.

	S	n	Z	d
Quartz sand (Senetakis et al., 2012)	8.74	0.49	0.62	-0.11
Volcanic clean sand (Senetakis et al., 2012)	4.53	0.54	0.52	-0.13
Volcanic silty sand (non-plastic fine)	8.70	0.43	1.89	-0.16
Volcanic silty sand (plastic fine)	1.07	0.74	3.40	-0.16

The data summarized in Table 2 show that:

– the percentage of fine content plays a key role on the small strain behaviour of volcanic soils. As a matter of fact, the nonplastic fine content (lower than 50%) increases the stiffness index, S and reduces the sensibility to stress state variation with respect to that measured on clean volcanic sand. On the other hand, the hard siliceous particles of quartz sand influence both the high values of the stiffness index and its sensibility to stress variation.
– the plasticity of fines (fine content higher than 50%) significantly reduces the initial shear stiffness of the volcanic soils if compared to that of pyroclastic soil with a fine percentage lower than 50% (non-plastic).
– fine content also influences the initial damping ratio of volcanic soils that is significantly higher than that of clean volcanic sand and quartz sand. The plasticity of the fine further enhances the dissipative properties of the pyroclastic soils tested.

The non-linear behaviour of the pyroclastic soils was analysed carrying out resonant column and torsional shear tests at increasing strain level.

Figure 7 shows the experimental results in terms of normalized shear modulus, $G/G_0(\gamma)$ curves (Figure 7a) and $D(\gamma)$ curves (Figure 7b). In these figures, the upper and lower bounds of the G/G_0 (γ) and $D(\gamma)$ curves proposed by Seed and Idriss (1970) for quartz granular soils are also drawn with a solid black line for comparison.

Data of the pyroclastic silty sand all plot outside of the literature curves with a remarkably more linear response of the volcanic soil samples in comparison to the corresponding behavior of quartzitic soils; this trend was very often observed in the case of volcanic soils (Senetakis, 2011; Orense et al., 2012). Beyond the linear threshold strain, γ_l, friction at particle contacts and particle rearrangement dominate the behavior of the granular assembly. These mechanisms contribute to the dissipation of energy, the reduction of shear stiffness and the increase in damping.

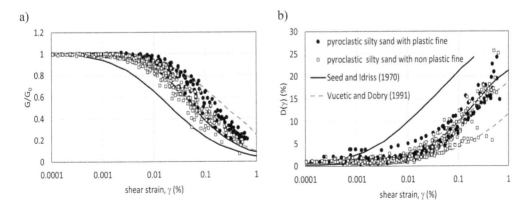

Figure 7.   Experimental data: a) normalized shear modulus and b) damping ratio vs. shear strain.

For the pyroclastic soils, the linear threshold (the shear strain corresponding to $G/G_0 = 0.95$) is shifted to higher strain levels in comparison to the quartz sand, and the volumetric threshold (the shear strain corresponding to $G/G_0 = 0.5$) as well. The more linear response of the pyroclastic soil can be related to interlocked packaging of their particles and to the highly irregular and jagged particles surface. Both these peculiar characteristics, often detected in the case of volcanic soil particles, can be responsible of micro-mechanisms that prevent significant particle rearrangement and thus the reduction of shear stiffness and the increase in damping.

The linear range is further enhanced by the plastic fine as shown by the black dots in Figure 7, where the literature curves proposed by Vucetic and Dobry (1991) for fine-grained plastic soil, for PI values of 15% and 50% are also plotted for comparison (dashed gray line). The linear threshold of silty sand with plastic fine exceeds that predicted by the literature curves confirming that plasticity of fine further increases the already more linear response of volcanic soils. As a matter of fact, the linear threshold increases with the applied confining stress ranging between 0.004% and 0.007% for silty sand with nonplastic fine and between 0.005% and 0.015% for silty sand with plastic fine.

Literature relationships neither catch the dissipative behaviour of the tested pyroclastic soils. Seed and Idriss (1970) curves predict a higher damping ratio at medium to high strain levels if compared to the experimental data. On the other hand, the relationships by Vucetic and Dobry (1991) underestimate the damping ratio with respect to that measured on pyroclastic silty sand with plastic fine.

Since the literature curves available are not able to describe the observed behaviour, the experimental data were interpreted adopting the modified hyperbolic model proposed by Dardendeli (2001), calibrating the model parameters on the experimental data set. The analytical expression of the modified hyperbolic model is given in Equation (3), where γ_{ref} is the reference strain that expresses the "linearity" and corresponds to a G/G_0 $(\gamma) = 0.5$; parameter (γ), namely the coefficient of curvature, expresses the overall slope of the G/G_0 (γ) curve. Parameter (α) was introduced by Darendeli for a best-fit of the hyperbolic model to the experimental data since, initially, in the hyperbolic model proposed by Vucetic and Dobry (1991), the only fitting parameter was the reference strain.

$$\frac{G}{G_0} = \frac{1}{1 + \left(\frac{\gamma}{\gamma_{ref}}\right)^\alpha} \tag{3}$$

The reference strain is given analytically through the empirical form of Eq. (4), where $A_{\gamma ref}$ is a constant that primarily depends on soil type and grading, β expresses the dependency of the non-linear response on soil plasticity, and n_γ is an exponent that synthesises the effect of confining stress on the reference strain.

$$\gamma_{ref} = \left(A_{\gamma_{ref}} + \beta PI\right)\sigma'^{n_\gamma} \tag{4}$$

The damping ratio curve, $D(\gamma)$, is modeled by modifying the Masing criteria adding an initial viscous damping, D_0 and introducing a function F aimed at reducing the damping ratio predicted by the Masing criteria at large strain levels. The equation expressing $D(\gamma)$ is shown herein together with that of the reducing function F.

$$D - D_0 = F\left(c_1 D_{Masing} + c_2 D^2_{Masing} + c_3 D^3_{Masing}\right) \tag{5}$$

$$F = f\left(\frac{G}{G_0}\right)^{0.1} \tag{6}$$

The initial damping dependency on plasticity index and stress state is taken into account as follows:

$$D_0 = (A_D + \delta PI)\sigma'^d \tag{7}$$

Using the modified hyperbolic model, the reference strain and coefficient of curvature of all specimens at variable mean effective stresses were evaluated by fitting the analytical model of Eq. (3) to the experimental data. The fitting parameters $A_{\gamma ref}$ and n_γ were then determined by plotting the reference strains against σ'. The results of the fitting procedure are shown in Figure 8. The mean values of the fitting parameters are summarized in Table 3 and compared with those obtained by Senetakis et al. (2013) on volcanic clean sand and quartz sand. The higher (α) values of the pyroclastic soils imply that the rate of G/G_0 decrease, below a threshold strain that corresponds to $G/G_0=0.50$, is less pronounced in comparison to the quartz soils. Regarding the constant $A_{\gamma ref}$, it can be seen in Table 3 that the pyroclastic soils exhibit systematically higher values in comparison to the quartz ones, a trend which supports the observations that the volcanic soils exhibit a significantly more linear shape of the $G/G_0(\gamma)$ curves and that in these soils there is a shift of the elastic and volumetric thresholds to higher strains.

a)   b)

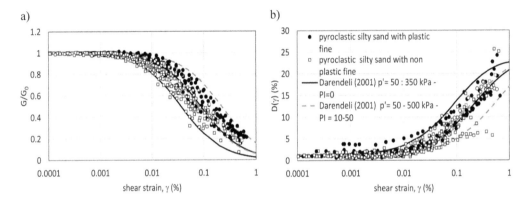

Figure 8. Experimental data fitted by Darendeli (2001) relationship with ad hoc calibrated parameters.

Table 3. Calibrated parameters of Darendeli (2001) relationships.

Darendeli (2001) parameters	α	$A_{\gamma ref}$	β	n	A_D	δ	d	f	c_1	c_2	c_3
Volcanic soil (Senetakis et al. 2013)	1.12	0.036		0.25	0.52		-0.13				
Quartzitic soil (Senetakis et al. 2013)	1	0.00527		0.47	0.62		-0.19				
Volcanic silty and sandy soils (this study)	1.12	0.030	0.001	0.18	1.50	0.013	-0.07	0.46	0.93	0.012	-0.0001

4 SEISMIC RESPONSE OF PYROCLASTIC SOILS

4.1 The geotechnical model

The influence of the above-described cyclic behaviour of Neapolitan pyroclastic soils on the seismic site response was here analysed comparing the numerical response of a well-characterized soil profile in the city of Naples with that obtained replacing an intermediate layer of pyroclastic soil of about 15 m, with a layer of Toyoura sand (TS). One-dimensional non-linear seismic response analyses in effective stress were carried out on both profiles.

The selected soil stratigraphy, drawn in Figure 8a, characterizes the site of a metro station located in Napoli city center, few meters from the shoreline (Licata, 2015; Fabozzi et al., 2017). Like in most of the seaside areas of Naples, the subsoil is characterized by alternating layers of volcanic and seashore deposits, lying above a sub-horizontal tufaceous bedrock. Starting from the ground surface, below a negligible thickness of the man-made ground cover, about 17 m of seashore sands (SS) lie upon 27 m of volcanic products, divided into 14 m of pyroclastic silty sand (Pyr), 10 m of pozzolana (Poz) and 3 m of the fractured facies of Neapolitan Yellow Tuff (Tuff1). The top of the bedrock (Tuff2) was intercepted at the depth of approximately 44 m. The groundwater level was detected within 1 m below the ground level.

The G_0 (z) profile based on the shear wave velocity measured in a Cross hole test is shown in Figure 9a. In the same figure, the $G_0(z)$ profiles adopted in the numerical analyses on the two different profiles are also drawn. It is worth noting that, in case of the Profile 1, for the pyroclastic soil layer (Pyr) the relationship here proposed to describe the shear modulus dependency of pyroclastic soils with nonplastic fine on current stress level was adopted. On the other hand, for Profile 2 the shear modulus variation with depth of Toyoura sand is modeled adopting the $G_o(p')$ relationship proposed by Asadi et al. (2020). To complete the subsoil model assumed in the dynamic analyses, Figure 9b reports the curves describing the variation of the equivalent shear stiffness, G, normalized with respect to the small-strain value, G_0, and of the damping ratio, D, with the shear strain amplitude, γ. For the pyroclastic layer (Pyr), the curves were determined by performing Resonant Column – Torsional Shear (RCTS) tests on undisturbed block samples taken at a depth of 18 m (Licata, 2015).

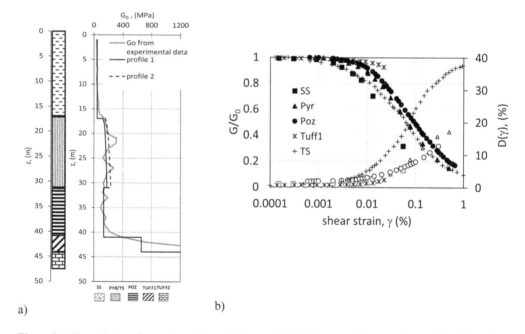

a)   b)

Figure 9.   Numerical analyses: a) stratigraphic log and initial shear modulus profiles and b) G/G₀ and D (γ) curves. (Marine sand, SS, Piroclastic silty sand, Pyr, Pozzolana, Poz, Toyoura sand, TS, fractured Yellow Tuff, Tuff1).

For the (SS), (Poz), and (Tuff1) formations, the curves were inferred from laboratory tests carried out on samples of the same lithotypes taken in other sites (Vinale, 1988). In the same figure, the shear modulus decay and damping curves of Toyoura sand (Iwasaki et al., 1978) are also reported.

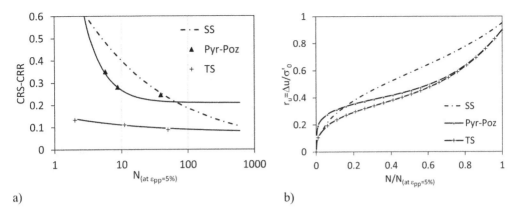

a)

b)

Figure 10. a) Cyclic resistance curves and b) pore pressure ratio curves.

Finally, the pore pressure buildup was modeled based on the results of undrained cyclic triaxial tests. Cyclic triaxial tests were carried out on samples retrieved from the saturated pyroclastic silty sand (Pyr), since the site was classified as susceptible to lique-faction. The block samples were retrieved at a depth of 18 m and are characterized by 30% of nonplastic fine and a relative density of 40%. All the undrained cyclic triaxial tests were carried out on undisturbed soil samples at a confining stress of 100 kPa. The experimental results are reported in Figure 10a in terms of cyclic resistance curve, obtained evaluating the number of cycles at liquefaction as that corresponding to the attainment of 5% double amplitude axial strain. Figure 10b shows the results in terms of pore pressure ratio, $r_u = \Delta u/\sigma'_0$, as a function of the number of cycles. In the same fig-ures, the cyclic response relevant to the marine sand (SS) and Toyoura sand (TS) is also reported. The cyclic resistance and pore pressure buildup curves of (SS) were obtained applying the procedure suggested by Chiaradonna et al. (2020), based on the results of the CPT test, while literature data by Orense and Pender (2016) are considered for Toyoura sand.

Coupled, non-linear analyses in the time domain were carried out using the code SCOSSA (Tropeano et al, 2016, 2019). In the code, the soil profile is modelled as a system of consistent lumped masses, connected by viscous dampers and springs with hysteretic behaviour. The non-linear soil behaviour is modelled by a hysteretic stress-strain relationship based on the modified MKZ model (Matasovic and Vucetic, 1993). corrected through to the procedure for strength compatibility proposed by Gingery and Elgamal (2013), to better predict the soil behaviour at large strains up to failure (Chiaradonna et al., 2018). Generalised Masing rules are adopted in the code to reproduce loading-unloading reloading cycles (Phillips and Hashash, 2009).

The build-up of excess pore water pressure in the saturated soils was simulated through the simplified model proposed by Chiaradonna et al. (2018). The model formu-lation is based on the definition of a 'damage parameter' which permits to synthetically express the seismic demand relevant to an irregular time-history of shear stress and to compare it to the cyclic strength of liquefiable soils, as measured in stress-controlled cyclic laboratory tests. As a result, the pore pressure ratio (i.e. the excess pore pressure normalized by the initial effective confining stress) can be straightforwardly computed as a function of the cumulated damage. The model adopted describes the cyclic resistance curve as follows:

$$\frac{(CSR - CSR_t)}{(CSR_r - CSR_t)} = \left(\frac{N_r}{N_l}\right)^{\frac{1}{\alpha}} \qquad (8)$$

213

where (N_r, CSR_r) are the coordinate of a reference point in the (N, CSR) plane, CSR_t represents a threshold value of CSR below which pore water pressure build-up does not occur and α describes the steepness of the cyclic resistance curve in a bi-logarithmic plot.

Moreover, the pore pressure model expresses the pore pressure ratio, r_u, as a function of the normalized number of cycles, N/N_L:

$$r_u = a \left(\frac{N}{N_l} \right)^b + c \left(\frac{N}{N_l} \right)^d \tag{9}$$

where *a, b, c* and *d* are curve-fitting parameters. The parameters of the pore pressure model calibrated for the different soil layers are reported in Table 4, and the related curves are drawn in Figure 10.

Table 4. Parameters of the pore pressure model.

Soil layer	CSR_t	α	a	b	c	d
SS	0.007	2.930	0.784	0.456	0.166	4
Pyr	0.21	0.667	0.487	0.202	0.413	4
TS	0.075	2.70	0.501	0.339	0.399	4

4.2 *The input motion*

The selection of seismic input used in the dynamic analyses was carried out following the probabilistic hazard approach suggested by the national building code. The reference value of peak ground acceleration (PGA = 0.168g) was evaluated referring to the life safety limit state, corresponding to a return period of 475 years. For the selected site, the reference acceleration for a 10% of exceedance within 50 years, is equal to 0.168g. The selection of seismic input used in the dynamic analyses was performed through *REXEL* code (Iervolino et al., 2009) by which 6 spectrum-compatible accelerograms belonging to a range of $5.5 < M_w < 7$ and $0 < R_{epi} < 30$km, in accordance with de-aggregation analysis, were scaled at $a_g = 0.168$g.

The main features of the selected accelerograms are reported in Table 5; all of them were recorded by European seismic stations on a stiff rock outcrop, i.e. a soil class 'A' according to the National Technical Code, to minimize the influence of local conditions of subsoil on amplitude, frequency content and duration of signals. The response spectra of the selected time histories and the compatibility with the target spectrum are illustrated in Figure 11.

Table 5. Parameters of the input motions.

Earthquake	Date	Mw	Fault Mechanism	R_{epi} [km]	PGA [g]	Site class
Bingol	01/05/2003	6.3	strike slip	14	0.292	A
Friuli	06/05/1976	6.5	thrust	23	0.356	A
Montenegro	15/04/1979	6.9	thrust	21	0.181	A
Umbria Marche	06/10/1997	5.5	normal	5	0.187	A
South Iceland	21/06/2000	6.4	strike slip	5	0.837	A
Val Comino	11/05/1984	5.5	normal	17.4	0.109	A

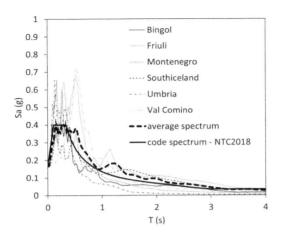

Figure 11. Acceleration spectra of the selected input motions.

4.3   *The analyses results*

Figure 12a-b-c-d and 12 e-f-g-h report the results of the 1D effective stress analyses carried out along the two soil profiles, in terms of maximum acceleration, shear strain, shear stress and pore pressure ratio vs. depth, obtained for the six input motions together with the average profiles. As shown in Figure 12h, liquefaction always occurs in the layer of Toyoura sand, often involving the whole layer. On the other hand, in the pyroclastic layer, no evidence of liquefaction was detected whatever the input is, as shown in Figure 12d.

The energy dissipated in the Toyoura sand layer prevents high pore pressure buildup in the overlaying layer of marine sand, with a pore pressure ratio, r_u, not exceeding 0.4 at the surface, whereas along the pyroclastic soil profile a continuously increasing pore pressure ratio is observed, in some case inducing the liquefaction of the first few meters of marine sand. Right above the spikes in the acceleration profiles (Figure 12a-e), the results show a significant attenuation of motion approaching the surface; this effect can be attributed to the degradation of stiffness and the increase of damping of the soil layer in which pore pressure build-up, which works like a kind of seismic isolation device, by limiting the acceleration. To better highlight this effect, Figure 13 reports the analyses results in terms of acceleration time histories referred to the reference input motion of South Iceland (Figure 13e) compared to those computed in the Toyoura sand layer and in the Pyroclastic layer (at 27.5 m from the ground level) and to those computed at the surface. It can be observed that as the earthquake waves propagate through the liquefiable layers, the acceleration undergoes different degrees of de-amplification accompanied by filtering of the high-frequency components. Those phenomena are associated with the pore pressure buildup in the intermediate layers, as can be observed from the time history of pore pressure ratio at the same depth, drawn in Figure 14 together with the time history of shear stress. Every time the shear stress overcomes the threshold value, $\tau_t = CSRt·\sigma'_v$ (black dashed line in Figure 14) defined by the pore pressure model adopted in the analyses, an excess pore pressure increment is generated. The threshold shear stress in the case of the pyroclastic layer is more than two times that of Toyoura sand, in which pore pressure starts to cumulate after the first 12 seconds and soon reaches the value of the effective stress at that depth. In the case of the pyroclastic soil layer, the pore pressure starts to increase at about the same time but, due to the higher value of threshold stress, it does not exceed 33% of the existing vertical effective stress at that depth.

To further verify the effect of pore pressure build-up in (Pyr) and (TS) soil layers, the shear stress– shear strain cycles computed at a depth of 27.5 m are plotted in Figure 15. Toyoura sand (black dashed line) undergoes stiffness degradation followed by the

215

development of a large deformation after only a few cyclic load applications. These large deformations and the reduction in stiffness of the grounds are indications of the occurrence of soil liquefaction. On the other hand, the pyroclastic soil (black solid line) exhibits limited degradation and deformations due to the reduced amount of pore pressure build-up.

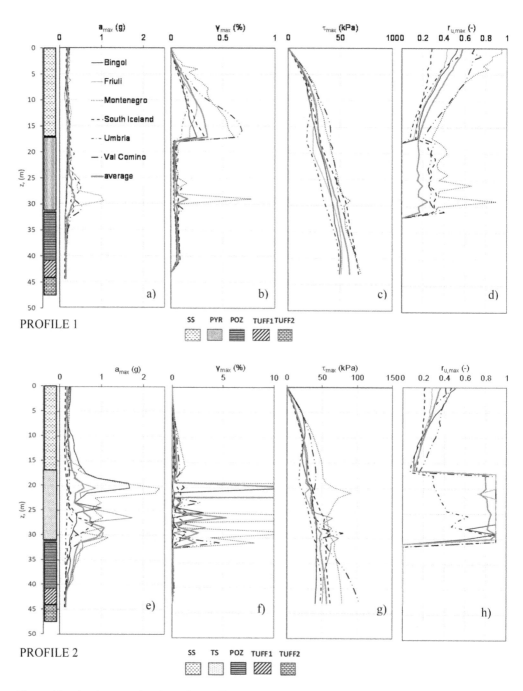

Figure 12.   Analyses results along the profile 1 and 2: a) and e) maximum acceleration, b) and f) maximum shear strains, c) and g) maximum shear stresses, d) and h) pore pressure ratio build-up.

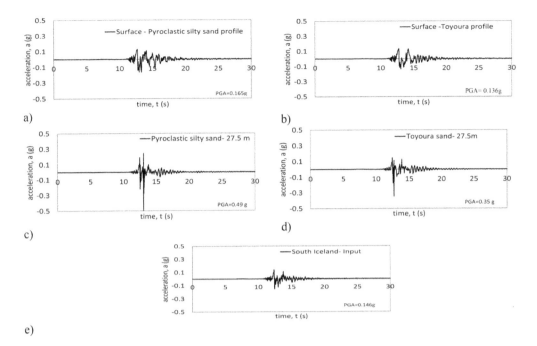

a)

b)

c)

d)

e)

Figure 13.   Acceleration time histories at surface of the a) profile 1 and b) profile2, at 27.5 m of c) profile 1 and d) profile 2, related to e) South Iceland input motion.

Finally, the post-cyclic consolidation settlement along the two profiles has been computed as the result of the dissipation of excess pore pressure, following the approach suggested by Chiaradonna et al. (2019), as follows:

$$\sum\nolimits_{i=1}^{N} \frac{\Delta\sigma'i}{E_{oed\ i}}$$ (10)

where $\Delta\sigma'i$ is the increment of vertical effective stress arising from the dissipation of the excess pore pressure, Δu; $E_{oed,i}$ is the constrained modulus and Δz_i is the thickness of the generic layer i that results from the discretization of the soil profile into N layers.

Assuming that the solid skeleton may be treated as an elastic medium characterized by Young's modulus, E', a shear modulus, G and Poisson's ratio, v', it follows that:

$$E_{oed} = \frac{2G(1-\nu')}{(1-2\nu')}$$ (11)

A Poisson's ratio equal to 0.3, and the degraded shear modulus G_{degr}, i.e. the mobilized modulus at the end of shaking along with the whole profile, were assumed to compute the constrained modulus.

An overall settlement of about 0.5 cm was computed in the case of the pyroclastic silty sand profile whereas it is more than double in the case of the Toyoura sand profile (1.7 cm).

Based on the results presented, it can be said that the pyroclastic soil would perform better than Toyoura sand when subjected to the specified earthquake excitation. This is consistent with the findings from cyclic triaxial tests that the liquefaction strength of pyroclastic soil is more than 2 times higher than that of Toyoura sand.

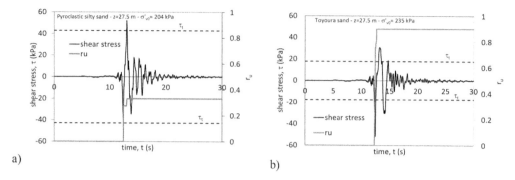

a)

b)

Figure 14. Time histories of shear stress and pore pressure build-up computed at 27.5 m of a) profile1 and b) profile 2 due to South-Iceland input motion.

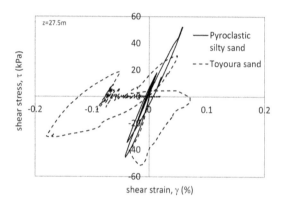

Figure 15. Stress-strain cycles computed at a depth of 27.5m within the pyroclastic soil layer and the Toyoura sand layer.

5 CONCLUSIONS

The paper contains a synthesis of a wide experimental campaign on pyroclastic soils retrieved at different sites in the Neapolitan metropolitan district aimed at investigating the main factors affecting their cyclic behaviour.

The results clearly highlight that these volcanic soils are characterized by higher linear threshold strain and initial damping ratio as well as higher liquefaction resistance when compared with hard-grained sands. This behavior can be justified at a particle scale looking at their irregular surface texture that induces significant interlocking effects.

The small and medium strain behaviour of pyroclastic soils seems to be ruled mainly by fine content. Samples exceeding 50% of fine content are plastic and 50% of fine content seems to be a threshold value beyond which soil behaviour is controlled by the fine matrix rather than by the granular skeleton.

The tested pyroclastic soils are also characterised by a liquefaction resistance higher than that of quarzitic sands due to both interlocked packaging of particles and high irregular and jagged particles surface.

To evaluate how the presence of a pyroclastic soil layer can affect the seismic ground response, the numerical response of a well characterized soil profile in the city of Napoli was compared to that obtained replacing an intermediate layer of pyroclastic soil of about 15 m, with a layer of Toyoura sand. One-dimensional non-linear seismic response analyses in effective stress were carried out on both profiles. The results clearly show that the pyroclastic soil layer exhibits lower excess pore water pressure build up when compared with that Toyoura sand subjected to the same level of input motion. Pore pressure build-up in the intermediate layer of both profiles implies a drastic reduction of the amplification at the surface as well as filtering of the high frequency content of the input. As a matter of fact, the degradation of stiffness and the increase of damping of the soil layer in which pore pressure build-up, make it like a kind of seismic isolation device. This effect has been observed on the response at the surface of both profiles, thus enhancing the beneficial role that can have a liquefiable layer at intermediate depth along with a soil profile. To further confirm this overall effect on the ground response at surface it has been necessary to evaluate the post cyclic settlement exhibited by the two soil columns subjected to the same input motion. To this aim a simplified procedure suggested by Chiaradonna et al. (2019) has been adopted, assuming that the settlement can be computed as the results of pore pressure dissipation in oedometric conditions. As it was expected the vertical settlement computed at the surface of the Toyoura sand profile is more than two times that computed in the case of the pyroclastic soil profile, due to the different amount of pore pressure build up caused by shaking. Therefore, it can be concluded that the presence of an intermediate liquifiable layer can be beneficial if it does not exhibit excessive pore pressure build up, as happened in the case of pyroclastic soil layer.

ACKNOWLEDGEMENTS

The collection and interpretation of the experimental data presented in this work were carried out as part of WP16.1 "Seismic response analysis and liquefaction" in the framework of the research programme funded by Italian Civil Protection through the ReLUIS Consortium (DPC-ReLuis 2019-2021).

REFERENCES

Asadi, M.B., Asadi, M.S., Orense, R.P. and Pender M.J., 2020, Small-Strain Stiffness of Natural Pumiceous Sand *J. Geotech. Geoenviron. Eng.*, 146(6), 06020006, DOI: 10.1061/(ASCE)GT.1943-5606.0002256.

Chiaradonna, A, Tropeano, G, d'Onofrio, A and Silvestri, F., 2018, Development of a simplifed model for pore water pressure build-up induced by cyclic loading. *Bulletin of Earthquake Engineering*, 16(9), 3627–3652. https://doi. org/10.1007/s10518-018-0354-4

Chiaradonna, A., d'Onofrio, A. and Bilotta, E., 2019, Assessment of post-liquefaction consolidation settlement - *Bulletin of Earthquake Engineering*, 17(11), 5825–5848, DOI: 10.1007/s10518-019-00695-0

Chiaradonna, A., Flora A., d'Onofrio, A., Bilotta, E. (2020) A pore water pressure model calibration based on in-situ test results *Soils and Foundations* Volume 60, Issue 2, April 2020, Pages 327–341 https://doi.org/10.1016/j.sandf.2019.12.010

Darendeli, M.B., 2001, Development of a new family of normalized modulus reduction and material damping curves. *Ph.D. Dissertation*, The University of Texas at Austin, 362 pp.

de Vita, S., Sansivero, F., Orsi, G., Marotta, E. and Piochi, M., 2010, Volcanological and structural evolution of the Ischia resurgent caldera (Italy) over the past 10 ka. *Geological Society of America, Special Paper*, 464, 193–239

d'Onofrio, A., Silvestri, F. and Vinale F., 1999, A new torsional shear device. *Geotechnical Testing Journal*, 22(2), 101–111

Evangelista, A., Nicotera, M.V. and Scotto di Santolo, A., 2002, Caratteristiche geotecniche dei terreni piroclastici della città di Napoli. *Proc. 21ˢᵗ Italian Geotechnical. Conference*, L'Aquila 11-14 September 2006, 45–52 (in Italian)

Evangelista, L., 2006. Suscettibilità alla liquefazione dinamica dei depositi sabbiosi in falda della città di Napoli. *Master thesis*, University of Napoli Federico II (in Italian).

Evangelista, L., 2011. A critical review of the MASW technique for site investigation in geotechnical engineering. *Ph.D. dissertation*, University of Napoli Federico II.

Evangelista, L. and Santucci de Magistris, F., 2011, Upgrading the simplified assessment of the liquefaction susceptibility for the city of naples, Italy. *Proc. 5th International Conference on Earthquake Geotechnical Engineering (Santiago, Chile)*

Fabozzi, S., Licata, V., Autuori, S., Bilotta, E., Russo, G. and Silvestri, F., 2017, Prediction of the seismic behavior of an underground railway station and a tunnel in Napoli (Italy). *Underground Space*, 2 (2), 88–105

Gingery, J.R. and Elgamal, A., 2013, Shear stress-strain curves based on the G/Gmax logic: a procedure for strength compatibility. *The 2nd IACGE International conference on Geotechnical and Earthquake Engineering*, Chengdu, China.

Hardin, B.O. and Drnevich, V.P., 1972, Shear modulus and damping in soils: design equations and curves. *Journal of Soil Mechanics and Foundations Division, ASCE*, 98(7), 667–691.

Iervolino, I., Galasso, C. and Cosenza, E., 2009, REXEL: computer aided record selection for code-based seismic structural analysis. *Bulletin of Earthquake Engineering*, 8, 339–362.

Iwasaki, T., Tatsuoka, F., Tokida, K. and Yasuda S., 1978, A practical method for assessing soil lique-faction potential based on case studies at various sites in Japan. *Proc. 2nd Int. Conf. on Microzonation for safer construction-research and application*, 2, 885–896.

Licata, V., 2015, A laboratory and field study on cyclic liquefaction of a pyroclastic soil. *PhD Dissertation*, University of Napoli Federico II, Napoli, Italy.

Licata V., d'Onofrio A., Silvestri F. (2018) Microstructural factors affecting the static and the cyclic resistance of a pyroclastic silty sand. *Geotechnique* 68 (5), 434–441 https://doi.org/10.1680/jgeot.16. P.319

Matasovic, N. and Vucetic, M., 1993, Cyclic characterization of liquefiable sands. *Journal of Geotechnical Engineering, ASCE*; 119 (11), 1805–1822.

Orense, R.P., Hyodo, M. and Kaneko, T., 2012, Dynamic deformation characteristics of pumice sand, *New Zealand society for earthquake engineering (2012 NZSEE) conference*. University of Canterbury; Christchurch, NZ.

Orense, R. P. and Pender, M.J., 2016, From micro to macro: An investigation of the geomechanical behaviour of pumice sand, *Volcanic Rocks and Soils* (Rotonda et al. *eds*), *Proceedings of the international workshop on volcanic rocks and soils*, Lacco Ameno, Ischia Island, Italy, 24-25, September 2016, 45–62.

Orsi, G., de Vita, S. and Di Vito, M., 1996, The restless, resurgent Campi Flegrei nested caldera (Italy): constraints on its evolution and configuration. *J. Volcanol. Geotherm. Res.* 74, 179–214

Phillips, C and Hashash, Y.M.A, 2009, Damping formulation for non-linear 1D site response analyses. *Soil Dynamics and Earthquake Engineering* 29(7), 1143–1158

Picarelli, L., Evangelista, A., Rolandi, G., Paone, A., Nicotera, M.V., Olivares, L., Scotto di Santolo, A., Lampitiello, S. and Rolandi M., 2006, Mechanical properties of pyroclastic soils in Campania Region. *Proceedings of the Second International Workshop on Characterisation and Engineering Properties of Natural Soils, Singapore*, 29 November-1 December 2006 Taylor and Francis, London (2006), 2331–2384

Santucci de Magistris, F. and Evangelista, L., 2007, Simplified assessment of the liquefaction susceptibility for the city of Naples, Italy. *Proc. 4th Intl Conf. on Earthq. Geotechnical Engineering*, Thessaloniki, Greece.

Senetakis, K., 2011, Dynamic properties of granular soils and mixtures of typical sands and gravels with recycled synthetic materials, *PhD dissertation*. Department of civil engineering: Aristotle University of Thessaloniki; Greece.

Senetakis, K., Anastasiadis, A. and Pitilakis K., 2012, The small-strain shear modulus and damping ratio of quartz and volcanic sands. *Geotechnical Testing Journal*, 35 (6)(ISSN: 1945–7545).

Senetakis, K., Anastasiadis, A. Pitilakis K. and Coop M.R., 2013 The dynamics of a pumice granular soil in dry state under isotropic resonant column testing *Soil Dynamics and Earthquake Engineering* 45, 70–79 http://dx.doi.org/10.1016/j.soildyn.2012.11.009

Seed, H.B. and Idriss, I.M., 1970, Soil Moduli and Damping Factors for Dynamic Response Analyses, Report EERC 70-10, *Earthquake Engineering Research Center*, University of California, Berkeley, CA.

Stucchi M., Meletti C., Montaldo V., Crowley H., Calvi G.M., Boschi E., 2011. Seismic Hazard Assessment (2003-2009) for the Italian Building Code. *Bull. Seismol. Soc. Am.* 101(4),1885–1911. DOI: 10.1785/0120100130

Tropeano, G., Chiaradonna, A., d'Onofrio, A. and Silvestri, F., 2016, An innovative computer code for 1D seismic response analysis including shear strength of soils. *Géotechnique*, 66(2), 95–105.

Tropeano, G., Chiaradonna, A., d'Onofrio, A. and Silvestri, F., 2019, A numerical model for non-linear coupled analysis on seismic response of liquefiable soils. *Computers and Geotechnics*, 105, 211–227. https://doi.org/10.1016/j.compgeo.2018.09.008

Vinale, F., 1988, Caratterizzazione del sottosuolo di un'area campione di Napoli ai fini di una microzonazione sismica. *Rivista Italiana di Geotecnica* 22 (2), 77–100(in Italian).

Vucetic, M. and Dobry, R., 1991, Effect of soil plasticity on cyclic response. *Journal of Geotechnical Engineering, ASCE*, 117(1), 89–107

Rock Mechanics and Engineering Geology in Volcanic Fields – Ohta, Ito & Osada (eds)
© 2023 copyright the Author(s), ISBN 978-1-032-27657-1

Numerical analysis of mechanical behavior of a tunnel in swelling volcanic rocks

Xiaodong Liu*, Shotaro Yamada & Takashi Kyoya
Tohoku University, Sendai, Japan

ABSTRACT: In this study, we propose an elasto-plastic-damage model to analyze mechanical behavior of a tunnel constructed in swelling volcanic rock mass. The elasto-plastic part of this model is based on the Modified Cam Clay (MCC) model, which can express the softening and hardening behavior of ground materials. The stiffness deterioration due to the damage that occurs with the process of water absorption is also taken into consideration. Furthermore, we formulate an implicit stress update algorithm (return mapping algorithm) to improve the computational accuracy and robustness of the computation for the model when it is implemented in finite element method (FEM). Several numerical simulations were conducted for a tunnel in swelling rocks called Sakazukiyama Tunnel, which is belong to Yamagata Expressway, to analyze mechanical behavior, containing the swelling displacement and damage progress with the process of swelling. The computation results show that stress increased remarkably as swelling happened; and the uplift of tunnel invert is larger when the damage due to the swelling of bedrock is considered, and that the simulation results with consideration of damage exhibit better approximation of the deformed state of Sakazukiyama Tunnel.

Keywords: Swelling Minerals, Constitutive Law, Damage, Return Mapping

1 INTRODUCTION

It is acknowledged that there are many volcanoes in Japan and most volcanic origin rocks contain swelling clay minerals (mainly smectite), such as mudstone, shale, tuff, rhyolite. It is reported that many mountainous tunnels in such volcanic rocks have been destroyed as the weathering and decomposition of igneous rocks with the process of water absorption. Therefore, it is an essential task to solve such severe engineering problems, which means making appropriate predictions of the mechanical behavior of the tunnel in swelling rocks so that rational operation and maintenance management can be conducted later. Although it is an urgent issue to establish a rational evaluation method to predict such a severe ground heaving phenomenon, unfortunately, no method to provide quantitative and accurate evaluation has been established.

This study proposes an elasto-plastic-damage model to analyze the Swelling Mechanical behavior of a tunnel constructed in swelling volcanic rock mass. It contributes to predicting the nonlinear mechanical response of the tunnel constructed in swelling rock mass, aimed at evaluating the stability of a tunnel constructed in swelling rocks after some years.

Such swelling deformation occurred at Sakazukiyama Tunnel, which locates between the Yamagata Zao IC and the Yamagata Kita IC on the Yamagata Expressway. It is reported that the geology around Sakazukiyama Tunnel is mainly Neogene Miocene tuff, intrusive rock rhyolite and basalt, and a huge ground heave occurred suddenly after 17 years of service

*Corresponding author: liu.xiaodong.r5@dc.tohoku.ac.jp

DOI: 10.1201/9781003293590-27

(Yuzo Okui et al., 2009). The numerical analyses for Sakazukiyama Tunnel are conducted by applying the method proposed in this study.

2 NUMERICAL ANALYSIS PROCESS

In this study, the following three-stage analyses are conducted to evaluate the stress state of the tunnel accurately with the process of expansion by water-absorption, which contains initial stress analysis, excavation analysis and swelling analysis.

2.1 Self-weight analysis

In order to carry out the excavation analysis explained in section 2.2, the stress state information of ground before excavation is required. And in order to obtain this stress state, the self-weight analysis is performed firstly. The boundary condition of the self-weight analysis is shown in Figure 1, and the stress distribution of the excavated area and non-excavated area are expressed as σ_{A0} and σ_{B0}, respectively.

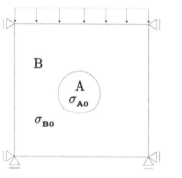
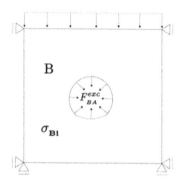

(a) Self-weight analysis   (b) Excavation analysis

Figure 1.   Boundary condition and stress distribution result.

2.2 Excavation analysis

After self-weight analysis, the excavation analysis is conducted by acting excavation vector F_{BA}^{exc} on the surface of excavation region A, which can be calculated by Equation (1) (David M. Potts and Lidija Zdravkovic, 1999).

$$\{F_{BA}\} = \int_{V_A} [B]^T \{\sigma_{A0}\} dV_A - \gamma \int_{V_A} [N]^T dV_A \tag{1}$$

B is the displacement-strain relationship matrix, **N** is the shape function matrix, γ is the bulk unit weight. The stress distribution of non-excavated area after excavation is express as σ_{Bi}, which is the initial stress state of swelling analysis introduced in next section.

2.3 Swelling analysis

In swelling analysis, we propose an elasto-plastic-damage model to evaluate the mechanical behavior of swelling. Furthermore, an implicated stress update algorithm is formulated for the accurate and robust computation in incremental analysis. And also, the consistent tangent modulus is derived in order to obtain the quadratic convergence rate in full Newton-Raphson iteration of FEM.

2.3.1 Swelling model

The basic formulas of the swelling model based on Modified Cam-Clay (MCC) is as follows.

The strain is decomposed into three parts as Eq. (2).

$$\varepsilon = \varepsilon^e + \varepsilon^p + \varepsilon^s \tag{2}$$

Where the upper index e, p and s are mean elastic, plastic and swelling, respectively. The hypo-elastic constitutive relation can be written as:

$$\dot{\sigma} = \mathbb{C}_{(\sigma;\varepsilon^s,\varepsilon^p)} : \dot{\varepsilon}^e \tag{3}$$

The elastic modulus tensor considered damage can be expressed as follows.

$$\mathbb{C}_{(\sigma;\varepsilon^s,\varepsilon^p)} = \left(1 - D_{(\varepsilon^s,\varepsilon^p)}\right)\overline{\overline{\mathbb{C}}}_{(\sigma)} \tag{4}$$

$\overline{\overline{\mathbb{C}}}_{(\sigma)}$ is the hypo-elastic modulus tensor in original MCC model, which depends on the average effective stress. Damage is taken into consideration in this model to evaluate the deterioration of rock stiffness with the process of swelling. The damage variable D is given by the following exponential function.

$$D_{(\varepsilon^s,\varepsilon^p)} = D_\infty[1 - \exp(-\chi)] \tag{5}$$

D_∞ is a parameter representing the maximum value of the damage. It is assumed that the damage parameter χ depends on the deviatoric plastic strain and volumetric swelling strain. Specifically, it can be expressed as follows.

$$\chi = a\sqrt{\frac{2}{3}}\|e^p\| - c\varepsilon_v^s \tag{6}$$

e^p is deviatoric plastic strain, ε_v^S is volumetric swelling strain, a and c are the coefficients representing the influence degree of damage progress relating deviatoric plastic strain and volumetric swelling strain, respectively. As the other basic equations and parameters such as yield function, flow rule, hardening parameter and so on, are equivalent to the MCC model, so the explanation is omitted here.

The volume swelling strain ε_v^S is determined by the following relationship.

$$\varepsilon_{v(p)}^s = m_s\,S_w\,\varepsilon_{v_{max}(p)}^s \tag{7}$$

m_s and S_w are the content of swelling minerals (montmorillonite, etc.) contained in unit volume rocks and swelling saturation, respectively. $\varepsilon_{v_{max}(p)}^s$ is the maximum volumetric swelling strain of swelling minerals, which dependents on mean stress state. Specifically, it can be expressed as eq. (8), and the relationship is shown in Figure 2 (Grob, 1972).

$$\varepsilon_{v_{max}(p)}^s = \begin{cases} \frac{\ln p_{s_c} - \ln p}{\ln p_{s_c} - \ln p_{s_0}}\varepsilon_{v_{max}}^{S_0} & (p \leq p_{s_c}) \\ 0 & (p > p_{s_c}) \end{cases} \tag{8}$$

p_{s_0} and p_{s_c} are the reference mean stress of swelling and the threshold of mean stress for the occurrence of swelling, respectively. $\varepsilon_{v_{max}}^{S_0}$ is the maximum volumetric swelling strain at the reference mean stress.

2.3.2 Return mapping algorithm

The return mapping algorithm for the model constructed above is formulated in this section. The algorithm was introduced by Simo et al., e.g., (Simo and Hughes, 1998). But, no

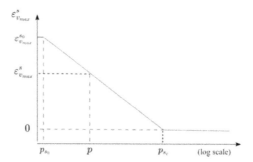

Figure 2. The relationship of maximum volumetric strain and mean stress.

formulation has been conducted for a model containing constrained pressure-dependent swelling, hypo-elastic and damage characteristics, as far as the authors know. For an implicit stress update algorithm, the equations to be solved can be summarized in Eq. (9).

$$
\begin{cases}
\boldsymbol{R}_{\sigma n+1}(\boldsymbol{\sigma},\varDelta\gamma) &= \boldsymbol{\sigma}_{n+1} - \boldsymbol{\sigma}_n - \mathbb{C}_{(\sigma\,;\,\varepsilon^s,\,\mathcal{B}^p)}:\varDelta\boldsymbol{\varepsilon}^e \\
R_{fn+1}(\boldsymbol{\sigma},\varDelta\gamma) &= MD\left[\ln\left(\frac{p}{p_{c_0}}\right) + \ln\left(\frac{M^2+\eta^2}{M^2}\right)\right] - \left(\varepsilon^p_{vn} + \varDelta\gamma\boldsymbol{I}_{(2)}:\frac{\partial f_{n+1}}{\partial\boldsymbol{\sigma}_{n+1}}\right)
\end{cases}
\tag{9}
$$

Eq. (9) is linearized as follows to obtain the variable corrections $\delta\boldsymbol{\sigma}$ and $\delta(\varDelta\gamma)$.

$$
\begin{cases}
\boldsymbol{R}_{\sigma n+1} + \dfrac{\partial \boldsymbol{R}_{\sigma n+1}}{\partial\boldsymbol{\sigma}_{n+1}} : \delta\boldsymbol{\sigma} + \dfrac{\partial \boldsymbol{R}_{\sigma n+1}}{\partial(\varDelta\gamma_{n+1})}\delta(\varDelta\gamma) = \boldsymbol{0} \\
R_{f\,n+1} + \dfrac{\partial R_{f\,n+1}}{\partial\boldsymbol{\sigma}_{n+1}} : \delta\boldsymbol{\sigma} + \dfrac{\partial R_{f\,n+1}}{\partial(\varDelta\gamma_{n+1})}\delta(\varDelta\gamma) = 0
\end{cases}
\tag{10}
$$

When solving Eq. (10), the two-pattern correction calculations are conducted for the case of elastic load or the plastic load.

In the case of the elastic state, since the increment of the plastic multiplier $\varDelta\gamma = 0$, the stress correction amount $\delta\boldsymbol{\sigma}$ can be obtained as follows simply.

$$
\delta\boldsymbol{\sigma} = -\left[\frac{\partial \boldsymbol{R}_{\sigma\,n+1}}{\partial\boldsymbol{\sigma}_{n+1}}\right]^{-1} : \boldsymbol{R}_{\sigma\,n+1}
\tag{11}
$$

In the case of a plastic load, the variable correction $\delta\boldsymbol{\sigma}$ and $\delta(\varDelta\gamma)$ are calculated as eq. (12).

$$
\begin{cases}
\delta(\varDelta\gamma) = \dfrac{R_{f\,n+1} - \frac{\partial R_{f\,n+1}}{\partial\boldsymbol{\sigma}\,n+1}:\left[\frac{\partial \boldsymbol{R}_{\sigma\,n+1}}{\partial\boldsymbol{\sigma}_{n+1}}\right]^{-1}:\boldsymbol{R}_{\sigma\,n+1}}{\frac{\partial R_{f\,n+1}}{\partial\boldsymbol{\sigma}\,n+1}:\left[\frac{\partial \boldsymbol{R}_{\sigma\,n+1}}{\partial\boldsymbol{\sigma}_{n+1}}\right]^{-1}:\frac{\partial \boldsymbol{R}_{\sigma\,n+1}}{\partial(\varDelta\gamma_{n+1})} - \frac{\partial R_{f\,n+1}}{\partial(\varDelta\gamma_{n+1})}} \\
\delta\boldsymbol{\sigma} = -\left[\dfrac{\partial \boldsymbol{R}_{\sigma\,n+1}}{\partial\boldsymbol{\sigma}_{n+1}}\right]^{-1} : \left[\boldsymbol{R}_{\sigma\,n+1} + \dfrac{\partial \boldsymbol{R}_{\sigma\,n+1}}{\partial(\varDelta\gamma_{n+1})}\delta(\varDelta\gamma)\right]
\end{cases}
\tag{12}
$$

2.3.3 Consistent tangent moduli

When solving the nonlinear equilibrium equation by the Newton-Raphson method in FEM, the consistent tangent modulus $\boldsymbol{C}^{ep}$ is required for obtaining the quadratic convergence rate (Simo, and Taylor, 1985). It is derived as follows.

$$C^{ep} = \Xi_{v+1} : \mathbb{C}_{n+1} - \frac{\Xi_{v+1} : \mathbb{C}_{n+1} : \beta_{n+1} \otimes \zeta_{n+1} : \Xi_{v+1} : \mathbb{C}_{n+1}}{I_{(2)} : \frac{\partial f_{n+1}}{\partial \sigma_{n+1}} + \zeta_{n+1} : \Xi_{v+1} : \mathbb{C}_{n+1} : \beta_{n+1}} \tag{13}$$

With

$$\begin{cases} \Xi_{v+1} = \left[\mathbb{I}_{(4)}^{sym} - \frac{\partial \mathbb{C}_{n+1}}{\partial \sigma_{n+1}} \underset{*}{*} \Delta\varepsilon^e + \Delta\gamma_{n+1} \mathbb{C}_{n+1} : \frac{\partial^2 f_{n+1}}{\partial \sigma_{n+1}\partial \sigma_{n+1}} - \frac{1}{9} w \mathbb{C}_{n+1} : I_{(2)} \otimes I_{(2)} \right. \\ \left. + \frac{1}{9} w \frac{\partial \mathbb{C}_{n+1}}{\partial \varepsilon_{n+1}^s} \underset{*}{*} \Delta\varepsilon^e : I_{(2)} \otimes I_{(2)} - \Delta\gamma_{n+1} \frac{\partial \mathbb{C}_{n+1}}{\partial \varepsilon_{n+1}^s} \underset{*}{*} \Delta\varepsilon^e : \frac{\partial^2 f_{n+1}}{\partial \sigma_{n+1}\partial \sigma_{n+1}} \right]^{-1} \\ w = \frac{m_s \, \varepsilon_{v_{max}}^{s0} \, S_{w\,n+1}}{\ln p_{sc} - \ln p_{s0}} \frac{1}{p_{n+1}} \\ \beta_{n+1} = \left[\frac{\partial f_{n+1}}{\partial \sigma_{n+1}} - \mathbb{C}_{n+1}^{-1} : \frac{\partial \mathbb{C}_{n+1}}{\partial \varepsilon_{n+1}^p} \underset{*}{*} \Delta\varepsilon^e : \frac{\partial f_{n+1}}{\partial \sigma_{n+1}} \right] \\ \zeta_{n+1} = \left[\frac{\partial f_{n+1}}{\partial \sigma_{n+1}} - \Delta\gamma_{n+1} I_{(2)} : \frac{\partial^2 f_{n+1}}{\partial \sigma_{n+1}\partial \sigma_{n+1}} \right] \end{cases} \tag{14}$$

The operator $\underset{*}{*}$ in the formula above means $(\blacksquare \underset{*}{*} \bullet)_{ijpq} = \blacksquare_{ijklpq} \bullet_{kl}$.

3 NUMERICAL ANALYSIS

The numerical simulation for a swelling deformed tunnel called Sakazukiyama Tunnel is carried out using the analysis method presented in section 2.

3.1 Self-weight analysis and excavation analysis

The boundary condition and the size of cross section of the model are shown in Figure 3. The earth surface load σ_y is set as 1000kN/m^2, equivalent to 50m overburden pressure, so the earth cover of the tunnel is about 100m, which corresponds to the topographical condition of Sakazukiyama Tunnel. The parameters used for self-weight analysis and excavation analysis are shown in Table 1.

Table 1. Parameters for self-weight analysis and excavation analysis.

ρ(t/m^3)	e_0	p_0(kPa)	p_c(kPa)	M	λ	κ	ν
1.73	1.26	200	3.0×10^9	1.55	0.43	0.05	0.3

The mean stress distribution of self-weight analysis is shown in Figure 4(a). It can be confirmed that the mean stress gradually increases in proportion to the depth due to the effect of its own weight. The stress obtained by the self-weight analysis was used as the initial stress of the bedrock, and excavation analysis was conducted later. The mean stress redistribution after excavation is shown in Figure 4(b). It can be confirmed that the stress release occurred, especially at the bottom and top side of the tunnel.

3.2 Swelling analysis

As it was reported that the swelling deformation of Sakazukiyama Tunnel is not only ground heave, but also extrusion of the side walls (squeezing) (Yuzo Okui et al., 2009). Therefore, the swelling domain is set as Figure 3(b), and the depth of the swelling domain is 20.0 m from the center of the tunnel to the end boundary of swelling domain, so the thickness of swelling

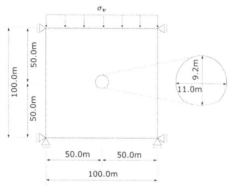

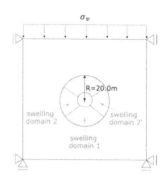

(a) Self-weight analysis and excavation analysis

(b) Swelling analysis

Figure 3.   Boundary condition and size of cross section.

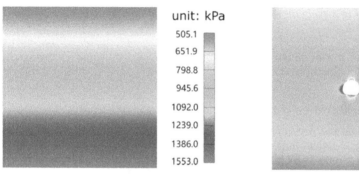

(a) Self-weight analysis

(b) Excavation analysis

Figure 4.   Mean stress distribution.

domain is about 15.0m as the radius of the tunnel is about 5.0m. The parameters used for swelling analysis in swelling domain are shown in Tables 2 and 3, and the parameters in other domains are the same as shown in Table 1.

Table 2.   Parameters for swelling analysis (1).

Swelling domain	e_0	p_0(kPa)	p_c(kPa)	M	λ	κ	ν
1	1.4	Result of	5.0×10^3	1.55	0.57	0.05	0.3
2	1.1	excavation	1.5×10^4	1.55	0.37	0.05	0.3
2'	1.3	analysis	1.5×10^4	1.55	0.34	0.05	0.3

Table 3.   Parameters for swelling analysis (2).

Swelling domain	D_∞	a	c	m_s	$\varepsilon_{v_{max}}^{s_0}$	p_{s_0} (kPa)	p_{s_c} (kPa)
1	0.95	30	30	0.40	-1.00	1.0	4.0×10^3
2	0.95	30	30	0.90	-0.33	1.0	4.0×10^4
2'	0.95	30	30	0.40	-0.25	1.0	4.0×10^4

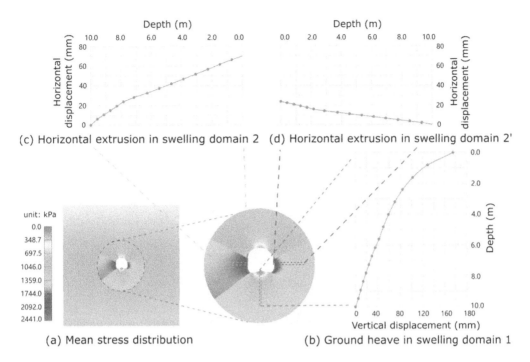

(c) Horizontal extrusion in swelling domain 2   (d) Horizontal extrusion in swelling domain 2'

(a) Mean stress distribution   (b) Ground heave in swelling domain 1

Figure 5. Simulation results of swelling analysis.

The mean stress distribution after swelling analysis is shown in Figure 5(a). It can be seen that the stress increases significantly in the swelling domain. The ground heave (vertical displacement) in swelling domain1 is shown in Figure 5(b), with the maximum reaches about 170mm. It needs to pay attention to that in order to compare the computation results with the measurement data in site, the computation results are modified by setting the point at a depth of 10m as a fixed point because the measured range of underground displacement measurement device is within 10m. Moreover, the horizontal extrusions in swelling domain2 and domain 2' are shown in Figure 5(c) and (d), respectively, with the maximums reaching about 70mm and 24mm. It is very close to the values reported in *Countermeasures Construction Report of Yamagata Expressway Sakazukiyama Tunnel*, which are 55mm and 28mm, respectively. The reason why the difference of the left and right side of the tunnel occurred is that the content of swelling clay minerals m_s is different in each area as shown in Table 3. In addition, the damage progress happened significantly around the tunnel, especially the region below the tunnel invert. Although it is desired to show more results of swelling analysis, there is no more space to show. So, the others will be discussed in detail at RMEGV2021.

4 CONCLUSIONS

In this study, we showed a multi-stage analysis process for evaluating mechanical behavior of tunnels constructed in swelling mountain. In addition, we proposed an elasto-plastic-damage constitutive law for the phenomenon of rigidity deterioration due to rock swelling, and formulated the return mapping algorithm for an accurate and robust computation. The method proposed in this study was validated that the quantitative evaluation is possible by showing the numerical simulations of Sakazukiyama Tunnel.

REFERENCES

Yuzo Okui, Takahisa Tsuruhara, Hiroyuki Ota and Satoru Sakuma, 2009, Analysis of behavior in saka-zukiyama road tunnel under use, Proceedings of tunnel engineering, JSCE, Vol.19, pp. 173–180.

David M. Potts and Lidija Zdravkovic, 1999, Finite element analysis in geotechnical engineering: theory, Thomas Telford.

Grob H., 1972, Schwelldruck im Belchentunnel (Swelling pressure in the Belchen tunnel), International Symposium for Tunneling, pp 99–119. (in German)

Simo, J.C. and Hughes, T.J.R., 1998, Computational Inelasticity, New York: Springer.

Simo, J.C., and Taylor, R.L., 1985, Consistent Tangent Operator for Rate-lndependent Elastoplasticity. Computer Meth. in Applied Mech. and Eng. Vol.48, 101–118.

Rock Mechanics and Engineering Geology in Volcanic Fields – Ohta, Ito & Osada (eds)
© 2023 copyright the Author(s), ISBN 978-1-032-27657-1

Time-dependent characteristics of tuff from creep test under different loading regimes

Takashi Ito* & Ömer Aydan
Department of Civil Engineering, University of the Ryukyus, Japan

ABSTRACT: Tuff, which is sedimentary rock consisting of volcanic ashes, has been used as building stone for centuries worldwide and some collapses of abandoned underground quarries occurred in recent years. The time-dependent response and monitoring of these abandoned underground quarries are of great importance. The authors have been carrying out some experimental studies on the creep behavior of tuffs from Japan and Turkey utilizing creep tests subjected to different loading regimes such as uniaxial or triaxial compression, Brazilian tensile, bending as conventional creep tests and impression creep tests as index test. The authors report some experimental results on time-dependent characteristics of tuffs during uniaxial and triaxial compression, Brazilian and impression creep experiments in this study. The experimental results are compared with each other for the suitability for creep characteristics and their implications are briefly discussed.

Keywords: Creep tests, brazilian tensile, uniaxial compression, impression, tuff, Oya tuff

1 INTRODUCTION

There are many underground quarries in Japan where tuff was quarried as building material. One of the well-known tuffs is Oya tuff (locally known as Oya-Ishi) and it was used as building stone in Imperial Hotel in Tokyo. They are mostly exploited using room and pillar technique (Figure 1). As discontinuities are few very, large underground openings created. Underground quarries are generally shallow with thin cover. The rock is either tuff or welded tuff.

(a) Oya   (b) Asuwayama   (c) Izu

Figure 1. Views of underground quarries.

*Corresponding author: takito@tec.u-ryukyu.ac.jp

DOI: 10.1201/9781003293590-28

As the quarrying is declining due to requirement stones as building stones in Japan, many of these underground quarries are now abandoned. Besides these underground quarries, there are also many abandoned lignite and coal mines. The presence of these abandoned quarries and mines is even forgotten. Nevertheless, the gravity acts on these abandoned quarries and mines. Furthermore, they may be immersed under groundwater. It is well-known that the strength of surrounding rock supporting the overburden above these abandoned quarries and mines decreases with immersion and absorption of groundwater. This reduction may be generally greater than 40-50%. The stresses in these abandoned quarries and mines create generally constant stress conditions if there is no yielding in pillars and roofs. However, it is well-known that some partial yielding and slabbing occurs in the pillars and roof in long-term and they may result in creep failure of pillars and roof so that some collapses over such areas occur from time to time. Figure 2 shows two examples of collapses occurred above abandoned tuff quarries.

(a) 1989 Collapse at Oya   (b) 2005 Collapse at Asuwayama

Figure 2.   Views of collapses at Oya (a) and Asuwayama.

The authors have been performing various experimental studies on the creep behavior of tuffs from Japan and Turkey utilizing different creep testing techniques, which involve uniaxial or triaxial compression, Brazilian tensile, bending as conventional creep tests and impression creep tests as index test. In this study, experimental results on time-dependent characteristics of tuffs during uniaxial and triaxial compression, as well as impression creep are described and they are compared with each other for the suitability for creep characteristics and their implications are discussed.

2   CREEP TESTING TECHNIQUES

Although servo-controlled testing devices are available for creep tests, the cantilever-type apparatus has been used in creep tests since early times and they are very economical. Furthermore, it is the most suitable apparatus for creep tests in practice because the load level can easily be kept constant with time. The load is applied onto samples by attaching deadweights to the lever. In addition, some three or four-point beam testing configurations are utilized for creep tests. In this section, a brief description of the devices utilized in this study is given.

2.1 Compression creep testing device

There are two 50 kN creep devices and the uniaxial compression creep experiments were mainly carried out by using a specially designed creep-loading device in the Rock Mechanics Laboratory of the University of the Ryukyus. The creep testing device is shown in Figure 3a. The device is capable of imposing the loads with an increment of 0.1kN and all loading operation is done manually so that no electric disturbance can affect loading operation. The axial displacement and related acoustic emission events are also monitored using measurement devices utilizing battery power source so that it is possible to carry out creep experiments without any conventional electricity sources. In addition, triaxial creep tests are possible through utilization of a specially designed cell.

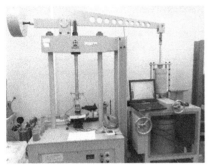 

(a) Compression creep device   (b) Brazilian creep device

Figure 3. Views of creep devices at the University of the Ryukyus.

2.2 Brazilian creep testing device

Figure 3b shows a cantilever type Brazilian creep experiment device at the Rock Mechanics laboratory of the University of the Ryukyus. The system is equipped with displacement transducer, acoustic emission sensor together with environmental parameters (temperature, humidity and air pressure). The data measured are stored in loggers, which operate entirely on batteries and experiments can be carried in any environment over a long-term (say months). This system induces loads on samples up to six times that imposed at the cantilever beam end.

2.3 Impression creep testing device

The impression creep experiment was first designed by Aydan et al. (2011) and the device was of a cantilever type equipped with battery operated displacement and acoustic emission devices. Impression creep experiments are relatively easy to perform and the capacity of loading equipment is relatively small compared to conventional creep experiments (Figure 4a). This system induces loads on samples up to ten times that imposed at the cantilever end and the maximum applied pressure could be up to 137 MPa. Figure 4b shows plastic yielding beneath the 3 mm flat-end needle. Aydan et al. (2008) theoretically showed that there would be at least three levels of end pressure to initiate the various yielding conditions (Figure 5). The first level would correspond to the tensile strength of rocks. However, the overall stiffness would be slightly higher after the yielding. The second yielding would correspond to uniaxial strength of rock and the overall stiffness would show strain hardening response. Once the strain level of rock beneath the needle reaches the residual strain level, the overall stiffness would drastically change. This level may be taken as the ultimate resistance.

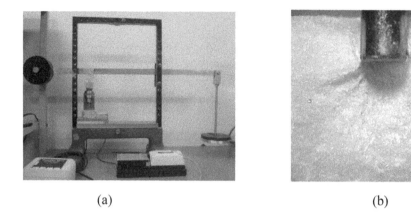

(a)   (b)

Figure 4.   (a) A view of impression creep device and (b) plastic yielding beneath the needle.

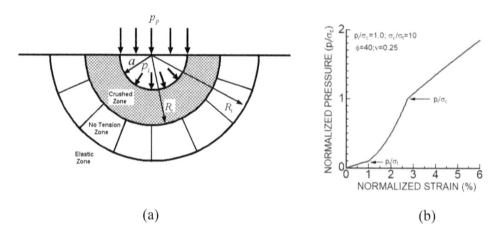

(a)   (b)

Figure 5.   (a) Yielding situations and (b) Yielding levels in penetration-axial stress response.

3   CREEP TESTS

3.1   *Uniaxial and triaxial compression creep tests*

Creep tests on Oya tuff and Tuff of Cappadocia carried out by Ito and Akagi (2001), Ito et al. (2016) and Ulusay et al. (2013) under dry conditions are plotted in Figure 6a, b, respectively. As noted from Figure 6, some of responses terminate with failure while the others become asymptotic to certain strain levels, depending the applied stress ratio (SR), which is defined as the ratio of applied stress to the short-term strength. The responses terminating in failure are generally divided into three stages. The transitions from the primary stage to the secondary stage and from the secondary stage to the tertiary stage are generally determined from the deviation of a linearly decreasing or increasing strain rate plotted in a logarithmic time space. Generally, it should, however, be noted that strain data must be smoothed before its interpretation. Direct derivation of strain data containing actual responses as well as electronic noise may produce entirely different results.

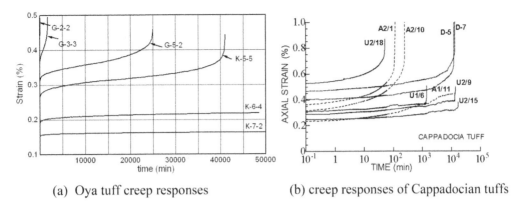

(a) Oya tuff creep responses (b) creep responses of Cappadocian tuffs

Figure 6. Creep responses from Oya tuff (Japan) and Cappadocian tuffs (Turkey).

When rocks have water absorption ability, their strength tends to decrease compared with that under dry condition. Particularly, the strength of soft rocks like tuffs decreases drastically and the strength reduction is generally greater than 60%. In some cases, soft rocks may disintegrate upon water absorption and the resulting strength reduction may be up to 100%. Several researchers investigated the effect of saturation on uniaxial compression creep tests (e.g. Ito and Akagi, 2001; Aydan et al. 2011). The effect of saturation during creep experiments should be regarded as the increase of the applied stress ratio with the consideration of strength of rock under saturated condition.

Triaxial compression creep experiments are quite limited as compared with uniaxial compression creep experiments due to sophistication of equipment and costs. Nevertheless, there were several attempts to conduct such tests (e.g. Ito and Akagi, 2011). Provided that friction angle is not rate-dependent, the stress ratios under triaxial compression creep test are defined in an analogy to that in uniaxial compression creep tests as:

$$SR = \frac{\sigma_1 - \sigma_3}{2c \cos \varphi + (\sigma_1 + \sigma_3) \sin \varphi} \tag{1}$$

where c, φ, σ_1 and σ_3 are cohesion, friction angle and maximum applied and confining stresses, respectively. If friction angle is rate-dependent, the ratio of the applied deviatoric stress to the deviatoric strength is used as stress ratio. However, the experimental results confirm that the rate-dependency of friction angle is negligible according to Aydan and Nawrocki (1998).

Figure 7 shows the creep response under a confining stress of 2 MPa and the failure time of compression creep tests under both uniaxial and triaxial compression environment. It is interesting to note that the overall tendency obtained in triaxial creep tests is basically similar to those of uniaxial compression creep tests irrespective of confining pressure.

3.2 Brazilian and bending creep tests

There are not many studies on tensile creep behavior of rocks using Brazilian creep tests. However, rock may be subjected to tensile stresses in nature such as cliffs with toe erosion and roof layers above underground openings excavated in sedimentary rocks. Aydan et al. (2011) and Ulusay et al. (2013) have recently reported some Brazilian tensile and bending creep tests on tuff samples.

Here we quote some experimental results from Ito et al. (2016) and Aydan et al. (2011). The diameter of samples was 46mm and their thickness ranged between 14 and 25 mm. All samples were subjected to creep loading level at a chosen period of time under dry conditions. After reaching the ultimate loading elevel, the samples were saturated. Figure 8a shows some of the measured response of a sample in Brazilian creep experiments on Oya tuff. Oya tuff sample numbered SN1-W3 was tested under fully saturated conditons at a stress ratio of 87%. As noted from the figures, acoustic emission occurs at each load increase, simultaneously.

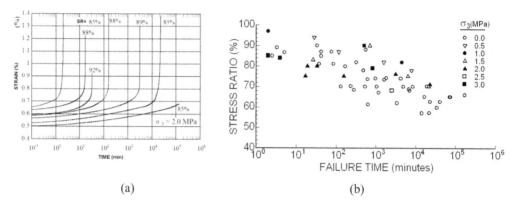

(a)   (b)

Figure 7.   (a) Creep response at a confining pressure of 2 MPa, and (b) Failure time versus stress ratio.

Ulusay et al (2013) reported a series of 3-point bending creep experiments on tuff sample at underground Avanos Museum under both dry and saturated conditions. During these tests, Acoustic Emission (AE) measurements were also conducted to investigate the fracturing phenomenon. An example of bending creep experiments is shown in Figure 8b. The samples were initially dry and later saturated. It was noted that the saturation of samples drastically reduces the creep failure time compared to the dry state.

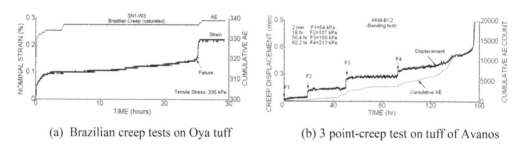

(a) Brazilian creep tests on Oya tuff   (b) 3 point-creep test on tuff of Avanos

Figure 8.   Brazilian and 3-point bending creep responses of tuff samples.

3.3   Impressions creep tests

Aydan et al. (2011) reported some impression creep tests on Oya tuff. These tests have been continued. Figure 9 shows two recent examples of impression creep experiments. The inspection of specimens after creep tests indicated that permanent displacement occurred and there was a cylindrical permanent hole together with radial fractures. The plastic zone radius is roughly twice the indenter radius. The tip pressure is also shown in Figure 9. It was also clearly noticed that the viscous behavior was stress-level dependent as seen in Figure 9. During experiments, an earthquake occurred and the ground shaking also had some effects on the creep response.

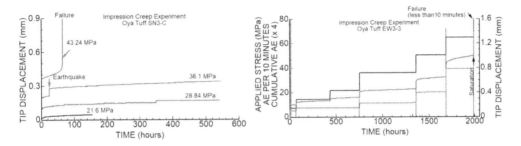

Figure 9. Impression creep responses of Oya tuff under different applied tip pressure.

4 COMPARISONS AND DISCUSSIONS

Figure 10 compares the failure time of Oy tuff samples tested in Brazilian, impression and uni-axial compression creep experiments under dry and saturated conditions. For comparing the impression creep experiments with other experiments, the stress ratio was defined by dividing the applied stress condition by 36 MPa. From experimental results, it is very interesting to note that if the stress ratio remains same, the failure time of dry and saturated samples are very close to each other. Furthermore, the failure times of samples tested under uniaxial compression and Brazilian creep experiments are also similar. These two important conclusions have strong implications in practice.

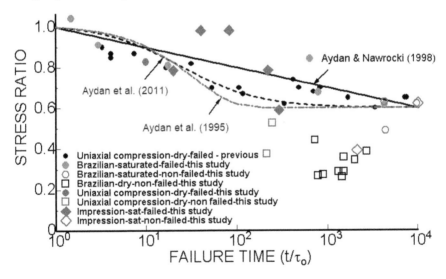

Figure 10. Comparison of failure time of various creep experiments and empirical relations by Aydan et al. (1995, 2011] and Aydan & Nawrocki (1998).

5 CONCLUSIONS

The experimental studies presented in this study have been still continuing. This study is also unique in a way that four different experimental techniques are utilized for the same rock. One of the main purposes is to explore the suitability of impression creep tests as an index test (Aydan et al. 2011) and the results are quite promising that there is a high possibility for the utilization of the impression creep as an index creep test. The preliminary experimental result clearly demonstrated that both conventional and index tests could be quite comparable and similar type responses are observed. As pointed out in previous section, the most critical aspect is the definition of the stress ratio in impression creep tests, which is easily obtained in conventional creep tests. If short-term tests on samples using the impression testing indenter are carried out, the results obtained from impression creep tests should yield almost the similar type material properties for time-dependent behavior of rocks.

REFERENCES

Aydan, Ö. and Nawrocki, P., 1998, Rate-dependent deformability and strength characteristics of rocks. *Int. Symp. On the Geotechnics of Hard Soils-Soft Rocks*, Napoli, 1, 403–411.

Aydan, Ö., Akagi, T. Ito, T., Ito, J. and Sato, J., 1995, Prediction of deformation behaviour of a tunnel in squeezing rock with time-dependent characteristics. NUMOG V, 463–469.

Aydan, Ö., Watanabe, S. Tokashiki, N. 2008, The inference of mechanical properties of rocks from penetration tests. *5th Asian Rock Mechanics Symposium (ARMS5)*, Tehran, 213–220.

Aydan, Ö, Rassouli, F. and Ito, T., 2011, Multi-parameter responses of Oya tuff during experiments on its time-dependent characteristics. *Proc. of the 45th US Rock Mechanics/Geomechanics Symposium*, San Francisco, ARMA 11–294

Ito, T. and Akagi, T. 2001, Methods to predict the time of creep failure (in Japanese). Proceedings of the31st Symposium on Rock Mechanics of Japan, 77–81.

Ito, T., Akagi, T., Aydan, Ö., Ulusay, R. and Seiki, T., 2016, Time-dependent properties of tuffs of Cappadocia, Turkey. EUROCK2016, Ürgüp, 229–234.

Ulusay, R., Aydan, Ö., Geniş, M and Tano, H., 2013, Assessment of stability conditions of an underground congress centre in soft tuffs through an integrated rock engineering method (Cappadocia, Turkey). Rock Mechanics and Rock Engineering, 46:1303–1321

Rock Mechanics and Engineering Geology in Volcanic Fields – Ohta, Ito & Osada (eds)
© 2023 copyright the Author(s), ISBN 978-1-032-27657-1

The inference of physico-mechanical properties of tuffs of Cappadocia Region and Phrygian Valley of Turkey from Needle Penetration Index (NPI)

Ömer Aydan*
Department of Civil Engineering, University of the Ryukyus, Okinawa, Japan

Halil Kumsar
Department of Geological Engineering, Pamukkale University, Denizli, Turkey

Resat Ulusay
Department of Geological Engineering, Hacettepe University, Ankara, Turkey

ABSTRACT: Needle penetration index (NPI) testing technique has been used for inferring the uniaxial compressive strength (UCS) of soft rocks particularly tunnelling through squeezing rocks in Japan. The authors have been using this device to infer the physico-mechanical properties of soft rocks for some time. In this study, the authors describe the characteristics of experimental results carried out on tuff samples from various parts of Cappadocia Region and Phrygian Valley in Turkey. The physico-mechanical properties have been estimated in terms of needle penetration index (NPI) and compared with experimental results on tuffs from Cappadocia and Phrygian Valley. Their implications are discussed in similar rock mass conditions with some considerations of archeological remains.

Keywords: Needle penetration index, tuff, physico-mechanical properties, empirical relations

1   INTRODUCTION

A practical index testing technique, which is called needle penetration index (NPI) test, has been used for inferring the uniaxial compressive strength (UCS) of soft rocks particularly in tunnelling through squeezing rocks in Japan. The device can measure the applied load and the penetration depth of the needle of the device. The ratio of applied load to penetration is called Needle Penetration Index (NPI). The authors have been using this device to infer the physico-mechanical properties of soft rocks (i.e. unit weight, elastic modulus, uniaxial compressive strength (UCS), Brazilian tensile strength, elastic wave velocity etc.) of soft rocks from Japan, Turkey and Egypt. As the damage caused to rock by penetration is quite negligible, the device can be effectively used in assessing the state of rocks and inferring its properties particularly in archeological structures excavated in/on rock mass. In this study, the authors describe the characteristics of the testing device and experimental results. Various empirical equations are used to infer the physico-mechanical properties in terms of needle penetration index (NPI) and compared with experimental results on tuffs from Cappadocia and Phrygian Valley in Turkey and their implications are discussed in similar rock mass conditions.

*Corresponding author: aydan@tec.u-ryukyu.c.jp

DOI: 10.1201/9781003293590-29

2 GEOGRAPHY AND GEOLOGY

2.1 *Cappadocia Region*

The Cappadocia Region is situated in central Anatolia forming a high plateau and surrounded by three old volcanoes (Erciyes, Melendiz, Hasandag; see Figure 1a. Kızılırmak River, flowing through the northern part of the region, is the principal architecture of the morphological features in the Cappadocia Region. Another important physiographic feature of the region is the fairy chimneys generally located along the valleys.

The Cappadocia Region is generally underlain by volcanic rocks of the Neogene-Quaternary period belonging to the Cappadocian Volcanic Province (CVP) (i.e. Toprak et al. 1994; Aydar et al. 2012). Basement rocks, Yesilhisar formation, Ürgüp formation and Quaternary deposits are the main units observed in the region. The Ürgüp formation, which shows the largest distribution in the region, includes a number of different tuff members and lava layers (Figure 1b). The underground and semi-underground rock structures and the fairy chimneys take place in this formation, particularly in its Kavak member with a thickness of between 10 and 150 m. The members of this formation are generally of dirty white, grey and pink colours and of clastic character, showing an alternation from the fine grained to coarse types with enclosed large and small lumps of pumice and obsidian.

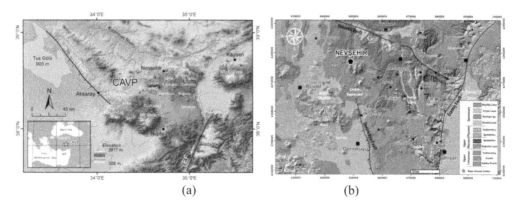

Figure 1.   (a) Geography and (b) geology of Cappadocia Region (from Aydar et al. 2012).

2.2 *Phrygian Valley*

Phrygians appeared in Anadolu (Turkey) about 3200 years ago following the collapse of Hittite Empire. The Phrygia is an area on the western end of the high Anatolian plateau. Phrygia begins in the northwest with an area of dry steppe with the Sakarya and Porsuk river system and the Phrygian capital was Gordion. Phrygia extends to the towns of Laodikeia on the Lycus and Hierapolis in western Turkey. The Frig(Phrygian) Valley is a rock-hewn valley and was inhabited by Phrygians this rock-hewn valley about 3200 years ago for about 500 years. It runs from Eskişehir through Afyon to Kütahya (Figure 2a, b. The Afyon-area is best-preserved among other areas.

3 PHYSICO-MECHANICAL PROPERTIES

3.1 *Cappadocia Region*

The Cappadocia tuffs are prone to atmospheric conditions, and therefore, susceptible to weathering. Mean values of some dry short-term geomechanical properties of the main

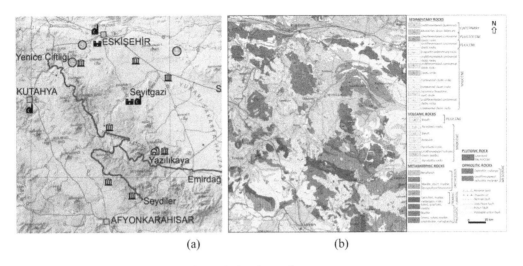

(a)   (b)

Figure 2.   (a) Geography and (b) geology of Phrygian Valley.

Cappadocia tuffs, in which rock structures are commonly observed, are given in Table 1. These tuffs are weak to very weak rocks and have high deformability. However, geomechanical properties of the Zelve tuff is lower than those of the other tuffs as noted from Figure 3a. Furthermore, the saturation drastically reduces the strength and deformation modulus. The results of UCS and modulus of elasticity of the tuffs measured parallel and perpendicular to bedding plane are scattered around line of anisotropy indicating that their mechanical properties should be fairly uniform (Aydan and Ulusay 2003, 2013).

Table 1.   Mechanical properties of the tuffs in Cappadocia.

Property	Gördeles Tuff (Derinkuyu)	Kavak Tuff (Ürgüp)	Zelve Tuff (Avanos)	Zelve Tuff (Zelve)
Unit Weight (kN/m³)	14.2	13.8	13.6	13.6
UCS (MPa)	9.4	5.4	3.9	3.6
Tensile Strength (MPa)	0.6	0.4	0.35	0.35
Elastic Modulus (GPa)	3.3	1.4	0.9	085
Poisson ratio	0.2	0.15	0.1	0.12
P-wave velocity (km/s)	2.0	1.9	1.6	1.5

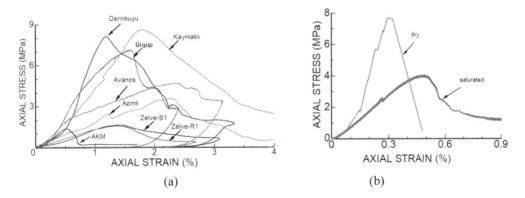

(a)   (b)

Figure 3.   (a) Strain-stress response and (b) the effect of saturation in uniaxial compression tests.

239

3.2 *Phrygian Valley*

Some rock samples were collected during the site investigations. Physical, mechanical and index properties of the tuffs in the valley were investigated under laboratory conditions. Physical properties involve unit weight, water absorption of rock samples. Mechanical properties involve elastic modulus, uniaxial compression and tensile strength of rock samples though 3-point bending and uniaxial compression experiments on dry and saturated rock samples. Figure 4 compares strain-stress response of coarse grain and fine grain tuff and the effect of saturation. Needle penetration and P-wave velocity measurements were also performed as index tests. Experimental results are summarized in Table 2.

Table 2. Physico-mechanical properties of rocks in the Phrygian Valley.

Property	Coarse-grained tuff		Fine-grained tuff	
	Dry	Sat.	Dry	Sat.
Unit Weight (kN/m³)	13.5-14.4	17.6	13-16	17-19
Water Absorbtion (%)	21.9-22.9		14.0-17.5	
Elastic Modulus (GPa)	0.8-1.2	0.3	1.5-1.7	
UCS (MPa)	5.2- 6.1	2.1-3.1	9.2-9.4	
Tensile Strength (MPa)	0.35	0.24	0.4	
P-wave velocity (km/s)	1.6-1.7	1.1-1.2	3.3-	2.3

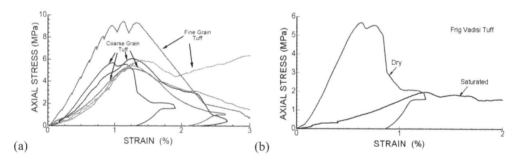

Figure 4. (a) Strain-stress response and (b) the effect of saturation in uniaxial compression tests.

4 NEEDLE PENETRATION INDEX TESTS

MARUTO Testing Machine Company (1999) produced a simple needle-penetration-index testing (NPI-T) device and it has been widely used in many engineering projects involving soft rocks in Japan (Aydan et al., 2012). It has been recognized by ISRM and it has been developed into an ISRM Suggested Method (SM) (Ulusay et al., 2014. The diameter of the needle is about 0.85 mm (Figure 5a). Aydan et al. (2008) has developed another penetration device for soft as well as medium and high strength rocks utilizing a simple low-capacity device (Figure 5b). The measured response can be analyzed using the theory developed by Aydan et al. (2008), which renders both deformation and strength characteristics of rocks through a single test. Aydan (2012) and Aydan et al. (2014) also utilized the conventional needle in the same device to evaluate the overall response during penetration and investigated the damage zone induced by the penetration device using both visual inspection as well as X-CT technique as shown in Figure 5c.

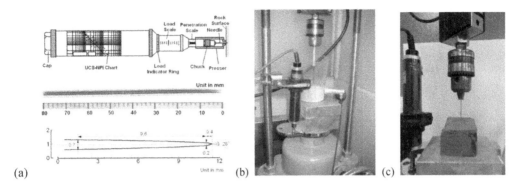

(a)   (b)   (c)

Figure 5. (a) Conventional NPI device and (b) Aydan et al. (2008) device (c) Aydan (2012)'s set-up.

4.1 *Cappadocia Region*

The set-ups shown in Figure 5b, c were utilized for tuff samples collected from various areas in Cappadocia Region. Figure 6a shows the response of tuff samples from the Derinkuyu Underground City including the one subjected to 270 freezing-thawing cycles. When responses of dry and saturated samples are compared, it is interesting to note that the initial parts are entirely different. However, penetration-load relation after a penetration of 1.6 mm shows quite similar stiffness. Regarding the stiffness of the sample subjected to freezing-thawing after 0.8 mm also show similar trend.

Figure 6b shows the penetration-load responses measured during tests for conventional needle and 2, 3mm needle configurations are different. Regarding 2 and 3 mm diameter needles with cylindrical flat tip-ends, the responses are similar and then they start to deviate from each other due to yielding of rock beneath the tip-end. The overall penetration-load responses become quite similar to that observed when the conventional needle is used. As pointed by Aydan et al. (2014), the conventional needle causes plastic yielding from the very beginning.

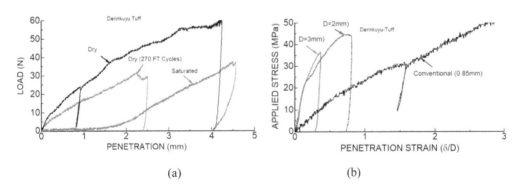

(a)   (b)

Figure 6. (a) penetration-load responses with conventional needle, (b) comparison of responses using different needle configurations.

4.2 *Phrygian Valley*

Similar tests were done on samples from Phrygian Valley and results are shown in Figure 7. Figure 7a shows the response of dry and saturated tuff samples. Figure 7b shows the penetration-load responses measured during tests for conventional needle, and 2 and 3mm needle configurations. Regarding 2 and 3 mm needles with cylindrical flat tip-ends, the responses are similar and then they start to deviate from each other due to yielding of rock beneath the tip-end. The overall penetration-load response becomes quite similar to that observed when the conventional needle is used due to plastic yielding. However, it should be noted that the yielding around needles is quite local and limited.

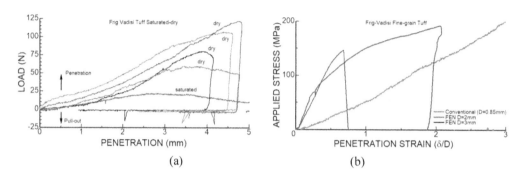

Figure 7. (a) Penetration-load responses with conventional needle, (b) comparison of responses using different needle configurations.

5 CORRELATIONS AMONG NPI AND PHYSICO-MECHANICAL PROPERTIES

Aydan (2012) and Aydan et al. (2014) established some empirical relations to estimate various physico-mechanical properties of soft rocks as a function of needle penetration index (NPI). Besides these relations, some empirical relations between NPI and uniaxial compressive strength (UCS) are also proposed. These empirical relations are summarized in the ISRM-SM for needle penetration index tests (see Ulusay et al., 2014). The NPI values versus uniaxial compressive strength (UCS), tensile strength (TS) and elastic modulus (EM) are shown in Figure 8. In the respective figures, results are differentiated for tuffs of Cappadocia Region and Phyrgian Valley tuffs. Although the results show some scattering due to non-homogeneity of the samples, the presence of a good correlations among the NPI and UCS, TS and EM reveals that the NPI can be useful for such soft rocks and applicable at historical sites under protection.

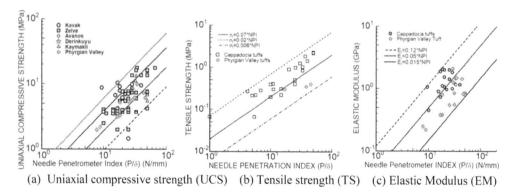

(a) Uniaxial compressive strength (UCS) (b) Tensile strength (TS) (c) Elastic Modulus (EM)

Figure 8. Correlation among NPI and UCS(a), TS(b) and EM(c).

6 CONCLUSIONS

The authors presented the results of the rock mechanics investigations on volcanic soft rocks from various parts of Cappadocia Region and the Frig (Phrygian) Valley. Needle penetration tests using the conventional needle and cylindrical rod-like needle of 3 mm dimater were also carried out and correlations among NPI and mechanical properties together with the use of empirical relations are compared and discussed. The good relationships among the physico-mechanical properties and NPI proposed by Aydan (2012) and Aydan et al. (2014) including those by others suggested that the needle penetration index test can be useful for such rocks and applicable at historical sites under protection. The strength of rock is reduced drastically under saturated conditions and wetting-drying processes, which are also clearly noted in the NPI results.

REFERENCES

Aydar, E., Schmitt, A.K., Çubukçu, H.E., Akin, L., Ersoy, O., Sen, E., Duncan, R.A. and Atici, G., 2012, Correlation of ignimbrites in the central Anatolian volcanic province using zircon and plagio-clase ages and zircon compositions. J. Vol. Geother. Res., 213-214, 83–97.

Aydan, Ö., 2012, The inference of physico-mechanical properties of soft rocks and the evaluation of the effect of water content and weathering on their mechanical properties from needle penetration tests. ARMA 12-639, 46th US Rock Mechanics/Geomechanics Symposium, Chicago, Paper No. 639, 10 (on CD).

Aydan, Ö. and Ulusay, R., 2003, Geotechnical and Geoenvironmental characteristics of man-made underground structures in Cappadocia, Turkey. Engineering Geology, 69, 245–272.

Aydan, Ö. and Ulusay, R., 2013, Geomechanical evaluation of Derinkuyu Antique Underground City and its implications in geoengineering. Rock Mech. and Rock Eng. 46, 731–754.

Aydan, Ö. and Kumsar H. 2016, A Geoengineering Evaluation of Antique Underground Rock Settlements in Frig (Phryrgian) Valley in the Afyon-Kütahya Region of Turkey. EUROCK2016, Ürgüp, 853–858.

Aydan Ö., Ulusay R, and Yüzer E. 1999, Man-made structures in Cappadocia, Turkey and their implications in rock mechanics and rock engineering. ISRM News Journal 6 (1), 63–73.

Aydan, Ö., Watanabe, S., and Tokashiki, N., 2008, The inference of mechanical properties of rocks from penetration tests. Proceedings of the 5th Asian Rock Mechanics Symposium (ARMS5), Tehran, pp. 213–220.

Aydan, Ö., Sato A., and Yagi. M., 2014, The Inference of Geo-Mechanical Properties of Soft Rocks and their Degradation from Needle Penetration Tests. Rock Mechanics and Rock Engineering, 47, 1867–1890.

MARUTO Testing Machine Company, 1999, Instructions for SH-70 Penetrometer (NANGAN-PENETOROKEI) (in Japanese), Tokyo, 4p.

Toprak, V., Keller, J., and Schumacher, R., 1994, Volcanotectonic features of the Cappadocian Volcanic Province. Volcanological Congress IAVCEI-Ankara, Excursion Guide, 58 pp.

Ulusay, R., Aydan, Ö., Erguler, Z.A., Ngan-Tillard, D.J.M., Seiki, T., Verwaal, W., Sasaki., Sato, A., 2014, ISRM suggested method for the needle penetration test. Rock Mech Rock Eng, 47, 1073–1085

Rock Mechanics and Engineering Geology in Volcanic Fields – Ohta, Ito & Osada (eds)
© 2023 copyright the Author(s), ISBN 978-1-032-27657-1

Exact solution of Richards' equation using heat transfer boundary and application to dry-induced deformation behavior of Tage tuff

Kazuki Mizuo
Formerly Saitama University, Saitama, Japan

YotaTogashi* & Masahiko Osada
Saitama University, Saitama, Japan

ABSTRACT: In geological disposal of radioactive waste and tunnel construction of high-speed railways, it is important to minimize the deformation of the rock to be excavated. When constructing such rock structures at sedimentary rock sites, it is important to consider the dry deformation of the rock masses. The authors have investigated changes in mechanical properties associated with continuous changes in the water content of green tuffs originating from submarine volcanoes widely distributed in eastern Japan (called Tage tuff). On the other hand, it is widely known that the water content in rock mass changes non-uniformly from the surface in contact with the atmosphere. Considering that the mechanical properties of rock mass change according to the water content, it is important to obtain the water content distribution. In this study, we show a method of linearizing the Richards' equation of a nonlinear partial differential equation by adding two conditions that the water diffusion coefficient is constant and the permeability coefficient does not depend on the coordinates, and the exact solution in which saturation changes depending on coordinates and time is obtained. Using the obtained exact solution, we simulated the drying experiment of Tage tuff conducted by the authors and examined some conditions based on the heat transfer boundary.

Keywords: Richards' equation, exact solution, unsaturated sedimentary rock, water content

1 INTRODUCTION

Changes in moisture content due to drying and wetting can be accompanied by very large deformations of sedimentary rocks. In some cases, a contraction strain that is one-third as large as the failure strain was observed in the process leading to drying (Osada 2014). Thus, when constructing such rock structures at sedimentary rock sites, drying deformation of such rock masses should be carefully considered. The authors have investigated mechanical properties green tuffs originating from submarine volcanoes widely distributed in eastern Japan (Togashi et al. 2018, 2019), and we also have estimated changes in mechanical properties associated with continuous changes in the water content of green tuffs originating from submarine volcanoes widely distributed in eastern Japan (Togashi et al. 2020, 2021a, 2021b).

On the other hand, it is widely known that the water content in rock mass changes non-uniformly from the surface in contact with the atmosphere. Considering that the mechanical properties of rock mass change according to the water content, it is important to obtain the water content distribution. The change in water content in the ground follows the Richards' equation (Richards 1931), but this equation is very non-linear, and even if it is solved by numerical analysis,

*Corresponding author: togashi@mail.saitama-u.ac.jp

DOI: 10.1201/9781003293590-30

a stable solution may not be obtained. Therefore, in recent years, many studies have been conducted to find an exact solution of the Richards equation under simplified conditions (Hooshyar and Wang 2016, Sugawara et al. 2018). Some methods have been proposed using a non-linear water retention curve (Broadbridge et al. 2017, Ekpoudom et al. 2017), but there are few cases where a heat transfer boundary is introduced.

In this study, we show a new method of linearizing the Richards equation of a nonlinear partial differential equation by adding two conditions that the water diffusion coefficient is constant and the permeability coefficient does not depend on the coordinates, and the exact solution in which saturation changes depending on coordinates and time is obtained. Using the obtained exact solution, we simulated the drying experiment of Tage tuff conducted by the authors and examined some conditions based on the heat transfer boundary.

2 EXACT SOLUTION OF RICHARDS' EQUATION

The following nonlinear partial differential equation have been proposed to predict changes in water content in unsaturated ground (Richards 1931).

$$\frac{\partial \theta}{\partial t} = \frac{\partial K}{\partial z}\left(\frac{\partial \psi}{\partial z} + 1\right). \tag{1}$$

Here, θ is the volumetric water content, t is the time, K is the unsaturated hydraulic conductivity, z is the coordinates, and ψ is the pressure head. The exact solution of this nonlinear PDE has not yet been found. In this study, we obtain the exact solution of this equation as follows. The Richards equation is transformed into the following.

$$\frac{\partial \theta}{\partial t} = \frac{\partial}{\partial z}\left(K\frac{\partial \psi}{\partial \theta}\frac{\partial \theta}{\partial z}\right) + \frac{\partial K}{\partial z}. \tag{2}$$

Here, the heat equation can be obtained by adding the condition that the moisture diffusion coefficient D, which is the slope of the water retention curve, is always constant ($D = K\partial\psi/(\partial\theta)$ = const.), and the condition that the unsaturated permeability coefficient does not depend on the coordinates ($\partial K/(\partial z) = 0$).

$$\frac{\partial \theta}{\partial t} = D\frac{\partial^2 \theta}{\partial z^2}. \tag{3}$$

The water retention curve is predominantly non-linear in the region close to saturation and dryness. However, the assumption that D is constant in the region where S is not too small and not too large holds. It is also rational to assume that K does not depend on coordinates if the stratum is uniform. Here, the following can be obtained by substituting the effective saturation $S = (\theta - \theta_r)/(\theta_s - \theta_r)$ into the above equation using the volume moisture content θs at saturation and the residual volume moisture content θ_r.

$$\frac{\partial S}{\partial t} = D\frac{\partial^2 S}{\partial z^2}. \tag{4}$$

Next, the initial conditions and boundary conditions will be described. First, the following equation is assumed as the initial condition.

$$S(z,0) = S_i. \tag{5}$$

245

We consider a closed interval where z is $(0, L)$, and S_i is a constant value. Here, the following Neumann boundary conditions are introduced in order to deal with various boundary conditions.

$$\frac{\partial S(0,t)}{\partial z} = 0, \quad -D\frac{\partial S(\pm L,t)}{\partial z} = h(S - S_t). \tag{6}$$

S_t is a constant value of terminal saturation. Since the exact solution cannot be obtained as it is, we introduce dimensionless saturation $s_d(z, t) = (S(z, t) - S_t)/(S_i - S_t)$ and rewrite the problem as follows.

$$\frac{\partial s_d}{\partial t} = D\frac{\partial^2 s_d}{\partial z^2}, \tag{7}$$

$$s_d(z,0) = \frac{S(z,t) - S_t}{S_i - S_t} = 1 \tag{8}$$

and

$$\frac{\partial s_d(0,t)}{\partial z} = 0, \quad -\frac{\partial s_d(\pm L,t)}{\partial z} = hs_d. \tag{9}$$

First, the general solution of Eq. 7 can be expressed as follows.

$$s_d = (A\cos pz + B\sin pz)Ce^{-Dp^2t} \tag{10}$$

A, B, and C are undetermined coefficients, and p is a non-zero positive real number. By differentiating this equation with z and substituting $z = 0$, the following is obtained from the boundary conditions Eq.9

$$(-Ap\sin pz + Bp\cos pz)Ce^{-Dp^2t}|_{z=0} = BpCe^{-Dp^2t} = 0 \tag{11}$$

When C is zero, s_d is always zero, so $B = 0$. Similarly, by substituting the boundary condition of $z = L$ in Eq. 9, the following is obtained.

$$-(-Ap\sin pz)Ce^{-Dp^2t}|_{z=L} = Ap(\sin pL)Ce^{-Dp^2t} = hA(\cos pL)Ce^{-Dp^2t} \tag{12}$$

Therefore, the following relational expression of p is obtained.

$$p\tan pL = h \tag{13}$$

If the solutions that satisfy Eq. 13 are $p_1, p_2, p_3 \cdots$, then their linear sum is also the solution, so s_d can be expressed as follows.

$$s_d = \sum_{n=1}^{\infty} (C_n\cos p_n z)e^{-Dp_n^2 t} \tag{14}$$

Substituting the initial condition Eq. 8 into this equation as

$$s_d(z,0) = 1 = \sum_{n=1}^{\infty} (C_n \cos p_n z) \tag{15}$$

To determine the Fourier coefficient C_n, the right-hand side of the above equation in the case of n by $\cos p_m$, $(m = 1, 2, \cdots)$ are multiplied and integrated. This integral takes a value only when $m = n$ due to orthogonality of trigonometric function, as shown below.

$$\int_0^L C_n \cos p_n z \cdot \cos p_m z dz = C_n \left(\frac{\sin(2p_n L)}{4p_n} + \frac{L}{2} \right) \tag{16}$$

Therefore, this equation is equal to the following equation.

$$\int_0^L 1 \cdot \cos p_m z dz = \frac{\sin(p_m L)}{p_m} \tag{17}$$

From the above, C_n can be obtained as follows.

$$C_n = \frac{4 \sin(p_n L)}{\sin(2p_n L) + 2p_n L} \tag{18}$$

Therefore, the exact solution of s_n is given as follows.

$$s_n = \sum_{n=1}^{\infty} \frac{4 \sin(p_n L)}{\sin(2p_n L) + 2p_n L} (\cos p_n z) e^{-Dp_n^2 t} \tag{19}$$

When the change of variables in Eq. (8) is taken back, an exact solution for saturation degree S can be obtained by setting $\beta_n = p n L$.

$$S(z,t) = S_t + (S_i - S_t) \sum_{n=1}^{\infty} \frac{4 \sin(\beta_n)}{\sin(2\beta_n) + 2\beta_n} (\cos \beta_n \, z/L) e^{-D\beta_n^2/L^2} \tag{20}$$

From Eq. 13, β_n is the solution of the following transcendental function, which is solved by the Newton- Rapson method.

$$\frac{\beta_n}{Lh} = \cos \beta_n \tag{21}$$

3 RESULTS AND SOLUTION VERIFICATION

First, the nature of the solution was verified using h as a parameter. h is called the heat transfer coefficient in the heat equation and changes the degree of heat transfer at the boundary. Figure 1 shows relationship changes between degree of saturation S and distance z due to h. At $h = 1$, it becomes an adiabatic boundary where Sr changes depending on the boundary conditions. At $h \approx 0$, S decreases uniformly with a constant saturation. This is close to the conditions assumed in the element test. When h shows a value different from 0 or 1, it is a heat transfer boundary. In this way, various boundary conditions can be expressed by the value of h. It is not known how the change in water content changes at the edges of the material as it dries. It is extremely important to prepare solutions according to various conditions in this way.

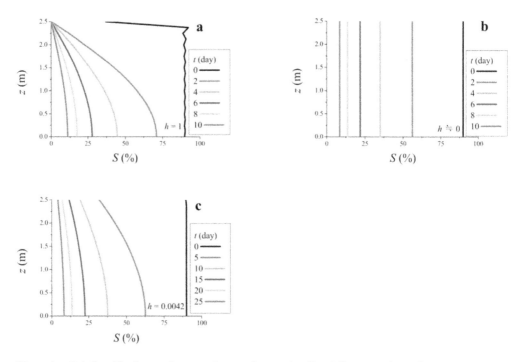

Figure 1.   Relationship changes between degree of saturation S and distance z due to h.

Next, the applicability of the exact solution shown here is verified by comparing it with the experimental results. The dry deformation experiment is shown in Figure 2. This is an experiment in which a wet specimen is dried in the air. The measurement items are deformation (strain) and mass change, and the water content change is specified from the mass change. Figure 3 compares the exact solution and the change in saturation due to the experimental results. The saturation calculated in the experiment is accurate because it takes into account the change in pore space due to deformation (Togashi et al. 2021a). It can be seen that the change in saturation in the drying phenomenon can be expressed fairly accurately. The values of h and D were adjusted appropriately here, but even if it is difficult to identify h, it is possible to identify D by conducting a water retention test. In that case, we think that any change in saturation can be expressed only by the value of h.

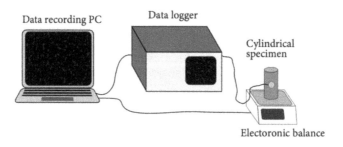

Figure 2.   Drying test for tuff (Togashi et al. 2021a).

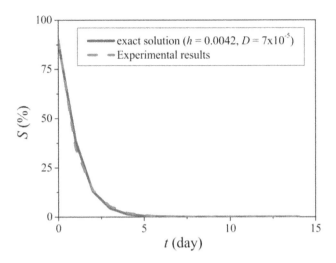

Figure 3. Comparison between test results and exact solution.

4 CONCLUSIONS

In this study, we derived the exact solution of the Richards equation of the heat transfer boundary and confirmed the nature of the exact solution. In addition, it was shown that the results of the dry deformation experiment based on previous studies can be appropriately expressed by the proposed exact solution. In the future, we plan to apply it to a method for estimating changes in saturation in the area affected by excavation.

ACKNOWLEDGEMENTS

This work was supported in part by a grant from the Ministry of Economy, Trade and Industry, Japan.

REFERENCES

Broadbridge P, Daly E and Goard J. 2017. Exact Solutions of the Richards Equation With Nonlinear Plant-Root Extraction. Water Resources Research. 51. https://doi.org/10.1002/2017WR021097

Ekpoudom II, Jacob AI and Itaketo UT. 2017. Analytical Solution of Richards'Equation for Application in Automatic Irrigation Systems. International Journal of Theoretical and Applied Mathematics. 3 (5),163–166.

Hooshyar M and Wang D. 2016. An analytical solution of Richards' equation providing the physical basis of SCS curve number method and its proportionality relationship. Water Resources Research. 52, 6611–6620.

Ricards LA. 1931. Capillary conduction of liquids through porous mediums. Physics. 1(5),318–333

Osada M. 2014. Drying-induced deformation characteristics of rocks. Proc ISRM Int Sym – ARMS8. ISRM-ARMS8-2014–001.

Sugawara K, Sayama K and Takara K. 2018. An analytical solution of Richards equation for soil with groundwater table. Proc. JSCE. 74(4),I1–I6. (In Japanese)

Togashi Y, Kikumoto M, Tani K, Hosoda K and Ogawa K. 2018. Detection of deformation anisotropy of tuff by a single triaxial test on a single specimen. International Journal of Rock Mechanics and Mining Sciences. 108, 23–36. https://doi.org/10.1016/j.ijrmms.2018.04.054

Togashi Y, Kikumoto M, Tani K, Hosoda K and Ogawa K. 2019. Non-axisymmetric and/or non- elementary response of anisotropic tuff in axisymmetric, elementary triaxial test. E3S Web of Conferences. 02008. https://doi.org/10.1051/e3sconf/20199202008

Togashi Y, Osada M and Yamabe T. 2020. Rotation of Three Principal Strains of Tuff during Drying Shrinkage. Proc ISRM Int Sym – EUROCK 2020. ISRM-EUROCK-2020-175.

Togashi Y, Imano T, Osada M, Hosoda K and Ogawa K. 2021a. Principal strain rotation of anisotropic tuff due to continuous water-content variation. International Journal of Rock Mechanics and Mining Sciences. 138, 104646. https://doi.org/10.1016/j.ijrmms.2021.104646

Togashi Y, Imano T and Osada M. 2021b. Deformation characteristics of sedimentary rock due to continuous changes of moisture content in wetting process. IOP Conference Series: Earth and Environmental Science. 703(1), 012021. https://doi.org/10.1088/1755-1315/703/1/012021

Rock Mechanics and Engineering Geology in Volcanic Fields – Ohta, Ito & Osada (eds)

Experimental study on the mechanical properties of past submarine landslide layers including volcaniclastic material

Yang Li, Hinako Hosono & Takato Takemura
Department of Earth and Environmental Sciences, Nihon University, Tokyo, Japan

Daisuke Asahina*
Geological Survey of Japan, AIST, Tsukuba, Japan

ABSTRACT: Past submarine landslides are often uplifted by tectonic movements and appear on land. The layer of the past submarine landslides are mechanically weaker than the surrounding strata and may cause landslides. The stability of a past submarine landslide is affected by (i) a load of overlying strata, (ii) pore pressure acting on the past submarine landslide surface, and (iii) the shear strength of the surrounding rock including the submarine landslides surface.

In this paper, we conduct a direct shear test and rebound hardness test to study the mechanical characteristics of past submarine landslides. The samples used in this study are tuffaceous sandstone (white pumiceous volcaniclastic material) and mudstone next to the weak surface of past submarine landslide in the Pleistocene Kiwada Formation of the Kazusa Group along the Isumi River in Otaki-cho, Chiba Prefecture, central Japan. We study the effect of the saturation of rock samples on friction strength and the relationship between hardness distribution and the weak surface was investigated. The proposed experimental approach shows that the present mechanical properties of the past submarine landslide surface can be used to evaluate the reactivation of the weak layer.

Keywords: Submarine landslides, Rebound hardness test, Friction test, Tuffaceous sandstone

1 INTRODUCTION

Some past submarine landslide sediments have been uplifted by tectonic movement and are now lying onshore. It is not known whether the slip surfaces of past submarine landslides still retain the mechanical properties as a weak surface. The stability of a past submarine landslide is affected by (i) a load of overlying strata, (ii) pore pressure acting on the past submarine landslide surface, and (iii) the shear strength of the surrounding rock including the submarine landslides surface.

The past submarine landslide sediments investigated in this study are the Pleistocene Kiwada Formation of the Kazusa Group along the Isumi River in Otaki-Cho, Chiba Prefecture, central Japan. The sliding surface of the submarine landslide is the boundary between mudstone and pumice layers. The cause of the submarine landslide is thought to be a decrease in shear strength due to an increase in pore pressure in the pumice layer (Utsunomiya, 2019). At the time of the past submarine landslide, the sediments were considered to be unconsolidated and the slope was low-gradient at a low angle.

*Corresponding author: d-asahina@aist.go.jp

DOI: 10.1201/9781003293590-31

Even if the angle of slope is low, submarine landslides are caused by the formation of a water film (e.g. Kawakita et al., 2020). However, due to the effects of consolidation and diagenesis over time, it is unlikely that past slip surfaces will continue to act as a slip surface. To evaluate the reactivity of past submarine landslides, we conducted in-situ rebound hardness tests and friction tests in the laboratory on the tuffaceous sandstone (white pumiceous volcaniclastic material) and mudstone.

2 MATERIALS AND METHODS

2.1 *Measuring and sampling location*

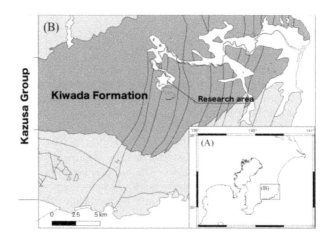

Figure 1.   Location map of the research area in Chiba prefecture, central Japan.

Past submarine landslide sediment can be observed in the Pleistocene Kiwada Formation of the Kazusa group along the Isumi river in Otaki-Cho, Chiba prefecture, central Japan (Figure 1). Here, the Kazusa group is a 3000 m-thick basin located on the western extension of the Kazusa basin which is terminated by the Preneogene accretionary complex deformed in response to the Izu–Bonin arc collision with the Honshu arc and is well exposed on the Boso Peninsula. The lower and middle Kazusa group consist mainly of deep-marine and upper-slope sediments. The Kiwada formation is in the middle Kazusa group and consists mainly of siltstone or mudstone in which thin turbidite sandstone and tuff beds are intercalated (Utsu-nomiya et al., 2019).

2.2 *Rebound hardness test*

We use the Equotip hardness tester (EHT) to measure the surface dynamic hardness. The EHT is a portable, electronic battery-operated NDT instrument, initially designed for measuring the surface rebound or dynamic hardness of metals and polymers. The EHT is designed base on the Leeb hardness test. This test is based on the principle of measuring the kinetic energy loss of an impact body after impacting on a test surface. During the impact, a permanent magnet integrated in the impact body passes through a coil in which voltage is induced by the forward and backward movement (Yilmaz & Goktan, 2019). The model of the EHT which we used in the rebound hardness test is Equotip 3-Portable Metal Hardness Tester. The impact device which we used is the basic type D impact. The unit of the surface hardness is the hardness value L with impact device D (HLD). The accuracy is ±4 HL. The value of hardness is the average value of 3 times at one point.

The in-situ surface hardness is measured 5 cm interval mesh. The measuring points are distributed in three successive layers including the sliding plane and showed in Figure 2. Here, Kb8A is the tuffaceous sandstone, and the sliding plane is between mudstone layers. The detail of the Kb8A tuff bed is described in Utsunomiya et al. (2019).

2.3 *Friction test*

The friction test of the tuffaceous sandstone and the mudstone which were sampled from the outcrop was conducted. The hand specimens of the tuffaceous sandstone and mudstone were taken from the Kb8A and Kb8B layers at the outcrop. The specimens were dried and ground into powder, and the powder was used as the specimen for the friction test. The friction test was conducted using a direct shear test apparatus and the powder specimens were placed between steel blocks with uneven surfaces.

The vertical load, shear load, and shear displacement were measured. The upper block was subjected to a vertical load of 15 kN, while the lower block was displaced at a shear rate of 1 μm/s. To make the sample fit the steel block, we keep the powder specimen under load for over 30 minutes before shearing. The friction tests were carried out under dry and saturated conditions. Here, the saturated condition was carried out in a water bath attached to the lower steel block, therefore, the test condition is a drainage condition in which no excess pore water pressure is generated.

3 RESULTS

The results of the in-situ rebound hardness test were shown as Figure 2. With the EHT surface hardness data, white tuffaceous sandstone is generally harder than the mudstone. In the mixing band, the part of tuffaceous shows the high value of hardness, the part of mudstone shows the value as the base mudstone.

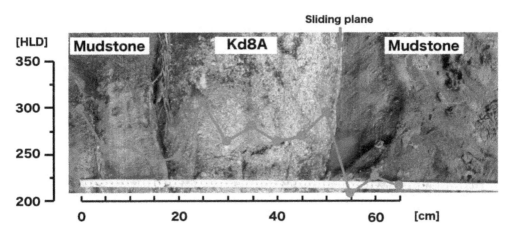

Figure 2. The distribution of the surface hardness which the measured in field. The red line is the hardness which measured by Equip-3. The circles show the measured points.

The results of the friction test are shown in Figure 3. The friction coefficient of the tuffaceous sandstone is smaller than the mudstone under the dry condition. Under the saturated condition, the friction coefficient of the tuffaceous sandstone is larger than the mudstone. The friction coefficients of the specimens were μMud =0.78 for the mudstone and μTuff =0.68 for the tuffaceous sandstone under dry conditions, and μ'Mud =0.42 for the mudstone and μ'Tuff =0.47 for the tuffaceous sandstone under saturated conditions, respectively. Friction coefficients differed greatly between dry and saturated conditions, and tended to differ between mudstone

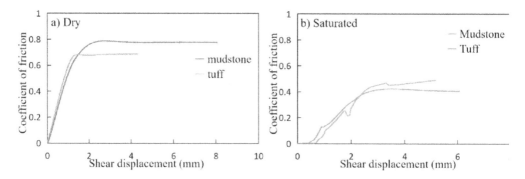

Figure 3. Results of friction tests, a) dry conditions, b) saturated conditions.

samples and tuff, that is, the friction coefficient is μMud >μTuff under dry conditions and μ'Mud < μ'Tuff under saturated conditions. The difference in friction coefficients between dry and saturated conditions may be attributed to the amount ratio of clay minerals.

4 DISCUSSION AND CONCLUSION

Submarine landslide sediments were unconsolidated and their cohesive force might be small, therefore, the friction coefficient under saturation conditions could control the sliding plane. From the results of friction tests, it is considered that the mudstone layer with the lower friction coefficient became the sliding plane. When considering the re-activation of landslides after tectonic uplift of submarine landslide sediments, it is necessary to consider not only the friction coefficient but also the degree of solidification. The friction coefficient is lower in tuffaceous sandstone than in mudstone, but the degree of solidification indicated EHT surface hardness is higher in tuffaceous sandstone than in mudstone, when the slip surface is unsaturated. It is suggested that both mudstone and tuffaceous sandstone can function as a sliding plane, depending on the balance between friction coefficient and solidity.

It is necessary to evaluate the degree of solidification of tuffaceous sediments appropriately and to discuss it together with the shear strength when considering the re-activation of a sliding plane of submarine landslide sediment after tectonic uplift. In addition, it will be important to accumulate these physical properties using in-situ and boring core samples, and to interpret them geologically based on the consolidation and diagenesis processes.

REFERENCES

Kawakita, S., Asahina, D., Takemura, T. Hosono, H., Kitajima, K., 2020, Effect of hydraulic and mechanical characteristics of sediment layers on water film formation in submarine landslides. Prog Earth Planet Sci 7, 62. https://doi.org/10.1186/s40645-020-00375-7

Utsunomiya, M., Noda, A., Otsubo, M., 2019, Preferential formation of a slide plane in translational submarine landslide deposits in a Pleistocene forearc basin fill exposed in east-central Japan. Geological Society London Special Publications 477 (1):SP477.3

Yilmaz, N. G., Goktan, R.M., 2019, Comparison and combination of two NDT methods with implications for compressive strength evaluation of selected masonry and building stones. Bull. Eng. Geol. Environ. 78, 4493–4503.

Rock Mechanics and Engineering Geology in Volcanic Fields – Ohta, Ito & Osada (eds)
© 2023 copyright the Author(s), ISBN 978-1-032-27657-1

Volcanic rocks characterization based on soft computing algorithms by using VRS empirical system

Luís Ribeiro e Sousa
Construct, School of Engineering, University of Porto, Portugal

Joaquim Tinoco
ISISE, University of Minho, Guimarães, Portugal

Rita Leal e Sousa*
Department of Civil, Environment and Ocean Engineering, Stevens Institute of Technology, Hoboken, USA

Karim Karam
Sarooj, Muscat, Oman

António Topa Gomes
Construct, School of Engineering, University of Porto, Portugal

ABSTRACT: This paper characterizes volcanic rocks through the application of an empirical geomechanical system designated as VRS (Volcanic Rock System). For the VRS, geotechnical information was collected from samples from several Atlantic Ocean islands that include Madeira and Canarias archipelagos, taking also into consideration data from other different sources. The new empirical system is based on the consideration of six geological-geotechnical parameters to which relative weights are attributed. The final VRS index value, which varies between 0 and 100, is obtained through the algebraic sum of these weights. Some representative correlations were obtained between VRS coefficients and RMR and GSI values. Correlations were obtained between deformability rock mass modulus and VRS with an exponential expression and also for each rock type. In addition, the volcanic rock formations were also analyzed using a decision tree algorithm and considering variables from the VRS and RMR systems.

Keywords: Volcanic rocks, Geomechanical characterization, VRS empirical system, Data Mining (DM)

1 INITIAL CONSIDERATIONS

Evaluation of the geomechanical parameters of rock masses can be carried out using the empirical classification systems. These systems consider, between others, the properties like the strength of the rock, density, condition and orientation of discontinuities, groundwater conditions and the stress state. To evaluate these properties, a numerical measure is given and, subsequently, a final geomechanical index is obtained by applying a numerical expression associated with the system. The result allows classifying the rock mass in a certain class associated with important information for the design like in some cases construction sequences, support needs and geomechanical parameters.

*Corresponding author: rsousa@stevens.edu

DOI: 10.1201/9781003293590-32

The purpose of this paper is to analyze the geomechanical behavior of volcanic rock formations, characterize them, and develop an empirical system, as well as to apply Data Mining (DM) techniques in order to develop new models. The empirical systems for volcanic rocks have been designated by VRS (Volcanic Rock System). Geotechnical information was collected from samples from several Atlantic Ocean islands that include Madeira, Azores and Canarias archipelagos, taking into consideration the data from different sources (Miranda et al., 2018). The various rock types are described with particular emphasis on the Madeira Island rock formations and their geomechanical properties.

Artificial Intelligence (AI) techniques are progressing very rapidly since 1956. AI today is labeled as a narrow when it is designed to perform a specific task or labeled general when designed to outperform humans at a very cognitive task (Sousa et al., 2018). The prediction of geotechnical formation behavior in geoengineering is complex because of the uncertainties in characterizing rock masses. In large projects, the real amount of geotechnical data that is generated and collected can be used to reduce uncertainties (Miranda and Sousa, 2012).

Data can hold valuable information such trends and patterns that can be used to improve decision making and optimize processes. Therefore, it is necessary to define standard ways of collecting, organizing and representing data. There are automatic tools from the field of AI and pattern recognition that enable one to analyze and interpret data using DM techniques (Witten et al., 2011, Leskove et al., 2014). DM is an area of computer science that lies at the intersection of statistics, machine learning, data management and databases, pattern recognition, artificial intelligence and other areas.

2   VRS EMPIRICAL SYSTEM TO VOLCANIC ROCKS

The VRS empirical system for volcanic rocks is an adaptation of the RMR system and includes a classification developed at São Paulo, for tunnels in basaltic formations (Ojima, 1981, Menezes et al. 2005). The new empirical system is based, like RMR system, on the consideration of six geological and geotechnical parameters to which relative weights are attributed. The final VRS index value, which varies between 0 and 100, is obtained through the algebraic sum of these weights (Miranda et al., 2018). With this index, it is possible to obtain strength properties, deformability moduli, and description of the rock mass quality, as well as recommendations for excavation and support needs and support loads, using correlations with other geomechanical indices.

The following geomechanical parameters were considered: P_1 - UCS; P_2 - rock weathering characteristics; P_3 - intensity of jointing; P_4 - discontinuity conditions; P_5 - presence of water; P_6 - disposition of blocks. Different weights are assigned to each parameter, as illustrated in Figure 1. In relation to RMR, the properties were identical for P_1, P_4 and P_5, but have different weights. The parameter due to discontinuities orientation P_6, introduced by Bieniawski (1989) as an adjustment of the sum of the remaining five parameters, was difficult to assign a weight, because it depends on groundwater conditions. Instead, it was substituted by another parameter related to the disposition of blocks. This parameter is considered to evaluate block stability. Four situations were considered: blocks of very favorable, favorable, acceptable and not acceptable which refer to the stability of the geotechnical structure. The VRS system considers for P_2 the rock weathering effect which is not considered by the RMR system, while P_3 is related to the joint intensity combining the effects of parameters P_2 (RQD) and P_3 (discontinuity spacing) considered by RMR system. The meaning of different parameters is given in Figure 1. The values that each parameter from P_1 to P_6 are indicated in Miranda et al. (2018).

A rock mass is classified into six classes. A rock mass designated as class VI has a behavior conditioned by the rock characteristics of deformability and strength, while

P₁	UCS	R₁	R₂	R₃	R₄	R₅
	(weight)	(15)	(9)	(6)	(3)	(1)
P_2	Rock weathering	A_1	A_2	A_3		
	(weight)	(20)	(12)	(4)		
P_3	Joint frequency	F_1	F_2	F_3	F_4	F_5
	(weight)	(25)	(20)	(15)	(10)	(5)
P_4	Joint surface conditions	B_1	B_2	B_3	B_4	B_5
	(weight)	(30)	(25)	(17)	(10)	(0)
P_5	Presence of water	C_1	C_2	C_3	C_4	
	(weight)	(10)	(7)	(4)	(0)	
P_6	Block position	D_1	D_2	D_3	D_4	
	(weight)	(0)	(-2)	(-5)	(-10)	

Figure 1. VRS classification and weights for the system.

a formation designated as class I behaves in accordance with the characteristics of the discontinuities. For rock masses with other classes, behavior is determined by the combination of both types of characteristics.

The collected data were organized and structured in a database composed of 108 records with 29 attributes (Miranda et al. 2018). The data were mainly obtained from Madeira Island (76%), with the rest from Canarias Islands (18%) and Mexico (6%). In the database, the deformability modulus of the rock mass (E_{RM}) was derived from the Serafim and Pereira (1983) formula, assuming the restriction of RMR<80. GSI was only calculated for RMR>23 according to the Hoek and Brown criterion (Hoek, 2007). The values of cohesion and internal friction angle were obtained through the software Roc-Data (Rocscience, 2015).

3 APPLICATION OF DM TECHNIQUES TO THE DATABASE

Prediction of geotechnical formation behavior in geoengineering is complex because of the uncertainties associated with the characterization of rock masses. The database can hold valuable information such as trends and patterns that can be used to improve decision making and optimization processes. It is however necessary to define standard ways of collecting, organizing and representing the data. AI tools and pattern recognition techniques enable one to analyze datasets to retrieve information there (Witten et al., 2011, Leskove et al., 2014). DM is an area of computer science that lies at the intersection of statistics, machine learning, data management and databases and pattern recognition. The formal and complete analysis process is called Knowledge Discovery from Databases (KDD) that defines the main procedures for transforming data into knowledge (Cortez, 2010).

In this work, a Decision Tree (DT) was adopted for volcanic rocks characterization. A DT is a direct and acyclic flow chart that represents a set of rules distinguishing classes or values in a hierarchical form (Berry and Linoff, 2000).). These rules are extracted from the data, using rule induction techniques, and appear in an "if-then" structure expressing a simple and conditional logic. Source data are split into subsets based on the attribute test values, and the process is repeated in a recursive manner.

All experiments were conducted using the R statistical environment (R Development Core Team, 2010) and supported through the *RMiner* package (Cortez, 2010), which facilitates the implementation of several DM algorithms (e.g., ANNs, SVMs or DTs algorithms), as well as different validation approaches such as cross-validation. For models' evaluation and comparison, three classification metrics were used based on the confusion matrix (Hastie et al., 2009, Miranda et al., 2018).

A hierarchical volcanic rock mass rating was developed based on a DT algorithm, taking as model inputs P_i (i=1,2 ...6) variables from the classification system of VRS (from here named HVR). A similar approach was followed but considering instead attributes P_i (i=1,2 ...6) variables from the RMR system (from here named HRMR). Figure 2 depicts the decision trees for HVR and HRMR systems.

Figure 3 shows the observed versus predicted classes using the HVR and HRMR models. For each observed class (x-axis), the percentage of each predicted class (y-axis) is shown. Figure 3a shows that the HVR model is unable to correctly identify class 1 and its best

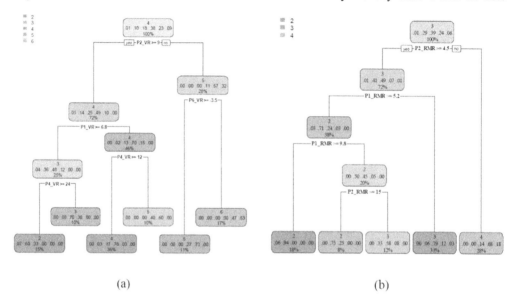

(a)   (b)

Figure 2. Decision trees for (a) HVR and (b) HRMR systems. Each node shows the predict class (first row), the predicted probability of each class (second row) and the percentage of observations in the node (third row).

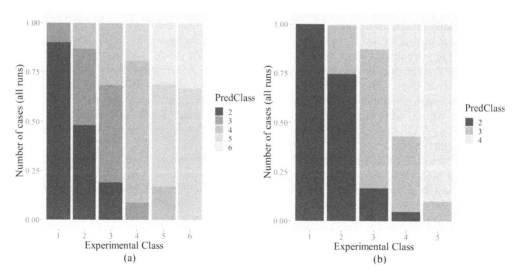

Figure 3. (a) HVR (a) and (b) HRMR performance.

performance is for class 4. Around 90% of VR identified as 1 (true condition) was classified by the HRV model as 2 and 10% classified as 3. Also, the HRMR model (Figure 3b) is unable to correctly identify VR class. The proposed DT is not able to identify classes 1 and 5. The best performance is observed for class 2, for which an F1-score around 73% was achieved. For classes 3 and 4, the proposed system seems to perform slightly well.

Finally, in Figure 4 is illustrated the relative importance of each input variable for both HVR and HRMR models. According to the proposed DT based on the HVR system, the three most relevant variables are P_2, P_4 and P_1 with an influence close to 30% each. Also, according to the DT based on the RMR system, the three most relevant variables are P_1, P_4 and P_2, with a total influence around 94%.

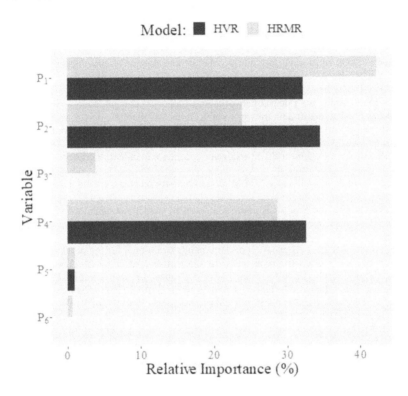

Figure 4.   HVR and HRMR relative importance.

4   CONCLUSIONS

The VRS empirical geomechanical classification system was developed specifically for volcanic rocks by adapting the more traditional RMR system. A database of volcanic rocks was created using mainly geomechanical information from different archipelagos. The VRS was calibrated and correlated with RMR system.

Decision Trees (DT) algorithm was applied to predicting VR classes. Parameter P_4 (discontinuity conditions) from the VRS is the most relevant variable and P_2 (rock weathering characteristics) is the second most influential parameter. Considering variables from the VRS and RMR systems, a better performance is achieved using attributes from the VRS.

According to hierarchical models, the HVR model was unable to identify correctly class 1 and its higher performance is observed for class 4. Also, the HRMR model was unable to correctly identify VR classes. Indeed, the proposed DT is not able to identify classes 1 and 5. The best performance is observed for class 2. For classes 3 and 4, the proposed system seems to perform slightly well. Although the initial classification attempt requires further improvements, this first attempt shows that the use of data mining tools in the study of volcanic rocks could be very useful, with important costs reduction. Moreover, the use of sensitivity analysis can help in the clarification (human understanding) of the high complexity of these models, in particular by measuring the relative importance of model attributes.

ACKNOWLEDGEMENTS

This work was financially supported by: Base Funding - UIDB/04708/2020 and Programmatic Funding - UIDP/04708/2020 of the CONSTRUCT - Instituto de I&D em Estruturas e Construções - funded by national funds through the FCT/MCTES (PIDDAC).

REFERENCES

Berry M, Linoff G 2000 Mastering data mining: the art and science of customer relationships management. New York: John Wiley and Sons.

Bieniawski Z 1989 Engineering rock mass classifications. John Wiley & Sons.

Cortez P. 2010 Data mining with neural networks and support vector machines using the R/miner tool. In: Proceedings of advances in data mining e applications and theoretical aspects; 572–583.

Hastie T, Tibshirani R and Friedman J. 2009 The elements of statistical learning: data mining, inference, and prediction. New York: Springer-Verlag.

Hoek E 2007 Practical rock engineering. https://www.rocscience.com/learning/hoeks-corner/course-notes-books.

Leskove J, Rajaraman A and Ullman J 2014 Mining of massive datasets, second edition, Cambridge University Press.

Menezes AT, Varela FM and Sousa LR 2005. Road tunnel Caniço-Camacha, Madeira Island. geotechnical study. In: Proceedings of the 2nd Portuguese-Spanish geotechnical symposium, 179–189 (in Portuguese).

Miranda T and Sousa LR 2012 Application of data mining techniques for the development og geomechanical characterization models for rock mass. In Innovative Numerical Modeling in Geomechanics, Eds. Sousa, Vargas, Fernandes & Azevedo, Taylor & Francis, London, Chapter 13, 245–264.

Miranda T, Sousa LR, Gomes AT, Tinoco J and Ferreira C 2018 Geomechanical characterization of volcanic rocks using empirical systems and data mining techniques. J. of Rock Mechanics and Geotechnical Engineering, 10, 138–150.

Ojima L 1981 Methodology of rock mass classifications for tunneling. LNEC Thesis. Lisbon (in Portuguese).

R Development Core Team (2010). R Foundation for Statistical Computing, Vienna.

Rocscience 2015 Rocscience. RocData, Version 5.003. 2015. https://www.rocscience.com.

Serafim JL and Pereira P 1983. Considerations on the geomechanics of Bieniawski. In: International symposium on engineering geology and underground construction, 33–42.

Sousa LR, Miranda T, Sousa LR and Tinoco J 2018 Deep underground engineering and the use of Artificial Intelligence techniques J. of Earth & Environmental Sciences, 3, 158.

Sousa LR, Miranda T, Sousa RL and Tinoco J 2021. The use of Data Mining techniques in rockburst assessment. J. Engineering, 3, 552–558.

Witten H, Frank E and Hall M 2011 Data Mining: Practical Machine Learning Tools and Techniques, 3rd edition, Morgan Kaufman Publishers.

Rock Mechanics and Engineering Geology in Volcanic Fields – Ohta, Ito & Osada (eds)
© 2023 copyright the Author(s), ISBN 978-1-032-27657-1

Defining the Hoek-Brown constant m_i for volcanic lithologies

Marlène C. Villeneuve*
Chair of Subsurface Engineering, Montanuniversität Leoben, Leoben, Austria

Michael J. Heap
CNRS, Institut Terre et Environnement de Strasbourg (ITES), Université de Strasbourg, France

Lauren N. Schaefer
U.S. Geological Survey, Golden, CO, USA

ABSTRACT: The empirical Hoek-Brown failure criterion is a well-known and commonly used failure criterion for intact rocks and rock masses, especially in geological engineering. The intact criterion is calculated using experimental triaxial compression test results on intact samples, whereas the rock mass criterion modifies the intact strength using quantified measures of the rock mass quality. The Hoek-Brown failure criterion includes a fitting constant for intact rocks, m_i, which controls the steepness and curvature of the failure envelope. Because of the existence of a table of m_i values for various rock categories, calculated using thousands of triaxial experiments, m_i values are often extracted from this table rather than fitted to the more time- and resource-intensive triaxial experiments.

Using hundreds of triaxial experiments on variously altered volcanic rocks from volcanoes around the world, we demonstrate that m_i varies dramatically based on a complex combination of alteration, lithology, and texture, for example ranging from 2-38 for andesite. In contrast, the tabulated values typically have small ranges, for example 25±5 for andesite. This means the failure criteria for volcanic rocks based on tabulated estimates could significantly over or under predict the intact strength, and thereby the rock mass strength, causing errors for stability and deformation assessments for a variety of volcanological and geological engineering purposes, from dome deformation or flank stability to excavation in volcanic rocks.

In this paper we not only highlight the high variability of m_i for volcanic rocks, but by building on published relationships between porosity and strength, we demonstrate that m_i too is sensitive to porosity. We propose a number of preliminary methods to constrain m_i values, including one using porosity.

Keywords: Rock mechanics, engineering geology, Hoek-Brown, m_i, failure criterion

1 INTRODUCTION

There is considerable literature on the mechanical behavior of intact, typically fresh/unaltered, commonly encountered lithologies, like granite and sandstone, while metamorphic and other sedimentary rocks, such as gneiss, schist and shale have also been reasonably well studied. In the last decade, numerous uniaxial and triaxial experiments conducted on fresh and altered (weathered and/or hydrothermally altered) extrusive

*Corresponding author: marlene.villeneuve@unileoben.ac.at

DOI: 10.1201/9781003293590-33

volcanic rocks have provided a dataset large enough to explore the impact of geological characteristics on mechanical behavior of these rocks (Heap and Violay, 2021).

Failure criteria, such as the Mohr-Coulomb or Hoek-Brown, are necessary for assessing brittle rock behavior under different stress conditions, such as slope stability, deformation and excavation analyses. The generalized Hoek-Brown failure criterion (Eberhardt, 2012) provides a brittle failure criterion for both the laboratory and the field scales, making it particularly applicable for assessing field-scale volcanic behavior, such as edifice deformation and instability. It can also be converted to Mohr-Coulomb parameters (Hoek and Brown, 1997) as necessary. While both of these failure criteria can be derived using a series of triaxial experiments, many practitioners and researchers do not have the necessary laboratory equipment, and depend instead on published values or rules of thumb. In the case of the Hoek-Brown failure criterion, the uniaxial compressive strength (UCS), an empirical constant, m_i, and the geological strength index (GSI) are required to derive the rock mass failure criterion. The UCS can be easily tested in the laboratory or estimated in the field, whereas the GSI can be observed in the field (Hoek et al., 2013) or estimated based on previous work (Heap et al., 2020). The empirical constant, m_i, however, is often taken from the rock category-based table in Hoek (2007). Despite the existence of a wide range of m_i values for volcanic lithologies (Richards and Read, 2011), m_i has not been studied in depth. While many methods for estimating m_i from mechanical experiments have been proposed (Cai, 2010; Richards and Read, 2013), we have sought to explore the dependence of m_i on geological characteristics to determine if a transfer function could be developed.

Recently published experimental data reviews by Heap and Violay (2021) and Heap and Villeneuve (2021) have highlighted that porosity is the dominant factor controlling strength and stiffness of extrusive volcanic rocks. Since porosity can be measured relatively easily and quickly in the laboratory and density in the field (e.g. Farquharson et al., 2015) using submersion techniques, porosity is an ideal candidate for developing a transfer function to estimate the Hoek-Brown failure criterion parameter m_i. We provide the results of an investigation of the effect of porosity on m_i and propose a method for using transfer functions based on porosity to estimate m_i when triaxial experimental results are not available.

2 DETERMINING M_I FROM TRIAXIAL EXPERIMENTS

The Hoek-Brown failure criterion for intact rock is an empirical curve fitting experimental triaxial deformation results, formulated as in Eq. (1).

$$\sigma'_1 = \sigma'_3 + C_o \left(m_i \frac{\sigma'_3}{C_o} + 1 \right)^{0.5} \tag{1}$$

where C_o is the fitted uniaxial compressive strength and m_i is an empirical fitting parameter, which are calculated according to Eq. (2) and Eq. (3), respectively (appendix in Hoek and Brown, 1997).

$$C_o = \sqrt{\frac{\Sigma y}{n} - \left[\frac{\Sigma xy - \left(\frac{\Sigma x \Sigma y}{n} \right)}{\Sigma x^2 - \left(\frac{(\Sigma x)^2}{n} \right)} \right] \frac{\Sigma x}{n}} \tag{2}$$

$$m_i = \frac{1}{C_o} \left[\frac{\Sigma xy - \left(\frac{\Sigma x \Sigma y}{n} \right)}{\Sigma x^2 - \left(\frac{(\Sigma x)^2}{n} \right)} \right] \tag{3}$$

where $x = \sigma_3$, $y = (\sigma_1 - \sigma_3)^2$ and n is the number of specimens tested.

We obtained wet and dry, room temperature experimental triaxial deformation results on mostly 20-mm diameter plugs, loaded mostly at a strain rate of 10^{-5} from the literature cited in Table 1. We derived the m_i values and compared them to average porosity (given as a fraction; measured via fluid saturation or gas pycnometry methods) for the sample series. Because of sparse data, we used a minimum of three test results per sample series; however, a minimum of five samples is the standard (Hoek and Brown, 1997). Where stress-strain plots of the deformation experiments were available, we used them to select only results exhibiting brittle deformation for the regression. Where these were not available, we used $\sigma_3 > 0.5 C_o$ (Hoek and Brown, 1997) and/or $\sigma_1 < 4.4\sigma_3$ (Mogi, 1966) to select only results with brittle deformation. We note that these guidelines are important when deriving the m_i value because the experimental results are only valid for brittle behavior. This means that some datasets could not be fully utilized (e.g. Weydt et al., 2020) because the confining pressure was too high to derive either the m_i for Hoek-Brown failure or a reliable cohesion and friction angle for Mohr-Coulomb failure. We also note that including laboratory UCS improves the reliability (as measured by the coefficient of determination of the m_i value (Hoek and Brown, 1997) but is not necessary.

3 RESULTS

We derived m_i values from experimental results in the literature and compiled them along with porosity, C_o, and a short description of the rock and the source (Table 1). A statistical analysis of the dataset shows that the m_i data are moderately normally distributed (Shapiro-Wilks W = 0.9489 > 0.9, p value = 0.02238 < 0.05), with an arithmetic mean of 20, a geometric mean of 17 and a standard deviation of 11. The standard deviation is higher than in the commonly used table in Hoek (2007) but is in accordance with the findings of Richards and Read (2011). The histogram has a slight skewness towards the lower values, and moderately negative kurtosis, meaning that it has short tails (Figure 1, left). Porosity, conversely, is not normally distributed (Shapiro-Wilks W = 0.8776 < 0.9, p value = 0.00000521 < 0.05) and is highly skewed towards lower porosity (Figure 1, right), as expected since very high porosity rocks (>0.35) are fairly uncommon, are difficult to sample and test, and will be ductile even at low confining pressure (Heap and Violay, 2021). When separated into the categories used in the Hoek (2007) table for lava, (auto)breccia/agglomerate, and tuff, our arithmetic mean m_i values (22±10, 16±8, and 3±1, respectively) are lower than the values in the table (25±5, 19±5, and 13±5, respectively).

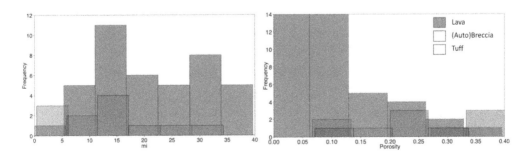

Figure 1.   Histograms of m_i (left) and porosity (right) for lava, (auto)breccia and tuff (see Table 1).

Table 1. Rock type, porosity fraction, fitted uniaxial compressive strength (C_o using Eq. 2) and the empirical Hoek-Brown constant (m_i using Eq. 3) of effusive volcanic rocks.

Rock Description	Porosity fraction	C_o (MPa)	m_i	Reference for source of experimental triaxial results
Altered Ruapehu andesite lava breccia	0.2	29	13	(Mordensky et al., 2022)
Unaltered Ruapehu andesite lava breccia	0.38	6	5	(Mordensky et al., 2022)
Altered Ruapehu andesite lava core	0.038	72	36	(Mordensky et al., 2022)
Unaltered Ruapehu andesite lava core	0.036	154	36	(Mordensky et al., 2022)
Volcan de Colima andesite	0.08	91	24	(Heap et al., 2015a)
Volcan de Colima andesite	0.11	68	34	(Heap et al., 2015a)
Mt. Etna basalt	0.04	189	22	(Heap et al., 2011)
Mt. Etna basalt	0.05	149	31	(Zhu et al., 2016)
Mt. Etna basalt	0.05	122	28	(Zhu et al., 2016)
Volvic trachyandesite	0.21	94	8	(Heap and Violay, 2021)
Mt. St. Helens dacite	0.18	48	19	(Heap et al., 2016)
Mt. St. Helens dacite	0.06	39	39	(Kennedy and Russell, 2012)
Mt. St. Helens dacite	0.10	16	16	(Smith et al., 2011)
Chaos Crags Dacite	0.12	49	34	(Ryan et al., 2020)
Whakaari dacite	0.06	97	18	(Heap et al., 2015b)
Whakaari altered tuff	0.30	53	4	(Heap et al., 2015b)
Altered Ruapehu andesite lava	0.15	103	16	(Schaefer et al., 2021)
Altered Ruapehu andesite lava	0.06	231	33	(Schaefer et al., 2021)
Altered Ruapehu andesite autobreccia	0.25	26	24	(Schaefer et al., 2021)
Altered Ruapehu andesite lava	0.06	166	39	(Schaefer et al., 2021)
Altered Ruapehu andesite autobreccia	0.40	10	12	(Schaefer et al., 2021)
Altered Ruapehu andesite lava	0.19	59	21	(Schaefer et al., 2021)
Altered Ruapehu andesite lava	0.09	112	25	(Schaefer et al., 2021)
Altered Ruapehu andesite lava	0.07	201	34	(Schaefer et al., 2021)
Altered Ruapehu andesite lava	0.09	109	29	(Schaefer et al., 2021)
Altered Ruapehu andesite lava	0.09	145	14	(Schaefer et al., 2021)
Altered andesite volcaniclastic (tephra)	0.48	10	21	(Schaefer et al., 2021)
Altered Ruapehu andesite lava	0.09	116	27	(Schaefer et al., 2021)
Unaltered Ruapehu andesite lava	0.06	179	48	(Schaefer et al., 2021)
Unaltered Ruapehu andesite lava	0.02	304	32	(Schaefer et al., 2021)
Unaltered Ruapehu andesite autobreccia	0.22	44	32	(Schaefer et al., 2021)
Altered Ruapehu andesite lava	0.07	147	31	(Schaefer et al., 2021)
Altered Ruapehu andesite lava	0.30	34	7	(Schaefer et al., 2021)
Altered andesite autobreccia	0.43	11	17	(Schaefer et al., 2021)
Altered Ruapehu andesite lava	0.20	80	22	(Schaefer et al., 2021)
Unaltered Ruapehu andesite autobreccia	0.33	48	6	(Schaefer et al., 2021)
Altered andesite lava (pyroclastic block)	0.04	90	13	(Schaefer et al., 2021)
Altered Ruapehu andesite lava	0.03	132	25	(Schaefer et al., 2021)
Altered Ruapehu andesite lava	0.17	65	14	(Schaefer et al., 2021)
Aquixtla altered andesite	0.21	56	8	(Weydt et al., 2020)
Aquixtla unaltered basaltic trachyandesite	0.12	197	13	(Weydt et al., 2020)

(Continued)

Table 1. *(Continued)*

Rock Description	Porosity fraction	C_o (MPa)	m_i	Reference for source of experimental triaxial results
Augile unaltered basaltic trachyandesite	0.009	344	10	(Weydt et al., 2020)
Pedernal altered rhyolitic lava	0.38	27	12	(Weydt et al., 2020)
Cuyoaco altered andesite	0.11	106	12	(Weydt et al., 2020)
Rotokawa altered andesite	0.096	100	15	(Siratovich et al., 2016)
Pinnacle ridge altered lava breccia	0.32	25	6	(Mordensky et al., 2018)
Volcan de Colima andesite	0.16	38.5	18	(Farquharson et al., 2016)
Emochi andesite	0.07	78	12	(Tang et al., 2018)
Tage Tuff	0.15	23	2	(Tang et al., 2018)
Ogino Tuff	0.11	31	3	(Tang et al., 2018)
Krafla altered basaltic hyaloclastite	0.13	44	10	(Eggertsson et al., 2020)
Krafla unaltered basaltic hyaloclastite	0.21	10	14	(Eggertsson et al., 2020)
Krafla unaltered basaltic hyaloclastite	0.12	45	15	(Eggertsson et al., 2020)

4 DISCUSSION

Rock type (Figure 2, left) or category (right) are not very useful for predicting m_i values; however, porosity/density can be. Heap et al. (2020) and Heap and Violay (2021) similarly found that the Young's modulus and uniaxial compressive strength of volcanic rocks depend more strongly on porosity than on rock type. The key observations are that m_i decreases with increased porosity, and that tuff and very high porosity rocks present as outliers. These have been removed for the development of a transfer function based on porosity (ϕ) for lava and (auto)breccia textures, given by Eq. (4):

$$m_i = 53.8e^{-1.4\phi} - 23.5; \text{when } \phi < 0.4 \tag{4}$$

Despite many relationships between mechanical properties and porosity following power functions (e.g. UCS), we also tested an exponential decay function. We used the Akaike information criterion (AIC, Akaike (1974)) to compare the functions because the r^2 goodness-of-fit test is not applicable to non-linear functions. Using the AIC, two models can be compared and the model with the lower AIC value is selected as the most suitable. We found that the exponential decay function (Figure 3) suits the dataset better, with a lower AIC than a power function (3451 vs. 3955).

The wide scatter in the dataset could arise from a number of areas, grouped into testing effects and natural variability. The data from the literature were tested under slightly different conditions, however we only selected those that were tested according to similar, minimum standards. More likely, based on the wide scatter observed in UCS and Young's modulus data for volcanics (e.g., Heap and Violay, 2021) even for tests conducted using identical methodology in a single lab (e.g., Schaefer et al., 2021) we attribute most of the scatter to natural variability. This suggests that more investigations into the impact of geological characteristics on m_i are needed, in addition to UCS and Young's modulus.

Given the wide scatter in the dataset, the transfer function should be used with an upper and lower bound of approximately $\pm$ 10 for samples with < 0.25 porosity and approximately $\pm$ 5 for samples with 0.25 < porosity < 0.4. Samples with very high porosity are generally extremely weak and will tend to compact (i.e. ductile behavior) rather than fail brittlely even at very low confining pressure and, thus, should not be characterized by the Hoek-Brown failure criterion. This dataset contains only three tuff samples, which cannot represent tuff in general, a rock type that is typically heterogeneous in terms of its physical and microstructural characteristics. Nevertheless, these samples appear to behave differently from lava and (auto) breccia and should be characterized by the Hoek-Brown failure criterion with care.

In the absence of porosity measurements, the transfer function in Eq. (4) can be used for field- or lab-derived bulk density, ρ_b, using the solid density, ρ_s, according to the method in Heap et al. (2020):

$$porosity = 1 - \frac{\rho_b}{\rho_s} \tag{5}$$

where the solid density has been measured as 2669, 2909, 2614 and 2307 kg/m³ for andesite, basalt, dacite and tuff, respectively (Heap et al., 2020). The solid density may deviate from these values for altered volcanics, depending on the type of alteration (i.e. alteration to low density clay versus dense epidote). We used this method to estimate porosity for altered samples from literature cited in Table 1 and found that the estimated porosity tended to be nearly identical for high-density alteration minerals and was typically within 0.05 of the reported porosity for low-density alteration minerals.

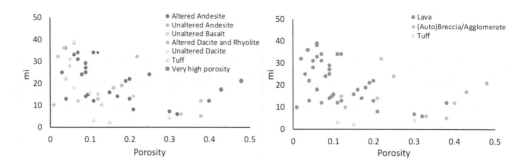

Figure 2.   Cross plots of m$_i$ as a function of porosity according to rock type (left) and category (right).

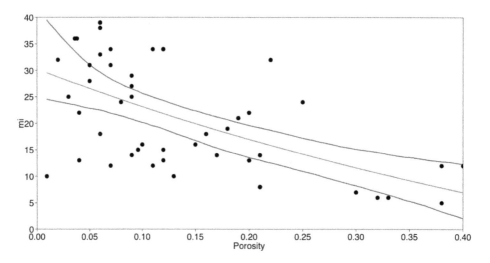

Figure 3.   M$_i$ as a function of porosity fraction with exponential transfer function (Eq. 4) (red) and 95% confidence interval (blue).

5 CONCLUSIONS

We present an analysis of recently published experimental triaxial deformation data that provides a large dataset with which to explore the potential for estimating the difficult to obtain, and often poorly estimated, empirical Hoek-Brown parameter m_i for volcanic rocks. First, we provide the simple regression equations for calculating m_i from experimental data. We then use these functions to produce the dataset and compare it with porosity. Finally, we provide a transfer function that uses porosity, an easily and commonly measured value in volcanic studies, to estimate m_i.

ACKNOWLEDGEMENTS

The authors thank all the collaborators who have worked in the field and laboratory to collect data and for their help developing simple solutions for parameterization of volcanic rocks as well as Stan Mordensky and Tamara Jeppson for their reviews.

REFERENCES

Akaike, H. (1974). A new look at the statistical model identification. IEEE T. Automat. Contr., 19(6), 716–723. https://doi.org/10.1109/TAC.1974.1100705

Cai, M., 2010. Practical estimates of tensile strength and Hoek–Brown strength parameter mi of brittle rocks. Rock Mech. and Rock Eng. 43(2),167–184. https://doi.org/10.1007/s00603-009-0053-1

Eberhardt, E., 2012. The Hoek–Brown failure criterion. Rock Mech. and Rock Eng. 45, 981–988.

Eggertsson, G.H., Kendrick, J.E., Weaver, J., Wallace, P.A., Utley, J.E.P., Bedford, J.D., Allen, M.J., Markússon, S.H., Worden, R.H., Faulkner, D.R., Lavallée, Y., 2020. Compaction of hyaloclastite from the active geothermal system at Krafla volcano, Iceland. Geofluids 2020, 1–17.

Farquharson, J., Heap, M.J., Baud, P., Reuschlé, T., Varley, N.R., 2016. Pore pressure embrittlement in a volcanic edifice. Bull. of Volcanol. 78, 6. https://doi.org/10.1007/s00445-015-0997-9

Farquharson, J., Heap, M.J., Varley, N.R., Baud, P., Reuschlé, T., 2015. Permeability and porosity relationships of edifice-forming andesites: A combined field and laboratory study. J Volcanol Geoth Res 297, 52–68. https://doi.org/10.1016/j.jvolgeores.2015.03.016

Heap, M.J., Baud, P., Meredith, P.G., Vinciguerra, S., Bell, A.F., Main, I.G., 2011. Brittle creep in basalt and its application to time-dependent volcano deformation. Earth Planet Sci. Lett. 307(1-2), 71–82.

Heap, MJ, Farquharson, J., Baud, P., Lavallée, Y., Reuschlé, T., 2015. Fracture and compaction of andesite in a volcanic edifice. Bull. Volcanol. 77, 55. https://doi.org/10.1007/s00445-015-0938-7

Heap, M., Kennedy, B., Pernin, N., Jacquemard, L., Baud, P., Farquharson, J., Scheu, B., Lavallée, Y., Gilg, A., Letham-Brake, M., Mayer, K., Jolly, A., Reuschlé, T., Dingwell, D., 2015. Mechanical behaviour and failure modes in the Whakaari (White Island volcano) hydrothermal system, New Zealand. J Volcanol Geoth Res 295, 26–42. https://doi.org/10.1016/j.jvolgeores.2015.02.012

Heap, M.J., Russell, J.K., Kennedy, L.A., 2016. Mechanical behaviour of dacite from Mount St. Helens (USA): A link between porosity and lava dome extrusion mechanism (dome or spine)? J. Volcanol. Geoth. Res. 328, 159–177. https://doi.org/10.1016/j.jvolgeores.2016.10.015

Heap, M.J., Villeneuve, M., Albino, F., Farquharson, J.I., Brothelande, E., Amelung, F., Got, J.-L., Baud, P., 2020. Towards more realistic values of elastic moduli for volcano modelling. J Volcanol Geoth Res 390, 106684. https://doi.org/10.1016/j.jvolgeores.2019.106684

Heap, M.J., Villeneuve, M.C., 2021 Calculating the cohesion and internal friction angle of volcanic rocks and rock masses. Volcanica 4(2),279–293. https://doi.org/10.30909/vol.04.02.279293

Heap, M.J., Violay, M.E.S., 2021. The mechanical behaviour and failure modes of volcanic rocks: a review. Bull. Volcanol. 83, 33. https://doi.org/10.1007/s00445-021-01447-2

Hoek, E., 2007. Practical Rock Engineering. Available online: https://www.rocscience.com/learning/hoeks-corner/course-notes-books

Hoek, E., Brown, E., 1997. Practical estimates of rock mass strength. Int. J. Rock Mech. Mining Sci. 34 (8),1165–1186. https://doi.org/10.1016/S1365-1609(97)80069-X

Hoek, E., Carter, T.G., Diederichs, M.S., 2013. Quantification of the geological strength index chart. 47th U.S. Rock Mechanics/Geomechanics Symposium. San Francisco, CA, USA, June 23- 26.

Kennedy, L.A., Russell, J.K., 2012. Cataclastic production of volcanic ash at Mount Saint Helens. Phys. Chem. Earth, Parts A/B/C 45-46, 40–49. https://doi.org/10.1016/j.pce.2011.07.052

Mogi, K., 1966. Pressure dependence of rock strength and transition from brittle fracture to ductile flow. Bull. Earthquake Res. Inst. 44, 215–232.

Mordensky, S.P., Villeneuve, M.C., Farquharson, J.I., Kennedy, B.M., Heap, M.J., Gravley, D.M., 2018. Rock mass properties and edifice strength data from Pinnacle Ridge, Mt. Ruapehu, New Zealand. J Volcanol. Geoth. Res. 367, 46–62. https://doi.org/10.1016/j.jvolgeores.2018.09.012

Mordensky, S.P., Villeneuve, M.C., Kennedy, B.M., Struthers, J. In review. Hydrothermally induced edifice destabilisation: the mechanical behaviour of rock mass surrounding a shallow intrusion in andesitic lavas, Pinnacle Ridge, Ruapehu (New Zealand). Eng. Geo.

Richards, L., Read, S., 2011. A comparison of methods for determining mi, the Hoek-Brown parameter for intact rock material. 45th U.S. Rock Mechanics/Geomechanics Symposium. San Francisco, CA, USA, June 26- 29.

Richards, L., Read, S., 2013. Estimation of Hoek-Brown parameter mi using Brazilian tensile test. 47th U.S. Rock Mechanics/Geomechanics Symposium. San Francisco, CA, USA, June 23- 26.

Ryan, A.G., Heap, M.J., Russell, J.K., Kennedy, L.A., Clynne, M.A., 2020. Cyclic shear zone cataclasis and sintering during lava dome extrusion: Insights from Chaos Crags, Lassen Volcanic Center (USA). J. Volcanol. Geoth. Res. 401, 106935.

Schaefer, L.N., Kereszturi, G., Villeneuve, M., Kennedy, B., 2021. Determining physical and mechanical volcanic rock properties via reflectance spectroscopy. J. Volcanol. Geoth. Res. 420, 107393. https://doi.org/10.1016/j.jvolgeores.2021.107393

Siratovich, P.A., Heap, M.J., Villeneuve, M.C., Cole, J.W., Kennedy, B.M., Davidson, J., Reuschlé, T., 2016. Mechanical behaviour of the Rotokawa Andesites (New Zealand): Insight into permeability evolution and stress-induced behaviour in an actively utilised geothermal reservoir. Geothermics 64, 163–179. https://doi.org/10.1016/j.geothermics.2016.05.005

Smith, R., Sammonds, P.R., Tuffen, H., Meredith, P.G., 2011. Evolution of the mechanics of the 2004–2008 Mt. St. Helens lava dome with time and temperature. Earth Plan. Sci. Lett. 307(1-2), 191–200.

Tang, Y., Okubo, S., Xu, J., Zhang, H., Peng, S., 2018. Loading-rate dependence of rocks in postfailure region under triaxial compression. Adv.in Materials Sci. Eng. 2018, 1496127.

Weydt, L.M., Ramírez-Guzmán, Á.A., Pola, A., Lepillier, B., Kummerow, J., Mandrone, G., Comina, C., Deb, P., Norini, G., Gonzalez-Partida, E., Avellán, D.R., Macías, J.L., Bär, K., Sass, I., 2020. Petrophysical and mechanical rock property database of the Los Humeros and Acoculco geothermal fields (Mexico). Earth Sys. Sci. Data 13, 571–598. https://doi.org/10.5194/essd-13-571-2021

Zhu, W., Baud, P., Vinciguerra, S., Wong, T., 2016. Micromechanics of brittle faulting and cataclastic flow in Mount Etna basalt. J. Geophys. Res.: Solid Earth 121(6),4268–4289.

Rock Mechanics and Engineering Geology in Volcanic Fields – Ohta, Ito & Osada (eds)
© 2023 copyright the Author(s), ISBN 978-1-032-27657-1

Bearing capacity of low-density volcanic rocks using the discontinuity layout optimization method

Ruben Galindo* & Miguel A. Millán
Universidad Politécnica de Madrid, Madrid, Spain

ABSTRACT: The failure criterion of low-density volcanic rocks differs radically from that of conventional rocks by manifesting collapse under isotropic stress. In this way, the shapes of the failure model do not reveal a continuously increasing growth of deviating stress with the isotropic stress, but they reach a maximum value after which they decrease until they vanish for the isotropic collapse pressure. As a consequence, engineering applications require the implementation of numerical codes and the resolution of associated numerical difficulties. This article presents the problem of the bearing capacity of a foundation on a low-density volcanic rock using the DLO (Discontinuity Layout Optimization) numerical method. The analysis of results shows the ability of the DLO method to solve the numerical difficulties associated with the complex failure criteria so that the convergence and stability of the solution can be achieved without generating high computational costs. Additionally, a discussion of the DLO results is also carried out, obtaining forms of failure on the ground following the real collapses in these volcanic materials. The bearing capacity values of the numerical analysis are also analyzed and compared with the expected values in this type of rock, resulting in a satisfactory comparison of results. In this way, an adequate and reliable resolution technique is provided to face the problem of bearing capacity in low-density volcanic rocks, overcoming limitations referred to in the technical literature regarding the difficulty of treating highly non-linear and non-monotonic numerical criteria, and that allows the introduction of isotropic collapse failure.

Keywords: Pyroclasts, Collapsible criterion, Bearing capacity, Shallow foundation, Discontinuity Layout Optimization

1 INTRODUCTION

One of the main difficulties of low-density volcanic rocks is that they are poorly understood in terms of their geomechanical properties. Under the action of an external load, at sufficiently "low" stress levels they behave as if it were a conventional rock, but when requested at higher stress levels, the bonds between their particles can break, leading to a sudden decrease in its volume and the reorganization of its particles, forming a more compact structure than the initial one. The previous process is known as mechanical collapse and involves a drastic change in the properties of the collapsible material, which can behave like a soil if its structure is completely destroyed.

There is a great interest for depicting a theoretical and practical framework to describe and to estimate the strength and deformability properties of the low- density

*Corresponding author: rubenangel.galindo@upm.es

DOI: 10.1201/9781003293590-34

volcanic pyroclasts in the field of civil engineering and building in volcanic areas. To face the design and calculation of the applications, it is necessary to use an adequate failure criterion, which correctly represents the behavior of low-density volcanic materials. Thus, the failure criterion of Serrano et al. (2016 and 2021), obtained from more than 250 samples of pyroclasts from Canary Islands, is applied to the problem of calculating a shallow foundation, establishing the stress state at each point of the ground to infer the rupture mechanism that occurs and for which the numerical formulation DLO (Discontinuity Layout Optimization) is applied. This failure criterion is implemented in FLAC [Fast Lagrangian Analysis of Continua, Itasca (1995)], in order to validate the numerical solution obtained with results calculated using the finite difference method.

2 FORMULATION OF THE FAILURE CRITERION

The parabolic criterion, in Cambridge variables p_{KR}^* and q_{KR}^*, for low-density volcanic pyroclastics suggested by Serrano et al. (2016 and 2021) is as follows:

$$q_{KR}^* = M(p_{KR}^* + t^*)\left(1 - \frac{p_{KR}^* + t^*}{P_c^* + t^*}\right)^{\lambda} \tag{1}$$

In law (1), represented in figure 1, there are four parameters (t^*, P_c^*, M and λ). t^* is the isotropic tensile strength, P_c^* is the isotropic compressive strength (the following notation is adopted: $P_{co}^* = P_c^* + t^*$), M is a frictional parameter, which is determined by triaxial tests, and λ is a mathematical adjustment parameter obtained from experimentation. The ratio t^*/P_{co}^* is called tensile coefficient ζ.

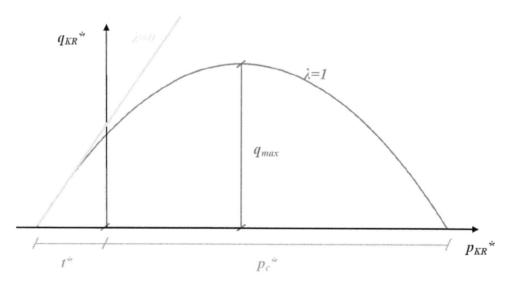

Figure 1.   Failure criterion of volcanic pyroclasts in Cambridge variables.

Conde (2013) and Serrano et al. (2016) have determined the parameters of the failure law (1) for a wide range of pyroclasts, as indicated in Table 1. From the results shown in the table, the following conclusions can be deduced: (a) For most pyroclasts the value of the parameter ζ is low, between 1-5%; (b) Almost all materials have a parameter λ very close to 1; (c) All materials, with some exception, have a parameter M between 1.5 and 3.

Table 1. Values of parameters of the failure law (1) obtained by Conde (2013) and Serrano et al. (2016).

Material	Tensile coef. ζ Mean (%)	Range (%)	Adjustment parameter λ Mean	Range	Frictional parameter M Mean	Range
Welded lapilli	9.1	1.8-17.1	0.98	0.93-1	2.43	1.78-2.97
Slightly welded lapilli	4.5	0-16.7	0.7	0.5-0.9	2.42	1.66-2.93
Welded pumice	18.4	5.1-25	0.84	0.63-1.02	2.2	1.68-2.73
Welded scoria	1.6	1.6	0.22	0.22	2.94	2.94
Disturbed welded lapilli	3.3	0.5-6.5	1	0.99-1	1.43	1.16-1.71
Disturbed welded pumice	2.9	0.3-5.9	1	0.98-1	2.1	1.51-2.75
Welded basaltic ashes	2.3	2.3	1	1	2.62	2.62
Slightly welded salic ashes	4.9	4.9	1	1	1.83	1.83
Disturbed welded salic ashes	1	1	1	1	2.94	2.94
Red pyroclastic flows	6.1	6.1	1	1	2.99	2.99
Yellow pyroclastic flows	22.7	22.7	0.89	0.81	1.77	1.77
Pozzolanna Nera	5.1	5.1	0.78	0.78	2.17	2.17
Fine-Grained pyroclastic flows	13	11.8-14.2	0.82	0.82	1.87	1.81-1.94

3 BEARING CAPACITY

The problem of a strip foundation resting a rock with collapsible strength criterion, assuming an associative flow law, collapsible failure criterion, perfect plasticity, weightless rock media and plane strain hypothesis is arised. The mathematical model to be solved is represented in Figure 2, where two boundaries are defined: the free boundary (boundary 1) and the boundary where the foundation supports (boundary 2). A distributed load is applied to boundary 1, which expressed dimensionlessly according to the same notation criteria that the variables is $f_1 = f_1^*/P_{co}^*$, while the foundation is directly represented by a distributed load on boundary 2 equal to $f_2 = f_2^*/P_{co}^*$. The boundaries are separated by a singular point, so that the free boundary has, in general, an inclination α with respect to the horizontal and the foundation is located at the end of its boundary 2 (that is, at the edge of the slope).

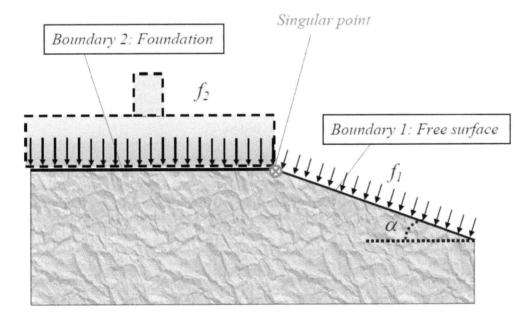

Figure 2. Definition of the bearing capacity problem to be solved.

4   DISCONTINUITY LAYOUT OPTIMIZATION METHOD

In this article, the Discontinuity Layout Optimization method (DLO) developed by Smith and Gilbert (2007) is used. This method is based on the traditional limit analysis, although it does not presuppose a certain mode of failure like that one. It requires the identification of all possible failure mechanisms in the medium, for a specific distribution of nodes, determining as critical the mechanism that is associated with the minimum energy.

This identification is achieved through an optimization method that identifies the critical arrangement of failure lines that leads to the collapse. The failure lines are generated by the connection of a series of nodes, which are distributed over the studied domain.

The problem is defined in terms of the relative displacement along each and every one of the possible failure lines, associating them with their corresponding tangential and normal relative displacements. When several lines converge in a node, displacement compatibility equations are applied.

Once the variables are defined, an objective function can be established that includes the total energy dissipated in the movement of the set of discontinuities. An optimization problem is arisen, whose solution will include a subset of discontinuities that form a compatible failure mode with minimum energy dissipation.

The accuracy of the solution is improved with a higher density of nodes, and the result can be considered as the upper bound solution (Smith et al., 2017).

In general, it can be concluded that the DLO method maintains the simplicity of the classical limit analysis, with an efficient identification of the discontinuities of the failure mechanism. However, it does not show the limitations of the theoretical problem regarding the identification of the mechanism before the calculation.

Furthermore, this method does not usually present problems of stability, blocking, or non-convergence of the solution when the problem is defined in a medium where the failure can occur without excessive confinement, Smith et al. (2017).

The aforementioned advantages and the low computational cost make the DLO a highly recommended method for the analysis of the bearing capacity of shallow foundations.

5   NUMERICAL CALCULATION AND VALIDATION

The commercial code GEO from LimitState (2019) is used in this research. The rock mass is defined in DLO models as a square domain with symmetry conditions along the y-axis (left edge), and encastre boundary conditions at the bottom and right edges. The free surface is assumed unloaded. The load is applied through a symmetric rigid foundation on top of the rock surface, with full connection at the contact nodes.

The rock domain size is specially defined for each model, depending on the extension of the failure mechanism. The bottom and right edges are located as far from the foundation as necessary to avoid interferences, usually by a trial-and-error process.

The discretization is chosen to balance the accuracy obtained using a larger number of points with the associated increase of the computational cost. A high density of points distributed in the domain (2000 nodes) is used, corresponding to an average nodal spacing of 1/15 of the foundation width. Those points are used to define all the possible failure lines and to apply the optimization procedure.

Eight analysis cases presented in Table 2 are studied, where the values of the adopted parameters and the bearing capacity results obtained by the DLO method are shown.

An isotropic collapse pressure value of $P_c^* = 1$ MPa, frictional parameter M equal to 1.5 and 2.5, tensile coefficients of 1 and 5%, and external load SC = 10 kPa and 500 kPa have been considered in all cases.

Table 2. Comparison of calculation cases between DLO and FLAC.

	Case 1	Case 2	Case 3	Case 4
DLO	M=1.5	M=1.5	M=2.5	M=2.5
FLAC	$\zeta = 0.05$	$\zeta = 0.05$	$\zeta = 0.05$	$\zeta = 0.05$
Error	$P_c^* = 1MPa$	$P_c^* = 1MPa$	$P_c^* = 1MPa$	$P_c^* = 1MPa$
	SC=10 kPa	SC=500 kPa	SC=10 kPa	SC=500 kPa
	0.76 MPa	1.01 MPa	1.1 MPa	1.1 MPa
	0.75 MPa	0.91 MPa	0.95 MPa	0.97 MPa
	1.3 %	9.9 %	13.6 %	11.8 %

	Case 5	Case 6	Case 7	Case 8
DLO	M=1.5	M=1.5	M=2.5	M=2.5
FLAC	$\zeta = 0.01$	$\zeta = 0.01$	$\zeta = 0.01$	$\zeta = 0.01$
Error	$P_c^* = 1MPa$	$P_c^* = 1MPa$	$P_c^* = 1MPa$	$P_c^* = 1MPa$
	SC=10 kPa	SC=500 kPa	SC=10 kPa	SC=500 kPa
	0.45MPa	1.01 MPa	1.09MPa	1.1MPa
	0.52 MPa	0.92 MPa	0.95 MPa	0.97 MPa
	13.5 %	8.9 %	12.8 %	11.8 %

Figure 3 Shows the results and breakage mechanisms obtained for cases 7 and 8; As can be seen, when the external overload increases, the failure mode changes, since the confinement increases and the destructuration is reached, locating the failure under the foundation. In case 7, the failure corresponds to the general mechanism of conventional soils and rocks as they have a low confining load, while case 8 shows the localized failure under the foundation, typical of low-density volcanic rocks according to the knowledge learned on the volcanic rocks.

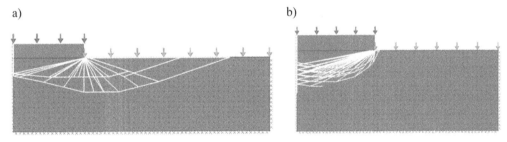

Figure 3. a) DLO model and breakage wedge for the calculation of a shallow foundation on rock with collapsible criteria: a) Case 7 indicated in Table 2; b) Case 8 indicated in Table 2.

In order to compare the results obtained applying the numerical method proposed in the present study, 2D models was used to calculated the cases by finite difference method (FDM), employing commercial code FLAC. Thus, Table 2 also shows the numerical results obtained applying finite differences. The numerical model was adopted in a way that the vertical load was directly applied on the ground surface (nodes), so that the characteristics of the foundation and the interaction with the ground surface did not influence the result. A symmetrical model, applying the plane strain condition, was used, where only half of the strip footing was represented (Figure 4a). In order to use the collapsible failure criterion defined in (1), it has been necessary to implement it through the user-defined programming module, which allows incorporating own constituent models using C++. In Figure 4b the failure mechanism located under the foundation for case 8 is shown.

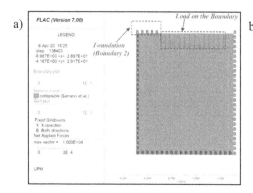
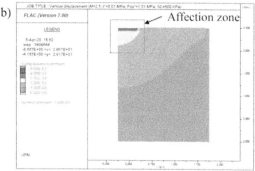

Figure 4.   a) FLAC model for the calculation of a shallow foundation on rock with collapsible criteria; b) Vertical displacements in the ground, located in a wedge of affection under the foundation, obtained numerically for the case 8 indicated in Table 2.

As can be seen, the values obtained are quite close, obtaining approximations between both methods of less than 14% in all the cases shown in Table 2.

6   CONCLUSIONS

From the research carried out in this work, the ultimate bearing capacity of a strip foundation with a collapsible strength criterion, assuming an associative flow law, perfect plasticity, weightless rock media and plane strain is studied using the DLO (Discontinuity Layout Optimization) numerical method. Two possible modes of breakage are identified in the ground under the foundation in collapsible rocks: plastic wedge and destructuring. In addition, numerical validation is performed by the finite difference method, using FLAC, obtaining good contrast results. It is also observed that the planned zones for destructuring coincide located under the foundation in both numerical models.

REFERENCES

Conde, M., 2013, Caracterización geotécnica de materiales volcánicos de baja densidad, *PhD Thesis*, Universidad Complutense de Madrid, Madrid (In Spanish).
Itasca, 1995, FLAC Version 7.0, User´s Manuals. Itasca Consulting Group, Minnesota, MN
LimitState, 2019, GEO Manual VERSION 3.5.d.
Serrano, A., Galindo, R., Perucho, A., 2021, Ultimate bearing capacity of low-density volcanic pyroclasts: application to shallow foundations. *Rock Mechanics and Rock Engineering*, 54, 1647–1670.
Serrano, A., Perucho, A., Conde, M. 2016, Yield criterion for low-density volcanic pyroclasts, *Int J Rock Mech Min*, 86, 194–203.
Smith, C.C., Gilbert, M., 2007, Application of discontinuity layout optimization to plane plasticity problems, *Proc. Royal Society A*, 463(2086),2461–2484.
Smith, C., González-Castejón, J. Charles, J., 2017, Enhanced interpretation of geotechnical limit analysis solutions using Discontinuity Layout Optimization. *19th International Conference on Soil Mechanics and Geotechnical Engineering*, Seoul.

Rock Mechanics and Engineering Geology in Volcanic Fields – Ohta, Ito & Osada (eds)
© 2023 copyright the Author(s), ISBN 978-1-032-27657-1

Studying the mechanical behavior of hard soils-soft rocks using numerical modeling

Khaled Abdelghafar*
Faculty of Engineering, Department of Mining, Petroleum, and Metallurgy, Cairo University, Giza, Egypt

Sherif Akl
Soil mechanics and foundations research laboratory, Faculty of engineering, Cairo University, Giza, Egypt

Hassan Imam & Samar Ahmed
Faculty of Engineering, Department of Mining, Petroleum, and Metallurgy, Cairo University, Giza, Egypt

ABSTRACT: Selecting a suitable mechanical model to represent the behavior of Hard Soils-Soft Rocks (HSSR) is a challenge for mining engineers. HSSR has intermediate properties between those of soils and hard rocks, it has a Uniaxial compressive strength (UCS) in the range 0.25 Mpa - 25 Mpa. Different constitutive models were proposed to represent the behavior of HSSR with a lack of certainty of their results. The applied mechanical models may be simple with few required input parameters or sophisticated with more input parameters. Using a simple model is easier than sophisticated one, but with lower accuracy. A compromise should be made. Two different mechanical models are used to study the effect of the mechanical model and the applicability of using more complex mechanical model. Mohr-Coulomb (M-C) as an example of simple models and Double Yield (DY) as a more complex one with multiple yield surfaces. The results show that the stresses around a circular opening excavated into a HSSR material begin to alter when using the more complex model at certain depth. At any depth smaller than this, there is no significant difference. This is related to the value of the resulting vertical stresses compared to the value of pre-consolidation pressure of the ground. Also, the DY model predicts higher safety factor. This sets a condition where using a more complex model is required.

Keywords: Numerical Modeling, HSSR, Soft rocks, Constitutive models, Mechanical behavior

1 INTRODUCTION

Dealing with Hard soil-Soft Rock materials (HSSR) seems to be difficult in mining engineering sector. HSSR differs from hard rocks, it lies between Soils and hard rocks with intermediate properties and transitional behavior. Neither rock mechanics side nor soil mechanics side consider these materials as their major specialty, because they are too soft to be tested in rock mechanics lab and too hard to be tested in soil mechanics lab, so this requires special testing apparatus (Kanji, 2014). These materials have low mechanical properties such as Uniaxial Compressive strength (UCS) which ranges from 0.25 Mpa to 25 Mpa (Kanji, 2014b). The main problems faced when dealing in such medium are the sampling and testing processes due the medium weakness. Moreover, the higher void ratio causes nonlinearity in the stress-strain curve, higher water content

*Corresponding author: Khaled.Abdelghafar93@cu.edu.eg

DOI: 10.1201/9781003293590-35

affects the strength and behavior, clay content induces the time dependent deformation, and tendency to transform from brittle to ductile at higher confining pressures, all are behavioral aspects of HSSR (Bailin, 1991). Also, HSSR shows multiple deformation mechanisms which should be predicted prior to any disturbance. Describing the behavior of HSSR requires a suitable mechanical model that can detect all behavioral aspects, so reasonable optimization should be done between model simplicity and result accuracy. Each model has limitations to work only within, otherwise it cannot give precise results. Simple models require few input parameters which means few properties and measurements while complex models require more inputs and more calculations and therefore higher accuracy. A mechanical model has three main components such as yield surface, flow rule, and hardening law. The yield surface defines the onset of plastic deformation, the flow rule determines the direction of the plastic flow, and finally the hardening law which determines if there is a change in the material properties with the developed plastic strain. Mohr-Coulomb (M-C), which assumes constant stiffness, cannot be used to model the time dependent behavior of rocks (Bonini, Debernardi, and Barla, 2009). M-C is still suitable for many geotechnical projects in which the material stiffness is kept constant. Moreover, bilinear M-C model, in which the material parameters change, is used to study the soft rock behavior (Bailin, 1991). Bailin reviewed a work in which Hoek-Brown (H-B) was claimed to be the most proper model to describe the rock response due to stresses and enchained this to the brittle behavior of harder rocks where ($\sigma_1 > 3.4 * \sigma_3$) at failure. Nova (Nova, 2006) claimed that soft rocks require using a model including hardening law.

Also, the Cam Clay model was applied to study the behavior of the normally consolidated clays (NC) (Itasca Consulting Group, n.d.) or the squeezing ground due to tunnelling activities (Shalabi, 2005). This model was criticized for being used to represent the behavior of overconsolidated (OC) clays due to strength overestimation (Yu, 2000).

Then, models with multiple yield surfaces were proposed to study the behavior of soft rocks such as marl, shale or weak sedimentary rocks with transitional behavior (Zhou and Zhu, 2010). Double Yield (DY) is one of these models which has two different yield surfaces; shear yield surface like that of M-C and volumetric yield surface. The volumetric yield surface is represented as a vertical line at the cap pressure value (pre-consolidation pressure) in (p-q) space (Figure 1). DY which was used for soft rocks and reached good agreement with the experimental results (Mochizuki, 2002). The DY model can represent the nonlinear, transitional behavior induced by the change of confining pressures and the strain softening or hardening behavior in which the material properties change with the plastic strain (Itasca Consulting Group., 2012). Another multiple-yield surface model, Cap Yield (CY), which has an ellipsoidal volumetric surface was applied besides M-C to study the lining behavior of a tunnel excavated into a soft clay material, the results showed that M-C predicts lower displacements than both the CY and the experimental observations (Do, Dias, Oreste, and Djeran-Maigre, 2013). Using these plastic hardening models gives more realistic outcomes (Cheng and Lucarelli, 2016) A summary of some mechanical models and their application is represented in Table 1.

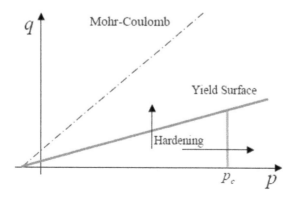

Figure 1. Yield surfaces of DY model.

Table 1. Summary of the mechanical models and applications.

Mechanical model	Application	Advantages	Disadvantages
Mohr-Coulomb (M-C)	Rocks and cohesive soils	Efficient in harder rocks and constant stiffness materials.	Cannot be efficient for soft ground deformations, needs hardening law (Bonini et al., 2009).
Hoek-Brown (H-B)	Hard rocks	Efficient for brittle rocks.	Not Efficient for ductile behavior, (Bailin, 1991).
Cam Clay	Clay, poorly consolidated materials.	Volumetric changes of rocks and soils. NC Clays.	Overestimated the strength of OC clays, (Vermeer & de Borst, 1984). Not efficient in detecting nonlinear elastic and predict unreal plastic shear strain in isotropic compression (Nova, 1986).
Modified Cam Clay (MCC)	Clay, soft rocks	Anisotropic behavior (Islam & Skalle, 2013),	Faced problems for Stiff clays (Hong, Pereira, Tang, & Cui, 2016).
Double Yield (DY)	Soft rocks, porous sandstone.	Volumetric changes, transitional behavior (Zhou & Zhu, 2010)	Needs cautious deal with input parameters.

2 METHODOLOGY

To test the applicability of using a mechanical model with multiple yield surfaces, a 2-D circular opening is numerically simulated in a typical HSSR material. Two different mechanical models were applied separately to test the effect of using the more complex model. Smectite shale, which was studied in Bailin's work (Bailin, 1991), is selected as a HSSR material with the following physical and mechanical properties represented in Table 2.

The numerical model represents a tunnel opening having 10-m diameter at two cases; Case 1: depth = 100 m, and case 2: Depth = 350 m. this is used to test the effect of depth on the results applying different mechanical models.

The model components include the problem geometry, the applied mechanical model, input parameters of medium properties, boundary conditions, and initial stresses. The varying depths are selected to test the change in vertical stresses due to depth. First of all, the geometry is represented with finer meshes to ensure reliable results. Then, two different mechanical models are separately applied using the required input parameters for both models. The first one is M-C which requires few inputs while the other model is the DY model which requires more input parameters. M-C represents the category of simple models with a single yield surface. On the other hand, DY represents a more complicated model with double yield surfaces as shown in Figure 1. The required input parameters of M-C are obtained through laboratory testing directly while the input parameters of the DY model need to be calibrated to ensure that these inputs match the material properties (Figure 3).

Regarding the initial stresses and the boundary conditions, the earth' surface is supposed to be a free surface ($\sigma_v = 0$, z=0), where σ_v is the vertical stresses due to gravity while the horizontal stresses are assumed to equal 85 % of the vertical stresses (k = 0.85). The model width is assigned to eliminate the effect of the fixed boundaries at both sides as shown in Figure 2. For simplicity and faster calculations, the model width is kept constant (100 m at each side of the opening) in Case 2 (depth: 350 m).

According to Table 2 and Table 3, the material has low physical and mechanical properties which increase the complexity of the problem. The selection of the mechanical model is not only the main problem but also the material nature and behavior.

For DY, the input parameters such as bulk and shear moduli are considered as the upper limits for their values if increased due to hardening, these values can be obtained through the

Table 2. Physical and Mechanical properties of shale (Bailin, 1991).

Water content	41.23%	young's modulus, E (Mpa)	230
Dry density	1100 Kg/m^3	Poisson's ratio (ν)	0.48
Saturated density	1720 Kg/m^3	Cohesion, C (Mpa)	1.3
Density	1565 Kg/m^3	friction angle (φ)	0
Porosity	62.5%	Pre-consolidation pressure, (Mpa)	4

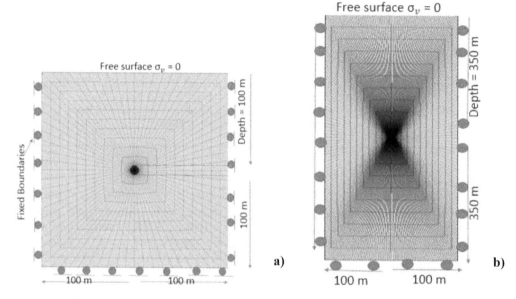

Figure 2.   Numerical model components a) case 1, b) case 2.

calibration process shown in Figure 3. Calibration is carried out by comparing the experimental stress-strain curve at a confining pressure of 2.5 MPa to the same curve obtained numerically with a continuous change of the inputs until reaching a reasonable match between both curves. then the final input parameters at matching state is chosen as the input parameters for DY model.

Finally, it is assumed that the medium behaves elastically before the excavation process and in the far field areas after excavating the opening. Only the surrounding area of the excavation behaves plastically according to the applied mechanical model.

3   RESULTS AND DISCUSSION

The results set a limiting minimum depth of construction below which there is no need to use more complex models. The developed stresses at depth lower than 261 m do not exceed the value of cap pressure or pre-consolidation pressure then the yield surfaces are not violated, so DY model predicts the same vertical stresses as M-C at the lower depth. At higher depth, DY predicts the lowest normal stress levels (Figure 4) which is predicted by having a look on Figure 1. Moreover, DY predicts higher vertical settlement due to the higher value of volumetric strain. Also, higher values of safety factor are predicted by DY due to the lower stress values and so lower strain increments (Figure 5). DY model has two surfaces as shown in Figure 1, the limits of the volumetric surface (vertical one) depart to the left direction which means softening behavior. In other words, Due to the developed volumetric strain, the bulk

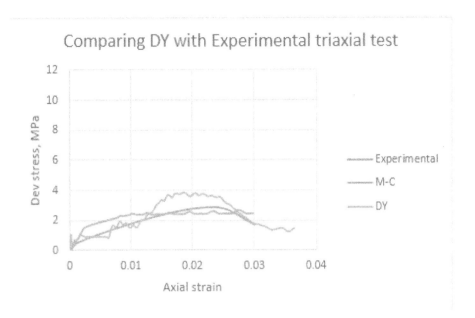

Figure 3.   Numerical Vs experimental stress- strain curves.

Table 3.   Calibrated input parameters for DY model.

Bulk modulus, Mpa	5000
Shear Modulus, Mpa	240
Young's modulus, Mpa	600
Poisson's ratio	0.48
Cap pressure, Mpa	4
Cohesion, Mpa	1.6
Friction angle	17°
R, multiplier	15

and shear moduli have lower values than the input values. Also, the cap pressure values decrease after excavation (2.1 MPa instead of 4 MPa) which means that the material softens in the volumetric yield surface direction. material anisotropy is not considered in this work, so it is considered as a limit for application.

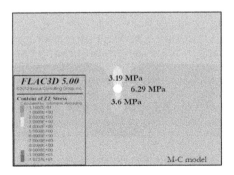 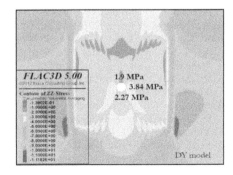

Figure 4.   Vertical stress distribution for both M-C and DY models, Case 2.

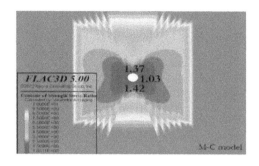 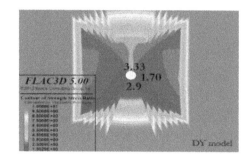

Figure 5. Distribution of Safety factor around the opening, case 2.

4 CONCLUSIONS

HSSR shows strain softening behavior observed through the calibration stress- strain curve and through the values of cap pressure after excavation. The main result obtained is the depth which is considered as a limiting condition for using more complex mechanical models, this depth is set to 261 m. At depth lower than this value, there is no need to use complex models where both simple and complex models give the same results as the yield surfaces are not violated. At higher depth, there is a significant change of the results where DY model predicts lower normal stresses, higher vertical displacement due to the higher volumetric deformation which are not predicted by simple mechanical models such as M-C. Also, DY predicts higher safety factor than M-C due to the lower total strain increments.

REFERENCES

Bailin, W. (1991). INVESTIGATIONS INTO THE MECHANICAL BEHAVIOUR OF SOFT ROCKS.

Bonini, M., Debernardi, D., Barla, M., & Barla, G. (2009). The mechanical behaviour of clay shales and implications on the design of tunnels. *Rock Mechanics and Rock Engineering*, *42*(2), 361–388. https://doi.org/10.1007/s00603-007-0147-6

Cheng, Z., & Lucarelli, A. (2016). Plastic hardening model II : Calibration and validation. *4th Itasca Symposium on Applied Numerical Modeling*, *2573* (Coduto2015), 393–402.

Do, N. A., Dias, D., Oreste, P., & Djeran-Maigre, I. (2013). 3D modelling for mechanized tunnelling in soft ground-influence of the constitutive model. *American Journal of Applied Sciences*, *10*(8), 863–875. https://doi.org/10.3844/ajassp.2013.863.875

Hong, P. Y., Pereira, J. M., Tang, A. M., & Cui, Y. J. (2016). A two-surface plasticity model for stiff clay. *Acta Geotechnica*, *11*(4), 871–885. https://doi.org/10.1007/s11440-015-0401-0

Islam, M. A., & Skalle, P. (2013). An experimental investigation of shale mechanical properties through drained and undrained test mechanisms. *Rock Mechanics and Rock Engineering*, *46*(6), 1391–1413. https://doi.org/10.1007/s00603-013-0377-8

Itasca Consulting Group. (2012). *FLAC 3D 5.0 manual. flac3D 5.0 manual.*

Itasca Consulting Group. (n.d.). CONSTITUTIVE MODELS: THEORY AND IMPLEMENTATION. (pp. 1–144). Retrieved from http://web.mst.edu/~norbert/ge5471/Assignments/Assign1-FLACI/FLAC manual/FLACmanual/f2d414.pdf

Kanji, M. A. (2014a). Critical issues in soft rocks. *Journal of Rock Mechanics and Geotechnical Engineering*, *6*(3), 186–195. https://doi.org/10.1016/j.jrmge.2014.04.002

Kanji, M. A. (2014b). Engineering works affected by soft rocks. *ISRM Conference on Rock Mechanics for Natural Resources and Infrastructure, SBMR 2014.*

Mochizuki, A. (2002). A New Numerical Model based on Double Yield Surface Concept and Validity Verification by FE-Analysis, (1994).

Nova, R. (1986). An extended Cam Clay model for soft anisotropic rocks. *Computers and Geotechnics*, *2*(2), 69–88. https://doi.org/10.1016/0266-352X(86)90005-4

Nova, R. (2006). Elastoplastic models for soils and soft rocks, 3651–3655. https://doi.org/10.3233/978-1-61499-656-9-3651

Shalabi, F. I. (2005). FE analysis of time-dependent behavior of tunneling in squeezing ground using two different creep models. *Tunnelling and Underground Space Technology*, *20*(3), 271–279. https://doi.org/10.1016/j.tust.2004.09.001

Vermeer, P. A., & de Borst, R. (1984). Non-Associated Plasticity for Soils, Concrete and Rock. *Heron*, *29*(3), 1–64. https://doi.org/10.1007/978-94-017-2653-5_10

Yu, H.-S. (2000). *Cavity Expansion Methods in Geomechanics. Cavity Expansion Methods in Geomechanics*. https://doi.org/10.1007/978-94-015-9596-4

Zhou, C. Y., & Zhu, F. X. (2010). An elasto-plastic damage constitutive model with double yield surfaces for saturated soft rock. *International Journal of Rock Mechanics and Mining Sciences*, *47*(3), 385–395. https://doi.org/10.1016/j.ijrmms.2010.01.002

Rock Mechanics and Engineering Geology in Volcanic Fields – Ohta, Ito & Osada (eds)
© 2023 copyright the Author(s), ISBN 978-1-032-27657-1

Effect of thermal shocks on physico-mechanical characteristics of volcanic rocks

Yuya Suda* & Ömer Aydan
Department of Civil Engineering, University of the Ryukyus, Nishihara, Japan

ABSTRACT: The authors collected some samples of tuff and welded tuff from Asuwayama region in Fukui Prefecture, Oya Region in Tochigi Prefecture of Japan and from Derinkuyu in Cappadocia Region of Turkey. Thermal shock was applied up to 6 hours and cool down under room temperature. The selected temperatures were 250, 500 and 800 Celsius. The geometries and physical and mechanical properties of samples were measured before and after thermal shocks. X-CT tomography technique is used to investigate internal damage due to thermal shocks in rock samples. Physical measurements involve density, p-wave and s-wave velocities while mechanical tests involve strain-stress responses under Brazilian and uniaxial compression tests. From these tests, tensile and compressive strength were determined under four different thermal regimes. These tests also allow us to determine triaxial yield characteristics as well as deformation properties under both elastic and non-elastic regimes at different selected temperature amplitudes. The authors report the outcomes of these experimental tests and discuss their implications.

Keywords: Thermal shock, volcanic rocks, physico-mechanical properties, X-CT Tomography

1 INTRODUCTION

Rock Engineering structures may be subjected to fires during their lifetime. Furthermore, there are many historical underground or semi-underground structures excavated in such as Cappadocia Region in Turkey, Nakhchivan and Kandovan in Azerbaijan, Kamakura and Sendai of Japan. Furthermore, masonry structures made of blocks of tuff or welded tuff are common worldwide. Therefore, the physico-mechanical characteristics or rocks subjected to thermal shocks are of great importance.

There are some experimental studies on the effect of thermal shocks on the mechanical properties of rocks. These experimental results clearly showed that the mechanical properties such as strength, deformation modulus decrease while their apparent porosity increases. It is pointed out that such variations are inferred to be due to internal damage caused by thermal shocks. The authors recently utilized X-Ray CT imaging technique to observed the internal damage caused by thermal shocks by varying temperatures from room temperature to 250, 500 and 800°C for about 6 hours and cooled down (Aydan et al. 2020). The experiments have been now continued using more rocks.

This study is a part of this experimental program. Physical characteristics were measured before and after thermal shocks. Furthermore, some mechanical properties of rocks were measured and compared with those of similar samples not subjected to thermal shocks. For some rocks, X-CT tomography technique is utilized to investigate the internal damage. The variation of mechanical properties is also assessed using the internal damage index. The authors report the outcomes of some of these experiments and discuss their implications in rock mechanics and rock engineering.

*Corresponding author: ysuda@tec.u-ryukyu.ac.jp

DOI: 10.1201/9781003293590-36

2 ROCK SAMPLE SITES

2.1 *Rock sampling sites in Japan*

Tuff samples are gathered from Asuwayama in Fukui Prefecture and Oya in Tochigi Prefecture in Japan. Asuwayama tuff has been extracted from the Asuwayama hill in Fukui City and it is commercially known as Shakutani stone. Shakudani or Asuwayama tuff is a welded tuff in Fukui Prefecture. Asuwayama formation is a result of volcanic activity during the formation of the Sea of Japan and it is formed during early Miocene of the Cenozoic Tertiary era. It is a pyroclastic flow deposit and it is characterized as dacite pumice lapilli. This tuff was used as building stone and other purposes (Aydan et al. 2014). There are many abandoned underground quarries, which collapse from time to time as happened in August, 2005.

Oya tuff formation is a Tertiary formation with the basement rocks such as chert, sandstone of the Paleozoic era, and the Mesozoic era. The Oya tuff was formed under marine environment about 20 million years ago and has a porous structure and bluish-green pumice in splashed patterns together with chunks of the clay mineral. Its clay mineral mainly consists of montmorillonite and zeolite. Oya tuff is a soft rock and it can be easily excavated. It is easily

     (a) Shakudani underground quarry   (b) Oya underground quarry

Figure 1.   Views of quarries of (a) Shakudani (saya) and (b) Oya tuff

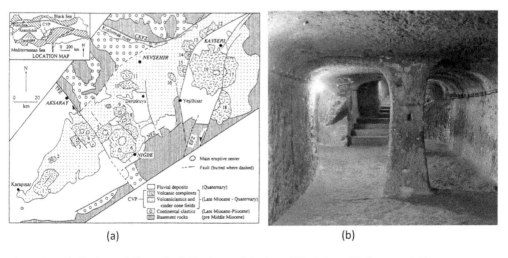

     (a)   (b)

Figure 2.   (a) Geology of Cappadocia Region and A view of Derinkuyu Underground City.

weathered and degraded. Especially the chunks of clay mineral in Oya tuff are easily washed away. Oya tuff or Oya stone, which is one of the most popular building stone materials in Japan, has been quarried in the Oya region, Utsunomiya, Japan. There were over 200 underground quarries in the past, and they also collapse from time to time as happened in February 10, 1989.

2.2 Rock sampling sites in Turkey

The Cappadocia area is situated in the Central Anatolia of Turkey and its altitude ranges between 1300 m and 1400 m, forming a high plateau. It has a triangular shape and it is surrounded by Erçiyes (3917 m), Melendiz (2935 m) and Hasandağ (3254 m) volcanoes. The Cappadocia Region is generally underlain by volcanic rocks of Neogene-Quaternary period belonging to the Cappadocian Volcanic Province (CVP) (Figure 2a). Basement rocks, Yeşilhisar formation, Ürgüp formation and Quaternary deposits are the main units observed in and around the area. The Ürgüp formation, unconformably overlying the Yeşilhisar formation, has widespread exposures in the area and the underground structures and the fairy chimneys are found in this formation. The formation is nearly horizontal and constitutes a 1100-1200m high plateau near Ürgüp and it is divided into 12 members in the region (Temel, 1992). These are, from older to younger, Çökek, Kavak, Zelve, Sarımaden, Damsa, Cemilköy, Tahar, Gördeles, Sofular, Topuzdağ, Kızılkaya and Kışladağ members. The Kavak member is the most commonly observed member of the Ürgüp formation and the underground cities are excavated in this unit (Aydan et al. 1999). Tuff of Derinkuyu Underground City (Figure 2b) and its close environment is located within the Gördeles member.

3 THERMAL SHOCK APPARATUS AND THERMAL REGIMES

The kiln is capable of applying thermal shock up to 1000 °C. The temperature of the kiln was elevated to selected temperature levels before setting samples (Figure 3) In this study, samples were subjected to selected temperature levels of 250, 500 and 800 °C for a duration of 6 hours. Then, the samples were taken out of the kiln and cooled down at room temperature.

   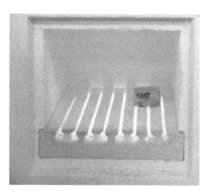

a) Overall view of the kiln   (b) Heated inside view

Figure 3.   An overall and inside view of the kiln used.

4 X-CT APPARATUS AND OBSERVATIONS

X-ray Computed Tomography (CT) device was produced by NIKON and its type is XT H320. The maximum size of the objects is 270x183x225 cm with a minimum focus diameter of $3\mu m$. Samples before and after thermal shocks were scanned by XT-H320 device and images were compared in order to see the internal damage (Figure 4). Figure 4b shows the view of samples before and after thermal shocks.

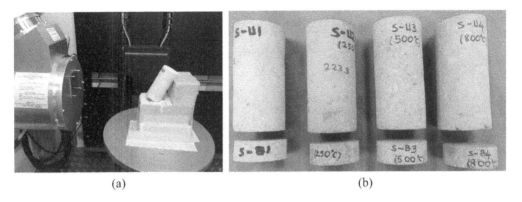

| (a) | (b) |

Figure 4. (a) A view of sample in XT H320 and (b) biews of Shakudani tuff samples after thermal shocks.

As noted from Figure 4b, there was no visible crack on the outer surface of samples while color changes were observed. However, a cross-section though the center of the sample in the X-CT images clearly indicated crack formation in the sample as shown in Figure 5(b). Figure 5(c) shows the volumetric crack opening in the sample.

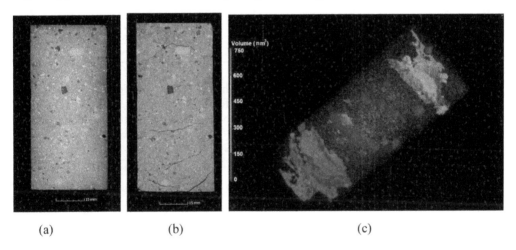

| (a) | (b) | (c) |

Figure 5. X-Ray images of S-U4 sample before(a) and after (b) thermal shock and (c) 3D visualization of internal damage in Sample S-U4.

5 EFFECT OF THERMAL SHOCKS ON PHYSICO-MECHANICAL PROPERTIES OF SELECTED VOLCANIC ROCKS

5.1 Shakudani (Asuwayama) tuff

Table 1 presents unit weight, p and s-wave velocity before (BTS) and after thermal shock (ATS) for Shakudani tuff samples subjected to uniaxial compression. As noted from the table, the thermal shock has a pronounced effect on the physical properties. The reduction of s-wave velocity is much higher than that of p-wave velocity of samples. Similarly, it can be said that physical properties decrease as the temperature increases. Furthermore, thermal shock causes some permanent straining in case of Shakudani tuff, particularly the sample S-U4 indicates great amount of permanent straining. Figure 6a shows strain-stress relation for each respective sample. It is interesting to note that deformation responses of samples S-U2 and S-U3 subjected to 500 °C or lower temperature levels imply that the increase of temperature increases result in stiffer response.

Table 1. Physical properties of Shakudani tuff compression samples.

No	γ (kN/m^3) BTS	ATS	V_p(km/s) BTS	ATS	V_s(km/s) BTS	ATS	ε_r (%)
S-U1	20.3	20.3	2.97	2.97	2.29	2.29	0
S-U2	19.9	19.9	3.04	2.97	2.30	2.13	0.15
S-U3	19.9	19.9	3.06	2.83	2.22	2.01	0.04
S-U4	19.9	19.4	3.06	1.51	2.21	0.89	1.63

Table 2 gives the physico-mechanical properties of Shakudani tuff samples for Brazilian tensile strength experiments. The overall tendency regarding the effect of thermal shocks is quite similar to those of compression experiments (Figure 6b). As noted from Table 2, some shrinkage type volumetric strain occurs for the sample subjected to 250 °C. Nevertheless, tuff samples, which are originally ash-fall deposits, become much stiffer and stronger for temperature levels less than 500 °C.

Table 2. Physico-mechanical properties of Shakudani tuff Brazilian test samples.

No	γ (kN/m^3) BTS	ATS	V_p (km/s) BTS	ATS	V_s (km/s) BTS	ATS	σ_t MPa	ε_r (%)
S-B1	19.2	19.2	3.21	3.21	2.32	2.32	4.16	0
S-B2	19.9	19.9	3.27	3.20	2.41	1.15	6.84	-0.83
S-B3	19.9	19.9	3.12	2.48	2.40	0.68	7.66	0.71
S-B4	19.9	18.9	3.01	1.71	2.36	0.66	3.37	2.66

Tensile strength of samples S-B2 and S-B3 were higher than S-B1 sample, which was not subjected to any thermal shock. However, higher temperature starts to induce some internal damage so that overall physico-mechanical properties tend to decrease. The deformation modulus and tensile strength increases may be related to this fact. However, the physico-mechanical properties of the sample subjected to a temperature shock of 800 °C starts to decrease, drastically.

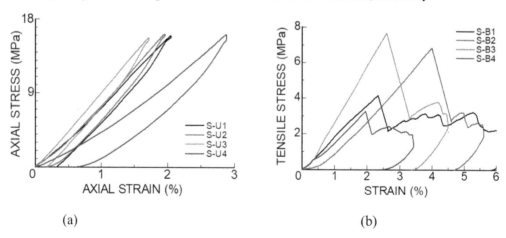

(a)   (b)

Figure 6. Strain-stress responses of Shakudani tuff samples in compression (a) and Brazilian (b) tests.

5.2 *Oya tuff*

Table 3 presents unit weight, p and s-wave velocity before (BTS) and after thermal shock (ATS) for Oya tuff samples. As noted from the table, the thermal shock has a pronounced effect on the physical properties. Furthermore, thermal shock causes some permanent straining in Oya tuff. Figure 7 shows strain-strain relation for each respective sample under compression and tensile regimes. It is interesting to note that deformation responses of samples subjected to 800 °C imply that the increase of temperature decreases the stiffness and strength for both loading regimes.

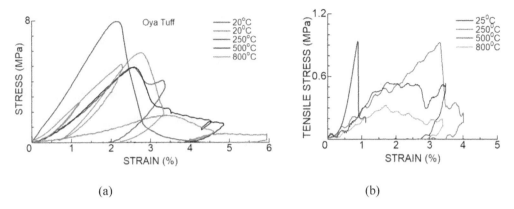

(a)   (b)

Figure 7.   Strain-stress responses of Oya tuff samples in compression (a) and Brazilian (b) tests.

Table 3.   Physical properties of Oya tuff compression samples.

No	γ (kN/m^3)		V_p(km/s)		V_s(km/s)		ε_r (%)
	BTS	ATS	BTS	ATS	BTS	ATS	
EWN-2	15.0	15.0	2.37	2.37	1.18	1.18	0
EW2-2	15.0	15.0	2.39		1.12		-2.31
WES-6	14.3	14.3	2.43		1.08		0.42
WE4S-5	15.0	13.8	2.25		1.23		0.17

5.3 Derinkuyu tuff

Table 4 presents unit weight, p and s-wave velocity before (BTS) and after thermal shock (ATH) for Derinkuyu tuff samples. As noted from the table, the thermal shock has a pronounced effect on the physical properties. The reduction of s-wave velocity is much higher than that of p-wave velocity of samples. Similarly, it can be said that physical properties decrease as the temperature increases. Furthermore, thermal shock causes some permanent straining of Derinkuyu tuff samples. Figure 8a shows strain-stress relation for each respective sample. It is interesting to note that deformation responses of samples subjected to 500 °C or higher temperature levels imply that the increase of temperature decreases stiffness and strength.

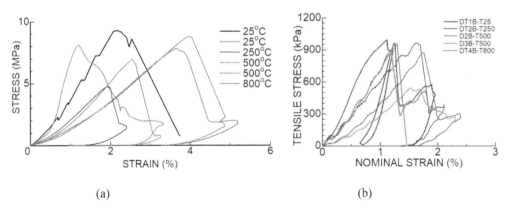

(a)   (b)

Figure 8.   Strain-stress responses of Derinkuyu tuff samples in compression (a) and Brazilian (b) tests.

Table 4. Physical properties of Derinkuyu tuff compression samples.

No	γ (kN/m³)		V_p(km/s)		V_s(km/s)		ε_r (%)
	BTS	ATH	BTS	ATH	BTH	ATH	
DT-U1	18.1	19.8	1.81	1.81	1.01	1.01	0
DT-U2	19.9	19.9	1.87	1.76	1.13	0.83	1.01
DT-U3	19.9	19.9	1.83	1.79	1.01	0.73	1.00
DT-U4	19.9	19.4	1.10	1.00	1.07	0.73	1.01

6 CONCLUSIONS

The authors have undertaken this study to investigate the effect of thermal shocks on physico-mechanical properties of volcanic rocks (tuff). The selected values of temperature were 250, 500 and 800 °C, for thermal shock tests for a duration of six hours. X-Ray CT images clearly showed the internal damage in sample and likely locations of damage. Inherent weakness planes within the sample eventually lead to internal damage and cracking. These damages in turn result in the reduction of physico-mechanical properties of tested rocks. However, it is also noted that the temperature up to certain levels may result in healing effects for tuffs. These tests also allowed us to determine triaxial yield characteristics as well as deformation properties under both elastic and non-elastic regimes at different selected temperature amplitudes.

ACKNOWLEDGEMENTS

The authors gratefully acknowledge Mr. T. Komuro of Nikon Corporation with their help visualization of internal damage of some rocks induced by thermal shocks and Prof. J. Tomiyama of the University of the Ryukyus for opinions on the interpretation of experimental results.

REFERENCES

Aydan, Ö. and Ulusay, R., 2003, Geotechnical and geoenvironmental characteristics of man-made underground structures in Cappadocia, Turkey. *Engineering Geology*, Vol. 69, 245–272.

Aydan, Ö., Sato A. and Yagi. M., 2014, The Inference of Geo-Mechanical Properties of Soft Rocks and their Degradation from Needle Penetration Tests. *Rock Mechanics and Rock Engineering*, 47:1867–1890.

Aydan, Ö., Tomiyama, J., Suda, Y. and Komuro, T., 2020, Visualization of internal damage of rocks subjected to thermal shocks up to 800□ by X-Ray CT imaging technique and their effect on mechanical properties. 47th Japan Rock Mechanics Symposium, JSCE, Tokyo, 214–219.

Temel, A., 1992, *Kapadokya eksplozif volkanizmasının petrolojik ve jeokimyasal özellikleri*. Ph.D. Thesis, Dept. of Geological Eng., Hacettepe University (in Turkish).

Rock Mechanics and Engineering Geology in Volcanic Fields – Ohta, Ito & Osada (eds)
© 2023 copyright the Author(s), ISBN 978-1-032-27657-1

Effects of clinopyroxene distribution by strong weathering of andesite

Eiji Nakata*
Central Research Institute of Electric Power Industry, Japan

Takehiro Ohta
Yamaguchi University, Yamaguchi, Japan

ABSTRACT: This study investigates weathering-related volume change by examining strong weathering of dense andesite. Andesite lava erupted on land during the Early Cretaceous, with the weathered rock consisting of a red-to-yellow-tinged matrix. The strongly weathered zone extends from 30 m to 40 m below the surface. Plagioclase disappears from the strongly weathered zone where kaolinite and halloysite form. In addition, the crack zone is several meters in width and lies between the strongly weathered zone and the fresh andesite zone.

Corestones consisting of fresh andesite have a diameter of about 20 cm and are often overlain in the upper part of the crack zone. In addition, rindlet-covered corestone 3 mm to 3 cm wide is observed around the fresh andesite zone. Na and Ca completely disappear from rindlet. Inversely, Ti and Zr are clearly concentrated in rindlet. These elements became mobile through weathering.

The clinopyroxene distribution uses the frequency ratio of high Fe-content pixels, indicating the same profiles for corestone and rindlet. The rock mass volume does not change through this weathering.

Ti increasing in rindlet is derived from the titanomagnetite dissolution. Plagioclase produces kaolinite and halloysite, and porosity increases with kaolinite and halloysite crystallization.

The results show that isovolumetric reaction occurring despite dissolution and crystallization with strong weathering. We consider that this weathering leads to surface slope changes.

Keywords: Volume change, Andesite, Spheroidal weathering, Saprolite

1 INTRODUCTION

As evidence of volume change, ground dug up for construction has been shown to upheave by +66 mm over seven years after unloading of the ground in the smectite-bearing crush zone (Tamura et al., 2007). Floor heave after completion has occurred in several mountain and coal mine tunnels in weak rock (Shimada et al., 1981). This phenomenon is a result of the unloading of ejector material or ground pressure. Grant (1986) revealed a 23% decreasing and 5% increasing volume for chemical reaction by the constant alumina isocon. Noe et al. (2007) studied house tilt and uplift via expansion of montmorillonite absorption of water as heaving bedrock. The house floors rise in the mudstone by crystallization of gypsum without an ejector (Oyama et al.,1998).

*Corresponding author: nakata@criepi.denken.or.jp

DOI: 10.1201/9781003293590-37

Alteration includes mineral formation and elemental motion. While it possible to determine the bulk density and porosity changes involved, it is not known whether the rock mass volume itself has changed. Weathering is mainly a dissolution reaction around the surface. Elements dissolved from rock-forming minerals because of rain infiltrating the ground from the surface. Kaolin minerals are crystallized to use residual ions. Smectite is an expansible clay mineral that comprises the commonly produced deeper area around the groundwater level to concentrate the infiltrating porewater with dissolved alkali and alkali earth ions (Anand and Paine, 2002). Kaolin and smectite formation may thus induce volume change. Volume changes in the regolith can be calculated from the ratios of the regolith to protolith densities and measured Ti concentrations in granodiorite (White, 2002). Changes in the volume are negligible, except in the shallow soils. The lack of volume changes is consistent with the preserved original bedrock texture (Velbel, 1990).

The volume of rock rarely changes by interaction, such as the formation of saprolite by weathering (White, 2002). The volume strain of rock is calculated on the assumption of immobile elements, including Ti, Zr and Nb (Brantley and White, 2009), as follows:

$$\varepsilon_j = c_i p_i / c_w p_w - 1 \tag{1}$$

c_w and ρ_w represent the concentration of immobile elements and bulk density of weathered rock, and *stor* the assumption of immobile elements, $\varepsilon_j = 0$ indicates isovolumetric weathering. Positive values of ε_j indicate expansion, whereas negative values indicate compression.

$c_x p_x$ is the concentration ff ox element per unit volume. If x elements are immobile, then $c_x p_x$ is consistent with the same value before and after alteration.

Strongly weathered andesite is distributed over a huge area in the southwest area of Yamaguchi Prefecture, Japan. Manganese hydroxide minerals fill in the joints of andesite, and slickenside is commonly confirmed on manganese minerals.

Spheroidal weathering, i.e., onion-skin weathering, of andesite initiates the transformation of andesite to saprolite. The rindlets have a diameter of about 20 cm and often occur in the upper part of the crack zone. Weathered rindlets 3 mm to 3 cm wide are observed around fresh andesite.

We introduce the volume change and transformation of elements of spheroidal andesite. Andesite lavas erupted during the Early Cretaceous. The weathered rock consists of a red-tinged, clay-rich matrix. The strongly weathered zone extends from 30–40 m below the surface. Plagioclase disappears from the strongly weathered zone where kaolinite and halloysite form. In addition, a crack zone several meters wide is distributed between the strongly weathered zone and the fresh andesite zone.

2   GEOLOGICAL SETTING AND OBSERVATIONS

Red-tinged matrix saprolite formed by andesite weathering is distributed in the southwest area of Japan. This andesite lava was deposited during the Early Cretaceous. Andesite contains a significant amount of titanomagnetite and clinopyroxene, lava overlie the conglomerate and sandstone alternation formation has 10 to 15° dipping toward the west. The saprolite has a maximum thickness of 30 to 40 m underlaid by an O-horizon in this area.

Typical weathering profiles show a currently under constructing road cut. Strongly weathered andesite (A-horizon; less than 1 m thick) is distributed below the organic topsoil (O-horizon; about 20 cm) and consists of dark red unstructured clay-rich saprolite. The B-horizon, which is 0.5 m thick and underlies the A-horizon, consists of yellow saprolite with manganese hydroxide minerals filling the joints, and contains clay-rich saprolite. The B-horizon is gradated to the C-horizon consisting of a red-to-yellow-tinged andesite. Porphyritic white clay mineral (kaolinite) and spheroidal weathered corestone are mainly confirmed in the C-horizon.

Samples are collected by the drilling of several holes through civil engineering works. Saprolite changes into unweathered andesite with a transition zone about 5 m thick. The weathering rindlet is along the joint with a corestone.

Joints are confirmed in the C-horizon outcrop. Almost all joints comprising manganese hydroxide minerals cross without displacement between saprolite and corestone.

3 INVESTIGATION METHODS

The samples were examined by XRD, XRF, EPMA and LA-ICP-MS, revealing mobile or immobile elements with weathering, as well as mercury porosimetry. An X-ray microscope (XGT) was also used to determine element transformation and mineral distribution ratio per pixel in rindlet and corestone to investigate volume change.

We used XGT-7200 (Horiba); UP213-AS (New Wave Research) for LA-ICP-MS; and X series II (Thermo Fisher Scientific). Samples for LA-ICP-MS comprised 0.1 g samples and a 0.2-g flux mixture of glass beads fused at 1,150 °C. LA-ICP-MS data of minor elements are calculated using the SiO_2 content by EPMA. We made the calibration curve fit the LA-ICP-MS intensity and contents of six standard (GSJ Geochemical Reference samples) fused glass beads.

4 RESULTS

The corestone (andesite) of the outcrop is shown in Figure 1. This corestone gradates into saprolite without a rindlet. The weathering front progresses continuously into the andesite. The numbers in the figure indicate magnetic susceptibility, which decreases in saprolite. This is in contrast to corestone, indicating that weathering affects magnetic mineral dissolution.

Figure 2 shows a cross-section of a spheroidal andesite gravel sample collected from a hill ridge. A rindlet 2 to 3 cm thick grows on the rim of the corestone. The boundary between the rindlet and the corestone (weathering front) is sharp and clear. Figure 2 also shows Ti, Fe and

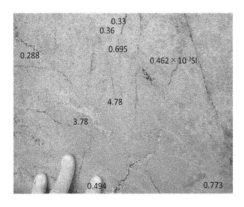

Figure 1. Saprolite originated by andesite in the C-horizon Numbers indicate magnetic susceptibility. Black film-filled cracks show manganese hydroxide minerals. Andesite (grey, center) in this figure retained the original structure.

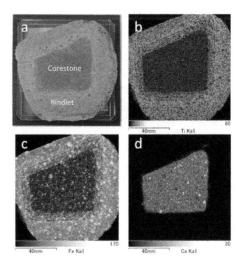

Figure 2. Spheroidal weathered andesite (a) Photograph of XGT sample, (b) Ti distribution, (c) Fe and (d) Ca. Rindlet has a high concentration of Ti and Fe.

Table 1. Chemical and physical properties.

	Saprolite	Rindlet	Corestone	Corestone	Rindlet	Titanomagnetite	
		A1	A6	Cpx	Cpx	fresh A	weathering B
SiO_2 (wt%)	57.29	60.07	58.01	51.21	29.02	0.95	14.74
TiO_2	0.86	2.95	1.08	0.14	0.07	21.75	0.30
Al_2O_3	31.14	18.82	18.30	2.10	16.73	1.84	8.41
Cr_2O_3	0.01	0.05	0.06	-	0.02	0.41	0.29
TFeO (Fe_2O_3)	9.06	20.58	8.10	17.37	25.48	(70.57)	(70.97)
MnO	0.67	0.30	0.15	0.58	0.35	3.58	0.11
MgO	0.52	0.56	3.61	12.63	14.27	0.10	4.19
CaO	0.06	0.05	7.41	12.32	0.47	0.00	0.04
Na_2O	0.03	0.02	3.07	0.17	0.04	0.03	0.00
K_2O	0.30	0.14	0.45	0.08	0.02	0.01	0.00
SO_3	0.02	0.02	0.02	0.01	-	0.02	0.03
P_2O_5	0.06	0.42	0.20	0.01	0.01	0.02	0.00
SrO	0.00	0.02	0.01	-	-	0.00	0.00
ZrO_2 (ppm)	118.63	406.99	144.39	-	-	0.00	0.00
Total	100.01	104.00	100.47	96.67	86.50	99.28	99.08
Porosity (%)	49.86	53.15	1.03	-	-	-	-
Bulk density (g/cm^3)	1.28	1.23	2.71	-	-	-	-
Median pore diameter (μm)	0.34	0.28	124.47	-	-	-	-

Ca distribution images by XGT analysis. Ti and Fe have accumulated in high concentrations in the brown rindlet to crystallize kaolinite and halloysite. Na, K, Mg and Ca are depleted from the rindlet. Na and K have the same tendency of Ca. Especially, Na and Ca completely disappeared in the rindlet. Al increases in the rindlet with gibbsite near the corestone. Conversely, Si decreases in the rindlet and is exchanged with Al.

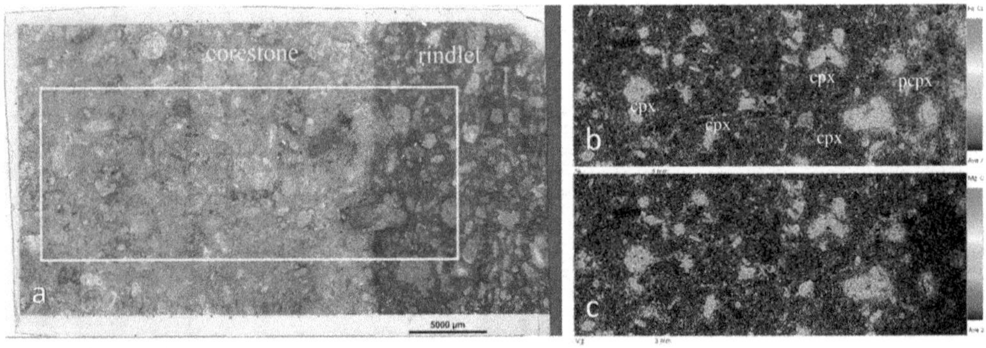

Figure 3. Reflection microscope and EPMA mapping data of corestone and rindlet (b) Fe, (c) Mg, analysis area is shown square in (a). cpx:clinopyroxene, pcpx:pseudo-clinopyroxene.

A close-up image was taken of structure of the corestone and the rindlet boundary using a thin section (Figure 3). Fe- and Mg-rich phenocrysts consist of clinopyroxene that exists in corestone and rindlet as pseudomorph in Figure 3. Clinopyroxene straddles the boundary between rindlet and corestone, which clearly consist of the same rock structure. The chemical contents and physical properties of saprolite, rindlet, corestone, clinopyroxene and titanomagnetite are shown in Table 1. Porosity increases with saprolite with about 50 vol.% compared with corestone, which has a porosity of 1 vol.%.

5 DISCUSSION

The magnetic susceptibility decreased in the saprolite (Figure 1). Imaoka and Nakamura (1982) reported Fe–Ti oxide minerals including Mn of Cretaceous volcanic rocks in southwest Japan. The formula of Fe–Ti oxide minerals, specifically, titanomagnetite, is $Mn_{0.2}Ti_{0.6}Fe_{2.3}O_4$ (Table 1), which is consistent with Imaoka and Nakamura. Reflectance microscope photographs of titanomagnetite in the rindlet and the corestone are shown in Figure 4. Titanomagnetite in the rindlet indicates coexistence in bright (A) and red reflectance corroded minerals (B), which do not transmit under polarizer transmitted light. Ti and Mn are clearly dissolved from bright reflectance mineral.

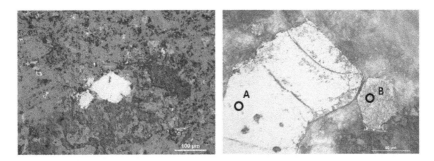

Figure 4. Reflection microscope of titanomagnetite Left: Titanomagnetite in corestone. Right: Titanomagnetite A in rindlet. A and B are EPMA quantitative analysis points (Table 1). Titanomagnetite B is weathered.

The results show that that Mn hydroxide mineral precipitated from Mn including titanomagnetite by dissolution. Ti and Fe also precipitated in the rindlet pore from this dissolved titanomagnetite. Titanomagnetite and clinopyroxene in the corestone are less altered than the rindlet and saprolite.

The porosity of the corestone is much lower than for rindlet and saprolite; also, the elemental distribution of the corestone is identical to that of the rim of the corestone. The weathering front of the spheroidal andesite is shown in Figure 2. These results indicate that many elements percolated through saprolite for the corestone rim by weathering. Also, Al percolation formed the enriched gibbsite zone around corestone. Porewater moved toward the corestone to dissolve feldspar, clinopyroxene, and titanomagnetite.

The progression of water with dissolution created the gibbsite supersaturated solution at the weathering front.

TiO_2, Na_2O and Zr contents per unit volume ($c_x p_x$) distribution are shown in Figure 5. Zr and TiO_2 contents per unit volume increase in the rindlet in contrast to corestone. Ti including magnetite dissolved by weathering. The decreased magnetic susceptibility of saprolite is consistent with the dissolution of titanomagnetite. If Zr is an immobile element through weathering, ε_j is calculated as –0.70 by Equation (1). ε_j for Ti also shows –0.66. This ε_j indicates compression; however, the density of the rindlet decreases with weathering. Ti and Zr are commonly recognized as immobile elements; however, the results show that these elements are mobile in this area's andesite by weathering. Mobilized Ti is derived from titanomagnetite by dissolution. Otherwise, the origin of Zr is unidentified in this study. Al also migrated from saprolite to the rindlet. Zr and Al are commonly persistent solubility elements dissolved under low pH solutions.

The occurrence of low pH solution is not identified in this area. One possibility is that this area is furnished with hydrothermal uranium ore. However, saprolitization progresses toward the surface and is distributed over a large area. We thus inferred that the cause of saprolite formation is weathering.

The results show that Zr and Ti are immobile elements even if the rindlet compressed 70% in volume. To investigate the volume changes with weathering, we examined the Fe distribution based on the XGT data. Fe including phenocrysts are distributed in the rindlet and

293

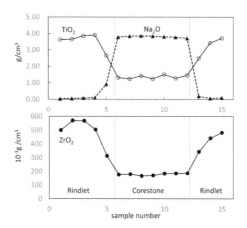

Figure 5.　Element concentration per unit volume of spheroidal andesite gravel.

corestone (Figure 3). The cpx in Table 1 shows the chemical composition of clinopyroxene and pseudo-clinopyroxene.

To estimate the volume change, we calculated the number of Fe (X-ray cts)-occupied pixels in Figure 6, which shows (1) Pixel counts for 5×5 pixels (0.970×0.970 mm). A pixel (0.194×0.194 mm) consisting of larger than 61cts ($Fe_2O_3 \fallingdotseq 25$ wt%) was selected for the XGT Fe counts. (2) Counts of the number in 0.970×0.970 mm for each pixel (x-axis: max. 25 pixels).

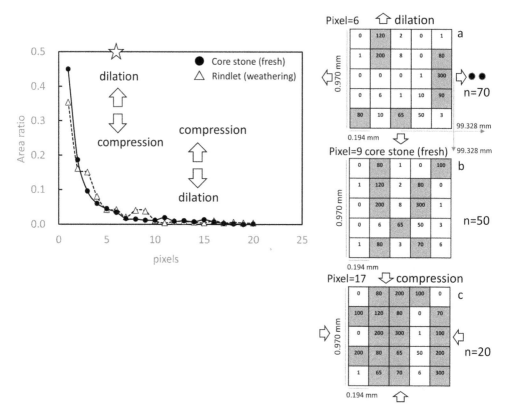

Figure 6.　The frequency ratio of clinopyroxene distribution in about 10 cm × 10 cm. The right figure shows a domain. Dilation (see right figure a): decreasing large numbers pixel. Compression (see right figure c): increasing large numbers pixel.

(3) Counts of the same pixel domain n. (4) Normalized (3) as the frequency ratio (y-axis). For example, Pixel =6 (Figure 6a) exists 70 (n) domains, an area ratio (Pixel=6) obtains 0.5 = 70/(70+50+20) (☆ in Figure 6). If a peak around the frequency ratio of 0.5 is increasing, this will show dilating. This figure shows the frequency ratio of the clinopyroxene distribution. When isovolumetric weathering occurs, the frequency ratio of clinopyroxene distribution shows the same curves for rindlet and corestone. These curves are quite similar to rindlet and corestone, and we inferred that isovolumetric weathering occurs in this area.

Our study indicates that Zr and Ti are mobilized by weathering caused by titanomagnetite dissolution. The rock volume is not changed andesite into saprolite by weathering and crystallization of kaoline clay with the dissolution of feldspar, even if the porosity increases from 1 vol% to 50 vol%.

6 CONCLUSIONS

Landslide and slope failures have occurred due to heavy rain at spheroidal weathering-developed slopes in Japan (Hirata and Chigira, 2019). Yamamoto et al. (2001) introduced that thin glossy black soil layers on discontinuous planes because amorphous Mn hydroxide precipitation causes decreased shear strength. Many slickenside planes develop on Mn hydroxide mineral-filled joints in this area. Small circular slides (5-m width) occurred on Mn hydroxide mineral-filled joint planes. Mn ions are derived from titanomagnetite dissolved with saprolitization.

Magnetite including Mn and Ti is dissolved by weathering, as shown in the reflection microscope and EPMA analyses. Ti and Zr are accumulated in rindlet. This study has shown that Ti and Zr should be specified as mobile elements.

The process of weathering progresses the dissolution of minerals with crystallization of kaolin minerals and a significant increase in porosity. We used the frequency ratio of Fe distribution to estimate the volume change from andesite corestone to rindlet. We could not show in this study that volume change occurs in the case of dense andesite that has undergone strong and long weathering.

REFERENCES

Anand, R. R. and Paine, M., 2002, Regolith geology of the Yilgarn Craton, Western Australia: Implications for exploration. *Australian Journal of Earth Sciences*, 49, 3–162.

Brantley, S. L. and White, A. F., 2009, Approaches to modeling weathered regolith. *Reviews in Mineralogy and Geochemistry*, **70**, 435–484.

Hirata, Y. and Chigira M., 2019, Landslides associated with spheroidally weathered mantle of granite porphyry induced by 2011 Typhoon Talas in the Kii Peninsula, Japan. *Engineering Geology*, **260**, 105217.

Imaoka, T. and Nakamura K., 1982, Iron-titanium oxide minerals of Cretaceous to Paleogene volcanic rocks in western Chugoku district, Southwest Japan – Special reference to manganese content of ilmenites-. *J. Japan. Assoc. Min. Petr. Econ. Geol.* 77, 235–255.

Noe, D. C., Higgins, J. D., and Olsen, H. W., 2007, Steeply Dipping Heaving Bedrock, Colorado: Part 1-Heave Features and Physical Geological Framework. *Environmental and Engineering Geoscience* **13**, 289–308.

Oyama, T., Chigira, M., Ohmura, N., and Watanabe, Y., 1998, Heave of House Foundation by the Chemical Weathering of Mudstone. *Journal of the Japan Society of Engineering Geology*, **39**, 261–272.

Shimada, S., Hokao, Z., and Sugiura, T., 1981, Experimental Study on the Floor Lift of Coal Mine Roadway Driven in the Weak Rock (1st Report). *Journal of the Mining and Metallurgical Institute of Japan*, **1126**, 1241–1244.

Tamura, E., Jyonai, A., Matuzaki, S., and Hasegawa S., 2007, Swelling Characteristic and Upheaval Mechanism of Smectite-bearing Crush Zone in Crystalline Schist. *Jour. Japan Soc. Eng. Geol.*, **48**, 80–89.

Velbel, M. A., 1990, Mechanisms of saprolitization, isovolumetric weathering, and pseudomorphous replacement during rock weathering – A review. *Chemical Geology*, **84**, 17–18.

White, A. F., 2002, Determining mineral weathering rates based on solid and solute weathering gradients and velocities: application to biotite weathering in saprolites. *Chemical Geology*, **190**, 69–89.

Yamamoto, T., Suzuki, M., Yoshiwara, Y., and Miyauchi, T., 2001, Slope failures due to thin glossy black soil layers on discontinuous planes and design strength parameter. *Journal of Japan Landslide Society*, **34**, 49–57.

Rock Mechanics and Engineering Geology in Volcanic Fields – Ohta, Ito & Osada (eds)
© 2023 copyright the Author(s), ISBN 978-1-032-27657-1

Seasonal change of a fracture aperture in porous tuffaceous rock with 0.01mm-order real-time monitoring

Masahiko Osada* & Kohei Funabiki
Saitama University, Saitama, Japan

ABSTRACT: For the purpose of capturing the long-term displacement behavior of a fracture, seasonal changes in fracture aperture were continuously monitored with 0.01 mm accuracy for an open fracture. The site is Yoshimi Hyaku-Ana, a nationally designated historic site, on a hill with a relative height of about 30 m. The target fracture found at the southwestern slope composed of a porous tuffaceous rock in Miocene epoch is open about 7 cm due to physical weathering caused by the intrusion of tree roots. The block above the fracture is unstable and is in risk to start sliding and falling. Therefore, three displacement sensors were arranged almost linearly in the direction of inclination of the fracture to measure the change in aperture width with the surrounding environment of temperature, relative humidity, air pressure and rainfall.

As a result of monitoring over one year, it was shown that the fracture is closed in the summer and opened in the winter, showing a high correlation with the temperature. After the rainy season of the year of observation, the width of fracture aperture was gradually shifted to open by less than 4 mm. Although sensor drift must be considered, the three sensors respond in the same way, but with different magnitudes.

We considered the change in aperture width with respect to temperature in two ways. One is the relationship between the daily average temperature and the aperture width. From this consideration, it was found that there is a correlation between the temperature and the aperture width before the rainy season, and that the aperture width becomes narrower as the temperature rises. It was also found that after the rainy season, all three sensors showed the same trend of aperture change.

Another consideration examined changes in temperature and aperture width during the day. As a result, it was found that the gradient in aperture width to temperature change at one day became smaller as the temperature decreased, and that the changes were clearly different before and after the rainy season.

These results imply that the upper block is moved to the dangerous side. In addition, we can propose that the observation of the diurnal change of the gradient is one of the effective indices for the signs of the change in the properties of the fracture at an early stage.

Keywords: tuffaceous rock, fracture aperture, seasonal change, real-time monitoring, Yoshimi Hyaku-Ana

1   INTRODUCTION

Rockfall is ultimately caused by the accumulation of displacement in the fractures that form the rock mass. In that sense, it is important to know how existing fratures open and close daily and how they respond to daily changes in temperature and humidity and rainfall.

*Corresponding author: osada@mail.saitama-u.ac.jp

DOI: 10.1201/9781003293590-38

However, such data are rarely reported, and there are some reports on Kuchitsu et al. (2005) and Osada et al. (2019).

Since there are many fractures in natural rocks and rockfalls often occur suddenly, it is impossible to measure all the opening and closing behaviors in each fracture. It is also important to observe the movement of unstable rock masses over a long period of time. In recent years, various measurements have been made with the progress of measurement technology and IoT technology (Janeras et al., 2017, Matsuoka, 2019).

It is important to capture both the behavior of the fracture itself and the behavior of the rock mass as a whole. In this paper, we report the measurement results for one year at Yoshimi Hyakuana in Saitama Prefecture, as an example of capturing the behavior of the fracture itself.

Figure 1.   Site, fracture details, and monitoring system.

2   METHODOLOGY

2.1   *The site: Yoshimi Hyaku-Ana*

Yoshimi Hyaku-Ana is a nationally designated historic site where more than 200 cave tombs exist on a hill with a relative height of about 30 m, as you can see in the lower right photo of Figure 1. The host rock is a porous tuffaceous rock (tuffaceous mudstone to sandstone) in Miocene epoch. In the cave tomb, chemical weathering is remarkable, and many studies have been conducted. (Oyama et al.,1998, Oyama et al.,1999, Kuchitsu and Ozaki, 1999, Oguchi et al., 2010, Takaya and Oguchi, 2011).

On the other hand, physical weathering is progressing on the southwestern slope where the cave tombs have been dug. The upper part of the slope, the more the rock mass deteriorates, and open fractures are formed at the entrance of many cave tombs as shown in the upper left of Figure 1. The roots of trees enter the fractures and lifting up the fractures apart. The fracture in Figure 1 is dipping 47° to the southwest on the rock slope with about 7 cm of the aperture. Although another side of the fracture is still connected to the bedrock, the block above the fracture is unstable and is in risk to start sliding and falling. Trees are currently being logged, but the roots remain as it is. The open fracture was confirmed in the winter of 2017, and as an emergency measure, a net has been installed to prevent falling and a sandbag has been placed at the bottom to suppress the impact of rockfall. While considering permanent countermeasures, displacement sensors are installed as shown in the lower left of Figure 1, and the change in fracture aperture is monitored to find out if the risk is getting worse.

2.2 Monitoring system

A displacement sensor manufactured by Tokyo Sokki Co., Ltd. (model: PI-5-50, resolution: 0.001 mm) was used to measure the change in aperture width. As shown in the lower left photograph of Figure 1, three displacement sensors were arranged almost linearly in the direction of inclination of the fracture. They are called S1, S2, and S3 from the right side, which is higher in position toward the fracture. This direction is almost along the root of the tree. Therefore, if the upper block above the fracture displaces with the root of the tree as the supporting points, the outputs of the three displacement sensors is expected to be almost the same, and if it rotates in the dipping direction, a systematic change of the outputs in magnitude is expected to occur.

The change in fracture aperture over time is affected by the surrounding temperature and humidity environment (Osada et al., 2019). An environmental sensor (BME280), by which temperature, humidity and air pressure can be measured with this one, is installed at a location as shown in the upper middle photo of Figure 1.

The obtained data are transferred to the cloud (using Ambient) via Wi-Fi at 1-minute intervals using the Raspberry Pi, which is a microcomputer board. In addition, a rain gauge OW-34-BP (manufactured by Ota keiki seisakusho co.) is installed inside the site and the record of the rainfall at 10-minute intervals is also transferred on the same cloud. It is a real-time monitoring system that can instantly visualize these data on the cloud.

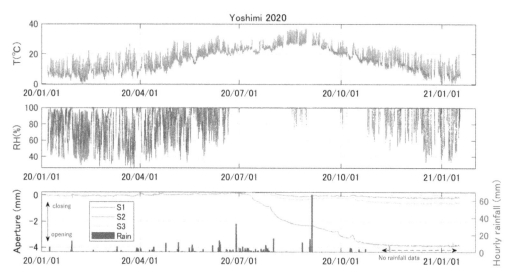

Figure 2.   Annual data of the fracture aperture with air temperature and humidity.

3   RESULTS

Figure 2 shows the data for one year from the first observation in January 2020. From top to bottom, the temperature, relative humidity, and the fracture aperture are shown. At the bottom of the figure, the 10-minute rainfall measured on the site is converted into 1-hour rainfall. The aperture indicates that it closes as the value increases.

Naturally, the temperature is high in summer and low in winter, with a maximum temperature of 37℃ and a minimum temperature of 0℃. The daily temperature range tends to be large on sunny days and small on cloudy days.

The relative humidity changes drastically from day to day, but after the rainy season, the relative humidity continues to be 100 %. According to the Japan Weather Association, the rainy season started around June 11th and ended around August 1st in this year. It can be said that the relative humidity tends to be 100% when the minimum temperature is 20°C or higher.

The fracture aperture did not change much until June but changed significantly after July. Although it is difficult to see on the scale of Figure 2, it has been observed that when it rains with a relatively large amount during the light rain season, the rainfall infiltrates into the upper block and the fracture is closed due to the weight increase (Osada et al., 2020). We think it is a characteristic behavior in a porous, soft rock.

In the changes after July, the change in S1 is particularly large, and it has moved to the opening side. Although it cannot be denied that the sensor may have drifted, the values of S2 and S3 are moving to the opening side. We will discuss it in detail later.

4 DISCUSSIONS

4.1 *The relationship between the temperature and the fracture aperture of daily average*

The relationship between the temperature and the fracture aperture was investigated. Since the temperature and aperture change throughout the day, the average temperature and average aperture are calculated and plotted for each day in Figure 3. The data in first half of the year are plotted in red, and the second half in blue. In addition, the regression line is drawn for the data in the first half, and the obtained gradient is described.

From the figure, it can be seen that there is a clear correlation between the temperature and the aperture in the first half of the year, and that the aperture tends to close as the temperature rises. The gradient at this time is about 0.01 mm/°C, which is almost the same value for the three displacement sensors. Such a clear result could not be obtained unless the measurement was performed on accuracy of less than 0.01 mm. Kuchitsu et al. (2005) reported that fracture aperture measurements were performed on the fractures found in Neogene tuffaceous sandstone and that there was a similar tendency. Although the details are not described, the value of 0.1 mm/°C is read from the figure, which is one order larger than our measurement.

In the latter half of the year, S1 suddenly changes the trend, that is, the aperture shows a large opening first irrespective to the temperature, and gradually return to the original trend. S2 and S3 also show the same route of the curves, although the absolute values in the change are small. From this result, we infer that there is something wrong with the upper block above the fracture. However, there was no systematic change with respect to the order of installation positions. A field check was conducted at the end of January 2021, but no visible changes such as the occurrence of new cracks were observed around S1 sensor.

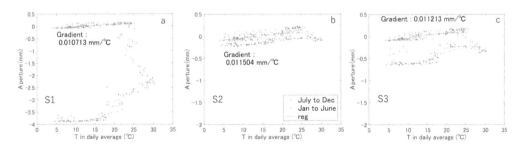

Figure 3.   Relation between temperature and aperture in average.

4.2　The diurnal change of the gradient

Figure 4 shows the measurement results of temperature and aperture on May 14, 2020. From this example, the aperture changes in correlation with the temperature even during the day. As mentioned above, the change in the aperture is affected by the latest rainfall, and there is a difference in the change tendency between cloudy and sunny. Therefore, we selected a day that satisfies the following two conditions and examined it.

- It is a sunny day (the temperature difference is large)
- There should be no heavy rain in the near future (preferably about a week)

Table 1 summarizes the changes in the aperture with respect to the temperature as a gradient. In the table, the temperature range of the day and its median, the individual gradients of S1 to S3, and their average value (S1 + S2 + S3)/3 and the value of (S2 + S3)/2 is described. Looking at this table, there is no significant change in S1 on the gradient. Therefore, the average values of the three displacement sensors are plotted against the median temperature in Figure 5. As a whole, it can be seen from the figure that the higher the temperature, the smaller the amount of change in the aperture per 1 °C. In the figure, the first half and the second half of the year are color-coded, but they are plotted on clearly different lines with the rainy season as the boundary, indicating that the latter half has a larger change against the same temperature. This implies that there was some change in the target fracture. In addition, the observation of the diurnal change of the gradient is one of the effective indices for the signs of the change in the properties of the fracture at an early stage.

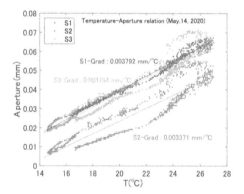

Figure 4.　Relation between temperature and aperture in one day as an example.

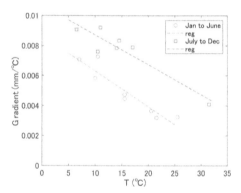

Figure 5.　The diurnal change of the gradient.

Table 1.　Summary of gradients in aperture width to temperature change at one day.

Date	T-Range °C	T-med°C	S1-grad mm/°C	S2-grad mm/°C	S3-grad mm/°C	S-avr mm/°C	(S2+S3)/2 mm/°C	diff mm/°C
2020/1/10	5–16	10.5	0.00655	0.00687	0.00840	0.00727	0.00764	-0.00036
2020/2/7	1–13	7	0.00650	0.00706	0.00772	0.00709	0.00739	-0.00030
2020/2/28	3–17	10	0.00534	0.00525	0.00692	0.00584	0.00609	-0.00025
2020/3/26	8–24	15.5	0.00422	0.00405	0.00513	0.00447	0.00459	-0.00012
2020/4/8	8–23	15.5	0.00453	0.00476	0.00494	0.00474	0.00485	-0.00011
2020/5/2	14–29	21.5	0.00344	0.00296	0.00320	0.00320	0.00308	-0.00012
2020/5/14	14–27	20.5	0.00379	0.00337	0.00379	0.00365	0.00358	-0.00007
2020/6/9	20–31	25.5	0.00360	0.00301	0.00317	0.00326	0.00309	-0.00017
2020/8/18	27–23	31.5	0.00373	0.00362	0.00492	0.00409	0.00427	-0.00018
2020/11/11	10–21	10.5	0.00619	0.00785	0.00879	0.00761	0.00832	-0.00071
2020/11/22	11–23	17	0.00652	0.00759	0.00957	0.00789	0.00858	-0.00069
2020/12/1	8–21	14.5	0.00677	0.00818	0.01007	0.00834	0.00913	-0.00079
2020/12/17	6–22	14	0.00641	0.00781	0.00929	0.00784	0.00855	-0.00071
2020/12/31	1–12	6.5	0.00734	0.00817	0.01172	0.00908	0.00995	-0.00087
2021/1/7	4–18	11	0.00795	0.00884	0.01083	0.00921	0.00984	-0.00063

5 CONCLUSIONS

By examining the data obtained from long-term observations over one year, the following was found.

1. The opening and closing behavior of porous, soft rock fracture can be captured by observing on the order of 0.01 mm in the field.
2. The fracture aperture shows a good correlation with the temperature and tends to narrow when the temperature rises and open when the temperature drops.
3. When the properties of fractures change, the correlation with temperature tends to change.
4. By observing changes in the aperture with respect to temperature changes on a daily basis, signs of changes in the properties of the fractures can be seen. It was shown that it may be possible to know at an early stage.

ACKNOWLEDGEMENTS

This work was supported by the staffs in the Buried Cultural Resource Center of Yoshimi Town. We would like to express our gratitude here.

REFERENCES

Janeras, M., Jara, J.-A., Roy´an, M. J., Vilaplana, J.-M., Aguasca, A., F`abregas, X., Gili, J. A. and Bux´o, P., 2017, Multi-technique approach to rockfall monitoring in the Montserrat massif (Catalonia, NE Spain), *Eng. Geol.*, 219, 4–20.

Kuchitsu, N., Jian, L. X., Seki, H., Morii, M. and Endo, T., 2005, Investigation of cracks for the preservation of Hyakkuhachi-Yagura tombs, Kamakkura City, Japan, *Science for conservation*, 44, 109–116.

Kuchitsu, N. and Ozaki, T., 1999, The evaporites observed at the Yoshimi Caves Historic Site, Japan, *Jour. Geol. Soc. Japan*, 105(4), 266–272.

Matsuoka, N., 2019, A multi-method monitoring of timing, magnitude and origin of rockfall activity in the Japanese Alps, *Geomorphology*, 336, 65–76.

Oguchi, C. T., Takaya, Y., Yamazaki, M., Ohnishi, R., Thidar, A. and Hatta, T., 2010, High acidic sulphate salt production on the cave wall in the Yoshimi Hyaku-Ana histric site, central Japan, *Proceedings of the XIX CBGA Congress*, Thessaloniki, Greece, 100, 413–419.

Osada, M., Hosokawa, K. and Miyamoto, S., 2019, Temperature dependent behavior of an open crack found at the top of the tunnel digged in Pleistocene soft rock, *Proceedings of the 46th Symposium on Rock Mechanics*, JSCE, 61–65.

Osada, M., Ozawa, K., Funabiki, K. and Yumi, A., 2020, Displacement behavior of open fracture in tuff with rainfall confirmed by long-term observation, *Proceedings of the 2020 Annual meeting of the Japan Soc. Eng. Geol.*, 145–146.

Oyama, T., Chigira, M., Ohmura, N., Sasaki, K. and Nagaoka, T., 1998, Weathering rate of mudstone on old unlined tunnel walls and the bacterial effect on it (1), *Abiko Research Laboratory Report*, CRIEPI, Vol. U98001.

Oyama, T., Sasaki, K. and Chigira, M., 1999 Weathering rate of tuff in old unlined tunnel walls and the bacterial effect, *Abiko Research Laboratory Report*, CRIEPI, Vol. U98046.

Takaya, Y. and Oguchi, C. T., 2011, Quantitative evaluation of debris production due to salt weathering of tuff in Yoshimi Hyaku-Ana, an historic site in central Japan, *Geographical Review of Japan*, 84(4), 369–376.

Groundwater and environmental problems in volcanic fields

Rock Mechanics and Engineering Geology in Volcanic Fields – Ohta, Ito & Osada (eds)
© 2023 copyright the Author(s), ISBN 978-1-032-27657-1

Hydrological methods for understanding the shallow groundwater flow in volcanic regions with abundant groundwater resources

Makoto Kagabu*

Institute of Integrated Science and Technology, Nagasaki University, Japan

ABSTRACT: Aso volcano is an active volcano in Kyushu Island, southwest Japan. This volcano is characterized by a large caldera, extending 18 km east–west and 25 km north–south, where a central cone is located. Aso volcano has abundant groundwater resources and is an area where many spring waters have been confirmed. Regional characteristics of spring discharge mechanisms could be found using hydrogeochemical techniques such as major ions, stable isotope ratios (δ^2H and $\delta^{18}O$), and age dating methods (^3H, CFCs, and SF_6) of the spring water. There is a clear difference in the type of major dissolved ions between the outer rim side (Ca-HCO_3 type) and the central volcano side (Ca-SO_4 type), suggesting differences in the groundwater flow system in each area. Among the springs in Aso caldera, the springs located in the area between Aso volcano and the two rivers were characterized by abundant dissolution of sulfate ions (SO_4^{2-}). This was considered an indicator of the influence of Aso volcano. Furthermore, such springs via the Aso volcano have a higher recharge elevation due to the lighter stable isotope ratios, and the age estimation method shows that these springs have a longer residence time. These hydrogeochemical results indicate the existence of large groundwater flow system. The understanding of groundwater flow mechanisms is a fundamental issue for the proper management of abundant groundwater resources in volcanic regions. Through the application of the methodology in this study revealed that there are springs with various flow systems in the Aso caldera.

Keywords: Groundwater flow system, Hydrology, Volcanic field, Stable isotopes, Age dating

1 INTRODUCTION

Aso caldera, located in the central part of Kyushu, is world-famous for its size as a caldera volcano and for the abundance of groundwater in its many springs. Elucidating the mechanisms of groundwater flow system (i.e., recharge, flow, and discharge mechanisms) in such a volcanic region will contribute to the conservation and sustainable use of local water resources.

The distribution of springs around the Aso caldera, their conditions of flow and water quality characteristics have been clarified by Shimano (1997). The regional characteristics of the springs have been investigated based on the concentration characteristics of dissolved major ions.

The hydrogen/oxygen stable isotope ratio of water also serves as a marker that is present in all water on earth. Water molecules such as $^1H^2H^{16}O$ and $^1H_2^{18}O$, in which hydrogen and oxygen atoms are replaced by heavier isotopes, are present in insignificant proportions. By measuring the degree of this proportion (i.e., isotope ratio), water can be classified, and the spatio-temporal distribution of water with different isotope ratios makes it possible to trace the water cycle. Additionally, young groundwater (residence time less than 50 years) is considered prevalent in Japan because of its abundant precipitation, and several age tracers have been proposed to estimate the young residence

*Corresponding author: kagabu@nagasaki-u.ac.jp

DOI: 10.1201/9781003293590-39

time (e.g., Plummer et al., 2000). By applying this method for estimating the residence time, we also attempted to understand the time scale of the regional water cycles.

This study presents a case study of groundwater flow mechanism in a caldera by estimating groundwater recharge elevation using hydrogen/oxygen stable isotope ratios and residence time using age tracers.

2 OVERVIEW OF THE STUDY AREA

Aso volcano is an active volcano in Kyushu Island, southwest Japan. This volcano is characterized by a large caldera, extending 18 km east-west and 25 km north-south, where a central cone is located (Figure 1). The caldera floor is divided into north and south parts called Aso-dani (Aso Valley) and Nangou-dani (Nangou Valley). Two large rivers, called Kurokawa and Shirakawa, respectively, flow on the lowest part of caldera floors at Aso-dani and Nangou-dani, respectively. Those rivers, which meet within the caldera, flow out at the western rim of the caldera. Aso volcano has abundant groundwater resources and is an area where many springs have been confirmed.

The average annual precipitation on Mt. Aso reaches about 3,250 mm (https://www.data.jma.go.jp/obd/stats/etrn/index.php), of which the average precipitation in June and July, the rainy season, reaches about 40% of the total annual precipitation. Because of the high permeability of the volcanic body, much of the precipitation is thought to recharge groundwater.

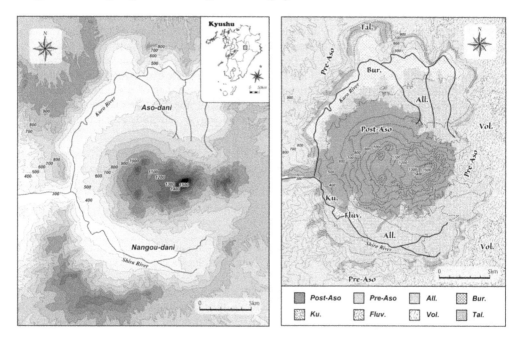

Figure 1.   Study area (Left) and geological settings (Right).
Bur.: Buried layer, Ku.: Kukino layer, Fluv.: Fluvial terrace, Vol: Volcanic ash, Tal.: Talus slope, Pre-Aso: Pre-Aso volcanic, Post-Aso: Post-Aso volcanic, All.: Alluvial (Modified from Shimano (1997)).

3 RESEARCH METHODS

3.1 *Major ions*

When clarifying the groundwater flow and the factors of water quality formation in a certain basin, it is easier to understand analytical results if they are shown in a diagram rather than simply comparing the concentration of each dissolved constituent (mg/L). This diagram is called a "Hexadiagram", and it shows the concentrations of the eight major dissolved constituents; Na^+, K^+, Mg^{2+}, Ca^{2+}, Cl^-, NO_3^-, SO_4^{2-}, and HCO_3^-. The concentrations of each of these constituents are shown as equivalent values (meq/L) in a hexagonal shape. The concentrations of the eight components are divided into anions and cations. The concentrations of each component are indicated by the distance from the centerline, so that it is possible to grasp which ions are dominant and to visually judge the water quality characteristics of the entire region.

3.2 *Hydrogen and oxygen isotope ratios (δD and δ^{18}O)*

Hydrogen and oxygen isotope ratios were expressed using δD and δ^{18}O, respectively, where $\delta = ((R_{sample}/R_{standard})-1)1000$ (‰) and R is the ratio of $D^{/1}H$ or $^{18}O^{/16}O$ in either the sample water (R_{sample}) or the Standard Mean Ocean Water ($R_{standard}$). We reported isotopic compositions in the standard δ-notation (‰) as deviations from the Vienna Standard Mean Ocean Water standard (V-SMOW). The analytical error was ±1.0‰ for δD and ±0.1‰ for δ^{18}O.

Since there is an "altitude effect" where the δ value of precipitation is lighter at higher elevations, and the δ value does not change during the groundwater flow process, the recharge elevation can be estimated by measuring the isotope ratio of spring water.

3.3 *Groundwater age tracer*

Measuring the age (residence time) of invisible groundwater makes it possible to define a series of time scales from recharge to discharge of groundwater, which will play an important role in appropriate groundwater management. It can be used to quantify the amount of groundwater resources in a region, to understand groundwater flow systems, and to predict future contamination. One of the age tracers is the radioisotope technique (^3H, ^{14}C, ^{36}Cl, etc.). These are methods that use the decrease in concentration of naturally occurring radioactive decay to estimate their age. Alternatively, there are methods to estimate the groundwater age by measuring artificially produced gases. These tracers include CFCs (chlorofluorocarbons), SF_6 (sulfur hexafluoride), ^{85}Kr (krypton 85), etc. From the measured dissolved gas concentrations in the groundwater, the atmospheric concentration at the time the groundwater was recharged can be obtained using the known solubility-recharge temperature relationship (Henry's law) and contrasting this value with the historical atmospheric concentration curve. In other words, the year of recharge is estimated and the residence time is determined by comparing when the concentration of the age tracer dissolved in the groundwater is equal to the atmospheric concentration.

4 RESULTS AND DISCUSSION

4.1 *Regional characteristics of spring water by major dissolved ions*

Based on Shimano's (1997) report on the characteristics of groundwater quality in the caldera and the results of the analyses in this study, Hexadiagrams were plotted on the map of water sampling points (Figure 2).

As a result, the samples on the outer rim side are mainly of Ca-HCO$_3$ type, and those on the central cone side are mainly of Ca-SO$_4$ type or intermediate type between Ca-SO$_4$ and Ca-HCO$_3$ type. The samples from the outer rim area have relatively low dissolved elements, and the higher the elevation, the lower the dissolved elements. This suggests that the contact between precipitation, which is a recharge source, and rocks and soils is less at higher elevations and the flow time is shorter (Shimano, 1997). On the other hand, the samples located on the central cone side have relatively high dissolved elements, indicating that the groundwater has been flowing through geology related to volcanic activity at the point where SO$_4$ elements are dominant. These regional differences in major ion types suggest differences in the groundwater flow system in each area.

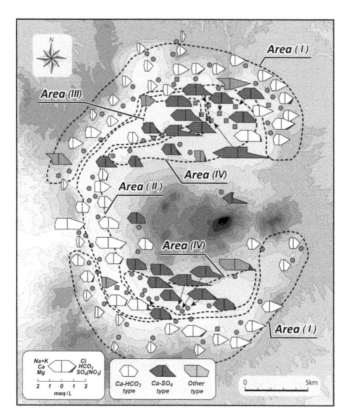

Figure 2. Spatial distribution of Hexadiagram in Aso caldera groundwater.
Four groundwater flow regions (I) to (IV) are also shown.

4.2 *Recharge elevation of spring water by hydrogen stable isotope ratios*

The planar distribution of hydrogen stable isotope ratios (δD) is shown in Figure 3. The caldera is geographically divided into the following four regions; (I) the outer rim of the mountain, (II) the central cone side, (III) the caldera lowland unconfined groundwater system, and (IV) the caldera lowland artesian system, are shown in Figure 3. The characteristics of the isotope ratios in each region are described below.

The lowest isotope ratios were observed in the (IV) area in the central Aso Valley, which averaged -55‰ (-57 to -53‰; N=22). The groundwater from these wells were recharged from relatively high elevation areas, considering the "altitude effect" of the isotope ratio. The isotope ratios in (II) area are slightly heavier than those in the artesian area, with an average value of -54‰ (-58 to -50‰; N=25). The isotope ratios in the (I) area are relatively heavy, averaging -52‰ (-54 to -47‰; N=31), suggesting that the groundwater is recharged from a lower elevation than that of the artesian area. Groundwater in the (III) area also shows heavy isotopic ratios, with an average of -52‰ (-55 to -46‰; N=15).

The groundwater recharge elevation was estimated based on the "altitude effect" of the isotope ratio (Kagabu et al., 2011). The altitude effect in Aso caldera (-1.64‰/100 m) is comparable to that of -1.95‰/100 m on the eastern slope of Mt. Minami-Yatsugatake (Kazahaya and Yasuhara, 1994) and -1.60‰/100 m on the northern slope of Mt. Fuji (Yasuhara and Kazahaya, 1995), which are in the same volcanic region in Japan as our study area.

The recharge elevations of (I), (II), (III), and (IV) area are estimated to be 753 m, 893 m, 759 m, and 972 m, respectively (Figure 4). Additionally, the estimated recharge elevation of the groundwater in (IV) area is higher than that of the outer rim of the Aso Valley (about 800 m above sea level). Therefore, the groundwater in the artesian area may be mainly recharged by precipitation on the central cone side.

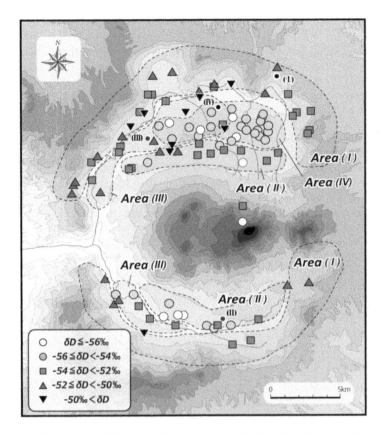

Figure 3. Spatial distribution of δD value. Four representative locations for groundwater age estimation (see in section 4.3) are also shown.

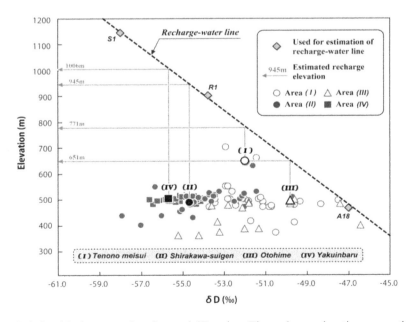

Figure 4. Relationship between elevation and δD value. The recharge elevation was estimated for four representative locations.

4.3   *Residence time of groundwater in the Aso caldera*

In this study, we selected groundwater-sampling points from each of the abovementioned four regions (See Figure 3), and conducted age tracer analysis to examine the residence time in the four regions with different flow systems.

The relatively long residence time was estimated to be around 28 years in areas (II) and (IV). On the other hand, the residence time was estimated to be around 21 years in areas (I) and (III), which was slightly shorter. The estimated residence time for CFCs was within the age range of the estimated residence time for ^{3}H (up to about 35 years) at all sites. So the CFC results were more limited in age.

When the measured CFC concentrations were converted to atmospheric concentrations, several samples were plotted at higher concentrations than the atmospheric concentration histories of each CFC shown in Figure 5 (three solid lines), which could not be compared with the atmospheric concentration histories. This may be due to excessive addition of CFCs, which cannot be explained by dissolution from the atmosphere. In particular, the CFC-11 and CFC-12 concentrations in the Otohime spring (area (III)), which is located in a residential area, were so high that the residence time could not be estimated. This is considered an anthropogenic addition of CFCs. Other studies were conducted to estimate groundwater residence time using SF$_6$ and ^{85}Kr age tracers in the Aso caldera area and the Kumamoto area located to the west of the caldera. Those results also estimated groundwater residence time of the same age scale (Kagabu et al., 2017).

5   GROUNDWATER FLOW MECHANISM IN THE ASO CALDERA

The groundwater flow system in the Aso caldera was classified on the basis of the hydrogeochemical characteristics, and the speed of the regional water cycle was revealed by the age tracer. A schematic diagram of the groundwater flow mechanism based on the results is shown in Figure 6.

In this study, the Aso caldera was divided into four regions; the outer rim of the mountain: region (I), the central cone side: region (II), the caldera lowland unconfined groundwater

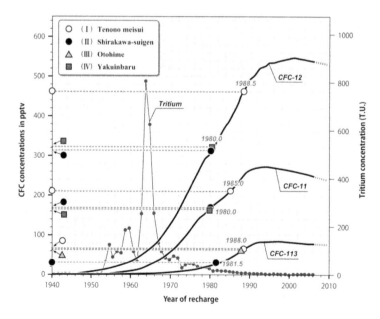

Figure 5.   Historical change of CFC-11, CFC-12 and CFC-113 (in the North American atmosphere) concentrations. Nine examples of the apparent recharge year in four sampling stations are shown.

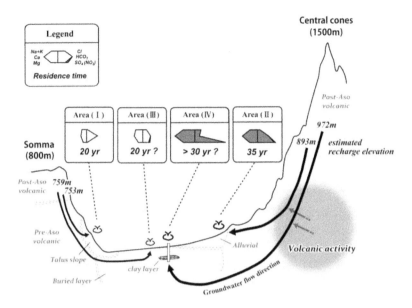

Figure 6. Schematic model showing the groundwater flow system in Aso caldera.

system: region (III), and the caldera lowland artesian well system: region (IV). The distribution characteristics of general water quality were Ca-HCO$_3$ type in region (I) and (III) with relatively low dissolved constituents, while Ca-SO$_4$ type in region (II) and (IV) with high dissolved constituents. The recharge elevation estimated from hydrogen stable isotope ratios is higher in regions (II) and (IV), which have a large groundwater flow system, while it is lower in regions (I) and (III), which is consistent with the low dissolved constituents. Furthermore, ^{3}H and CFC analyses, which are age tracers, were conducted on representative springs and groundwater samples from each region to evaluate the residence time. The groundwater residence time of about 20 to 35 years estimated in this study reveals the regional hydrological cycle scale.

REFERENCES

Kagabu M., Shimada J., Shimano Y., Higuchi S. and Noda S., 2011, Groundwater flow system in Aso caldera. *Journal of Japanese Association of Hydrological Sciences*, 41, 1–17. (in Japanese)

Kagabu M., Matsunaga M., Ide K., Momoshima N. and Shimada J., 2017, Groundwater age determination using ^{85}Kr and multiple age tracers (SF$_6$, CFCs, and ^{3}H) to elucidate regional groundwater flow systems. *Journal of Hydrology: Regional Studies*, 12, 165–180.

Kazahaya K. and Yasuhara M., 1994, Recharge and flow processes of groundwater in the Mt. Yatsugatake from the viewpoint of hydrogen stable isotope ratios of spring water. *Journal of Japanese Association of Hydrological Sciences*, 24, 107–119. (in Japanese)

Plummer L. N., Rupert M. G., Busenberg E. and Schlosser P., 2000, Age of Irrigation Water in Ground Water from the Eastern Snake River Plain Aquifer, South-Central Idaho. *Groundwater*, 38, 264–283.

Shimano Y. (1997): Hydro-chemical study of spring waters in the Aso caldera, central Kyushu. *Bunsei bulletin*, 8, 43–67. (in Japanese)

Yasuhara M., and Kazahaya K., 1997, Stable isotopic composition of groundwater from Mt. Yatsugatake and Mt. Fuji, Japan. *Proc. of the Rabat Symp.*, *IAHS Publ.* 244, 335–344.

Rock Mechanics and Engineering Geology in Volcanic Fields – Ohta, Ito & Osada (eds)
© 2023 copyright the Author(s), ISBN 978-1-032-27657-1

Relationship between the chemical composition of groundwater and the geological structure around tunnel

Kei Yamashita*, Suguru Shirasagi & Kazuhiko Masumoto
Kajima Corporation, Tokyo, Japan

Takehiro Ohta
Yamaguchi University, Yamaguchi, Japan

Hajime Sanjo & Hiroaki Yoshino
Japan Railway Construction, Transport and Technology Agency (JRTT), Kanagawa, Japan

ABSTRACT: It's important to predict in advance the volume and recharge area of discharge water in the tunnel. These are related to the structure of the aquifer, and the degree of connection between groundwater and surface water (well water, river water) is a key for prediction in advance. Previous studies (Kasama et al., 1972), (Ii et al., 1994) have suggested that the changes in the water quality (Electrical Conductivity (EC), pH, redox potential, chemical composition etc.) of the discharge water in the tunnel could be related to the changes in the groundwater flow and geology around the tunnel. This paper discusses about the applicability of the method to evaluate the degree of mixing of surface water in the tunnel, which focuses on the difference in water quality between surface water and groundwater.

To verify this applicability of the evaluation method, chemical composition of the discharge water collected from the advancing boring in the Yamaguchi section of the Central Alps Tunnel, wells and river water around the tunnel was analyzed. In this area, multiple fracture zones of the Atera Fault, which is one of the largest active faults in Japan, are distributed around tunnel and almost orthogonal to the tunnel.

As a result of this survey, discharge water in the tunnel showed pH, EC, and bicarbonate concentration higher than the surface water. This difference means that the discharge water in the tunnel stayed at the aquifer longer than the surface water. Measuring of discharge water from the advancing boring in some areas showed that EC increased with amount of discharge water but in other area, EC decreased. The former is due to the inflow of groundwater with long residence time, and the latter is due to the inflow of surface water.

It was found that analysis of water quality from advancing boring can help assess whether groundwater is connected to surface water or not.

Keywords: tunnel, fault, water-rock interaction, groundwater, geological structure

1 INTRODUCTION

In the construction of tunnels, discharge water in the tunnel causes troubles in the construction process and affects the surrounding environment such as drought on the surface.

*Corresponding author: keiy@kajima.com

DOI: 10.1201/9781003293590-40

Therefore, it is important to predict in advance the volume and recharge area of discharge water in the tunnel. These are related to the structure of the aquifer, and the degree of connection between groundwater and surface water (well water, river water) is a key for prediction in advance.

We investigate the applicability of the evaluation method that utilizes the difference in the quality of surface water and groundwater to understand the degree of contamination of surface water. In this study, chemical composition of the discharge water collected from the advancing boring in the Yamaguchi section of the Central Alps Tunnel, wells and river water around the tunnel was analyzed. This paper reports the method of evaluating the degree of connection of surface water from the discharge water in the tunnel.

2 GEOLOGY AROUND THE TUNNEL AND SAMPLING AREA

The geology of study area is based on Mesozoic Cretaceous Nohi rhyolite, which is intruded by granite porphyry and Inagawa granite at the same period. This area is located in the Atera Fault zone, which is one of the most active faults in Japan. According to the results of the geological survey, 14 major faults are recognized, and the Atera Fault Zone is reported to 208k300m to 206k840m at Central Alps tunnel (Figure 1). Water samples were collected from well and river around the tunnel (Figure 2) and long advancing boring, which had 450m length and was at 100 to 120m depth from the ground surface. In addition, there is an observation well where the groundwater level is continuously measured. This well located at about the middle of the advancing boring (yellow square in Figure 2), and the changes in the groundwater level were observed while the advancing boring were drilling.

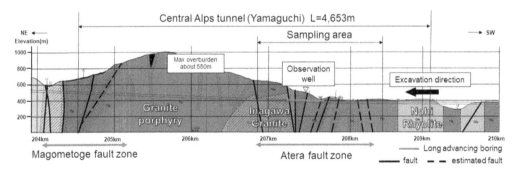

Figure 1. Geological Cross Section and Sampling area.

3 CHEMICAL COMPOSITION OF SAMPLES FROM LONG ADVANCING BORINGS, WELLS AND RIVERS

The results of chemical analysis of the discharge water from the advancing boring, the wells and river water are shown in Table 1 and Figure 3. As a result of comparing these water quality, pH, EC, bicarbonate concentration and Na/Ca ratio of the long advancing boring are higher than the surface water. This suggests that the discharge water from the long advancing boring is the groundwater with longer residence time, where water and rock are more reactive than the surface water. Thus, it was found that the quality of the water varies greatly depending on the discharging point.

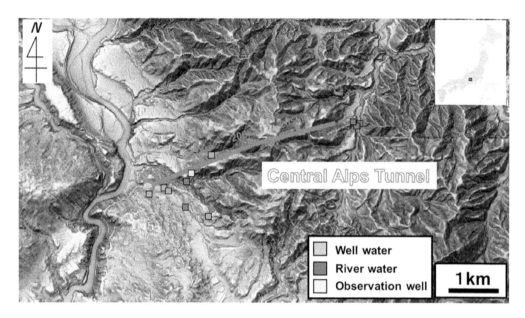

Figure 2.　Location of the Central Alps Tunnel.

Table 1.　pH, EC, bicarbonate concentration of sample from long advancing boring, wells and rivers.

	pH	EC (mS/m)	Bicarbonate concentration (ppm)
River water	7.3 ~ 7.6	3.5 ~ 6.4	12 ~ 26
Well water	5.6 ~ 8.6	4.0 ~ 13.5	12 ~ 53
Long advancing boring	8.3 ~ 8.7	26.1 ~ 31.4	120 ~ 164

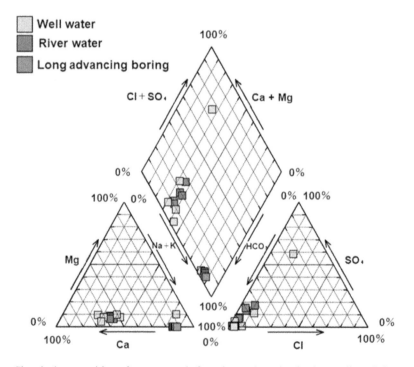

Figure 3.　Chemical composition of water sample from long advancing boring, wells and rivers.

4   EVALUATION OF DISCHARGE WATER FROM THE LONG ADVANCING BORING

Figure 4. shows the EC and the amount of discharge water during the drilling of the long advancing boring at length of 100 to 450 m. The amount of discharge water increased significantly during drilling at four locations, i.e. A, B, C, and D in Figure 4., and therefore the distribution of the aquifer was predicted at these locations. On the other hand, the EC decreased at A, B, and D, and increased at C. This suggests that surface water with a short residence time was mixed at A, B, and D, while groundwater with a long residence time was mixed at C. Moreover, the distribution of the faults is almost the same as A, B, and D where the amount of discharge water increased. Figure 5. shows the temporal change of the water level in the observation well located beside the drilling area of the long advancing boring (around B in Figure 4). When the long advancing boring passed near the observation well, the water level in the observation well decreased significantly and the amount of discharge water in the long advancing boring increased by about 100 L/min (B in Figure 4). It suggests that surface water flowed into the long advancing boring.

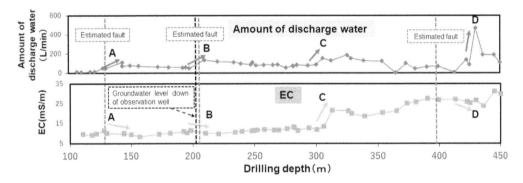

Figure 4.   EC and the amount of water inflow during the drilling of the long advancing boring at length of 100-450m.

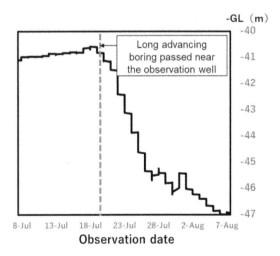

Figure 5.   Changes in water level at observation well located around the drilling area of long advancing boring.

Figure 6 shows the changes of the amount of the discharge water and EC after drilling the long advancing boring. The discharge water includes both surface water with short residence time and groundwater with long residence time. After drilling, the amount of discharge water decreased while the EC increases gradually. This means that a large amount of surface water was initially drawn in, but its ratio decreased due to the lowering of the water level, then the ratio of groundwater with long residence time increased. In this way, we were able to show the possibility of evaluating the influence of the surface water by measuring and analyzing water quality from advancing boring.

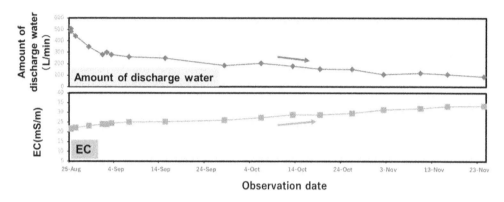

Figure 6.   Changes in amount of discharge water and EC after drilling long advancing boring.

5   CONCLUSIONS

The results of chemical analysis of the discharge water from the advancing boring, pH, EC, bicarbonate concentration and Na/Ca ratio of the long advancing boring are higher than the surface water. This suggests that the discharge water from the long advancing boring is the groundwater with longer residence time, where water and rock are more reactive than the surface water.

The discharge water from the long advancing boring shows that surface water with a short residence time came into the groundwater at the point where EC decreased, and groundwater with a long residence time came into the groundwater at the point where EC increased. As a result of observation of the discharge water from the long advancing boring, a large amount of surface water with a short residence time was initially drawn in, but as time passed, the proportion decreased and finally the proportion of groundwater with a long residence time increased.

We will continue to carry out measurement and chemical analyses of the groundwater from advancing borings in order to validate the method proposed in this paper.

REFERENCES

Kasama, T., Tsurumaki, M., 1972, Occurrence and Quality of Groundwater in the Tunnels of the Rokko Mountain Area, Kinki District, Japan, *Eng. Geol.*, 12, 16–28.
Ii, H., Misawa, S., 1994, The Groundwater Chemistry of the Matsumoto Tunnel and Surrounding Area, Journal of Ground water Hydrogy., 36, 13–29

Rock Mechanics and Engineering Geology in Volcanic Fields – Ohta, Ito & Osada (eds)
© 2023 copyright the Author(s), ISBN 978-1-032-27657-1

Selection of excavation method and recent progress at Yotei Tunnel on the Hokkaido Shinkansen line (from Shin-Hakodate-Hokuto to Sapporo)

Kyohei Ambai* & Kento Narita
Japan Railway Construction, Transport and Technology Agency (JRTT), Niseko Railway Construction Office, Hokkaido, Japan

Naoki Nishimura
Japan Railway Construction, Transport and Technology Agency (JRTT), Hokkaido Shinkansen Construction Bureau, Hokkaido, Japan

Masashi Negoro
Okumura Corporation, Shinkansen Yotei Tunnel (Hirafu) Site office, Hokkaido, Japan

ABSTRACT: Yotei Tunnel is the tunnel in the Hokkaido Shinkansen (bullet train line from Shin-Hakodate-Hokuto to Sapporo) running between Niseko town and Kutchan town (Abuta District, Hokkaido) and 9,750 m in length.

Yotei Tunnel is surrounded by Yotei Volcano on the east side and Niseko Volcanic Group on the northwest side. The maximum overburden is approximately 57 meters and the excavating ground is mainly composed of debris-avalanche deposits which means volcanic deposits produced by the orogenic movements and the eruptive activity from Yotei Volcano. This unconsolidated and sandy or silty layer sometimes contains high compressive strength boulders formed by andesite over 1 meter in diameter. Furthermore, from the additional survey, we estimate that the groundwater level for the entire length almost reaches the ground surface. Therefore, with NATM under high hydraulic pressure and unconsolidated ground conditions, there are risks of tunnel collapse and drought exerted by a lot of spring water from the tunnel cutting face. As a measure for these risks, a casting support tunneling system using TBM (SENS) is adopted for Yotei Tunnel because there are tunneling excavation cases under unconsolidated ground conditions. However, it is important to discuss the countermeasures for the high hydraulic pressure and high compressive strength boulders because there are few cases under those conditions.

This paper discusses the geological survey around Yotei Tunnel and the examination of construction methods for high hydraulic pressure and boulders. Afterward the research introduces the present work in the section of Yotei Tunnel in Sapporo side (Yotei Tunnel Hirafu Section).

Keywords: Tunnel, SENS, Unconsolidated Ground, High Hydraulic Pressure, Boulder

1 INTRODUCTION

The Hokkaido Shinkansen (bullet train line from Shin-Hakodate-Hokuto to Sapporo) starts at Shin-Hakodate-Hokuto Station and extends 212km to the north. The subject of this report, the Yotei Tunnel, locates between Oshamanbe Station and Kutchan Station and lie on both the town of

*Corresponding author: ambai.kyo-k7f6@jrtt.go.jp

DOI: 10.1201/9781003293590-41

Niseko and the town of Kutchan (Abuta District, Hokkaido) at 277 km 560 m – 287 km 310 m from the origin point of Shin-Aomori Station (9,750 m in length). It is under construction in 2 sections: Arishima Section (4,181 m in length) on the origin side and Hirafu Section (5,569 m in length) on the terminal side (Figure 1).

The Yotei Tunnel is surrounded by Yotei Volcano on the east side and Niseko Volcanic Group on the northwest side. At the designing stage, the excavation with NATM was considered, because the excavating ground was good enough although the ground water level was high. However, as a result of detailed survey, if excavating with NATM, it would be difficult for the cutting face to stand steadily, and there would be concerns that a large-scale drought might occur due to excavation.

In this paper, the geological survey around the Yotei Tunnel and examinations of the construction methods at the design stage are discussed. In the latter part of this paper, the present construction situation of Hirafu Section which started the excavation in April 2019 is also introduced.

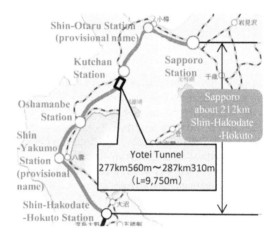

Figure 1.   Location of the Yotei Tunnel.

2   SELECTION OF EXCAVATION METHOD BASED ON GEOLOGICAL CONDITIONS

2.1   Geological condition

In the design of this tunnel, geological surveys such as boring or electrical prospecting were carried out. Figure 2 shows a geological profile of Yotei Tunnel based on the geological survey.

As shown in Figure 2, the excavating ground of this tunnel consists of the debris-avalanche deposits from the tunnel section to the ground surface. These debris-avalanche deposits are composed of andesite and volcanic ash sand due to the volcanic activity of Yotei Volcano. These debris-avalanche deposits are assorted into 4 types as shown in Table 1.

As shown in Figure 2, the excavating ground of the Yotei Tunnel was expected to contain a lot of andesite in part. Therefore, to investigate the detail of andesite, prospecting boring survey was carried out in the section where andesite was anticipated. By the survey, it was confirmed that andesite boulders over 1 m in diameter were included in the excavating ground. On the other hand, in Unit II of debris-avalanche deposit where sand is predominant, the core easily collapsed in the boring survey. Then it was assumed that debris-avalanche deposits had a low degree of consolidation when sand was dominant. Therefore, it was expected that the excavating ground would be mixed with hard and soft ground.

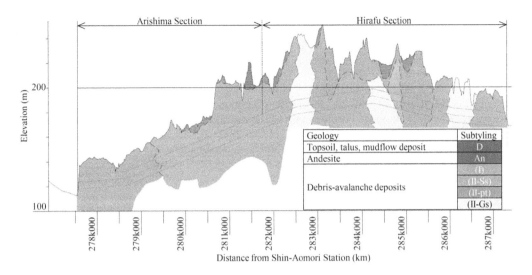

Figure 2. The Yotei Tunnel geological section.

Table 1. Assortment of debris-avalanche deposits.

Subtyping	Profile
I	Debris-avalanche blocks containing hard andesite blocks are predominant and continuously distributed. This unit irregularly contains silt to andesite gravel.
II-Ss	Containing debris-avalanche block and andesite gravel, and mainly composed of silt to sand as substrate.
II-pt	Containing grayish white pumice and volcanic ash.
II-Gs	Containing homogeneous grayish white sand.

2.2 Groundwater condition

Figure 3 shows the relationship between the overburden and head of groundwater of the Yotei Tunnel based on geological survey.

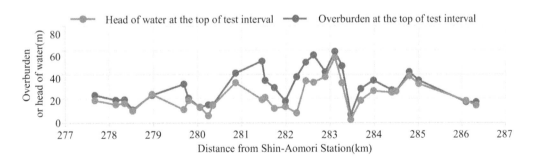

Figure 3. Relationship between overburden and head of groundwater.

From the Figure 3, it was assumed that there would be several places of confined groundwater along the tunnel and that the maximum water pressure of approximately 0.5 MPa would act during tunnel excavation.

319

2.3 Selection of excavation method

Considering the excavation with NATM as a general tunnel excavation method, it is thought to be difficult that the tunnel face stands steadily by itself if excavating ground is composed of debris-avalanche deposits described in **2.1**.

Moreover, if groundwater is abundant in excavating ground of debris-avalanche deposit as described in **2.2**, it would be further difficult that the tunnel face stands by itself. The lowering of groundwater level by drainage boring is one of the measures for groundwater problem in tunnel excavation. However, the excavating ground of the Yotei Tunnel has debris- avalanche deposit reaching almost to the ground surface, that is, there is no impermeable layer.

Therefore, when the groundwater level is lowered, a large-scale drought in the ground was worried. Therefore, in this construction, SENS, which had been used in unconsolidated ground, was adopted instead of conventional NATM.

3 EXAMINATION OF SENS ADAPTATION FOR YOTEI TUNNEL

3.1 Overview of SENS and problems for adaptation of SENS

SENS (a cast-in-site lining system using TBM) is a tunnel excavation method developed in the construction of the Sanbongihara Tunnel on the Tohoku Shinkansen bullet train line. This is a system in which excavation is conducted by keeping stability of tunnel cutting face with closed shield machine, while excavating ground is supported by the extruded concrete lining (ECL) as a primary support constructed in the shield machine tail section. After confirmation of stabilization of primary support, concrete is casted like a secondary lining in NATM. SENS is an abbreviation of Shield, ECL, and NATM System.

Table 2 shows construction conditions of the Yotei Tunnel with comparison to former construction experiences. As shown in the table, all of SENS tunneling have been adopted in unconsolidated ground. However, there is no construction case of SENS under high water pressure condition or in the excavating ground which contains boulders and is composed of hard and soft layers like the Yotei Tunnel, whereas there are some cases of the shield method in such conditions, e.g., (Omori et al., 1992). Therefore, it needed to examine countermeasures against groundwater and boulders.

Table 2. Compilation of construction conditions.

	Yotei Tunnel	Tsugaru-Yomogida Tunnel	Sanbongihara Tunnel
Geology	Sand, gravel, silt and andesite	Sandy soil	Sandy soil, loam, cohesive soil and pumice
Nature	Unconsolidated		
Water pressure	0.5 MPa	0.4 MPa	0.2 MPa
overburden	57 m	93 m	32 m

3.2 Examination against groundwater

In this tunnel, existence of abundant groundwater had been recognized from the preliminary investigation. Since SENS machine covers excavating ground, it is expected that it can prevent drought in the ground caused by a large-scale water flood into the tunnel. However, in the shield excavation, there is a possibility of the blowoff, which is caused by the loss of the balance between the earth pressure and water pressure of the ground and mud pressure of the shield machine, and the loss of the plastic flow state of the excavated soil. Therefore, in order to secure waterproofness of excavated soil and earth pressure at tunnel cutting face, it was decided to install a two-staged screw conveyor with a shaft in the second stage, shown in Figure 4(a).

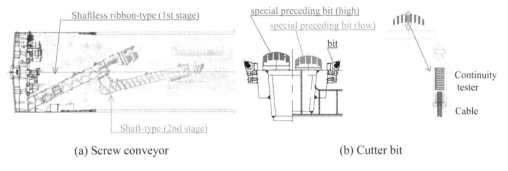

(a) Screw conveyor   (b) Cutter bit

Figure 4.   Countermeasures for the Yotei Tunnel's excavating ground.

Nevertheless, it is not sufficient under this tunnel's condition, shown in Figure 3, because the two-staged screw conveyor in this machine is theoretically available against water pressure up to 0.4 MPa. Therefore, in an environment where the water pressure exceeds 0.4 MPa, it was decided to use the blowoff inhibit additive to ensure the plastic flow state of slurry. The additive improves shear resistance and frictional resistance by aggregating soil and water. In adding the additive, most suitable formulation to the geology of the Yotei Tunnel was examined and selected (Kamagaya et al., 2020).

3.3   Measures against boulders

In this tunnel, the existence of boulders has been confirmed by the preliminary investigation. The maximum diameter of gravel which appear during excavation is estimated as approximately 1,200 mm, but it is difficult to specify where such large gravels would appear. Therefore, the following countermeasures in the shield machine, not from the ground, were examined.

(1) Rock ingesting

When a gravelly ground is excavated by a shield machine, two ways are considered in general: a crushing type in which a rock block is broken into small pieces by a roller bit, and a capturing type in which a rock block is taken in with being large. Since the rock masses are scattered in the unconsolidated ground in this tunnel, in the case of crushing type, there is a concern that the roller bit cannot be rotated due to the soil bitten around the roller bit, and thus the capturing type was adopted. The screw conveyor has two stages as described in **3.2** in order to maintain the tunnel face earth pressure, and in order to take in large boulders, the first stage was equipped with a shaftless ribbon-type screw conveyor in which the gravel diameter of 1,200 mm assumed from the past survey can be taken in, as shown in Figure4(a).

(2) Improvement of wear resistance of cutter bit

Shown in Figure 4(b), as a measure against wearing of cutter bit, a step is provided in the special preceding bit, so that the preceding bit functions for a long time, and the load to the teeth bit is reduced, thereby improving the drillable distance by the equipped bits.

(3) Check of the bit wear amount

A continuity tester embedded in the bits constantly monitor the wear state of the bit. Thus, wear of the bits can be checked, and necessity of maintenance such as replacement of the bits can be estimated.

(4) In case of emergency

In the unlikely event such that a screw conveyor is blocked, or unexpected bit damage occurs, the following measures are being considered.

(a) Gravel destruction equipment

When gravel is caught in a screw conveyor intake part, the gravel is crushed by a gravel crushing device (breakers, rock drills, etc.) from a gravel crusher insertion valve.

(b) Bit inspection and replacement shaft

When bit replacement is necessary due to wear of the bits, a shaft is constructed by a liner plate, and the bits are replaced from the opening.

(c) Ground improvement from the ground or from inside the shield machine

At first, watertight zone is formed in front of the shield machine. When excavation reached the watertight zone, change of bit or removal of gravel are performed in a chamber.

4   CONSTRUCTION OF THE HIRAFU SECTION

The Hirafu Section began excavation in April 2019 based on the consideration of Chapter 3. As of April 1, 2021, the excavation has progressed at 1,918Rings (2,874 m).

4.1   *Actual measures to groundwater*

Figure 5 shows the results of ground water pressure and injection of blowoff inhibit agent.

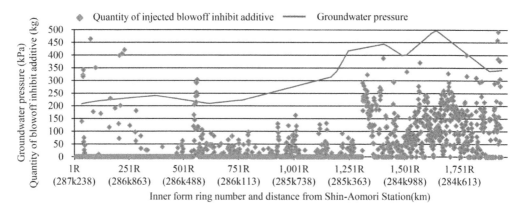

Figure 5.   Ground water pressure and injection of blowoff inhibit additive.

As shown in the figure, ground water pressure generated at the tunnel face has changed with the progress of excavation, reaching 0.5 MPa at the maximum. On the other hand, large blow-off prohibiting excavation has not been occurred yet by injecting the additive appropriately.

Figure 6 shows the estimation and measurement of spring water in tunnel in March 2021.

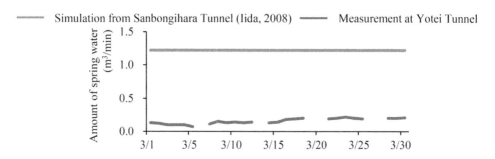

Figure 6.   Spring water in tunnel (March 2021).

The estimate was set from the past measurement at Sanbongihara Tunnel. The figure shows that the amount of spring water in the Hirafu Section is less than 1/5 of the estimated amount. This suggests that the construction is proceeding without deteriorating the surrounding ground water environment.

322

4.2 Actual measures to boulders

Figure 7 shows the particle size distribution and maximum gravel diameter of slurry.

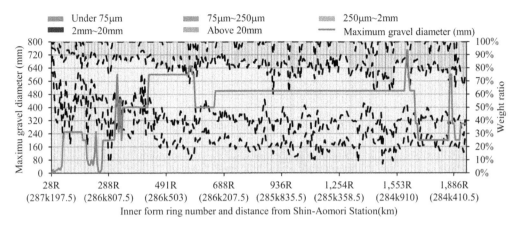

Figure 7. Particle size distribution and maximum gravel diameter of slurry.

From the figure, in the section currently excavated, the alternating layers centered on sand with a fine-grained soil content of around 20% are being excavated. The excavating ground is not uniform in geology, and especially after 850R, the ratio of gravel and sand changes greatly, but uniformity coefficient is high, $U_c = 8 \sim 26$, and it can be said that the soil has a good particle size distribution. From 1,140R to 1,207R and from 1,436R to 1,550R, these are the sections where the existence of large gravel was expected in the preliminary survey. The proportion of gravel and rock mass discharged from the screw conveyor tended to increase slightly, but the size was up to 750 mm in diameter. A fresh crack which can be assumed to have been broken by a cutter or a spoke was recognized in the discharged rock mass, but there was no trace of a rock over 1 m from its shape. When the discharged gravel was examined with a rock Schmidt hammer, it was $75 \sim 275$ MPa in strength, and comparatively soft and hard mass were mixed. However, there has been no trouble that the excavation stopped due to the existence of gravel.

5 CONCLUSIONS

As a result of geological survey in designing the Yotei Tunnel, it was proven that the excavating ground was composed of debris-avalanche deposit with abundant ground water. If excavation was carried out with NATM under such excavating ground, the collapse of the tunnel cutting face and droughts on the ground by the mine flood were likely to occur. Therefore, SENS, which has been used in unconsolidated ground, was adopted in this construction. In addition, to cope with ground water and boulders in the Yotei Tunnel, improvement of the screw conveyor and the bit, and the blow-off inhibit additive were adopted. Currently as of April 1, 2021, the Hirafu Section has been excavated to 2,874 m.

It is expected that boulders will appear in the sections ahead, and that further high water pressure will also act, but we will steadily proceed with the excavation safely.

REFERENCES

Iida, H., 2008, Study on Casting Support Tunneling System using TBM in Un-consolidated Soil with High Groundwater Level, Waseda University, 2008, Ph.D. thesis (in Japanese).
Kamagaya, Y., Negoro, M., Nakamura, M. and Narita, K., 2020, *Examination of foaming material and mud additive in large-scale mud pressure shielding method, Proceedings of 75th JSCE Annual Meeting,* Japan, 9–11, September 2020, VI–936 (in Japanese).
Omori, K., Tomobe, M., Takahira, H. and Soeda, S.,1992, River crossing by shields running in parallel. Sapporo City Rapid Rail Toho Line, *Tunnels and Underground*, 23(12), 1015–1024 (in Japanese).

Rock Mechanics and Engineering Geology in Volcanic Fields – Ohta, Ito & Osada (eds)
© 2023 copyright the Author(s), ISBN 978-1-032-27657-1

Groundwater chemistry controlled by geologic conditions at Yotei Volcano, Hokkaido, Northern Japan

Takehiro Ohta* & Yuka Kubo
Yamaguchi University, Yamaguchi, Japan

Chihiro Kawakami
Yamaguchi Prefectural Institute of Public Health and Environment, Yamaguchi, Japan

Kento Narita
Japan Railway Construction, Transport and Technology Agency, Sapporo, Japan

ABSTRACT: There are huge volume and good quality groundwater in volcanic fields, therefore the volcanic fields are estimated as suitable water resource area. However, because geological structures of aquifer in volcanic fields are often heterogeneous, it is not easy to estimate groundwater distribution and groundwater flow path. Therefore, it is difficult to predict water quality degradation and depletion of water resource by anthropogenic activity such as welling, tunneling and agricultural activity. It is too expensive to use isotope tracer to assess the impact of anthropogenic activity to groundwater resources. If the temperature, pH, EC, Eh and ion concentrations can predict the change of groundwater conditions, it become easy to assess the groundwater condition changes by anthropogenic activity. Therefore, we try to understand the relationship between volcanic geology and groundwater chemistry at Yotei Volcano at Hokkaido, Northern Japan.

Hydrogen and oxygen stable isotope ratios show that all groundwater in Yotei Volcano is derived from meteoric water. The groundwaters on Yotei Volcano were classified into five groups base on geochemical analysis and hierarchical cluster analysis (HCA). Principal Component Analysis (PCA) shows that water-rock interaction is most advanced in groundwater in several lava flows presumably composed of basalt and low-flow rate rocks, followed by groundwater from fine-grained pyroclastic rocks and debris avalanche deposits. From PCA, the geochemical process at some springs seems to organic matter decomposition and anthropogenic pollution.

Keywords: Groundwater Chemistry, Geologic Conditions, Multivariate statistical analysis, Yotei Volcano

1 INTRODUCTION

In volcanic fields, groundwater is abundant and in many cases of good quality, making it an excellent water resource. However, due to the heterogeneity of the geology in volcanic regions, it is difficult to understand the distribution and flow of groundwater. Therefore, for example, it is difficult to predict the deterioration of water quality and depletion of water sources due to volcanic and hydrothermal activities (Aiuppu et al., 2003) and anthropogenic activities

*Corresponding author: takohta@yamaguchi-u.ac.jp

DOI: 10.1201/9781003293590-42

such as pumping by wells, tunnel construction works, and farming (Shimada, 2013, etc.). Predicting and reducing the impact of such groundwater flow and water quality changes on water sources is an urgent issue. In volcanic fields, it is not easy in finance to understand the groundwater flow by using tracers such as stable isotopes. On the other hand, because groundwater quality, especially major ion concentrations, are determined by water-rock reactions in the flow path, it become easy to estimate groundwater flow if these parameters can be used as the tracers. Therefore, it is necessarily to make clear the availability of these parameters as the tracer for hydrology.

This study compares groundwater quality to volcanic geology on Yotei Volcano (Figure 1) as the first step to confirm the availability of the pH, electric conductivity (EC), temperature, and ion concentration as the hydrological tracers. At Yotei Volcano volcanic gases and fluids activity are not remarkable, then it is able to understand the water quality formation mechanism by simple reaction between recharge water and volcanic rocks. This report describes the characteristics of groundwater quality and the relation between the water quality and volcanic geology at Yotei Volcano.

Figure 1. Location, geomorphology of Yotei Volcano and rock facies distribution related sampling wells and springs (based on Katsui, 1956 and Uesawa et al, 2011).

2 STUDY AREA

2.1 Geologic and hydrogeologic settings

Yotei Volcano, which is located at central of southern western Hokkaido, rises from an elevation of about 200 m at its base to about 1800m at the summit, and the volcano is about 11 km across at its base (Figure 1, Uesawa et al. 2016). On the west base of the volcano lava flows, lava domes, and pyroclastic cones, which ejected from summit crater and/or adventive crater are distributed (Figure 1, Katsui, 1956). Alluvial fan deposits are widely distributed around the base of the volcano, and a debris avalanche deposit from this volcano is located along its southwest base (Figure 1, Moriya, 1983). The southeast base of this volcano is covered debris avalanche deposit derived from nearby volcano (Figure 1, Moriya, 1983). The eruptions of Yotei Volcano were divided into three stages, the ejecta of each stage consists of lavas and pyroclastic rocks which is pyroxene andesite (Katsui, 1956).

Hokkaido regional development bureau (1984) illustrated the hydrogeologic map of Yotei Volcano. At 450 to 500 m in elevation, the surface aquifer consists of pyroclastic rocks, the talus deposits play permeable layer at base of the volcano (Hokkaido regional development bureau, 1984).

2.2 *Hydrologic setting*

Tsurumaki (1989) reported the groundwater geochemistry of spring waters around Yotei Volcano. However, temperature and water quality of each spring is constant, ion concentration differ from each spring, which reflects the geology of aquifer (Tsurumaki, 1989). Furthermore, NO_3^- increased since 1970's, that suggested the influences of fertilization at farmland (Tsurumaki, 1989).

3 METHODOLOGY

3.1 *Field investigation*

A total of 34 groundwater samples were collected from 8 wells and 20 springs around Yotei Volcano from October 2016 to November 2021 (Figure 1). Temperature, pH, electric conductivity (EC), and redox potential (Eh) of groundwater were measured at the fields. At the boring wells, water tables were measured before sampling.

3.2 *Chemical analysis*

The major cations (Ca^{2+}, Mg^{2+}, Na^+ and K^+) were analyzed by inductively coupled plasma optical emission spectrometry (ICP-OES: ICP-OES 5110, Agilent Technologies). The major anions (Cl^-, SO_4^{2-}, NO_3^-, F^-, PO_4^{2-} and Br^-) were measured by ion chromatograph (IA-300, DKK-TOA), and HCO_3^- was determined as total alkalinity by sulfuric acid titration to the endpoint of pH=4.5. Charge balance errors of almost samples were less than 5%.

3.3 *Stable isotope ratios*

Stable isotope ratios are useful tracers for understanding groundwater recharge. In this study, the ratios verified the origin and recharge altitude of each groundwater. Stable isotope ratios, $d^{18}O$ and dD were measured by laser cavity ring-down spectroscopy (CRDS: LWIA DLT-100, Los Gatos Research).

3.4 *Multivariate statistical analysis*

Hierarchical cluster analysis (HCA) was used to classify the groundwater samples. HCA is a valuable tool for understanding spatial and temporal pattern of groundwater chemistry and for estimating controlling factors of that (Jiang et al., 2015). In this study HCA is applied to 12 geochemical parameters (major cations and anions) of 34 groundwater geochemical measurements.

Principal component analysis (PCA) was adopted to assess the geochemical processes of each groundwater. PCA is a powerful tool for quantitative investigations into geochemical processes affecting groundwater chemistry such as contamination (Jiang et al., 2015) and water-rock interactions (Koh et al., 2016). The principal components were extracted from the correlation matrix of 12 geochemical parameters (major cations and anions).

Table 1. Results of geochemical analysis of groundwater samples.

No.	Temperature (°C)	pH	EC (mS/m)	Eh (mV)	$\delta^{18}O$ (‰)	δD (‰)	F (mg/L)	Cl	NO_2	Br	NO_3	SO_4	PO_4	HCO_3	Ca	K	Mg	Na	Charge balance error (%)
B-164-1	11.5	7.98	23.40	256	-11.3	-69.3	0.58	10.60	0.03	0.05	0.49	0.38	2.61	124.73	7.44	3.27	3.52	39.26	-0.43
B-165-1	10.8	7.82	11.12	403	-11.0	-67.6	0.06	8.30	-	0.04	0.96	3.01	-	44.60	6.68	2.45	3.43	9.68	2.43
B-212	7.5	7.41	13.30	427	-11.4	-68.2	0.08	10.90	-	0.07	1.05	11.90	-	40.41	9.35	4.23	2.69	11.58	2.32
B-213	8.0	7.47	8.74	440	-11.5	-71.1	0.04	11.10	-	0.03	0.10	3.69	-	25.77	5.68	1.12	2.00	8.32	1.35
B-223	9.6	6.55	15.03	394	-10.1	-63.7	0.05	18.30	-	0.10	0.06	6.67	-	32.70	6.81	8.74	2.84	10.08	1.63
B-225	10.0	7.48	15.65	225	-11.4	-70.3	0.11	11.60	-	0.02	0.68	24.30	-	34.16	9.32	1.27	4.53	13.14	1.11
B-226-1	10.2	7.17	11.09	367	-11.0	-67.9	0.06	12.00	-	0.04	0.04	10.80	-	25.77	6.47	1.64	3.04	8.88	0.55
B-226-2	9.3	7.13	11.07	300	-11.0	-68.7	0.07	10.60	-	0.03	0.05	10.60	-	29.96	6.53	1.69	3.15	8.81	-0.22
B-227	7.5	6.62	47.00	462	-11.4	-69.8	0.02	123.54	-	0.05	2.42	5.81	-	15.32	20.00	3.40	7.06	53.29	1.09
F-8	9.1	7.56	33.00	295	-11.9	-73.1	0.19	14.00	-	-	-	9.37	0.24	170.80	26.35	3.39	11.94	29.19	3.49
G-9	8.2	7.79	7.62	357	-	-	0.03	12.40	-	0.04	-	3.25	-	16.01	3.55	1.39	1.55	8.36	1.59
H-1	5.2	7.91	9.28	421	-11.1	-69.5	0.06	10.50	-	0.02	2.34	4.35	0.22	24.40	5.40	1.40	2.08	8.39	0.39
H-13	10.2	7.57	35.70	444	-11.7	-74.1	0.23	14.80	-	0.03	1.15	10.70	0.31	187.49	23.40	3.76	12.92	31.00	-1.07
HR-1	7.0	7.83	19.29	595	-	-	0.13	6.93	0.01	0.01	2.50	2.69	0.22	19.52	4.87	1.53	1.29	7.43	6.40
HR-1	6.8	7.65	8.09	462	-	-	0.15	6.51	-	0.01	2.61	2.56	0.31	29.28	5.62	1.87	1.55	7.31	-0.20
HR-2	6.6	7.69	8.02	456	-	-	0.15	6.33	-	0.01	2.61	2.53	0.21	26.49	5.55	1.85	1.54	7.29	3.06
K-6	7.5	6.61	19.57	465	-	-	0.17	7.28	-	0.02	1.60	4.62	0.24	97.60	18.52	2.49	5.87	14.75	4.16
NE-1	6.9	6.97	10.17	459	-	-	0.12	6.13	-	0.02	4.85	5.96	0.26	34.16	7.65	2.10	2.31	8.52	2.38
NE-2	6.7	4.92	5.00	532	-	-	0.11	6.59	0.03	0.02	4.40	5.94	0.23	29.28	6.53	2.42	1.90	8.45	2.14
NE-2	6.6	7.84	9.15	458	-	-	0.13	5.36	-	0.03	3.58	4.73	0.29	29.98	6.69	1.89	2.05	7.74	4.26
NS-Y-21-2	9.2	6.76	32.10	435	-	-	0.27	9.28	-	0.02	1.33	5.21	0.23	161.04	28.41	3.16	10.77	22.52	4.84
P-8	5.0	7.94	7.94	465	-11.3	-69.5	0.06	10.30	-	0.03	1.78	3.86	0.08	19.52	4.35	1.52	1.61	7.57	-0.54
R-11	6.2	5.59	5.43	502	-12.4	-76.3	0.18	7.35	0.00	0.01	1.77	3.69	0.32	34.16	7.04	3.48	2.64	9.44	8.95
R-11	6.4	7.53	10.77	446	-12.3	-75.6	0.19	6.81	-	0.02	1.92	3.60	0.32	39.04	7.49	1.54	2.98	9.17	4.91
S-1	6.2	6.23	6.30	399	-12.0	-73.4	0.18	7.63	0.00	0.02	1.90	10.78	0.27	34.16	9.94	4.08	3.12	9.12	8.91
S-1	5.9	7.50	13.90	397	-	-	0.21	7.26	-	0.03	2.07	12.60	0.33	44.60	11.17	1.81	3.88	9.33	2.92
SE-1	7.3	7.60	13.30	421	-	-	0.10	9.37	-	0.03	6.39	6.78	0.12	43.92	11.14	3.01	3.63	9.12	3.52
SW-1	9.2	6.11	19.82	554	-12.0	-72.7	0.11	8.90	0.02	0.02	0.55	3.29	0.23	87.84	9.94	7.11	5.23	14.07	-1.75
SW-1	7.5	6.92	20.80	432	-	-	0.14	9.18	-	0.03	0.94	3.28	0.22	105.90	20.96	2.54	6.32	14.24	3.62
SW-2	14.4	8.30	10.78	521	-	-	0.01	10.98	-	0.02	0.74	3.56	-	14.64	4.20	1.84	1.70	7.47	6.25
SW-3-2	17.8	7.69	22.20	476	-	-	0.02	7.57	0.03	0.04	-	4.21	-	19.52	5.56	1.26	2.07	6.52	10.16
TR-2	10.4	5.32	11.82	516	-	-	0.00	9.19	0.02	0.04	5.97	3.22	-	19.52	4.89	0.24	2.08	7.52	0.37
W-2	9.5	7.81	8.19	334	-	-	0.01	9.28	-	0.03	0.65	5.24	-	24.40	5.28	1.87	1.94	6.21	-2.70
W-3	15.8	7.17	11.61	505	-	-	0.02	8.87	-	0.05	1.28	4.27	-	34.16	9.59	0.84	2.96	8.21	8.88

4 RESULTS AND DISCUSSION

4.1 *Recharge of groundwater at Yotei Volcano*

Oxygen and hydrogen stable isotope ratios of the groundwater in Yotei Volcano indicate that the origin of the groundwater is meteoric water (Table 1, Figure 2). It is estimated that the recharge elevations of these groundwaters are ranged from 580 m to 1525 m, mainly above 1000 m, from the $d^{18}O$ values (Mizutani and Satake, 1997). From these, it is considered that the groundwaters in Yotei Volcano are the past meteoric water recharged at from midslope to nearby summit of this volcanic body without contamination of volcanic water and hydrothermal water. Therefore, it is assumed that geochemistry of the groundwaters is formed by chemical reaction between meteoric water and rock of this volcano.

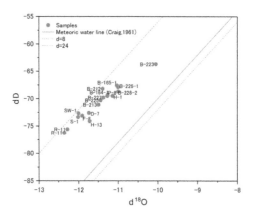

Figure 2. $d^{18}O$ and dD values of groundwaters from Yotei Volcano.

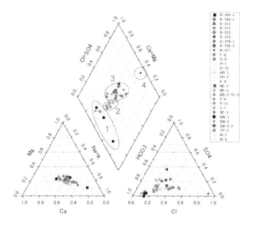

Figure 3.   Piper diagram of groundwaters from Yotei Volcano.

4.2 *Classification of groundwater around Yotei Volcano using major elements and HCA*

To understand the geochemical processes of the groundwater, the groundwaters were classified based on the piper diagram (Figure 3) qualitatively and on HCA (Figure 4) in statistically. The groundwaters of Yotei Volcano showed $Ca-HCO_3$ type mainly, $Na-HCO_3$ type and Na-Cl type in minor. Those were divided into four groups on piper diagram i.e., 1: the group showing the trend from $Ca-HCO_3$ to $Na-HCO_3$ type with highest HCO_3, 2: the group of $Ca-HCO_3$ type with moderate HCO_3, 3: the group of $Ca-HCO_3$ type with low HCO_3, 4: the group of Na-Cl type (Figure 3).

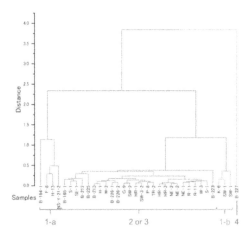

Figure 4.   Dendrogram of groundwaters from Yotei Volcano.

HCA is powerful tool to categorize the many groundwater samples quantitatively because based on statistical theory. As mentioned above the groundwaters could be divided into four groups qualitatively on piper diagram, therefore it was tried that the groundwaters were classified in four clusters by HCA. Figure 4 is the HCA dendrogram using group average method with Euclidian distance. The resalt indicated that group 4 isolate from others remarkably, and group 2 and 3 cannot divide. On the other and, group 1 could be segregated from other groups, furthermore that was divided into two sub-groups i.e., 1-a and 1-b.

From these results, the groundwaters on Yotei Volcano were classified into four or five groups. The samples including each group are listed as below.

Group 1-a: B-164-1, F-8, H-13, NS-Y-21-2

Group 1-b: K-6, SW-1

Group 2: B-165-1, HR-1, HR-2, NE-1, NE-2, R-11

Group 3: B-212, B-213, B-223, B-225, B-226-1, B-226-2, G-9, H-1, P-8, S-1, SE-1, SW-2, SW-3-2, TR-2, W-2, W-3

Group 4: B-227

Group 1 samples excepting B-164-1 are situated nearby lava flows or eruption center at eastern base of the volcano. Group 2 other than B-165-1 are distributed at northern and northeastern base of the volcano nearby lava lows. Group 4, B-227, is locater beside main road which is Root 5.

4.3   *Geochemical process assuming by PCA*

From the principal component analysis of 12 major elements in groundwater, four principal components (PCs) were extracted using a Kaiser criterion, which explained 77% of the total variance (Table in Figure 5). PC1 and PC2 seems to main components in groundwater because these two explain over 50 % of total variance (Table in Figure 5). PC1 had higher positive loadings on HCO_3, Ca, Mg ad Na. PC1 scores in Group 1-a showed highest and followed by Group 4 and 1-b (Figure 1 (a)(b)). Group 2 and 3 had equal to or below zero in PC1 scores (Figure 1 (a)(b)). PC2 had highest positive loading in PO_4, followed by NO_2 and F and had slightly higher negative loading in SO_4 and Cl. Only one of Group 1-a, B-164-1, showed higher score in PC2, and Group 4 had highest negative in PC2 score (Figure 2(a)(c)). PC3 had highest positive loading in Br, followed by Cl. PC3 scores did not relate with groundwater groups, showed higher in B-164-1, B-223 and B-227 (Figure 2(b)(c)). However, Group 1-a excepting B-164-1 had slightly lower values in PC3 scores than other samples. In PC4, NO_3 and Cl were higher positive loadings.

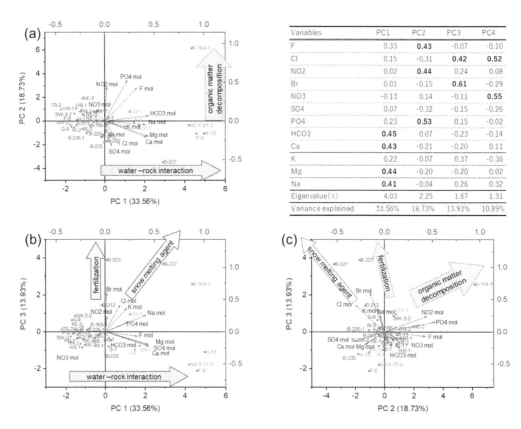

Variables	PC1	PC2	PC3	PC4
F	0.33	0.43	-0.07	-0.10
Cl	0.15	-0.31	0.42	0.52
NO2	0.02	0.44	0.24	0.08
Br	0.01	-0.15	0.61	-0.29
NO3	-0.13	0.14	-0.11	0.55
SO4	0.07	-0.32	-0.15	-0.26
PO4	0.23	0.53	0.15	-0.02
HCO3	0.45	0.07	-0.23	-0.14
Ca	0.43	-0.21	-0.20	0.11
K	0.22	-0.07	0.37	-0.36
Mg	0.44	-0.20	-0.20	0.02
Na	0.41	-0.04	0.26	0.32
Eigenvalue(λ)	4.03	2.25	1.67	1.31
Variance explained	33.56%	18.73%	13.93%	10.89%

Figure 5. Principal component loadings, and loadings and scores of principal components (PC). (a) PC1 loadings versus PC2 loadings, (b) PC1 loadings versus PC2 loadings, (c) PC2 loadings versus PC3 loadings. The colors of labels in each diagram are same as Figure 4.

As PC1 loadings shows higher in HCO_3, Ca, Mg and Na, which are discharged by the reaction between minerals in aquifer and water containing CO_2 gas, PC1 scores seems to illustrate the water-rock interaction in aquifer. Because positive value of these loadings if PC1 scores become higher the degree of water-rock interaction more prograde. From Figure 5 (a)(b) it is considered that the water-rock interaction is most prograde in the groundwater in Group 1-a and followed by Group 4 and Group 1-b. The groundwater in Group 1-a spring from toe of lava flows situated at eastern to southeastern base of Yotei Volcano, and Group 1-b spring from intra part of pyroclastic rocks or debris avalanche deposits. It considered that some lave flows in Yotei Volcano are under conditions at which the water-rock interaction progress in easy, such as basaltic rocks, low flow rate rocks. On the other hand, it is reasonable that water-rock interaction progress in pyroclastic rocks and debris avalanche deposits because those will have large surface areas due to consisting of many fine grain materials.

The highest positive variable in PC2 loadings is PO_4, therefore PO_4 seems to influence to PC2 scores. The source of PO_4 in groundwater seems to anthropogenic contamination, decomposition of organic matter in soils and so on (Ioka et al., 2015). As PO_4 loading in PC2 is most positive, if PC2 score of the groundwater shows higher, the water seems to be affected PO_4 pollution. It is only B-164-1, which is one of Group 1-a, to be shown high PC2 score. This point is well at intra of debris avalanche deposits, and the groundwater at this point showed remarkable low value in Eh. Therefore, the ground at this point may contain organic matter originated past plants involving debris avalanche and is under reduction environment. As the organic matter will decomposed in reduced environment, it is estimated that the groundwater of B-164-1 contains amount of PO_4 from organic matter decomposition. While

Group 4, B-227, shows negative score of PC2, therefore groundwater of Group 4 is affected by Cl and SO_4 which show negative loading.

PC3 loading shows higher in Br and Cl. It is considered that Br in groundwater will be derived from agricultural chemicals. On the other hand, the origin of Cl in groundwater assumes many cases such as fossil seawater. High value of PC3 scores were calculated in only three groundwater i.e., B-164-1, B-223 and B-227. B-227 is located beside main road and is plotted at same direction of synthetic vector of Cl and K loadings in Figure 5(b)(c). Therefore, it is assessed that the groundwater of B-227 are polluted by snow melting agent consist of KCl, which are given the road at winter. In Figure 5(b) B-223, which is in the area that anthropogenic activity is high, is plotted at extension direction of Br and NO_2 loadings. From this, the groundwater of B-223 seems to be affected by anthropogenic pollution such as agricultural activity. The groundwater of B-164-1 has high PC3 score, however that is not plotted at extension direction of Br and Cl in Figure 5(c). Therefore, it seems that the groundwater of B-164-1 is affected anthropogenic pollution only limited and influenced from organic matter decomposition mainly as above mentioned.

5 CONCLUSIONS

This report described the characteristics of groundwater quality and the relation between the water quality and volcanic geology at Yotei Volcano base on the geochemical analysis, stable isotope ratios and multivariate statistical analysis. The results are concluded as follows.

i.   The groundwaters in Yotei Volcano are the past meteoric water recharged at from mid slope to nearby summit of this volcanic body without contamination of volcanic water and hydrothermal water.
ii. The groundwaters on Yotei Volcano were classified into five groups i.e., Group 1-a: B-164-1, F-8, H-13, NS-Y-21-2; Group 1-b: K-6, SW-1; Group 2: B-165-1, HR-1, HR-2, NE-1, NE-2, R-11; Group 3: B-212, B-213, B-223, B-225, B-226-1, B-226-2, G-9, H-1, P-8, S-1, SE-1, SW-2, SW-3-2, TR-2, W-2, W-3; Group 4: B-227
iii. The water-rock interaction is most prograde in the groundwater in Group 1-a, which estimated to consist of basaltic and/or low permeable rocks, and followed by Group 1-b, which consists of fine-grained pyroclastic rocks or debris avalanche deposits.
iv. The geochemical process at some springs seems to organic matter decomposition and anthropogenic pollution.

REFERENCES

Aiuppa, A., Bellomo, S., Brusca, L., D'Alessandro, W., Federico, C., 2003. Natural and anthropogenic factors affecting groundwater quality of an active volcano (Mt. Etna, Italy), *Appl. Geochem.*, 18, 863–882.
Hokkaido regional development bureau, 1984, Hydrogeologic map of Hokkaido (in Japanese).
Ioka, S., Onodera, S. and Muraoka, H., 2015, The behavior of phosphorus in groundwater, *Chikyu Kankyo*, 20, 47–54.
Jiang, Y., Guo, H., Jia, Y., Cao, Y. and Hu, C., 2015, Principal component analysis and hierarchical cluster analyses ofarsenic groundwater geochemistry in the Hetao basin, Inner Mongolia, *Chemie der Erde*, 75, 197–205.
Katsui, Y., 1956, Geology and Petrology of Yotei volcano. Explanatory Text of the Geological Map of Japan (Scale 1:50,000). Rusutsu, Appendix, 1–8 (in Japanese).
Koh, D-C., Chae, G-T., Ryu J-S., Lee, S-G. and Ko, K-S., 2016, Occurrence and mobility of major and trace elements in groundwater from pristine volcanic aquifers in Jeju Island, Korea, *Applied Geochemistry*, 65, 87–102.
Mizutani, Y. and Satake, H, 1997, Hydrogen and Oxygen Isotope Compositions of River Waters as an Index of the Source of Groundwaters, *Journal of Groundwater Hydrology*, 39, 287–297 (in Japanese with English abstract).
Moriya, I., 1983, Topographies of Japanese Volcanoes. University of Tokyo Press (135 p. (in Japanese)).

Shimada, J., 2013, Recent Challenges for sustainable groundwater management at Kumamoto Area, besed on the regional groundwater flow system – pumping permission for the regional groundwater management -, *Journal of Groundwater Hydrology*, 55, 157–164 (in Japanese with English abstract).

Tsurumaki, M., 1989, Visit to valuable water spring (7), *Journal of Groundwater Hydrology*, 31, 165–173 (In Japanese).

Uesawa, S., Nakagawa, M. and Egusa, M., 2011, Reinvestigation of Holocene eruptive history of Yotei Volcano, southwest Hokkaido, *Japan. Bull. Volcanol. Soc. Jpn.* 56, 51–63 (in Japanese with English abstract).

Uesawa, S., Nakagawa, M. and Umetsu, A., 2016, Explosive eruptive activity and temporal magmatic changes at Yotei Volcano during the last 50,000 years, southwest Hokkaido, Japan, *Journal of Volcanology and Geothermal Research*, 325, 27–44.

Rock Mechanics and Engineering Geology in Volcanic Fields – Ohta, Ito & Osada (eds)
© 2023 copyright the Author(s), ISBN 978-1-032-27657-1

Environmental geological study for existence mode and leaching mechanisms of hazardous elements in terrestrial sediments in volcanic arcs

Shinichi Atsuta*
Asano Taiseikiso Engineering Co., Ltd., Tokyo, Japan,

Takehiro Ohta
Yamaguchi University, Yamaguchi, Japan

ABSTRACT: Some troubles of arsenic leaching and acid water drainage from natural ground have been reported at mainly terrestrial sediments in volcanic arcs. Arsenic is abundant in volcanic ejecta on the forearc side of the volcanic front. Volcanic terrestrial sediments have accumulated on the marine sediment, and recycling arsenic has been confirmed in the subduction zone. The interaction of arsenic and marine sediment clay particles are also important to understand soil contamination processes on land by arsenic. The sediments in the Kazusa Group, which were well explored stratigraphically volcanic ash stratum as the key bed, were marine sediments from the late Pliocene to the middle Pleistocene. The concentrations of sulfur and some heavy metal elements in these were related to sedimentation depth.

The concentrations of arsenic in leachate and of arsenic and sulfur in rock samples were analyzed, and these maximum concentrations were 0.08 mg/L, 84 mg/kg and 1.8 wt.% respectively. There is a remarkable relationship between stratigraphy and sulfur content, recognizing the geochemical differences between fresh and weathered sediments. The sulfur and arsenic are distributed in framboidal pyrite crossing to pumice particles in the sediments. Pyrites included in the sediments were divided in three types. Framboidal pyrites in open space had highest arsenic concentration. We could confirm the change of arsenic leaching ratio and pH in leachate water from sediments using thermo-dynamic simulator "PRHEEQC". This result suggests that arsenic leaching and acid water drainage from sediments accrue from dissolution of pyrite. The proposed method could predict the phenomenon caused by dissolution of framboidal pyrite such as arsenic leaching and acid water drainage. The Sustainable Development Goals (SDGs) involve a goal, for efforts to reduce soil pollution, and the importance of effectively utilizing the generated soil as a construction by-product is increasing.

Keywords: Volcanic Terrestrial Sediments, Arsenic, Acid Water, Framboidal Pyrite, PRHEEQC

1   INTRODUCTION

Some troubles of arsenic leaching and acid water drainage from natural ground have been reported at mainly terrestrial sediments in volcanic arcs (Ohta et al., 2013). Arsenic is abundant in volcanic ejecta on the forearc side of the volcanic front. Volcanic terrestrial sediments have accumulated on the marine sediment, and recycling arsenic has been confirmed in the subduction zone.

*Corresponding author: s.atsuta@atk-eng.jp

DOI: 10.1201/9781003293590-43

The emplacement mode of arsenic in marine sediment are also important to understand soil contamination processes on land by arsenic.

In this study, it will be investigated that the existence mode and leaching mechanisms of hazardous elements in Kazusa Group sediments using geochemical and mineralogical methods. Moreover, the mineral assemblage and these amounts contained in the sediments were determined by mineralogical method, then the adsorption and desorption of arsenic on minerals were verified by using of thermo-dynamic simulator "PRHEEQC". The results of environmental geological studies on the modeling of the existence form and dissolution mechanism of arsenic in the Kazusa Group sediments are described. In this paper, the leaching characteristics of natural elements and water quality of leachate from construction soils was predicted based on the thermo-dynamic simulation, and the availability of this simulation was evaluated.

2 METHODOLOGY

2.1 Details of experimental and study samples

The sediments in the Kazusa Group, which were well explored stratigraphically volcanic ash stratum as the key bed, were marine sediments from the late Pliocene to the middle Pleistocene. The concentrations of sulfur and some heavy metal elements in these were related to sedimentation depth. The concentrations of arsenic in leachate and of arsenic and sulfur in rock samples were analyzed by (ICP-MS 7700, Agilent Technologies), and these maximum concentrations were 0.08 mg/L, 84 mg/kg and 1.8 wt.% respectively (Atsuta and Ohta, 2016) (Table 1).

Table 1. Various analysis results of Kazusa Group sediments (after Atsuta and Ohta, 2016).

Analysis items	unit	Result	Lower limit of quantification	Standard value
Arsenic leaching amount	mg/L	0.003 ~ 0.080	0.001	0.01
Bulk composition: Arsenic	mg/kg	N.D. ~ 84	1	39
Bulk composition: Sulfer	wt.%	N.D. ~ 1.8	0.02	2.0
pH(H_2O)	—	6.7 ~ 9.2	—	5.8 ~ 8.6
pH(H_2O_2)	—	2.4 ~ 9.9	—	3.5

(N.D.: Not detected)

Thin section samples of Kazusa Group sediments were prepared and their mineralogy characteristics were examined. Atsuta and Ohta (2016) reported that the mineral assemblage of the Kazusa Group sediments consists of plagioclase and quartz as the major minerals, with biotite and potassium feldspar. The matrix of the sediments is filled with amorphous iron hydroxide, calcite, smectite, illite, pyrite, and "framboidal pyrite" (Rust G.W., 1935) showing a raspberry-like appearance.

The mineral chemistry was analyzing a wavelength dispersive electron probe microanalyzer (JEOL, JXA-8230: EPMA) at the Center for Instrumental Analysis, Yamaguchi University. The analysis conditions were an acceleration voltage of 15 kV, a sample current of 2.0×10^8 A, a beam diameter of about 1 μm, and the correction method was the ZAF method. Table 2 shows the standard samples used for the measurement.

Table 2. Standard samples for mineral chemical analysis.

Fe	Hematite	(Fe2O3)	Fe2O3	99.90 wt.%
S	Sphalerite	(ZnS)	ZnS	32.90 wt.%
Zn	Sphalerite	(ZnS)	ZnS	66.89 wt.%
As	Gallium Arsenide	(GaAs)	GaAs	51.80 wt.%
Pb	PbVGeOxide	(PbVGeOxide)	PbO	79.29 wt.%
Cu	Cu-metal	Cu	Cu	100.00 wt.%
Hg	HgTe	(HgTe)	HgTe	50.00 wt.%

Thermo-dynamic simulation using the simulator "PHREEQC" (Parkhurst and Appelo, 2013) was used to verify the leaching process of natural elements in sediments and the change in water quality of leachate.

The simulation was conducted assuming a virtual embankment of about 10 m × 10 m × 10 m = 1,000 m³, consisting of the Kazusa Group sediments, which are gave off from the construction sites. The ideal mineral content was calculated from the bulk chemical composition of Kazusa Group sediments by a method using C.I.P.W. norm calculation. For the particle size of the mineral, measured values in microscopic such as plagioclase (200 μm) and calcite (200 μm) were used.

PHREEQC can calculate the equilibrium constant of the adsorption reaction, the hydrogen ion, cations, anions of the solution, the surface hydroxyl group concentration of each mineral, the pH condition of the surface mineral and the adsorption amount of each ion. For the reaction formula and equilibrium constant of adsorption of each ion to goethite and gibbsite, the thermal database (wateq4f) was used.

The arsenic leaching and acid water drainage phenomenon in the virtual embankment was set "KINETICS" keywords. The rate equations for silicate minerals were based on the method after Ohta et al. (2013), and rate equation of carbonate minerals used the coefficient of calcite minerals. These rate equations were set with the "Rate" keywords, and set with the "EQUILIBRIUM_PHASES" keywords.

3 RESULT

3.1 *Result of experiments*

In the mineral analysis, pyrite (FeS_2) and framboidal pyrite (FeS_2) were analyzed using EPMA to investigate the leaching phenomenon of arsenic and acidic water. Pyrite could be classified into the following three types according to its morphology and production status (Figure 1).

Type	Framboidal Pyrite		Pyrite
Condition	(1) Open Space Type	(2) In Pumice Type	(3) Cubic Type
BSE image	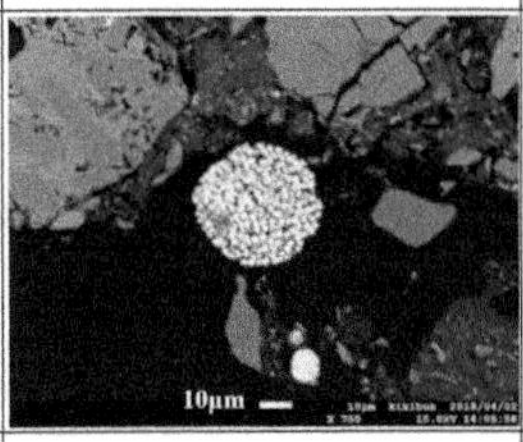	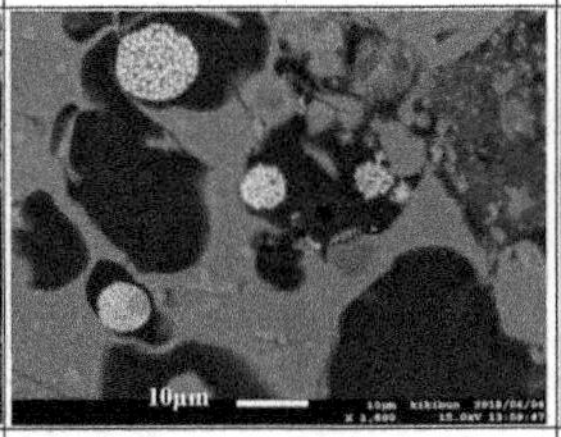	
As (wt.%): Maximum	1.1	0.53	0.46
As (wt.%): Average[*]	0.28	0.18	0.18
As (wt.%): Median[*]	0.22	0.16	0.12
Population	n=40	n=36	n=15

([*]: The average and median values were calculated by removing the data below the lower limit of quantification.)

Figure 1. Classification of pyrite and arsenic content in the Kazusa Group sediments.

(1) Open Space Type: Framboidal pyrite morphology, in which the crystal were not surrounded by other crystals.
(2) In Pumice Type: Framboidal pyrite morphology, crystallized in the voids of pumice particles in the sediments.
(3) Cubic Type: Pyrite's original cubic shape type (euhedral).

Comparing the three types of pyrite, type (1) tends to have a higher arsenic content (Figure 2). On the other hand, no difference was confirmed in the frequency distribution of arsenic content between type (2) and type (3).

3.2 *Result of thermo-dynamic simulation*

Figure 3 shows the long-term prediction results of the leaching phenomenon of Kazusa Group sediments by thermo-dynamic simulation. The pH of the leachate is initially acidic, on the other hand rises to near neutral over time. As the main component element, calcium ion (Ca^{2+}) showed a gradual increase. It was also confirmed that ferrous ion (Fe^{2+}) and sulfate ion (SO_4^{2-}) increased at similar concentration levels, indicating the effect of dissolution of pyrite.

On the other hand, it was confirmed that the concentration of arsenic (As(V)) gradually increases in the acidic region of the leachate, and conversely decreases when the pH value changes to neutral (Figure 3). This is thought to be due to the acidification caused by the dissolution of pyrite at earlier stage, which accelerated the leaching of arsenic from the pyrite. When neutralizing minerals such as calcite and plagioclase dissolve and the pH becomes neutral, arsenic adsorbs on the surface of the amorphous iron hydroxide produced by the dissolution of pyrite, and the arsenic ion concentration is thought to have decreased.

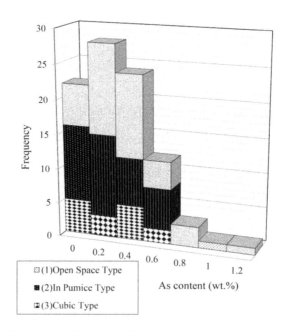

Figure 2. Pyrite classification and frequency distribution of arsenic content in the Kazusa Group sediments.

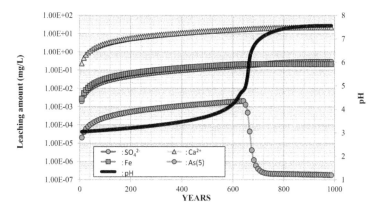

Figure 3. Thermo-dynamic simulation results of Kazusa Group sediments.

4 DISCUSSION

The results of mineral analysis and thermo-dynamic simulation suggest the mechanisms of arsenic leaching of Kazusa Group sediments. It was suggested that arsenic may be dissolved at the same times as acidification by promoting the dissolution of arsenic-containing framboidal pyrite (Eq.1). The proposed method could predict the phenomenon caused by dissolution of framboidal pyrite such as arsenic leaching and acid water drainage.

$$2Fe(S, As)S + 7O_2 + 2H_2O \rightarrow 2H_2SO_4{}^- + 2(S, As)O_4^{2-} + 2Fe^{3+} \tag{1}$$

Sawlowicz (1993) suggested that diagenesis-metamorphism causes the growth of framboidal pyrite to euhedral pyrite. Because the sediments of the Kazusa Group have not been subjected to a high degree of diagenesis, the pyrite exists as "framboidal pyrite" with a large surface area of crystals. This promoted the leaching of arsenic from the pyrite (Figure 4).

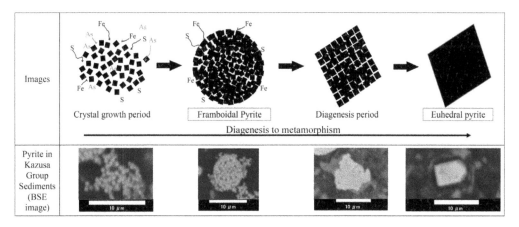

Figure 4. Hypothetical pathways for the formation of euhedral pyrite via framboids in the Kazusa Group sediments.

5 CONCLUSIONS

Pyrite could be classified into three types based on morphology and mineralogical analysis. The results showed that the concentration and frequency of arsenic content varied by types.

Comparing the three types of pyrite, type (1) tends to have a higher arsenic content (Figure 2). On the other hand, no difference was confirmed in the frequency distribution of arsenic content between type (2) and type (3). Comparing the arsenic content of type (1) and type (3), Type (1) is higher. This difference is thought to be due to the leveling of arsenic by diffusion during the crystal growth of "framboidal pyrite" which became a more stable euhedral pyrite during the process of diagenesis (Figure 4).

We could confirm the change of arsenic leachate and pH in leachate water from sediments using thermo-dynamic simulator "PRHEEQC". This result suggests that arsenic leaching and acid water drainage from sediments accrue from dissolution of pyrite. Therefore, it seems that the simulating method could predict the phenomenon caused by dissolution of framboidal pyrite such as arsenic leaching and acid water drainage.

In recent years, the Sustainable Development Goals (SDGs) involve a goal, for efforts to reduce soil pollution, and the importance of effectively utilizing the generated soil as a construction by-product is increasing. When considering construction soil as a geomaterial resource, it is significant that it is now possible to predict, based on geological and petrological characteristics, the leaching mechanisms of naturally occurring heavy metals and other substances due to weathering processes such as water-rock reactions that will occur in the future and medium- to long-term behavior such as changes in redox state due to construction. By predicting the leaching mechanisms based on geological and petrological characteristics, it will be possible to correctly assess the geological risk. Therefore, the method will contribute to SDG's goals.

REFERENCES

Atsuta, S. and Ohta, T., 2016, Existence form and leaching mechanism of arsenic of natural origin of submarine sediments of the Kazusa Group sediments, *22nd Symposium on Soil and Groundwater Contamination and Remediation*, 498–502

Ohta, T., Marumo, K. and Atsuta, S., 2013, Thermo-dynamic simulation of batch leaching tests for soil and rock samples, *13th Japan Symposium on Rock Mechanics & 6th Japan-Korea Joint Symposium on Rock Engineering*, 601–606

Parkhurst, D.L. and Appelo, C.A.J ., 2013, Description of Input and Examples for PHREEQC version3 - A Computer Program for Speciation, Batch-Reaction, One-Dimensional Transport, and Inverse Geochemical Calculations, U.S. Geological Survey Techniques and Methods, book 6, chap. A43, 497p.

Rust G.W., 1935, Colloidal primary copper ores at Cornwall mines, southeastern Missouri. *Journal of Geology*, No.43, 398–426

Sawlowicz, Z., 1993, Pyrite framboids and their development: a new conceptual mechanism, *Geologische Rundschau*, Volume 82, 1, 148–156

Rock Mechanics and Engineering Geology in Volcanic Fields – Ohta, Ito & Osada (eds)
© 2023 copyright the Author(s), ISBN 978-1-032-27657-1

Estimation of groundwater quality forming mechanism in Kuju volcano based on principal component analysis and stable isotope rations

Yuka Kubo* & Takehiro Ohta
Yamaguchi University, Yamaguchi, Japan

Chihiro Kawakami
Yamaguchi Prefectual Institute of Public Health and Environment, Yamaguchi, Japan

ABSTRACT: The volcano stores a large amount of groundwater in the mountain body. Because the heated water coming from a magma activity around a volcano is rise from deeper ground, chemical variety is recognized in groundwater and heated water. On the other hand, it is assumed by experiments and simulations that the groundwater chemistry is controlled by water-rock interaction between groundwater and rocks neighboring. Therefore, it is necessary to consider water - rock interaction for examination of the quality of the water formation. Furthermore, the influence by the magma activity must be considered in the volcano area.

The Kuju volcano is an active volcano located in a volcanic front of Middle Kyushu and the Beppu - Shimabara rift valley. Various springs exist around the Kuju volcano, but the water quality formation mechanism as a whole is not examined. Therefore, this study clarifies a characteristic of the springs and the heated waters around this volcano and examines water - rock interaction in this volcano area. And it is intended to clarify it about the water quality formation mechanism of the whole Kuju volcano based on them.

We performed the quality of the water analysis of the sample, and principal component analysis based on the result. Moreover, a stable isotope rations of water samples were analyzed.

It is thought from the results of the principal component analysis that the quality of the water which is rich in Ca and Mg was formed by water-rock interaction with the andesite, and that the quality of Na and K rich water is caused by water-rock interaction with dacite. Furthermore, it is thought that the origin of the groundwater is rainfall from the result of the isotope analysis.

Keywords: Kuju volcano, stable isotope rations, principal component analysis

1 INTRODUCTION

Volcanoes store large amounts of water in their bodies, and there are many spring ponds at the foot of the mountains (Yamamoto,1995). In the vicinity of volcanoes, hydrothermal fluids originating from magmatic activity also rise. Therefore, it is thought that the groundwater and springs in volcanic bodies show chemical diversity because of influence by magma-derived hydrothermal fluids and volcanic gases (Ohta, 2006). However, these studies have not examined a model of water quality formation in the groundwater of volcanic bodies based on water-rock interactions. On the other hand, it has been reported from batch-leaching tests and

*Corresponding author: a041vcu@yamaguchi-u.ac.jp

DOI: 10.1201/9781003293590-44

simulations using PHREEQC that groundwater quality is formed as a result of water-rock interaction with surrounding rocks (Ohta et al., 2019).

In the Kuju volcano, local water quality formation has been studied for carbonate spring at the eastern foot of the volcano (Yamada et al., 2007) and hydrothermal water at the Komatsu Jigoku in the western part of the volcano (Kiyosaki et al., 2006), but there has been no systematic study on water-rock interaction in the entire Kuju volcano. The purpose of this study is to clarify the relationship between volcanic geology and water-rock interaction around Kuju volcano, where springs and hydrothermal springs are affected by volcanic activity and those that are not, by conducting multivariate analysis for the water chemistry.

2 STUDY AREA

The study area is Kuju volcano, which is an active volcano located in a Beppu-Shimabara Graben leading from Beppu Bay to Ariake Sea, and which makes up volcanic front of Middle Kyushu with Yufu-Tsurumi volcano and the Aso volcano (Kawanabe et al., 2015). Kuju volcano is divided into three areas; i.e. the eastern area, the central area, the western area geomorphologically (Kawanabe et al., 2015). This volcano is comprised several small stratovolcanoes and lava domes, which consist of andesite to dacite rock (Figure 1, Kawanabe et al., 2015).

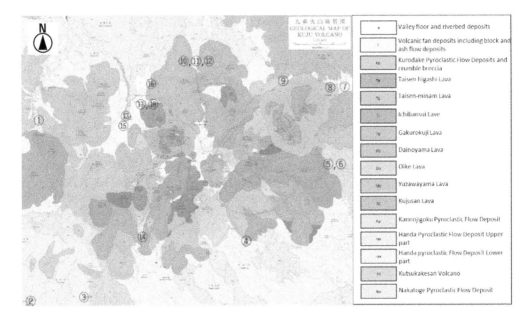

Figure 1.   Geology and sampling sites (based on Kawanabe et al., 2015).

3 METHODS

Water samples were collected from 11 spring waters, 1 river water,1 hydrothermal water, and 3 hot spring waters in March 2019 to September 2020 (Table 1). The temperature, pH, Electrical Conductivity (EC), Oxidation-Reduction Potential (ORP) of water samples were measured at sampling sites. And samples were collected when these parameters were stable.

Table 1. Characteristics of sample collection sites.

No	sample	Type	Geology around the spring※	Remarks
1	Komatsu jigoku	hydrothermal water (flowing)	Handa Pyroclastic Flow Deposit (d)	
2	Ikeyama	spring	Handa Pyroclastic Flow Deposit (d)	
3	Yamabuki	spring	Handa Pyroclastic Flow Deposit (d)	
4	Itibanmizu	ground water	Ichibansui Lave(d)	
5	Imamizu	ground water	Gakurokuji Lava(a)	carbonate spring
6	Imamizu river	river		
7	Yoiyana	ground water	Kurodake Lava Dome(a)	carbonate spring
8	Kurotakesou	ground water	Kurodake Lava Dome(a)	carbonate spring
9	Oike	spring	Kurodake Lava Dome(a),Dainoyama Lava(a)	
10	Naruko river	river		downstream of Hokkein Onsen
11	Borehole	flowing	Yuzawayama Lava(a),Dainoyama Lava(a)	
12	Drain hole	ground water	Yuzawayama Lava(a),Dainoyama Lava(a)	cut slope
13	Kannojigoku	cold spring water	Kannojigoku Pyroclastic Flow Deposit, Handa Pyroclastic Flow Deposit (d)	
14	Akagawa Onsen	hot spring water	Kujusan Lava(a)	
15	Hossyo Onsen	cold spring water	Kannojigoku Pyroclastic Flow Deposit, Handa Pyroclastic Flow Deposit (d)	
16	Yuzawa	spring	Yuzawayama Lava(a)	
17·18	Yubiyama	spring	Yubiyama Lava(a)	

※(a) shows andesitic and (d) shows dacitic

The major anions (Cl^-, SO_4^{2-}, and NO_3^-) were measured using ion chromatography (ThermoFisher SCIENTIFIC ICS-1600). HCO_3^- was determined by sulfuric acid titration. The major cations (Ca^{2+}, Mg^{2+}, Na^+, and K^+) were measured by inductively coupled plasma optical emission spectrometry (ICP-OES; Aglient Technologies 5110 ICP-OES). Trace elements were measured by Inductively Coupled Plasma Spectrometry (ICP-MS; Agilent 7500cx ICP-MS).

In order to clarify the origin of water, isotope analysis was conducted. Stable isotope ratios, $\delta^{18}O$ and δD were measured by laser cavity ring-down spectroscopy (CRDS: LWIA DLT-100, Los Gatos Research).

Principal Component Analysis was carried out at 10 sites using metallic elements (B, Al, V, Cr, Mn, Fe, Co, Ni, Cu, Zn, Ge, As, Se, Rb, Sr, Mo, Cd, Sb, Pb, U) and major elements (Na, K, Mg, Ca, HCO3, SO4, Cl, NO3, F).

Principal Component Analysis is a technique for summarizing data consisting of multiple variables into a small number of Principal Component. A principal component is a first-order combination of the original variables multiplied by their coefficients and added together and is mainly used to characterize a sample (Shimizu,2015).

4 RESULTS AND DISCUSSION

4.1 Classification of water samples around Kuju Volcano by major elements

The results of the water quality analysis are shown in Table 2 and Figure 2. Komatsu Jigoku, KannoJigoku, and Akagawa Onsen are plotted for Type I, Ichibanmizu for Type V, and others for Type II. It is thought that Ichibanmizu is plotted in type V.

Table 2. Results of water quality analysis (from December 2019 to November 2020).

Sample		1. Komatsu jigoku	2. Ikeyama	3. Yamabuki	4. Ichibanmizu	7. Yoiyana	8. Kurotakesou	9.Oike	11. Borihole	12.Drain hole	13. Kannojigoku	14.Akagawa onsen	15. Hossyo Onsen	16. Yuzawa	17. Yubiyama①	18. Yubiyama②	18. Yubiyama②
Sp.Date		20200917	20200917	20200917	20200917	20201111	20201112	20201112	20200916	20201112	20201112	20191201	20200916	20200917	20200917	20200916	20200917
pH		2.87	7.67	7.67	—	7.42	5.53	4.95	5.92	6.28	6.17	4.63	5.24	4.73	5.74	6.33	5.69
EC	(mS/cm)	93.8	9.18	13.04	—	11.13	39.9	14.01	26.3	178.6	19.58	32.00	186	38.1	34.4	18.08	18.04
ORP	(mV)	358	125	143	—	233	272	289	238	31	202	-4	-178	-65	172	236	207
Temperature	(°C)	62.4	14.3	14.7	—	11.6	11.5	11.0	13.1	16.7	12.0	13.3	24.4	19.0	16.4	13.3	13.4
Na	(mg/L)	2.65	1.57	2.06	—	5.61	13.05	7.65	2.76	58.83	8.12	4.10	66.49	5.71	6.32	3.03	2.57
K		8.07	2.93	2.99	—	2.12	3.99	1.39	2.51	10.67	2.60	4.31	17.94	7.22	4.56	2.76	2.19
Mg		5.15	2.17	3.97	—	2.38	15.43	1.97	8.68	79.86	5.76	7.18	53.30	11.92	11.91	5.21	4.01
Ca		16.95	7.80	13.02	—	10.31	52.04	18.23	36.63	242.19	37.98	43.57	488.65	41.92	31.61	16.45	17.31
HCO_3		0.00	39.04	43.92	—	29.28	195.20	41.13	80.52	1169.11	97.60	0.00	119.91	0	52.46	15.34	4.88
SO_4		244.3	7.4	19.3	—	26.2	31.2	5.3	47	93	11.7	116.24	1000.6	147.2	81.6	46.9	55.1
Cl		1.7	2.9	2.8	—	2.9	4.1	1.6	4.7	30.9	3.9	12.19	15.3	23.1	25.1	10.4	8.1
NO_3		0.08	0.59	0.3	—	0.93	1.94	1.51	1.67	0	1.23	0	<0.05	<0.05	1.25	0.87	1.29
F		0.07	0.05	0.07	—	0.09	0.45	0.44	0.12	0.02	0.03	0.184	0.46	0.10	0.14	0.19	0.21
B	(µg/L)	7	6	—	—	18	29	10	30	119	28	51	53	—	—	—	—
Al		8014	5	—	—	2	82	733	21	8	39	811	104	—	—	—	—
V		5.7	5.3	—	—	9.5	22.6	31.1	12.2	<0.1	8.2	1.7	0.3	—	—	—	—
Cr		1.7	0.2	—	—	<0.1	<0.1	<0.1	0.5	<0.1	<0.1	<0.1	<0.1	—	—	—	—
Mn		431	<0.1	—	—	0.8	11.9	105.3	0.3	803.1	0.9	294.7	2700	—	—	—	—
Fe		12098	2	—	—	<1	<1	2	4	269	12	4	25	—	—	—	—
Co		6.3	<0.5	—	—	<0.5	<0.5	<0.5	<0.5	8.0	<0.1	<0.5	<0.5	—	—	—	—
Ni		4.3	<0.5	—	—	<0.5	<0.5	<0.5	<0.5	0.8	<0.5	<0.5	<0.5	—	—	—	—
Cu		8.3	<0.5	—	—	<0.5	34	8	<0.5	<0.5	<0.5	10	<0.5	—	—	—	—
Zn		45	8	—	—	<1	17	1	6	<1	3	<1	<1	—	—	—	—
Ge		1.0	<0.1	—	—	<0.1	<0.1	<0.1	<0.1	<0.1	<0.1	0.1	0.8	—	—	—	—
As		<0.1	0.2	—	—	<0.1	0.1	<0.1	0.2	0.2	0.3	<0.1	40	—	—	—	—
Se		<0.1	<0.1	—	—	<0.1	<0.1	<0.1	<0.1	<0.1	<0.1	<0.1	<0.1	—	—	—	—
Rb		23.0	8.4	—	—	6.7	9.9	3.2	8.3	20	9.1	12.4	34.9	—	—	—	—
Sr		75	62	—	—	78	283	124	175	857	169	149	1126	—	—	—	—
Mo		<0.1	0.2	—	—	<0.1	0.3	1.2	<0.1	0.4	0.1	<0.1	<0.1	—	—	—	—
Cd		0.08	<0.025	—	—	<0.025	0.029	0.059	<0.025	<0.025	<0.025	<0.025	<0.025	—	—	—	—
Sb		<0.1	<0.1	—	—	<0.1	<0.1	<0.1	<0.1	<0.1	<0.1	<0.1	<0.1	—	—	—	—
Pb		<0.1	<0.1	—	—	<0.1	0.9	0.1	<0.1	<0.1	<0.1	<0.1	<0.1	—	—	—	—
U		0.224	0.035	—	—	<0.025	0.339	0.029	0.071	0.836	0.067	<0.025	<0.025	—	—	—	—
δD	(‰)	—	—	—	—	—	—	—	—	—	—	-58.9	—	-59.2	-60.4	-60.0	-61.0
$\delta^{18}O$		—	—	—	—	—	—	—	—	—	—	-9.1	—	-9.1	-9.1	-9.1	-9.1

4.2 Oxygen and hydrogen isotope analysis

We collected precipitation at three sites in the northwestern part of Kuju and analyzed it to obtain $\delta D = 7.99*\delta^{18}O + 10.16$, which is almost the same as the formula of the world's meteoric water line $\delta D = 8*\delta^{18}O + 10$ (Craig, 1961). The results of 13 water samples (Table 3) were compared with the meteoric water line, and all of samples except for Komatsu Jigoku are near the meteoric water line, indicating precipitation origin. In the case of Komatsu Jigoku, δD is almost the same as that of groundwater, but $\delta^{18}O$ is higher than that of groundwater, suggesting that groundwater reacted with rocks and/or mixed with volcanic gases under high temperature underground. The recharge elevation was calculated according to $\delta D = -0.014H - 41.7$ (Kazahaya and Yasuhara, 1994), and it was found that all the points except Komatsu Jigoku were recharged on the hillside near the outflow point.

4.3 Principal component analysis

Kubo et al., (2020) performed principal component analysis using ion concentrations and ion concentration ratios for Na, K, Mg, Ca, HCO_3, SO_4, Cl, NO_3, and F (Figure 4). They classified the springs in Kuju volcano into the following five categories, 1) those affected by volcanic gases mainly H_2S and SO_2 (No.1,10,13,14), 2) those affected by volcanic gases mainly volcanic CO_2 (No.5,7,8,11), 3) formed by the reaction with felsic minerals (No.2,3,4), 4) formed by the reaction with mafic minerals, and 5) possibly anthropogenic contamination (No.6).

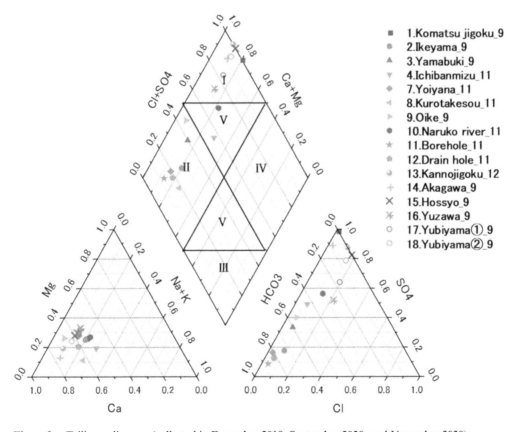

Figure 2. Trilinear diagram (collected in December 2019, September 2020, and November 2020).

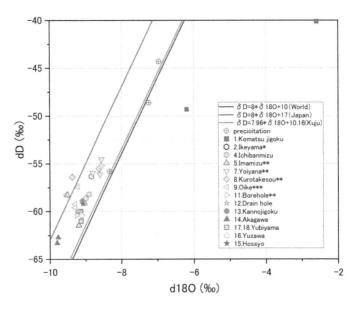

Figure 3.   Stable isotope ration.

Table 3.   Isotope data.

Sample	Sp.Date	δD(‰)	$\delta^{18}O$(‰)
1.Komatsu jigoku	201903	-49.3	-6.2
	201911	-40.1	-2.6
2.Ikeyama*	201609	-56.31	-8.84
4.Ichibanmizu	201911	-58.2	-8.9
5.Imamizu**	201703	-58.28	-9.52
	201704	-59.13	-9.03
	201708	-61.45	-9.19
7.Yoiyana**	201703	-54.53	-8.55
	201704	-55.95	-8.80
	201708	-56.11	-8.59
8.Kurotakesou**	201703	-56.39	-9.37
	201704	-58.64	-9.00
	201708	-55.51	-8.62
9.Oike	201609*	-59.24	-9.29
	201703**	-57.41	-9.23
	201704**	-58.25	-9.53
	201708**	-59.59	-9.30
11.Borehole**	201704	-60.38	-9.24
	201708	-55.19	-8.55
14.Akagawa Onsen	201911	-63.3	-9.80
15.Hossyo Onsen	20200917	-9.08	-59.2
16.Yuzawa	20200917	-9.11	-60.4
17.Yubiyama①	20200916	-9.14	-60.0
18.Yubiyama②	20200917	-9.13	-61.0
Makinoto Parking	20200613	-44.3	-7.0
Makinoto Prospect Terrace	20200613	-55.8	-8.3
Chojabaru visitor center Parking	20200614	-48.6	-7.2

* Ohta (2017)
** Yamashita (2018)

In this study, trace elements were newly added and the principal component analysis was conducted. The results of the principal component analysis are shown in Figure 5.

From Figure 5, Akagawa Onsen (14) is plotted in the first quadrant, borehole (11) and Kannojigoku (13) are plotted in the second quadrant, (2), (4), (7), (8), (9), and (12) are plotted in the third quadrant, and Komatsu jigoku (1) is plotted in the fourth quadrant. Principal component 1 had higher variations of Fe, Ni, Al, Cr, Zn, Ge, and Co, and principal component 2 had higher variations of Mn, Sr, Rb, As, and B positively, and V, Cu, Cd, Pb, and Se negatively. The water temperature was highest at Komatsu Jigoku, followed by Akagawa Onsen and the borehole, these only three water samples had positive PC1, and the temperature was lower at the rest of the sites, suggesting that the principal component 1 represents the elements dissolved depending on the water temperature.

Yoiyana and Kurotakeso distributed around Kurodake Lava dome which consist of basaltic andesite, have high V content and large negative PC2. Kubo et al., (2020) illustrated that the water quality of these two derived from interaction with basic andesite from the principal component analysis for major ion concentrations. Therefore it seems that negative PC2 values represents water-rock interaction with basic andesite. This is supported by that the amount of V in igneous rocks varies significantly with the amount of silicon dioxide (Koshimizu and Kyotani, 2002). The other side, Akagawa Onsen and Borehole, which are influenced by volcanic activity from estimation of principal component analysis for ion concentration, had high positive PC2 value. The positive value of PC2 therefore may show degree of volcanic activity.

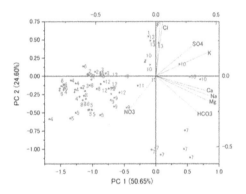 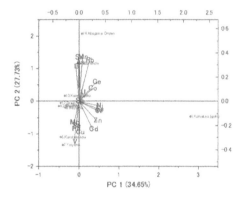

Figure 4. Principal component analysis for major ion concentration. Red plots are PCA scores for each sample. Numbers are same as Table 1. Blue lines show synthetic vectors of components for each variable.

Figure 5. Principal component analysis for trace elements. Red plots are PCA scores for each sample. Blue lines show synthetic vectors of components for each variable.

From this, we can classify them into four groups: 1) high water temperature group, 2) affected by volcanic activity, 3) markedly reacted with rocks, and 4) not much affected.

5   CONCLUSIONS

The results of this study can be summarized as follows.

(1) According to the water quality analysis, the springs waters and hot spring waters around Kuju volcano were distinguished as follows; Type 1: Komatsu Jigoku, KannoJigoku, Akagawa Onsen, Type 5: Ichibanmizu, and Type 2: Others.
(2) According to the oxygen/hydrogen isotope ratio, the groundwater at the points other than Komatsu Jigoku is assumed to originate from the meteoric water recharged in the

surrounding volcanic body. The groundwater in Komatsu Jigoku is thought to have reacted with rocks under high temperature and mixed with volcanic gases.

(3) From the principal component analysis, we can classify them into four groups: 1) high water temperature group, 2) affected by volcanic gases, 3) markedly reacted with rocks, and 4) not much affected by water-rock interaction and/or volcanic activity.

REFERENCES

Craig, H. (1961): Isotopic variations in meteoric waters. *Science*, 133, 1701–1703.

Kawanabe, Y., Hoshizumi, H., Itoh, J. and Yamasaki. S. (2015): Geological Map of Kuju Volcano. Geological Survey of Japan, AIST. (*in Japanese*)

Kazahaya and Yasuhara (1994): Flow process of groundwater recharge in Yatsugatake from the viewpoint of hydrogen isotope ratio of spring water. *Hydrology*, Vol.24, No.2, 107–119. (*in Japanese with English*)

Kiyosaki, J., Oikawa, K., Taguchi, S. and Chiba, H. (2006): Geochemistry of Hot Spring Water from Komatsu Jigoku Steaming Ground in Hatchobaru Geothermal Field, Oita Prefecture, Japan. *Fukuoka University Journal of Science*, Vol.36, No.1, 15–23. (*in Japanese*)

Koshimizu, S. and Kyotani, T. (2002): Geochemical behaviors of multi-elements in water samples from the Fuji and Sagami Rivers, Central Japan, using vanadium as an effective indicator. *Japanese Journal of Limnology*, Vol.63, 113–124. (*in Japanese*)

Kubo, Y. and Ohta, T. (2020): Study on Mechanisms forming groundwater/hotwater quality at the Kuju volcano by principal component analysis. *Proceeding of the 2020 Annual meeting of the Japan Society of Engineering Geology*. (*in Japanese*)

Ohta, K. (2006): Hydrothermal System at Unzen Volcano and its Geological Background. *Journal of the Geothermal Research Society of Japan*, Vol.28, No.4, 337–346.

Ohta, T. (2017): Study on Mechanisms forming Groundwater Quality at Volcanic Rock Area, *Proceeding of the 2017 Annual meeting of Japan Society of Engineering Geology*. (*in Japanese*)

Ohta, T., Hattori, S., Kikuchi, Y. and Shimofusa, D. (2019): Experimental and Numerical Study on to form the Groundwater Quality in Altered Volcanic Rock Area, *IAEG/AEG Annual Meeting Proceeding*, Volume 4, 89–95.

Shimizu, K. (2015): Theory and practice of multivariate analysis for practical use. *JUSE Press*, Tokyo, pp.149 (*in Japanese with English*)

Yamada, M., Amita, K and Ohsawa, S. (2007): Isotope-hydrological Study on Formation Mechanism Carbonate Springs at the Southeastern Foothills of Kuju Volcano, Central Kyushu, Japan., *J. Hot Spring Sci.*, Vol.54, No.4, 163–172. (*in Japanese*)

Yamamoto, S. (1995): volcano body springs in japan, *Ko-konshoin*, Tokyo, pp.264 (*in Japanese*)

Yamashita, K. (2018): Geochemical study on mechanism of groundwater forming at the eastern part of Kuju Volcano, Oita Prefecture, Yamaguchi University Bachelor Thesis, 80p. (*in Japanse*)

Rock Mechanics and Engineering Geology in Volcanic Fields – Ohta, Ito & Osada (eds)
© 2023 copyright the Author(s), ISBN 978-1-032-27657-1

Rapid evaluation of leaching potential of heavy metal elements from tunnel excavated rocks using handheld X-ray fluorescence spectrometer

Shusaku Yamazaki* & Toshiyuki Kurahashi
Civil Engineering Research Institute for Cold Region, Japan

Kenta Tsuruya
Japan Railway Construction, Transport and Technology Agency, Japan

Naoto Okamoto
Kiso-Jiban Consultants Co., Ltd., Japan

ABSTRACT: The effective assessment of elution of toxic heavy metal elements such as arsenic, selenium and lead from excavated rocks is one of the challenges in the safety evaluation in the tunnel construction in the volcanic field. We measured whole rock compositions of multi elements including As using Si-PIN type handheld-XRF instrument on 230 m long boring core from a tunnel construction where the elution of heavy metal elements was a concern. The measurement conditions are determined by pellet standards as 30 seconds for trace element of Cu, Zn, As, Se, Sb, Pb, Rb, Sr and Ba and 15 seconds for major elements of S, K, Ca, Ti, Fe and Mn. Under these conditions, the handheld-XRF measurement was carried out about 500 points at intervals of about 50 cm. The accuracy and detection limit were roughly about 10-20 mg/kg and 20% for trace elements and a few thousands mg/kg and 3-30 % for major elements. The result shows that changes in lithological description about color, weathering and hydrothermal alteration are represented by changes in covariation of heavy metal elements with other elements, such as between As and S, Fe and Mn. The horizontal variation of As composition by handheld-XRF of the core is consistent with the high As leaching (>0.3 mg/L) zone detected in a leaching test using a mixture of five samples taken at 10 m intervals from the core. This indicates that the handheld-XRF method on boring core sample is effective as a higher-resolution and rapid method for estimating or screening the leaching amount and trend of heavy metals from hydrothermally altered excavated rocks in volcanic field.

Keywords: Heavy Metal Elution, Handheld XRF, Boring Core, Tunnel Construction

1 INTRODUCTION

Toxic heavy metal elements such as arsenic (As) and lead (Pb) are occasionally concentrated in and leached from soil and rock in a volcanic field due to regional or local hydrothermal alteration and ore mineralization. In a tunnel construction, the leaching of heavy metals from excavated rocks to environment is often a problem. In a general method, a boring core sample taken from the tunnel route before excavation, and the elution of heavy metals is estimated by leaching test by bottle shaking and column test

*Corresponding author: yamazaki@ceri.go.jp

DOI: 10.1201/9781003293590-45

(e.g. Tabeline et al., 2018). Then, the amount of countermeasure volume is estimated and the safety treatment method is designed. However, these leaching tests are limited in the number of case due to cost and require a several days to a week from sampling to result. Therefore, there is a demand for methods that can quickly determine or screen the amount of elution from rock.

Handheld X-ray fluorescence analysis (hXRF) is a method that can directly analyze the multi element concentrations, including the toxic heavy metals, on the sample surface such as a boring core in a short time without pre-treatment (e.g., Rowe et al., 2012). This method is widely used for on-site screening of polluted soils (e.g. ISO 13196). However, it is not well understanding about the correspondence with the heavy metal leaching test and hXRF results and its applicability to simple and rapid risk assessment for tunnel excavation using boring core samples. In this study, we verified the measurement accuracy of hXRF method when applied to rock standards. Next, we conducted it as multi element analyses with about 50 cm intervals on the As-elusible boring core sample from a tunnel route in a volcanic filed that has been hydrothermal alteration and mineralization. In addition, the effectiveness of the hXRF method as an indirectly estimating method for the elution of the heavy metals from rock samples was examined.

2   SAMPLE AND METHODOLOGY

2.1   Boring core sample

The boring core sample with 230 m long and about 60 mm in diameter was horizontally drilled in the Kutchan Group sediments, which mainly consists of andesitic volcaniclastic rocks of Late Miocene to Pliocene age, distributed in the Akaigawa village, Hokkaido island, Japan. The sampled core mainly consists of matrix-supported conglomerates with black to green color hydrothermally altered andesitic to rhyolitic gravels and white to red silicified gravels that are 1-20 cm in diameters. Cubic pyrite with a maximum diameter of 2 mm is abundant in the silicified matrix throughout the core. Black mineral veins and brown clayey veins were partly observed in the matrix. At first 20 m of the core, the conglomerate matrix is colored brown to yellowish brown indicating zone of surface weathering and oxidation. Beyond 20 m, the matrix color changes to dark gray to greenish ash gray (Figure 4a, b). The tuffaceous sand layers are observed around 110, 180 and toward 210 to 230 m. Around 180-190m is poor recovery and consists of clayey drilling slime with gravels of few cm in size. The large part of matrix, gravels and tuffaceous sand layers are partially or completely soft and clayey by alteration. The secondary mineral assemblage by XRD analysis is characterized by illite, smectite, kaolinite and pyrite, indicating that the mineralization is caused by acidic hydrothermal alteration (Utada, 1981). In addition, jarosite was detected in the surface weathering zone which would be caused by oxidation of sulfide minerals.

2.2   Handheld X-ray Fluorescence (hXRF) analysis

The handheld-type X-ray fluorescence spectrometer (hXRF) used in this study is Innov-X System α-4000 equipped Ta-tube with a maximum power of 2 W (40 kV, 50 µA), spot size of 6 mm and Si-PIN detector. The measurement conditions and detectable elements were 15 kV and 7.5 µA with 0.1 mm thick Al-filter for the elements of P, S, Cl, K, Ca, K, Ti, Cr, Mn, Fe, I and Ba, and 40 kV and 8.5 µA with 2 mm thick Al-filter for trace elements of Co, Ni, Cu, Zn, As, Se, Rb, Sr, Zr, Mo, Hg, Pb, Ag, Cd, Sn and Sb. The measurement results of each sets of elements are calculated as the quantitative value ± error in mg/kg (ppm) unit based on the calibration curves at factory setting. In this study, the elements of P, Cl, I, Co, Zr, Mo, Hg, Ag, Cd, Sn were excluded the result because of large error or undetected in the standards and the boring core.

At first, in order to verify the suitable measurement time and accuracy required for a core analysis, total 32 pressurized pellets of standard samples of rock (GSJ and IAEGA), soil (JSAC) and cement (JCA) materials were measured by the hXRF. The pellet with about 2 mm in thickness was formed by hand-press of 1.2 g powder sample using 2 cm diameter aluminum ring at 30 MPa for 2 min. Measurement error, detection limit, and accuracy of calibration curve were verified in variation of measurement times with 5, 10, 15, 30 and 45 seconds.

Next, the hXRF measurement of the boring core samples was conducted by 15 and 30 seconds (total 45s) on core sample stored in a box, which after pre-treatment of cutting, surface washing and naturally dried in box at outside temperature at winter. The conglomerate matrix, soil and tuffaceous sand were measured at about 50 cm intervals and gravel, mineral vein, and slime as supplementary. Total number of analysis was at 500 points. During the measurement, the detector calibration was carried out every 1 to 2 hours. The detection limits were calculated as the mean error (background error) of data when the target element was not detected (Table 1).

2.3 *Elution test for toxic heavy metal elements*

The data of leaching test for heavy metal elements was provided by a tunnel construction project of Japan Railway Construction, Transport and Technology Agency (JRTT). The sample for the leaching test was collected two ways (Figure 4c). One is 23 samples by a 5-point mixing sample from 10 m core with 2 m intervals. Another is 34 samples from common piece to the hXRF measurement. The core samples were crushed to less than 2 mm and then it was shaken in bottle with water/rock ratio of 10:1 at 200 rpm for 6 hours. The elemental concentration of the leachate was measured by ICP-MS after filtration of 0.45μm. The whole rock arsenic concentration for the common sample was also measured using the desktop-XRF to verify the measurement accuracy of hXRF on the boring core.

Table 1. Detection limits of hXRF measurement on the boring core.

Element	Detection limit (mg/kg)	Element	Detection limit (mg/kg)	Element	Detection limit (mg/kg)
S	1847	Mn	69	Pb	18
Ca	766	Sb	55	Sr	9
K	591	Cu	54	Se	7
Ba	508	Hg	41	As	7
Fe	235	Zn	22	Rb	6

3 RESULT AND DISCUSSION

3.1 *Investigation of reasonable condition of hXRF measurement for geological sample*

The measurement results of the standard pellets are shown in Figure 1 and 2. As representative result for trace elements, the analytical error of As systematically decreased with increasing measurement time or concentration (Figure 1), suggesting that the measurement time would be 30 seconds for the trace elements, assuming that the sufficient accuracy is within ±20%. If the concentration of As exceeds 50 mg/kg, the error will decrease below 10% at 30 seconds. Similarly, using the result of S, which has the larger error in the major elements, we determined 15 seconds for the major elements in terms of balance between the total analytical time, error and the detection limit.

The accuracy of the quantitative results of hXRF was checked by the reference values of the standards in the analytical time of 30 and 15 seconds (Figure 2). The results showed strong correlations, on or around 1:1 line, in the elements of As, Fe, Mn and Pb. Although Rb, Sr, Zn, and Cr are not illustrated in Figure 2, these elements also showed a strong liner correlation. The result of S also shows a liner correlation, but

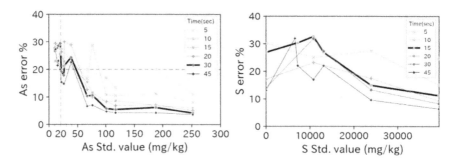

Figure 1. Measurement error (%) variation with analytical time of the hXRF in the standard materials.

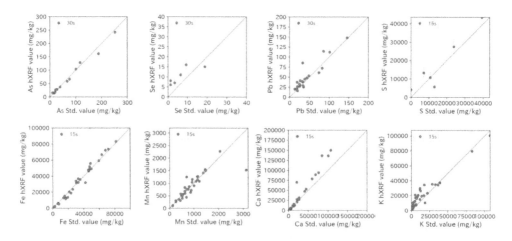

Figure 2. Relationship between the hXRF results and reference values of the standard materials.

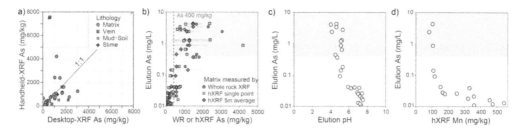

Figure 3. (a) Comparison between spot analysis using hXRF and whole rock analysis using desktop-XRF, and (b) comparisons of As concentrations between the whole rock analysis and the single spot and 5m average by hXRF, (c) pH vs. elusion As, and (d) elution As vs. Mn concentrations by hXRF.

they plotted slightly lower side. Se also tends to slightly high. Because they have 6 and 5 reference data, the quantitative correlation could not be verified sufficiently. For the results of Ca and Ti, the correlation of a quadratic curve apart to lower values than those of reference was observed, indicating that additional matrix correction is required to treat as quantitative values. For the elements of Sb and Hg, there is not sufficient numbers and range of the reference values. The collection of appropriate standards for these elements is an issue for the future. On the other hand, the elements of Zr, Ba, and K did not show well linear correlation. As the results, it is considered that the quantitative measurement using hXRF has sufficient accuracy to understand concentrations of heavy metals such as As, Se and Pb and also the associated elements of S, Fe, Mn, Cu, Zn, Sr, which are expected to relatively mobile during hydrothermal alteration of volcanic rocks.

The detection limits of the heavy metal elements As, Se, and Pb by hXRF were calculated as 7, 7, and 18 mg/kg, respectively (Table 1). Since the measurement of actual boring cores depending on the lithological types and sample matrix and surface conditions, the realistic detection limit would be estimated to be 2 or 3 times higher as around 10-20 mg/kg for the heavy metal elements. According to the Soil Contamination Countermeasures Law of the Ministry of the Environment, Japan and its guideline, the alarming standard value of whole rock concentrations for As, Se, and Pb in soil is set at >150 mg/kg. Therefore, the detection limit by this hXRF method is sufficiently small relative to the standard value. It is known that the S concentration is considered as an indicator of acidification of an excavated rock and as amount of sulfide mineral accretion. It is reported that the acidification of the leachate from excavated rocks in hydrothermally altered volcanic rocks becomes promoted if the sulfur content exceeds 0.3 wt% (3000 mg/kg; Hattori et al., 2003). The detection limit of the S content by this study is estimated about 2000 mg/kg. It is considered to be sufficient for use as a screening index to detect a kind of acidification and sulfide mineralization in a boring core sample.

Figure 3a and 3b show a comparison of arsenic concentrations of the whole rock powder using desktop XRF and those of surface measurement using hXRF for the common samples taken from the core. Since the hXRF is a spot analysis of the core surface, the heterogeneity of the lithology, mineral grains, water content and the shape and roughness of sample surface is affected to the X-ray fluorescence intensity in principle. The scattered difference between the hXRF spot analysis of mineral veins of mm-scale and some part of the matrix and the whole rock analysis is interpreted to reflect this heterogeneity (Figure 3a). The chemical heterogeneity is expected to be appeared in the conglomerate matrix, which is composed of coarse-grained sand and gravel consisting of various minerals and rocks.

However, the spot measured results of the conglomerate matrix correlate well with the results of the dried and pulverized whole-rock XRF analysis for most of the samples, especially for As concentrations below 2000 mg/kg. It can be concluded that the hXRF spot analysis targeting heavy metal elements of the boring core consisting of hydrothermally altered volcaniclastic rocks provides a practical geochemical signature as same as the whole-rock analysis.

3.2 Relationships between hXRF geochemistry, the elution test results and lithology

In order to estimate the whole volume of heavy metal polluted rocks in tunnel excavation, leaching test for heavy metal elements using the mixed samples for 10 m sections by 5-point mixing of 2 m interval were conducted using the boring core samples covering 230 m of the tunnel route. The results of the leaching test of the 10 m section and the hXRF common samples are shown in Figure 3 and 4c. For the heavy metal elements, arsenic was the only detected element from entire the core, which are the concentrations of As ranging from 0.001 to 5.9 mg/L. The elution of As in the 80-180 m section were much higher than the environmental quality standards of 0.01 mg/L and also 0.3 mg/L of the higher secondary standard (the Ministry of the Environment, Japan), indicating that the section requires appropriate treatment and management during and after the tunnel construction project (Figure 4c). The amount of As elution increased with decreasing pH of the leachate (Figure 3c).

In order to reveal a various geochemical heterogeneity from local to regional scale, a sufficiently small sampling interval comparable to the change of target geology is required. However, the elution test requires to take several days to obtain the results, and even if the measurement is carried out at 10 m intervals, which is considered to be highly frequent in a construction project, the analytical resolution is too low to understand the details of mineralization and alteration of volcanic rocks, which can often cause significant changes in tens cm to several meters. In addition, it is difficult to interpret the detailed mechanisms of element elution

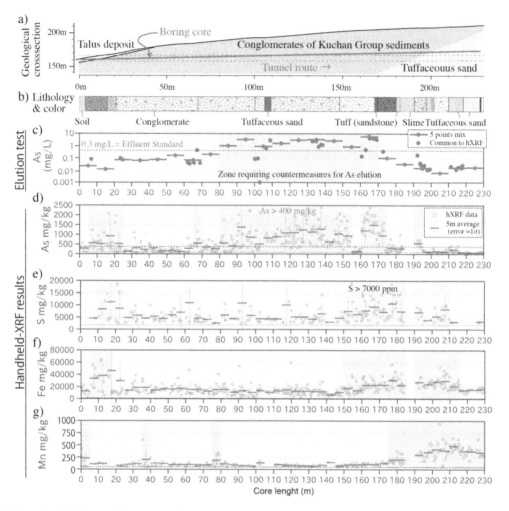

Figure 4. (a) Geological cross-section and sampling location, (b) color and lithological description of the core, (c) As leaching test results, and concentrations of (d) As, (e) S, (f) Fe and (g) Mn from handheld-XRF analysis of the boring core.

in water-rock interaction for excavated rocks, because of the limitation of analytical targets, which are generally the regulated toxic heavy metal elements of As, Se, Pb, Cr^{6+}, Cd, F and B and pH, electrical conductivity and redox state. In comparison, the hXRF core analysis, which enables rapid and simultaneous measurement of multi elements including heavy metal elements of As, Se and Pb, can operate immediately after core drilling. In this study, the hXRF measurements in the tunnel construction site could be performed at a rate of about 100 points per hour (20-30 m/h).

The results of hXRF analysis of As concentration in the conglomerate matrix show a concentration range of 17-5526 mg/kg, showing large variation even within 1-2 m of the measurement locations (Figure 4d). This indicates that the analytical results of single spot, 6 mm in diameter, on the core surface with 50 cm intervals sensitively reflects the sub meter scale heterogeneity of As distribution in the matrix. Therefore, highly-As elution zone detected by the leaching tests can also be detected more detailed resolution and understand As leaching mechanism by composite usage of multi element covariations between As and S, Fe and Mn concentrations measured by rapid and easy hXRF method on a boring core sample, consisting of volcaniclastic rocks experienced hydrothermal alteration and sulfide mineralization. On the other hand, the average value of As concentration in length of 5 m shows smooth variation through the core and distinctive high-As zone around 80-180 m, which seems to be comparable to the change of lithology and the zone of high-As elution (Figure 4a-d). In more detail, the reddish to brownish oxidized (weathered) section without surface soil in the 5-20 m is identified as a higher As concentration zone. The rapid decreasing of As after 20 m is corresponding to the change of color to greenish gray of the matrix. The section of 150-160 m is also identified as a section where As concentration is distinctively low, however the corresponding change of lithology and color is inapparent. In the 190-200 m, a gradual decreasing of As concentration was observed in the boundary between the conglomerate layer and the lower tuffaceous sandstone layer. As the relationship between the hXRF results and core description, the weathering zone with the formation of jarosite in 5-20 m also corresponds to the zone with enrichments of S and Fe (Figure 4e and d). It is expected that the relatively lower elution of As in the weathering zone, which would occurred by oxidation of sulfide minerals, could be explained by adsorption and/or coprecipitates onto jarosite or other iron hydro oxides (Tomioka et al., 2005). The localized spikes in S and Fe concentrations observed in the 50-150 m may correspond to the presence of focusing sulfide mineralization and the distribution of mineral veins, however there was no clear relationship with the As elution. The higher As concentrations over 400 mg/kg at 80-150 m is considered to be originated local sulfide mineralization or later hydrothermal circulation in the conglomerate matrix. At 150-180 m of the gray to ash gray zone in the lower part of the conglomerate are recognized as highly silicification zone containing quartz veins and infiltration and pyrite in the matrix, which well consistent with the distribution of high S and Fe concentrations (Figure 4e and 4f). Beneath the gray to dark gray conglomerate layer, the tuffaceous sandstones after 180 m characterized by Mn enrichment (Figure 4a and 4g). This implies that the stage of hydrothermal mineralization was different from those of the conglomerate layer. As the combine analysis between hXRF and the elution test, it is notable characteristics that there was a clear negative correlation between the Mn concentration and amount of As elution (Figure3d, 4c and 4g). This negative correlation is presumably due to the influence of either or both factors: changes of sediment material and difference of hydrothermal circulation or sulfide ore mineralization mechanisms. This interpretation is still not clear, and it will require further mineralogical and petrographic analysis.

As summarize above, in the study area, where consists of volcaniclastic rocks experienced hydrothermal alteration and sulfide mineralization during the Miocene to Pleistocene age, the highly elution zone of As can also be detected more detailed resolution and understand As leaching mechanism by composite usage of multi element covariations between As and S, Fe and Mn concentrations measured by rapid and easy hXRF method on a boring core sample.

4 CONCLUSIONS

Using pellet-formed standards, it is evidenced that the hXRF method for the toxic heavy metal elements of As, Se and Pb performs the detection limits of 10-20 mg/kg with error about 20% by 30 seconds measurement. The hXRF analysis at 50 cm interval on the boring core was possible to identify multi element geochemical variations such as As and Mn in high resolution, corresponding to changes of lithology, alteration and ore mineralization. In addition, the easy and rapid measurement by hXRF can also support to identify and explain a cryptic lithological variation that are difficult to understand by naked-eye description on core sample. It is concluded that handheld XRF analysis of borehole cores at 0.5 m intervals in volcaniclastic rocks with hydrothermal alteration and mineralization can be used to determine, screen and evaluate the high leaching zone of heavy metal elements exceeding environmental standard concentrations.

ACKNOWLEDGEMENTS

The analysis of the boring core samples and information about the boring core and tunnel construction site and data set of leaching test result were fully supported by the Japan Railway Construction, Transport and Technology Agency (JRTT).

REFERENCES

Hattori, S., Ohta, T., and Kiya, H., 2003, Engineering geological study on exudation of acid water from rock mucks -evaluation methods of rocks at the Hakkouda Tunnel near mine area -, *Jour. Japan Soc. Eng. Geol.*, 43(6), 359–371.

ISO, 2013, ISO 13196: Soil quality — Screening soils for selected elements by energy-dispersive X-ray fluorescence spectrometry using a handheld or portable instrument. *International Organization for Standardization.*

Rowe, H., Hughes, N., Robinson, K.,2012, The quantification and application of handheld energy-dispersive x-ray fluorescence (ED-XRF) in mudrock chemostratigraphy and geochemistry, *Chem. Geol.* 324-324, 122–131.

Tabelin, C. B, Igarashi, T., Villacorte-Tabeline, M., Park, I., Opiso, E. M., Ito, M., Hiroyoshi, N., 2018, Arsenic, selenium, boron, lead, cadmium, copper, and zinc in naturally contaminated rocks: A review of their sources, modes of enrichment, mechanisms of release, and mitigation strategies, *Sci. of the Total Envir.*, 645, 1522–1553.

Tomioka, Y., Hiroyoshi, N., Kubo, Y., and Tsunekawa, M., 2005, Effect of jarosite on the removal of arsenic ions in sulfuric acid solution, *Shigen-to-Sozai*, 121, 597–602.

Utada, M., 1981, Hydrothermal alterations related to igneous activity in Cretaceous and Neogene formations of Japan, *Mining Geol. Spec.* Issue 8, 67–83.

Geoengineering and infrastructures in volcanic fields

Rock Mechanics and Engineering Geology in Volcanic Fields – Ohta, Ito & Osada (eds)
© 2023 copyright the Author(s), ISBN 978-1-032-27657-1

Pozzolanic soils: A natural resource for soil improvement techniques

Manuela Cecconi*
Department of Engineering, University of Perugia, Italy

ABSTRACT: Soil improvement techniques with binders are used in several geotechnical engineering applications, as a sustainable solution for the reuse of waste soils. An experimental research started some years ago on lime-treated medium/coarse-grained pyroclastic soils from volcanic districts of Central Italy, whose geotechnical characterization was previously investigated in the past.

The suitability to lime stabilisation was broadly investigated and verified through laboratory geotechnical testing. The mechanical improvement induced by lime addition highlighted the development of a structured behavior linked to the chemo-mineralogical evolution of the system, while showing at the same time the effectiveness of the improvement technique to pozzolanic soils. While for clayey soils, the fine fraction is mainly responsible for ion exchange and pozzolanic reactions induced by lime, for the investigated pyroclastic soils pozzolanic reactions are dominant processes due to low quantity of fine fraction along with abundance of aluminates and silicates. The amorphous phase is found to play a fundamental role in the development of pozzolanic reactions and in the formation of new cementitious compounds.

The research has been performed following a multi-scale approach i) involving some special experimental techniques in order to observe the chemo-physical evolution of the system over time, ii) looking at the microstructure of the solid skeleton through mercury porosimeter tests and scanning electron microscopy, iii) aimed at investigating the mechanical behavior of lime-treated materials, by means of more conventional geotechnical laboratory techniques. The mechanical response of raw and lime-treated pozzolanic soils has been investigated in terms of 1D compressibility and shear strain behaviour. Through this multiscale and multidisciplinary approach - from mineralogy to geochemistry to geotechnics - it has been possible to explore the link between the phenomena detected at the microscale level and the macroscopic behaviour of the treated materials.

The research aims also to lay the basis for reconsidering the use of pozzolanic soils as resources and innovative materials. Due to amorphous structure and abundance of silicon and aluminum, pozzolanic soils can be considered suitable natural materials to act as precursors for alkaline activated binders in soil treatment.

Keywords: Pozzolanic soils, Lime-treatment, Microstructure, Mechanical behaviour upon shear, Compressibility

1 INTRODUCTION

Soil-treatment with lime addition is certainly one of the most widespread methods used worldwide to improve the geotechnical properties of soils, mainly clayey soils (Bell, 1996; Boardman et al., 2001; Locat et al., 1996; Cambi et al., 2012; Vitale et al., 2016, Russo, 2019). This technique appears to be very effective for the reuse of soils that are not suitable for earthworks in

*Corresponding author: manuela.cecconi@unipg.it

DOI: 10.1201/9781003293590-46

their natural state. During the construction of large infrastructures (road or railways embankments, dams, etc.), the excess of not-suitable soil can be a frequent occurrence, yielding to the corresponding problem of the allocation of excavated material in dedicated areas. This is a crucial and up-to-date issue since, from a sustainable point of view, there is an increasing an urgent need to re-use "local soils" in geo-environmental and geotechnical works. The reuse of excavated soil can be therefore considered as an important and sustainable opportunity, aimed at reducing the environmental impact.

When lime is added to soil, cation exchange and pozzolanic reactions take place, thus modifying relevantly the microstructure and improving the mechanical properties of the treated soil (Bell, 1996). These two mechanisms, referred to as modification and stabilization of treated soils respectively (Rogers & Glendinning, 1996), develop simultaneously but with different time scale. The cation exchange between calcium ions and clayey minerals takes place very quickly, inducing the flocculation of the fine-grained fraction. At a macroscopic scale, the grain size distribution of the treated soil shows a reduction of the finer fraction, and the plasticity index is then considerably reduced. High pH value of pore water (approaching the target value of 12.4, being the pH value of calcium saturated water) is induced by portlandite ($Ca(OH)_2$) dissociation in water, promoting dissolution of silicon and aluminum from clay minerals. The subsequent reaction of silica and alumina with the available calcium ions in solution (see Figure 1) triggers pozzolanic reactions with formation of stable cementing secondary phases, calcium-silicate-hydrates C-S-H and calcium-aluminate-hydrates C-A-H (Diamond & Kinter, 1966; Ingles & Metcalf, 1972; Little, 1996; James et al., 2008, Russo, 2019). The development of pozzolanic reactions, with the formation of these stable compounds is slower than cation exchange on time scale, and it is responsible for the improvement of the mechanical properties of the treated soil in terms of increase of shear strength and reduction of soil compressibility (Locat et al., 1996; Sivapullaiah et al., 2000; Tremblay et al., 2001; Rao & Shivananda, 2005; Vitale et al. 2017). Pozzolanic reactions are time and temperature dependent; the production of secondary hydrated phases is favored by prolonged curing times and high temperatures (Bell, 1996).

The possibility of using lime treatment technique for pozzolanic soils, generally unsuitable for earthworks in their natural state, and the analysis of the effectiveness of the treatment in terms of soil improvement represent the first goal of a recent wide experimental research, whose results are discussed and summarized in this paper. The majority of presented data are already published on Journal papers, whose references are all acknowledged. The achieved results may lead to a second, but no less important long-term goal, standing in reconsidering pozzolanic soils as resources and innovative materials, acting as precursors for alkaline activated binders in soil treatment. This aspect is only addressed in the paper, since the experimental work is currently under progress.

Pozzolanas are silica or aluminosilicate rich materials very reactive to calcium ions dissolved in pore water forming cementing compounds such as calcium-silicate-hydrate indicated as C-S-H or calcium-aluminate-hydrate C-A-H. The term "pozzolana" comes from Latin word *lapis puteolanus* and – in the Italian geotechnical literature – is generally attributed to incoherent pyroclastic deposits diffused in the Campanian Region (Pozzuoli, Naples - Italy). On the other hand, the pozzolanas widely spread in the Latium region are generally not cohesionless materials but, rather, they exhibit pronounced pozzolanic properties as well. The pozzolanic activity is therefore possible when high amounts of silica in an amorphous or glassy state (non-crystalline) are present, provided at the same time that the material has a high specific surface area. When mixed with water and lime, pozzolanic materials are capable of curing, at ambient temperatures, both in air and under water, resulting in hydration products which are rather insoluble and therefore also resistant to the washout action of water.

On this premise, it is clear why, from the ancient Roman times, natural pozzolanic soils were used to produce hydraulic binders for the construction of structures exposed to the action of rainwater, for water containment structures, i.e. tanks, canals, aqueducts as well as for maritime works. Although the chemical reactions are much slower, hydration products are similar to those of cement. Therefore, the capability of a pozzolanic material to combine with lime and produce water-insoluble compounds has been advantageously exploited in the production of cements consisting of mixtures of Portland cement and pozzolanic materials of

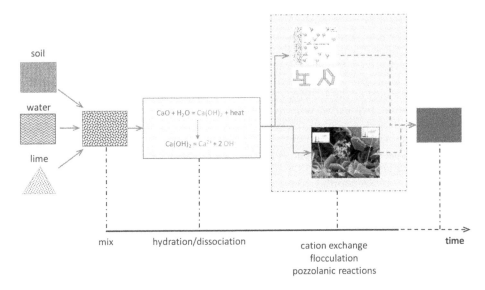

soil

water

lime

$$CaO + H_2O = Ca(OH)_2 + heat$$

$$Ca(OH)_2 = Ca^{2+} + 2 OH^-$$

mix   hydration/dissociation

cation exchange
flocculation
pozzolanic reactions

time

Figure 1.   Effects of lime addition to a soil-water system (by courtesy of Russo, 2019).

various kinds. Artificial pozzolanas can be obtained by electrostatic or mechanical precipitation of dust-like particles during coal combustion (*fly ash*), or can be obtained by condensation of silica vapors during the production of silicon and iron-silicon alloys (*silica fume*). The use as natural additives in cement industry is well documented (Blanco Valera et al., 2006; Martínez-Ramírez et al., 2006; Mertens et al., 2009).

In this paper, the attention focuses on lime treatment and improvement of natural pozzolanas. These are very diffuse in Central and Southern Italy, covering a total extent of 8000÷9000 km². Most of them are of pyroclastic origin. The cities of Roma and Napoli are both located in the heart of a wide volcanic region mainly formed by the volcanic complex of Colli Albani at the south-east of Roma, by the Mounts Vulsini and Mounts Sabatini volcanic Complexes at North, North-West and by the Somma Vesuvio and the Phlegrean Fields districts in the area of Naples, at the south-east and west of the town respectively. Due to their dissemination over large areas densely settled since historic time, the close interaction between pyroclastic deposits and geotechnical works is not limited to present-day engineering projects. Italian pyroclastic materials have been exploited since the roman period, to be used in the construction industry, e.g. as building blocks, to produce hydraulic mortars and cement and, even today, as the underlying stratum of road pavements and tennis courts. Prolonged mining activities in pozzolanic deposits have resulted in complex networks of underground cavities in urbanized areas, which have frequently collapsed with the formation of subsidence troughs or, in the most severe cases, sinkholes at the ground surface. Open quarries of pozzolanas are frequent in the south east of Roma, with subvertical excavation fronts which can be as high as 20 – 25 m. There might be also underground cavities not yet surveyed and mapped; in Central Italy pyroclastic rocks often form steep slopes at the margin of historical towns, affected by progressive erosion, often occurring through sudden failures, and thus exposed to significant hazard (Bernabini et al., 1966, Lembo Fazio et al., 1984; Tommasi & Ribacchi, 1998; Rotonda et al., 2002, Cecconi et al., 2020).

As with most pyroclastic soils, these materials are considered as problematic soils due to their nature, heterogeneity, microstructure features, and their complex mechanical behavior upon loading, in both saturated and unsaturated conditions. Therefore, while of great importance from the engineering point of view, the mechanical characterization of pyroclastic soils is rather complex, firstly because of the difficulties of retrieving intact undisturbed samples. For some of the investigated coarse-grained pozzolanic soils typical of the subsoil of Roma (Colli Albani volcanic complex), the observed mechanical behavior is generally stiff and brittle at low confining pressures (rock-like behavior) and become softer and more ductile at larger

359

confining stress (soil-like behavior). The range of stress where the behavior is rock-like is often controlled by inter-granular bond strength and microstructural features of the material and of the single constituents, rather than by the sole initial voids ratio, stress state and history, as for soils (Cecconi & Viggiani, 2001). A comparison of compressibility and strength properties of different pyroclastic materials from Latium and Umbria in Central Italy, was presented by Cecconi et al. 2011, 2016.

All this considered, pyroclastic soils are not very commonly used for lime stabilization and - to the Author's knowledge - until the last decade, there was a lack of systematic studies on this topic which motivated the current research.

In order to assess the improvement of physical and mechanical properties induced by lime, a large experimental work on lime stabilization of different – for origin and nature - Italian pyroclastic soils, was developed. Cecconi et al. (2011a, 2011b, 2011c) and Cecconi et al. (2013) showed the suitability to the treatment of two pyroclastic soils and the effects of treatment parameters on the engineering properties of the treated materials. Cecconi & Russo (2013) highlighted the effects of treatment parameters, such as type and per cent by weight of lime, type of binder (CaO or Ca(OH)$_2$), initial water content, curing time.

Cambi et al. (2016) evidenced that in the case of volcanic soils with low amount of exchanging minerals, cationic exchange only plays a secondary role. While for clayey soils fine fraction is mainly responsible for ion exchange and pozzolanic reactions induced by lime (Vitale et al. 2016, Vitale et al. 2017), for pyroclastic soils pozzolanic reactions are dominant processes due to low quantity of fine fraction along with abundance of aluminates and silicates, especially when exchanging minerals such as zeolites are absent. This was confirmed by the following studies on a zeolite rich pyroclastic soil belonging to a deposit of the Orvieto cliff in the Umbrian Region (Central Italy). Guidobaldi et al. (2017) showed the high reactivity of the system exhibited by the quick precipitation of hydrates as tangible products of the pozzolanic reactions.

In this paper the attention is focused on three peculiar coarse-grained pyroclastic soils belonging to different volcanic districts in Central Italy, an ignimbrite from the volcanic complex of Mounts Vulsini, Orvieto in the Latium region (Cappelletti et al., 1999; Gentili et al., 2014; Tommasi et al., 2015; Verrucci et al., 2015), a pozzolanic soil from the Colli Albani volcanic complex at the outskirts of Roma (Cecconi & Viggiani, 2001; Cattoni et al., 2007; Cecconi et al., 2010) and, moving towards south, a pozzolana belonging to the stratigraphic succession of Somma-Vesuvius eruptions products in the Campanian region (Papa, 2007; Russo et al., 2016; Cecconi et al., 2013).

The invitation to submit a keynote lecture to this international workshop provides the opportunity to summarize and rethink the work done to date and define the area which still motivates some further research, with the objective of addressing relevant engineering applications and find sustainable solutions even in the context of cultural heritage, whereas the most advanced methodologies and techniques to reduce degradation through soil improvement are required. If the main engineering interest in the geotechnical behavior of the investigated pozzolanic soils was primarily derived by the need to assess the stability of sub-vertical cuts and underground cavities opened for exploitation, the advancement of research on the topic of lime treatment and the comprehension of the mechanical behavior of such lime-treated soils has opened a new perspective in the possibility of a reuse of pozzolanic materials, to be considered as natural resources for soil improvement techniques.

2 INVESTIGATED SOILS

In the present study, as a first step of discussion, the results of experimental investigations on three not-treated pozzolanas from different pyroclastic deposits from the Mounts Vulsini, Colli Albani and Somma-Vesuvius volcanic complexes (see their location on the map in Figure 2) are compared, focusing the attention on their microstructural features since these may affect their geotechnical properties under different loading conditions. These soils belong to deposits including fallout and flow pyroclastic soils.

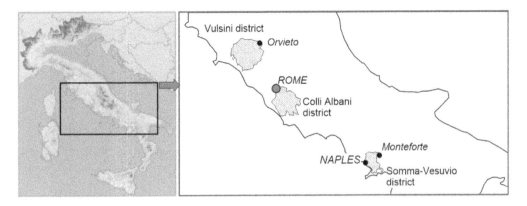

Figure 2.   On the map, location of the examined volcanic districts.

2.1   *Geological origin*

The first investigated pyroclastic material belongs to a thick ignimbrite deposit erupted about 330,000 years ago in the Vulsini volcanic complex (Cappelletti et al., 1999) and locally known as "Orvieto-Bagnoregio Ignimbrite". The volcanic activity of the Vulsini District is dominated by pyroclastic deposits covering an area of about 2200 km^2. The four main eruptive centers show a similar pattern: a strombolian and effusive activity at the beginning, followed by caldera-forming eruption with ignimbrite associate deposits and finally an eruptive phase marked by strombolian magmatic and hydromagmatic events. The pyroclastic soil was originated by the Bolsena eruptive center about 330 ky ago (Santi, 1991) and mainly consists of pyroclastic products generating thick ignimbrite deposits, often overlying stiff overconsolidated clays. These deposits form large areas of the eastern part of the volcanic plateau where intense erosion isolated mesas, buttes and inselbergs made of pyroclastic materials, which, since Etruscan times, were chosen for establishing naturally defended human settings (Sciotti, 1981). During the Middle Ages, most of them developed into villages and towns that now treasure important and artistic heritage. The most representative town is Orvieto (from Latin *Urbs Vetus*) sited 320 m above sea level and about 180 m above the alluvial plain of the Paglia River, at the foot of the hill (Bozzano, et al., 2008). As built on the top of a mesa formed by the Orvieto and Bagnoregio Tuff, the Orvieto cliff consists of a lithic facies (tuff) and a slightly coherent facies (pozzolana) resulting from distinct flow units (Nappi et al., 1994), some of which experienced a widespread zeolitization (Gentili et al., 2014). At the study site, most of the cliff is formed by pozzolanas, made of a weaker upper layer and a lower more consistent one (Rotonda et al., 2002). The experimental results discussed in this paper were obtained on samples belonging to the pozzolana facies retrieved from the North-Eastern corner of the Orvieto hill, at the base of the cliff, right beneath the Albornoz fortress (see Figure 3). The investigated pozzolana is denoted as OR soil.

The second investigated pozzolana is a coarse-grained pyroclastic weak rock, locally known as Pozzolana Nera (PN) typical of the area south-east of Rome, originated from the third cycle of an intense activity of the volcanic complex of the Colli Albani dating to the Upper-Middle Pleistocene (De Rita et al., 1992), and belonging to the unit belonging to the Vulcano Laziale lithesome (Giordano et al., 2006). The geotechnical properties of this material have been extensively investigated in the past and the outcomes of the study have been already published (Cecconi & Viggiani, 2001; Viggiani & Cecconi, 2007), including some papers presented to previous editions of International Workshop on Volcanic Soils and Rocks (Madeira Island – Portugal, 2002; Ischia Island- Italy 2015). The Colli Albani volcanic complex, also known as Vulcano Laziale, is located about 20 km south-east of the city of Rome. The volcanic products extend over an area of about 1500 km^2. The most important structure is the central caldera (Giordano et al. 2006) which was formed about 600,000 years ago (De Rita et al. 1995). Different periods of activity have been recognized, which resulted in the

Figure 3. At top left, outcrops areas of the Orvieto-Bagnoregio ignimbrite (after Nappi et al. 1994, modified); at bottom right sampling site at the base of the cliff.

deposition of 300 km^3 of pyroclastic materials, which outcrop extensively to the south of the Aniene river and east of the Tiber river (Figure 4). Most of the southern part of the city of Rome have been built on these pyroclastic products. The presence of magmatic and lahar deposits indicates that the volcano is still active although quiescent, with fumarole activity (Funiciello et al. 2003). The unit of Pozzolana Nera (PN soil) can be clearly identified in Figure 4b) which shows a quarry located south east of Rome. The deposit is massive and chaotic, with maximum thickness of about approximately 20 m.

The third investigated material, the Monteforte Irpino (MF) pyroclastic soil is a weathered and humified ashy soil belonging to the stratigraphic succession of Somma-Vesuvius eruptions products, at the north side of Mount Faggeto. At the site, the acclivity of the slopes reach values of about 30°-35°; in these areas these are considered typical values for mudflows landslide initiation (Papa, 2007; de Riso et al., 1999).

The MF soil characterizes the upper layers of the succession of about 40 km northwest of the volcano Somma-Vesuvius (Figure 5a) The entire stratigraphic succession overlies a fractured and karstified limestone of the Mesozoic and locally on strips of Miocene flysch. The material belongs to the shallower more recent unit, i.e. fallout deposits including millimetric greenish pumices, monogranular and strongly porphyritic with leucite and biotite crystals (layer 2 in orange, in Figure 5b). Lavic and carbonate lithic fragments are abundant and increase upward. In the studied area, this unit presents a marked lateral discontinuity and, often, a slight thickness (Di Crescenzo et al., 2007; Papa, 2007). The relevant hydro-mechanical properties of this soil are discussed in Papa (2007), Nicotera & Papa (2007).

2.2 *Microstructure and physical properties*

To get a full comprehension of the microstructural and chemo-mineralogical features of natural/raw and lime treated studied materials (see §3), several chemical and microstructural

Figure 4. a) Approximate western and northern limit (dash and dot line) of Pozzolana Nera outcrops. Dotted line encloses the urbanized Rome area (Cecconi et al., 2016); b) sub-vertical cut in a quarry of pozzolana (Via di Fioranello, Ardeatina, South-East Rome).

Figure 5. Monteforte Irpino site: a) location and b) simplified geological profile (adapted from Nicotera & Papa, 2007).

analyses were performed, some of them less common in soil mechanics and geotechnics, while more specific in the fields of mineralogy and geochemistry: Micro-X Ray Fluorescence (μ-XRF), X Ray Diffraction (XRD), Thermogravimetric Analysis (TGA), Fourier-Transformed Infra-Red (FTIR) spectroscopy, 29Si Nuclear Magnetic Resonance Spectroscopy (29Si NMR) analyses, nitrogen adsorption measurements (BET analyses), mercury porosimeter tests and SEM analyses.

The analyses were carried out with special equipment available at different institutions and Research Centers (University of Perugia, University of Cassino and Southern Lazio, Italy; IMN - Université de Nantes and IFSTTAR, France).

The investigated *Orvieto* pozzolanic soil (OR soil), is a zeolitized pyroclastic silty sand. Crystals of chabazite and amorphous phases were detected at SEM, the amorphous phase recognized in its characteristic glassy state (Figure 6), consistently with data obtained from the X-ray powder diffraction (see XRD spectrum in Figure 7) which clearly shows the predominant presence of a calcium enriched zeolite Ca-Chabazite, whose amount is furtherly proved by DTG and NMR spectra (see Guidobaldi et al., 2017, 2018 for details). Pore size distributions

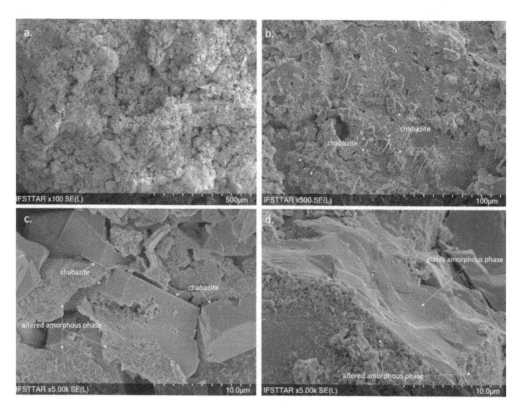

Figure 6. OR soil: SEM micrographs (Guidobaldi et al., 2018).

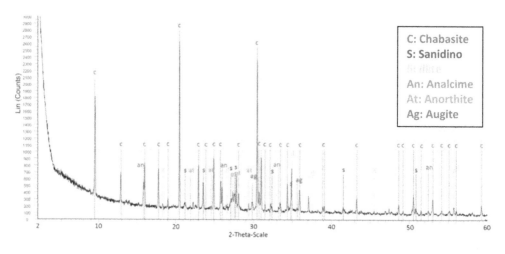

Figure 7. OR soil: X-ray diffraction spectrum (after Guidobaldi et al., 2017).

of not-treated OR soil are shown in Figure 8. A bimodal pore size distribution is noted, characterized by two main classes of pores ranging between 0.3 and 3 μm and 3–12 μm.

Micro-scale analyses of *Pozzolana Nera* from Rome (PN soil) revealed a matrix consisting mainly of dark grey to black fine ash shards, and crystals such as leucite, clinopyroxene and biotite (Giordano et al. 2006). The texture of natural PN soil is intermediate between granular- and

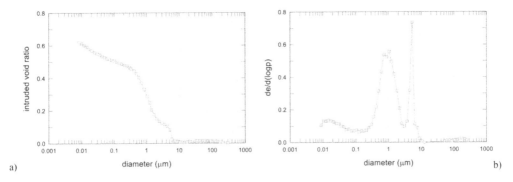

Figure 8. OR soil: Pore size distribution a) intruded void ratio and b) frequency of pore entrance diameters.

Figure 9. PN soil: SEM micrographs: a) 72×; b) 200× and c) 4500×.

matrix-sustained with randomly oriented clasts (isotropic clast orientation, according to Cecconi et al. 2010).

As recently documented and discussed in some other papers (e.g.: Viggiani & Cecconi, 2007), the microstructure of such material consists of sub-angular grains and grain aggregates of very variable size with a rough surface, even if it is not possible to discern grains from aggregates of grains; in some cases it is hard to identify individual particles at all (Figure 9). intrinsic inter-particle bonds, probably due to the original material continuity during deposition, are made of the same constituents of grains and aggregates.

According to Cambi et al. (2016), no zeolites have been found in PN soil. This was confirmed by X-ray powder diffraction analyses, which allowed to define the mineralogical composition of the soil (Figure 10). It will be shown that the absence of ion-exchanging minerals controls the whole reactivity of the system when PN comes into contact with lime, since ion exchange processes involving Ca^{2+} cations are not expected. At the same time, the high reactivity of natural PN soil is enhanced by high values of surface area, varying from 35 to 25 m^2/g as grain size increases (Figure 11a). According to the grain size distribution shown in Figure 11b), the selected material is classified as a medium sand.

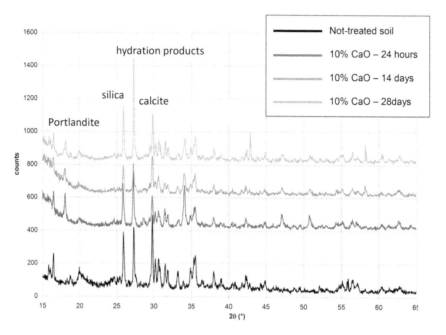

Figure 10.   PN soil: X-ray diffraction pattern for not-treated soil and after 10%CaO lime treatment at increasing curing times (after Cecconi & Russo, 2013).

Moving to *Monteforte* pyroclastic soil (MF), its bulk mineralogy was determined by X-ray powder diffraction (XRD), on randomly oriented samples (Figure 12). Monteforte soil mainly contains amorphous phases with small amount of Quartz, Kaolinite and Muscovite/Illite.

At SEM, the pores and grains shape and size variability is observed (Figure 13). In terms of grain size distribution, MF soil is a silty sand (Figure 14).

The elemental composition and mineralogical composition of OR, PN and MF soils are summarized in Tables 1 and 2, while the physical properties (mean values) of the three pozzolanas are reported in Table 3.

3   LIME TREATMENT BENEFICIAL EFFECTS

In engineering practice, the soil suitability to lime treatment is mainly based on its grain size distribution and plasticity index properties (e.g.: CNR, 1973, 2002; British Standard Institutions, 1990, 2006, French guidelines SETRA/LCPC, 2000). However, considerations only based on grain-dimensions are not sufficient, while investigating the distinct role of cation exchange and pozzolanic reactions is crucial for the assessment of the soil suitability to lime treatment. PN and MF samples for laboratory testing were prepared selecting the passing percent to the 2 mm sieve of the oven dried soil, while OR sample were prepared with the soil fraction passing the sieve 450 μm. The percentage of 7% by weight of powdered quicklime for MF and 10% for the PN provided the maintenance of a stable pH for long curing times, allowing the full

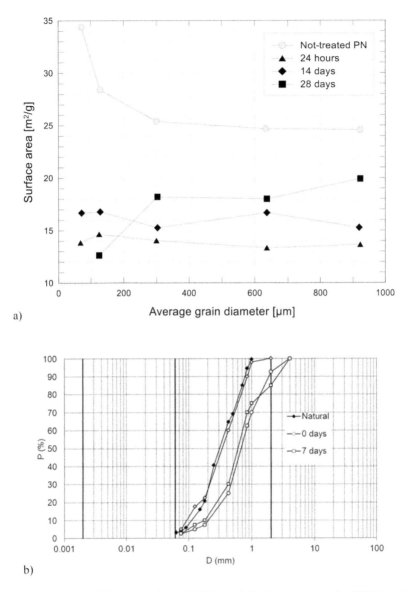

a)

b)

Figure 11. PN soil: a) specific surface from B.E.T. analysis (Brunauer et al., 1938) based on data obtained from nitrogen adsorption measurements on 10%CaO lime-treated PN soil (after Cecconi & Russo, 2013); b) grain size distributions.

development of pozzolanic reactions to take place (Figure 15). Due to the poor clay fraction of the investigated soils, the required of amount of lime required is generally high.

For OR soil, calcium hydroxide $Ca(OH)_2$ (i.e. hydrated lime) was mixed in proportion of 2 and 5% by dry weight of soil. Lime treated samples were then prepared by hand mixing the soil with the chosen fixed amount of quicklime powder and distilled water, allowing the quicklime to hydrate for 24 hours; hydrated lime containing more than 95% of portlandite was used for the treatment of OR soil.

Once obtained a homogenous powder, 20% by dry weight of soil of distilled water was added to the PN and OR soils-hydrated lime mixture, and 30% for MF soil.

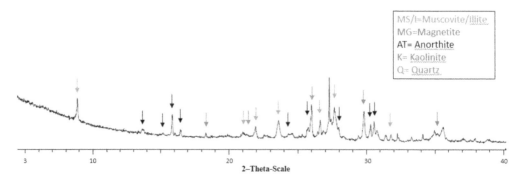

Figure 12.   MF soil: X-ray diffraction pattern (by courtesy of Russo et al., 2016).

Figure 13.   MF soil: SEM micrograph 6400× (by courtesy of Russo et al., 2016).

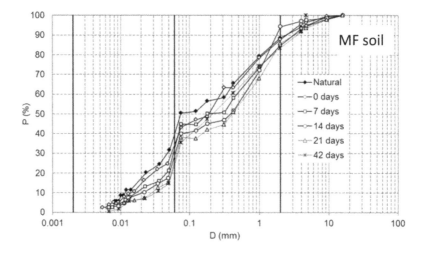

Figure 14.   MF soil: Grading curves for natural and treated samples (after Cecconi et al., 2013).

Table 1. Chemical composition of OR, PN and MF soils.

OR soil		PN soil		MF soil[(*)]	
component	weight (%)	component	weight (%)	component	weight (%)
MnO	0.14	Na_2O	0.57	MnO	0.26
SiO_2	55.06	SiO_2	51.46	SiO_2	50.68
Al_2O_3	15.00	Al_2O_3	22.82	Al_2O_3	25.23
MgO	0.64	MgO	1.603	MgO	1.04
K_2O	10.47	K_2O	6.45	K_2O	5.52
CaO	10.96	CaO	5.29	CaO	6.77
FeO	5.94	FeO	10.80	FeO	10.36
		TiO_2	0.99	PbO	0.02
				Cr_2O_3	0.02
				CuO	0.02
				ZnO	0.02

(*) by courtesy of Russo, 2021

Table 2. Mineralogical composition of OR and PN soils.

OR soil		PN soil	
component	weight (%)	component	weight (%)
Chabasite	23.90	Biotite	2.57
Analcime	0.90	Analcime	4.80
Anothoclase	4.50	Leucite	27.22
Phillipsite	1.00	Muscovite	5.26
Sanidinie	10.10	Sanidine	1.82
Calcite	2.40	Calcite	0.11
Augite	1.90	Augite	16.18
Amorphous	55.30	Albite	6.70
		Anortite	2.70
		Illite	4.07
		Ggypsum	1.87
		Amorphous	26.70

Table 3. Physical properties of raw OR, PN and MF soils (average values).

pozzolana	w/c (%)	G_s (-)	γ (kN/m^3)	γ_d (kN/m^3)	n (%)
OR	15	2.43	15.8	13.7	44
PN	13	2.69	16.5	14.6	45
MF	50	2.66	12.69	8.46	68

Grain size distributions were determined for both PN and MF soils at increasing curing times (namely 0, 7, 14, 21, 42 days for MF and 0, 7 days for PN). Conventionally, a curing time of 0 days corresponds to 24 hours after the addition of quicklime and water to the oven dried soil. Figures 11b), 14) show that the effects of lime addition are instantaneous, leading to the reduction of the fine fraction and the increase of the coarser fraction.

In the following sub-sections, the microstructural changes induced by lime treatment, detected from different several special laboratory tests, are examined (§3.1) while, as a direct effect and result, some aspects of the improved mechanical behaviour of lime-treated investigated soils are described and discussed (§ 3.2, 3.3).

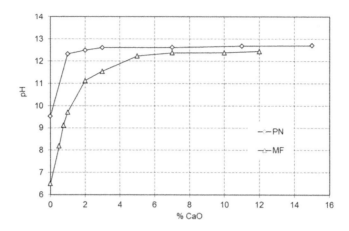

Figure 15. pH measurements for lime treated PN and MF soils (Cecconi et al., 2013).

3.1 *Microstructure changes*

Microstructural investigations and geochemical analyses were performed with the purpose of linking the chemo-physical evolution of the lime-soil-water system to the macroscopic mechanical improvements obtained from lime-treatment.

For Orvieto soil, soil and lime were preliminary mixed in the same weight proportions, in order to enhance the ongoing modifications at particle level. From the X-ray diffractograms (Figure 16) it is noted that the peaks of Portlandite, which appear after 24 h form lime addition, decrease with time indicating that the ongoing reactions involve a significative amount of lime. Even the peaks ascribed to zeolite Chabazite reduce with time, while a new mineralogical phase, namely Monocarboaluminate hydrate (C4ACH11), is recognizable since 7 days of curing and the intensity of the peak steadily increases with time.

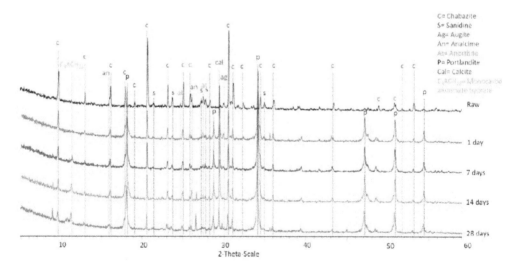

Figure 16. Diffractrograms of lime treated OR soil up to 28 days (Guidobaldi et al., 2017).

Hydrated lime $Ca(OH)_2$ of 5% by dry weight of soil was then mixed with OR soil and 20% water content. The results of X-ray diffraction analyses on treated OR did not show any relevant modification in the soil mineralogical composition after treatment; this suggests the minor contribution of crystalline phases to the generation of secondary products, while confirms that secondary products principally form at the expense of the amorphous phase. This phase is consumed by pozzolanic reactions, representing the main source of silica and alumina for the formation of secondary phases. The amount of original amorphous phase consumed by the reactions is replaced by the new compounds, detected as not crystalline by XRPD analyses. This hypothesis is supported by the fact that the quantitative interpretation of the XRPD patterns does not highlight significant variations on the abundance of mineralogical phases (see Guidobaldi et al., 2018). When hydrated lime is added to OR natural soil in amounts compatible with practical purposes, the role played by crystalline phases (including chabazite) in the formation of C-S-H and C-A-H is secondary with respect to that of the amorphous. The results obtained from TGA allow to interpret the features of the system evolution (Figure 17), i.e. a noticeable decrease of Portlandite mass loss in the very short term and - in the temperature range of hydrates phases - the increase of mass losses corresponding to a progressive increase of the amount of hydrates, with particular high rate in the short term. In the short time, the cementitious compounds derived from the pozzolanic activity are not at crystalline state but mostly amorphous, as confirmed by the X-ray analyses.

Therefore, the consumption of existing mineralogical phases is detectable since the very short term while – simultaneously - their peaks intensity decreases with the formation of hydrated phases. The observations suggest that Chabazite and the amorphous phase play an active role in the evolution of the lime-soil system and hydrates formation.

As concerns Monteforte soil, a sharp decrease of Portlandite mass loss is observed corresponding to a specular abrupt increase of hydrates phases resulting in cementitious products (see Figure 17b).

This evidence is also confirmed by Mercury Intrusion Porosimetry (MIP) results performed on treated OR soil at increasing curing times (Figure 18) showing an increase of frequency in the smallest pores, with pore entrance diameter <0.1 , due to the formation of secondary hydrates phases in the very short term (24 hours of curing). Increasing the curing time, a progressive filling of cavities and voids causes the reduction of frequency characterizing each class of pores, being particularly evident for the pore class with entrance diameter in the 0.3 – 3 range which is progressively vanished within the first 28 days of curing.

SEM analyses were performed by Guidobaldi et al. (2018) on lime-treated OR soil. Figure 19 shows he presence of gel hydrates after 28 days from lime addition. Gel phases are characterized by the smallest pore range (i.e. pore diameters lesser than 0.01 μm).

From the above results, it is clear that – for the investigated pozzolanic soils - the formation of cementitious compounds derived from the pozzolanic activity is the most relevant effect due to lime addition. Differently for clayey soils, where cation exchange develop in the very short time and are then followed by pozzolanic reactions, for these materials cation exchange and pozzolanic reactions progress simultaneously, the consumption of lime occurring fast while new hydrated phases are formed. Due to the absence of ion-exchanging minerals, pozzolanic reactions involve the entire lime molecule immediately after lime addition (Cambi et al. 2016).

3.2 *Behaviour upon shearing and soil dilatancy*

In this paragraph, the attention is focused on the behaviour upon shearing of not treated and lime-treated OR and PN soils which was investigated through direct shear tests, by adopting a standard shear box apparatus (ASTM 3080-90, 1994). Lime treated samples were prepared by hand mixing the soil with a fixed amount of quicklime powder, precisely 10% CaO for PN, 2% and 5% $Ca(OH)_2$ for OR soil, and distilled water (w/c = 20%) for both PN and OR soils. Increasing curing times were considered for both treated pyroclastic soils (namely 24 hours, 7, 28 days). The soil-lime-water mixture was then dynamically compacted inside standard 60 × 60 mm shear text boxes in n.3 thin layers by using a hollow cylindrical mallet of mass 850 g, sliding along a vertical bar from a height of 0.40 m (n.10 consecutive blows). All tests on OR and PN samples were performed in not-saturated conditions, but no suction measurements were performed before and during the tests.

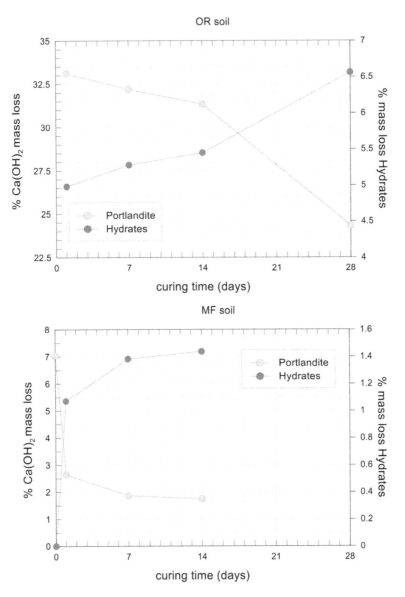

Figure 17. Quantitative analyses of thermogravimetric results: mass loss of secondary hydrated phases and portlandite at increasing curing time: a) OR and b) MF soils.

The stress-strain behaviour observed for PN and OR soils is shown in Figures 20, 21 in terms of shear stress, τ, vs. horizontal displacement δx, and measured vertical displacements δy vs. δx. Direct shear tests were performed at different increasing stress levels (50 - 800 kPa).

At low stress level (50 kPa) and 7 days-curing time, the brittle behaviour of PN soil - associated to soil dilatancy (i.e. negative volume strains) - is indicated by an abrupt decrease of shear stress, which occurs immediately after the peak strength mobilization. On the other hand, the stress-strain behaviour for raw, not-treated PN soil is undoubtedly more ductile and associated to positive volume strains (Figure 20a).

Figure 21 shows the stress-strain behaviour observed for OR treated compacted samples after curing times of 24 hours and 7 days. Again, it is rather evident that the observed

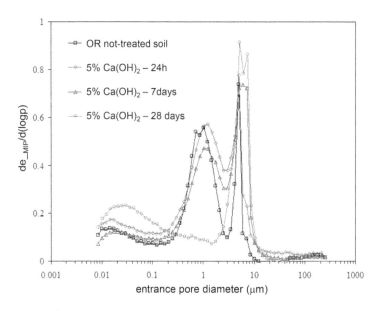

Figure 18. Pore size distribution evolution for raw and lime treated OR soil over curing time (adapted from Guidobaldi et al., 2018).

a)   b)

Figure 19. SEM micrographs of 5% Ca(OH)$_2$ lime-treated OR soil after 28 days from lime addition; a) 1000×; b) 5000 ×.

behaviour depends strongly on lime percentage and curing time, as well as on the applied stress level (in the range 50 - 200 kPa).

For both PN and OR soils it is worth noted that at "the end of test" conditions (horizontal displacements, δx ≅7 mm), when a stable condition is attained in terms of mobilized shear stress, even if volumetric strains still develop, both the raw samples and the treated samples exhibit – approximately – the same shear strength. This is true for all the investigated stress levels within 7 days from lime addition.

By indicating with τ_{max} the maximum shear stress attained during the test, the maximum stress ratio (τ_{max}/σ_v) exhibited upon shearing by raw and lime-treated PN and OR soils are shown in Figure 22 as a function of the applied vertical stress (σ_v). At low vertical stress, the larger are the lime-percentages and curing times, the higher is the maximum stress ratio. The trend of data suggests that the brittleness of the material reduces with the applied stress level (as expected) whereas, especially for PN soil, the effect of lime - percentage and curing time seems to vanish. Although the data for OR soil are still rather limited, a similar trend can be expected.

The role of lime percentage and curing time is also investigated in terms of soil dilatancy behaviour. Here, dilatancy is simply calculated as the ratio – δy/δx, i.e. negative values of δy lead to positive values of dilatancy. Regarding this point, the results obtained for both PN and OR samples are again plotted together in Figure 23. At small values of dilatancy, data reveal that OR 5%-lime-treated samples exhibit higher stress ratios, larger than those mobilized for PN 10%-lime-treated soil. On the other hand, larger PN soil dilatancy is needed to attain the same stress ratio of OR soil. When dilatancy is approximately null, i.e. at the end of test, all data from both soils seem approximately to overlap.

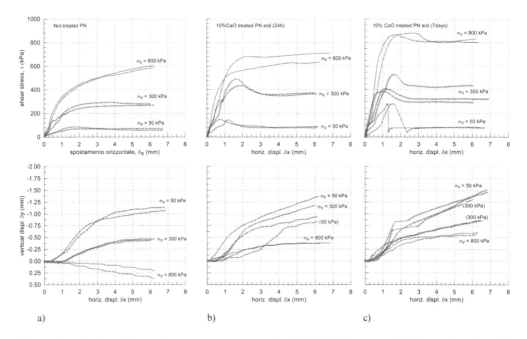

a)   b)   c)

Figure 20.   Results of direct shear tests on not-treated and 10%CaO lime treated PN samples at low (50 kPa), intermediate (300 kPa) and high (800 kPa) vertical stress, at different curing time (after Cecconi et al., 2011c).

Such experimental evidences of a shear strength gain after the addition of lime detected for both PN and OR soil can be related to the chemo-mineralogical evolution of the system induced by addition of lime. The high pH environment (larger than 12.4, see Figure 15) induced by lime addition enables the dissolution of the aluminosilicate constituents of both OR and PN pozzolanic soils.

The X Ray patterns for 10% CaO lime-treated PN soil - at increasing curing times (see previous Figure 10) - confirm again this interpretation. Signal of portlandite ($Ca(OH)_2$) is well detected in the system immediately after treatment (i.e. 24 hours) due to addition of lime, whereas a significant consumption detected for samples cured 7 and 28 days highlights the ongoing of pozzolanic reactions over time.

A further validation of the chemical evolution of the system is shown in Figure 24 whereas the quantitative interpretation of TGA analyses on treated OR soil (see also previous Figure 17a), showing the time dependent increase of mass loss in the temperature interval 0-330°C - compatible with chabazite dehydration (mainly occurring below 140°C), portlandite dehydroxylation, and precipitation of secondary hydrated phases (CSH, CSAH, CAH) - is coupled with the maximum shear stress increase.

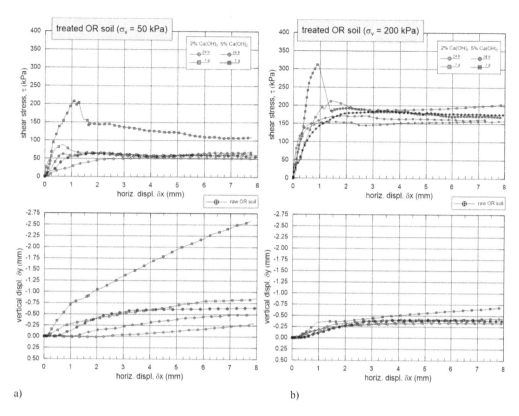

Figure 21. Results of direct shear tests on lime treated OR samples at a) low (50 kPa) and b) intermediate (200 kPa) vertical stress (after Guidobaldi et al., 2017).

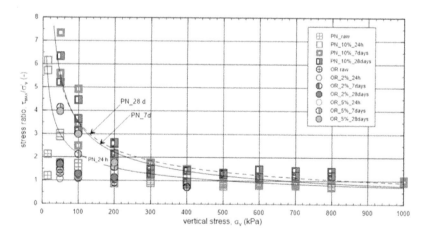

Figure 22. Maximum stress ratio (τ_{max}/σ_v) vs. vertical stress for not-treated and lime treated OR and PN soil.

3.3 Compressibility in 1D-oedometer conditions

A following stage of experimental work has been focused on the 1D-compressibility behaviour of not treated and stabilised pozzolanas upon oedometer conditions, with regard to both partially and fully saturated conditions. Details of the mechanical testing procedures and apparatuses are given in Cecconi et al. (2011c, 2013). During oedometer compression, vertical stress

375

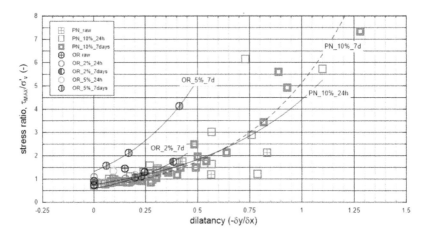

Figure 23.   Maximum stress ratio vs. soil dilatancy for not-treated and 7 days-lime treated OR and PN soil.

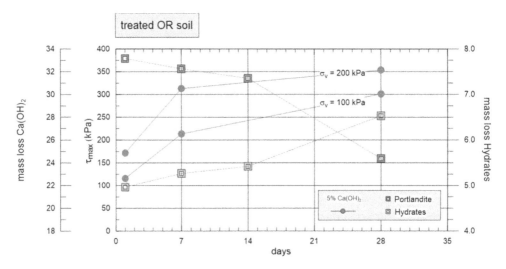

Figure 24.   Lime treated OR soil: quantitative analyses of TGA results and shear strength increase vs. curing time.

was conventionally applied in successive steps; stress increments were applied each hour, in view of the fact that the 90% of settlements systematically took place in less than one minute.

The compressibility curves of compacted unsaturated not treated Monteforte soil samples (MF) are reported in Figures 25a) and b). Figure 25a) shows the influence of degree of saturation on not treated soil prepared at low initial void ratio ($e_0=1.3$) and increasing degrees of saturation (namely 59, 62 and 68 %). The lower the degree of saturation, the larger the observed soil stiffness at stress level before yielding. In Figure 25b) the combined effects of partial saturation and lime stabilisation on the compressibility of looser samples ($e_0 = 1.6$) are shown. A reduced compressibility is observed for partially saturated not treated samples ($S_r = 68\%$) and a further increase in stiffness pertains to treated samples (21 days of curing, M21 in the plot); apparently, the effects of partial saturation on soil compressibility are negligible compared with those induced by lime addition. The relevant increase of yield stresses due to lime addition is shown in Figure 25c), where a slight dependency of yield stresses on S_r is highlighted.

Some other special tests were performed aimed at investigating the soil collapse upon wetting and how this is inhibited by lime treatment. Wetting paths were performed by submerging MF soil samples with distilled water few minutes after the load increment application at three reference stress levels, namely 200 kPa, 600 kPa and 1200 kPa (corresponding to pre-yield, yield and post-yield stresses). As shown by the time - settlements curves upon wetting reported in Figure 26, collapse was detected for each stress level, with lower settlements for the sample wetted at the pre-yield stress level. Lime addition induced a relevant modification of the observed phenomenon. The increase of the saturation degree upon wetting (performed at 1200 kPa stress level, corresponding to the yield stress level of treated MF soil) did not trigger the soil collapse, differently from the observed behaviour of not-treated material.

By considering the reduced curing times of the tested specimens (namely 0 and 7 days), it can be deduced that the cementitious products of pozzolanic reactions are effective immediately after lime addition. This result is consistent with one observed for lime-treated OR soil, whereas the soil compressibility was found to decrease since the very short term (see Figure 27a), consistently with the progressive bonding of aggregates due to the formation of secondary phases (Figures 17). The progressive attainment of a yield condition, which characterizes the compressibility curves for longer curing times, is undoubtedly consistent with the formation of cementitious compounds in a not crystalline state.

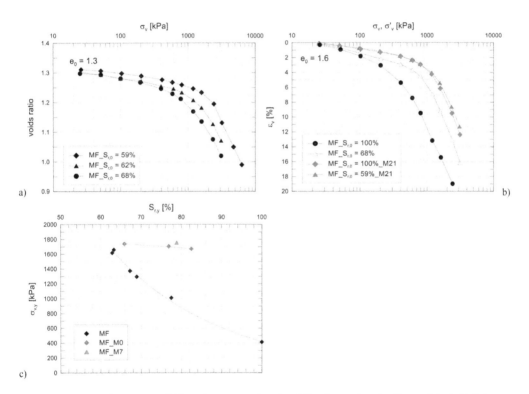

Figure 25. MF soil: compressibility of not treated and stabilized samples: a) effects of the initial degree of saturation; b) combined effects of degree of saturation and lime addition; c) effects of degree of saturation at yielding and lime addition, after 24h and 7 days of curing, on yield stresses (after Cecconi et al., 2013).

A further comparison, in this case between the 1D-compressibility of Orvieto soil and Pozzolana Nera is shown in Figure 27, at different curing times. For lime treated OR samples, the improvement in terms of one dimensional stiffness is relevant for curing times larger than 7 days, but an increase is detected also for samples tested after 24 h. The improvement in the

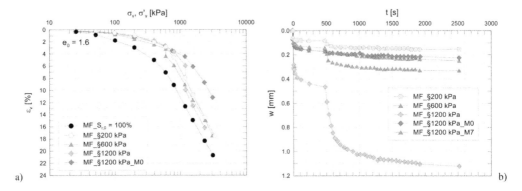

Figure 26. MF soil: collapse upon wetting of the not treated and lime stabilised MF samples: a) compressibility curves; b) settlements vs. time (after Cecconi et al., 2013).

mechanical response is more pronounced after 28 days of curing, when compressibility strongly decreases as well as the yield vertical stress increases. This behaviour persists at longer times, as it is confirmed by the test results at 90 days of curing, which are approximately overlapping those observed after 28 days.

PN soil is characterized by a larger compressibility. Also in this case, the mechanical improvement due to lime addition is detectable after 7 days from treatment and increases further after 21 days of curing, when compressibility before yield strongly decreases and, at the same time, the yield vertical stress increases.

For stress levels larger than yield stresses, both treated OR and PN samples show a higher compressibility index due to destructuration induced by loading. However, the complete destructuration of treated samples have not been achieved for the applied stress levels.

4 POZZOLANIC SOIL AS RESOURCES FOR NON-TRADITIONAL BINDERS

The potential development of soil stabilisation techniques using alkaline activation for geotechnical engineering purposes is very attractive since this technique could be an alternative to the ordinary use of lime and/or Portland cement for soil stabilisation (Coudert et al. 2019, Abdullah et al. 2020). The production processes of Portland cement and lime are energy intensive and emit a large quantity of CO_2. In the construction field, for which cement production contributes to at least 5–8% of global carbon dioxide emissions, alternative industrial by-products (e.g. high-calcium fly ash, rice husk ash, silica fume) have been successfully used as cementing agents in soil improvement resulting in environmental and economic benefits. In the low carbon agenda, it is then of outmost relevance the development of cost- and carbon-efficient technologies. Earlier studies have shown that alkali-activated materials develop relatively high strength, require low energy consumption and produce low CO_2 emissions during synthesis. That is why alkali activated materials could represent a viable sustainable alternative to traditional binders for soil stabilisation.

Alkali Activated Materials (AAM) are generally synthesised from amorphous reactive aluminosilicates (such as metakaolin or fly ash) reacting with an alkaline solution (made of sodium hydroxide NaOH and sodium silicate Na_2SiO_3). At room temperature, they give a hardened material with mechanical properties potentially suitable for traditional binders replacement. The type of aluminosilicate material needed in the alkali activation process varies as well. It is widely accepted the suitability of artificial pozzolanas (fly ash, ground granulated blast furnace slags, etc.) and calcined clays (e.g., metakaolin) for AAM synthesis.

At the same time, increasing attention is being recently paid to the possibility of using natural pyroclastic soils as precursors for alkaline activation. Many studies have been carried out to demonstrate the possibility of using these materials as precursors for binders with low environmental impact (Kamseu et al. 2014, Lemougna et al., 2011, Takeda et al., 2014, Djobo

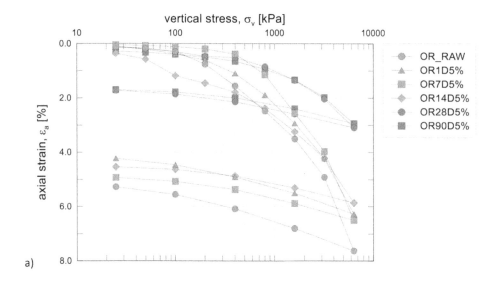

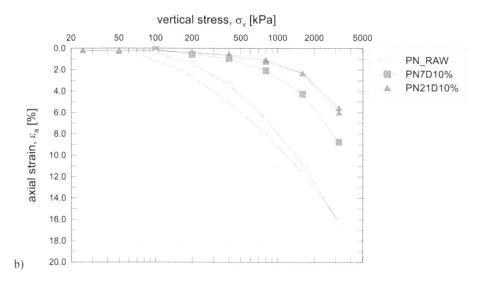

Figure 27. Compressibility curves for 1D oedometer compression tests on not-treated pozzolanas (raw) and a) 5%Ca(OH)$_2$ lime-treated OR soil, b) 10%CaOH lime-treated PN soil.

et al. 2014, Lemougna et al. 2020), and the results from literature show that volcanic ash-based alkali-activated cements/concretes develop good physical and mechanical properties as well as excellent durability compared to ordinary Portland cements.

However, the practical application of alkali-activated pyroclastic soils in soil stabilization projects has been quite limited so far. Verdolotti et al. (2008) showed the effectiveness of alkali-activation of natural pozzolana from Phlegrean Fields (Pozzuoli, Napoli) in terms of mechanical improvement, suggesting deep-mixing technique as viable solution for in situ alkali-activation of natural pozzolanas.

Zhou et al (2021) explored the feasibility of preparing eco-friendly grouting materials based on volcanic ash and metakaolin based alkali activated materials and investigated the synergistic effects of mechanical strength and rheological characteristics. The potential of using pyroclastic soil-based AAM for clayey soil stabilization compared to ordinary Portland cement has been showed (Ghadir and Ranjbar, 2018) for different initial water conditions of treated samples. Hossain and Mol (2011) explored the use of lime and pozzolana to enhance the strength properties of marly soils. The possibility of partial replacement of more widely used precursors as fly ash with volcanic ash as binders for soil stabilisation has been recently explored by Tigue et al. (2018).

Laboratory testing on the investigated pozzolanic soils to be used as precursors for alkaline activation will be the main focus of the development of the research presented so far.

5 CONCLUSIONS

Lime treatment, which is recognized to be a very common and consolidated technique for mechanical improvement of fine grained clayey soils, has proved to be also effective for soil improvement of pozzolanic soils. Due to some important and peculiar aspects of the mechanical behaviour of these materials, such as collapse during saturation, and the serious limit that this behaviour may represent for their use in geotechnical engineering applications, these soils might be discarded and not used as construction materials. The possibility of their reuse through lime stabilisation is therefore very interesting from a technical point of view. In this framework, pozzolanic soils represent a natural resource.

The results of a recent extensive experimental investigation on the effects of lime addition on the microstructure and geotechnical features of three pozzolanic soils, different for nature and mineralogical composition, are described and discussed in the paper in an organic and comparative manner.

The investigated materials, the Orvieto pozzolana (OR), the Pozzolana Nera (PN) and the Monteforte pozzolana (MF) belong to pyroclastic deposits, respectively originated from the Italian volcanic district of Mounts Vulsini (Orvieto), Colli Albani (South East of Rome), Somma Vesuvius (Naples). In their natural condition, due to their peculiar microstructure and mechanical behaviour, these materials had been diffusely studied in the past.

A summary of the main microstructural features has been firstly outlined in this paper. All the investigated silty-sand pozzolanas contain abundant percentages of $SiO2$ of about 51-55% and moderately high percentage of $Al2O3$ of about 23-25% for PN and MF soils and 15% for OR soil. Minerals of Leucite are predominant in PN soil ($\cong$ 24%), while OR soil is enriched in zeolites, mainly Chabasite ($\cong$ 24%). Significative is the presence of Amorphous glassy component, 55% and 27% in weight for OR and PN soil respectively. This composition together with large values of specific surface area favors the high reactivity of the soil-lime system as soon as lime is added to soil. The suitability of investigated materials to lime stabilisation was verified by means of pH measurements at various lime percentages and curing times. Pozzolanic samples were then lime-treated with CaO or $Ca(OH)_2$ in different weight percentage and mixed with a fixed amount of water content.

Physical characterization, minero-chemical development and microstructural analyses were performed on untreated and lime-treated samples in order to highlight the effectiveness of the treatment during the development of the lime-induced reactions. The experimental research involved some experimental techniques typical of geotechnics and geochemistry, from the micro scale at which microstructure and mineralogical composition are analysed to the scale of the laboratory volume sample. This multiscale approach allowed to understand the link between chemical reactions, microstructure modifications and observed mechanical behaviour. With regard to this point, the evidence of treated OR soil is indicative: when hydrated lime (5% of $Ca(OH)_2$) is added to the natural OR soil, the role played by crystalline phases, including zeolites, in the formation of C-S-H and C-A-H is secondary with respect to that of the amorphous phases and, rather, the amorphous component was found to be the main phase involved in pozzolanic reactions, while crystalline phases show a negligible contribution to the formation of hydrated secondary products.

The microstructural features induced by lime addition clearly affect the mechanical properties of lime-treated pozzolanic soils. The results of oedometer and direct shear tests have been interpreted as a function of curing time and, therefore, of the evolution of the system.

In terms of mechanical behaviour upon shearing, the response of lime-treated soils depends on applied stress level, amount of added lime and curing time. The addition of lime to the Orvieto pozzolana and Pozzolana Nera yields to an increase of shear strength (peak) but, at the same time an increase of soil brittleness and dilatancy; these features are more evident for larger lime contents and longer curing times. The formation of newly hydrated phases are responsible for this observed behaviour. The increase of soil dilatancy is reasonably due to bonding between grains/ grain aggregates inferred by new cementitious products. Data from pore size distributions of OR soil also reveal filling of pores and the formation of a new small class of pores, again confirming the progressive formation of secondary hydrates phases within 28 days since lime addition.

The effect of lime addition due to the progressive development of stable bonds between aggregates deriving from the products of pozzolanic reactions has been also observed in terms of reduced compressibility and increase of yield stress upon 1D oedometer loading conditions. For the three investigated pozzolanas, the improvement in terms of one dimensional stiffness is relevant for curing times larger than 7 days. As concerns the Monteforte pozzolana, which in its natural state is characterized by a collapse behavior upon wetting, it is found that lime treatment, with proper parameters, inhibit the collapsible behavior from the very short time after lime addition. This result is relevant in the perspective of embankment construction by employing pozzolanic soils.

It can be concluded that the improved mechanical behaviour of investigated pozzolanic soils, observed after the addition of lime, i.e. the reduced compressibility, increased yield stress, inhibited collapse upon wetting, increased peak shear strength – all of which are directly linked to the chemo-mineralogical evolution of the system – represents an interesting result for practical application of lime-treatment techniques in the fields of engineering and geology/ geotechnical engineering. Through a wide and long experimental research, the study has shown the effectiveness of lime-treatment technique - originally devoted only to fine-grained materials – and its applicability to medium/coarse grained pozzolanic soils.

ACKNOWLEDGEMENTS

The research presented in this paper is the result of a fruitful collaboration with many Researchers working at different research Italian and French institutions – Department of Engineering and Department of Physics and Geology, University of Perugia (Italy); Department of Civil and Mechanical Engineering, University of Cassino and Southern Lazio (Italy), Institut Français des Sciences et des Technologies des Transports (IFSTTAR), Institut des Matériaux Jean Rouxel (IMN), Université de Nantes (France), whose support is gratefully acknowledged.

I wish to thank Prof. Giacomo Russo (University of Naples Federico II) very much for the innumerable constructive discussions we had regarding data interpretation. Many thanks to Dr. Costanza Cambi and Prof. Paola Comodi (University of Perugia) for sharing the lab. experimental investigation.

The scientific contribution of Dr. Giulia Guidobaldi and Dr. Enza Vitale who performed most of chemical and geotechnical tests during their Ph.D. and post-Doc research activities is much appreciated.

I would also like to thank Dr. Fabio Marmottini (University of Perugia) and Dr. Paolo Tommasi (CNR - Institute for Environmental Geology and Geo-Engineering, Rome, Italy) for their valuable suggestions, and Eng. Diego Bellavita for his help in editing some graphs.

REFERENCES

Abdullah H.H., Shahin M.A., Walske M. L., 2020, Review of Fly-Ash-Based Geopolymers for Soil Stabilisation with Special Reference to Clay, *Geosciences* 2020, 10, 249, doi:10.3390/geosciences10070249.

Bell, F. G., 1996, Lime stabilisation of clay minerals and soils, *Engineering Geology*, 42, 223–237.

Bernabini, M., Esu, F., Martinetti, S., Ribacchi, R., 1966, On the stability of the pillars in an underground quarry worked through soft pyroclastic rocks, *Proc. I Congr. Int. Soc. of Rock Mechanics*, Lisboa, 2: 285–291.

Blanco Valera, M., Martínez Ramírez, S., Ereña, I., Gener, M., & Carmona, P., 2006, Characterization and pozzolanicity of zeolitic rocks from two Cuban deposits. *Applied Clay Science*, 33, 149–159.

Boardman, D.I., Glendinning, S. and Rogers, C.D.F., 2001, Development of stabilisation and solidification in lime-clay mixes, *Géotechnique*, 51, 533–543.

Bozzano, F., Bretschneider, A., & Martino, S., 2008, Stress–strain history from the geological evolution of the Orvieto and Radicofani cliff slopes (Italy), *Landslides*, 5: 351–366.

Brunauer, S., Emmett, P. H. & Teller, E. 1938, Adsorption of gases in multi molecular layers. *J. Am. Chem. Soc.* 60, 309–319.

BSI (1990) BS 1924-2: Stabilized materials for civil engineering purposes. Part 2: Methods of test for cement-stabilized and lime-stabilized materials. BSI, London, UK.

BSI (2006) BS EN 14227-11: Hydraulically bound mixtures – specifications. Part 11: Soil treated by lime. BSI, London, UK.

Cambi, C., Carrisi, S. and Comodi, P., 2012, Use of the Methylene Blue Stain Test to Evaluate the Efficiency of Lime Treatment on Selected Clayey Soils, *Journal of Geotechnical and Geoenvironmental Engineering*, 138(9),pp.1147–1150.

Cambi, C., Guidobaldi, G., Cecconi, M, Comodi, P., Russo, G., 2016, On the ICL test in soil stabilization. *Int. Proceedings of the 1st IMEKO TC4 International Workshop on Metrology for Geotechnics*, Benevento, Italy, 17–18 March 2016, 31–34.

Cappelletti, P., Langella, A., Colella, A., De Gennaro, R. 1999, Mineralogical and technical features of zeolites deposits from northern Latium volcanic district. *Periodico di Mineralogia* 1999, 68, 127–144.

Cattoni E., Cecconi M., Pane V., 2007, Geotechnical properties of an unsaturated pyroclastic soil from Roma, *Bull.of Eng.Geol. and the Env.*, 66, 403–414.

Cecconi, M. and Viggiani, G.MB., 2001, Structural features and mechanical behaviour of a pyroclastic weak rock, *Int. J. Numer. Anal. Meth. Geomech.*, 25, 1525–1557.

Cecconi, M., Scarapazzi, M. and Viggiani, G. MB., 2010, On the geology and the geotechnical properties of pyroclastic flow deposits of the Colli Albani, *Bull.of Eng.Geol. and the Env.*, 69, 185–206

Cecconi M., Pane V., Marmottini F., Russo G., Croce, P., dal Vecchio S., 2011a, Lime stabilisation of pyroclastic soils, in: *Proceedings of the 5th International Conference on Unsaturated Soils*, Barcelona, Spain, 6–8 September 2010; Volume 1, pp.537–541.

Cecconi M., Rotonda T., Tommasi P., Viggiani G. M.B. 2011b, Microstructural features and compressibility of volcanic deposits from Central Italy, In: *V International Symposium on Deformation Characteristics of Geomaterials*, vol. II, p. 884–891, Chung CK., Jung YH., Kim HK., Lee JS., Kim D.S Editors, Seoul, Korea, 1-3 September 2011.

Cecconi M., Ferretti A., Russo G., Capotosto A. 2011c, Mechanical properties of two lime stabilized pyroclastic soils. in: *Int. Symposium on Deformation Characteristics of Geomaterials*, vol. II, p. 772–778, Chung CK., Jung YH., Kim HK., Lee JS., Kim D.S Editors, Seoul, Korea, 1-3 September 2011.

Cecconi, M., & Russo, G., 2013, Microstructural features of lime-stabilised pyroclastic soils. *Géotechnique Letters*, 3, 124–129.

Cecconi M., Capotosto A., Russo G., 2013. Mechanical behaviour of two lime stabilised pyroclastic soils. *TerDOUEST 2013 Seminar*, 18-19 June, Marne La Vallee (France), 13–18.

Cecconi, M., Rotonda, T., Verrucci, L., Tommasi, P., & Viggiani, G., 2016, Microstructural features and strength properties of weak pyroclastic rocks from Central Italy, in: *Int. Workshop on Volcanic Rocks and Soils*, Rotonda et al. Eds, Lacco Ameno, Ischia Island, Italy, 24-25 sept 2015.

Cecconi M, Cambi C, Carrisi S, Deneele D, Vitale E, Russo G., 2020, Sustainable Improvement of Zeolitic Pyroclastic Soils for the Preservation of Historical Sites. *Applied Sciences* 2020, 10(3): 899, doi:10.3390/app10030899.

Coudert E., Paris M., Deneele D., Russo G., Tarantino A., 2019. Use of alkali activated high-calcium fly ash binder for kaolin clay soil stabilisation: Physicochemical evolution. Construction and Building Materials. Volume 201, 20 March 2019, p. 539–552.

CNR (1973). Stabilizzazione delle terre con calce. Bollettino Ufficiale del 21.02.1973n.36, Parte IV, Norme Tecniche.

CNR-UNI10006. (2002). Costruzione e manutenzione delle strade. Tecniche di impiego delle terre.

De Rita, D. Funiciello, R., Rosa, C., 1992, Volcanic Activity and drainage network evolution of the Alban Hills area, *Acta Vulcanologica*, Marinelli, 2, 185–198.

De Rita, D., Faccenna, C., Funiciello, R. and Rosa, C., 1995, Stratigraphy and volcano-tectonics, in: *The volcano of the Alban Hills*, Trigila ed., SGS Roma.

Diamond, S., & Kinter, E., 1966, Adsorption of calcium hydroxide by montmorillonite and kaolinite. *Journal of Colloid and Interface Science*, 22, 240–249.

de Riso R., Budetta P., Calcaterra D., Santo A., 1999. Le colate rapide in terreni piroclastici del territorio campano. *Atti della conferenza Previsione e Prevenzione di Movimenti Franosi Rapidi*, Trento, p. 133–150.

Di Crescenzo, G., Rotella, M. & Santo. A, 2007. Il contributo della geologia per lo studio dei meccanismi di innesco di colate rapide di fango al campo sperimentale di Monteforte Irpino (AV). In: *C. Nunziata (ed.) Piattaforme Evolute di Telecomunicazioni e di Information Technology per l'Offerta di Servizi al settore Ambiente Petit-Osa*: 263–272. Rome, Aracne.

Djobo J. N. Y., Elimbi A., Tchakouté H. K., and Kumar S. 2017, Volcanic ash-based geopolymer cements/concretes: the current state of the art and perspectives. *Environ. Sci. Pollut. Res. Int 2017*; 24 (5):4433–4446.

Funiciello R., Giordano G., De Rita D., 2003, The Albano maar lake (Colli Albani Volcano, Italy): recent volcanic activity and evidence of pre-Roman Age catastrophic lahar events, *Journal of Volcanology and Geothermal Research*, 123, Issues 1–2, 2003, 43–61.

Ghadir P., and Ranjbar N., 2018. Clayey soil stabilization using geopolymer and Portland cement. *Constr. Build. Mater.* 2018; 188: 361–371.

Gentili, S., Comodi, P., Nazzareni, S., & Zucchini, A., 2014, The Orvieto-Bagnoregio Ignimbrite: Pyroxene crystal-chemistry and bulk phase composition of pyroclastic deposits, a tool to identify syn- and post-depositional processes. *European Journal of Mineralogy*, 26, 6, 743–756.

Giordano G., De Benedetti A.A., Diana A., Diano G., Gaudioso F., Marasco F., Miceli M., Mollo S., Cas R.A.F., Funiciello R., 2006, The Colli Albani mafic caldera (Roma, Italy): Stratigraphy, structure and petrology, *Journal of Volcanology and Geothermal Research*, 155, Issues 1–2, 2006, 49–80.

Guidobaldi, G., Cambi, C., Cecconi, M., Deneele, D., Paris, M., Russo, G., Vitale, E., 2017, Multi-scale analysis of the mechanical improvement induced by lime addition on a pyroclastic soil. *Eng. Geol.* 2017, 221, 193–201.

Guidobaldi, G., Cambi, C., Cecconi, M., Comodi, P., Deneele, D., Paris, M., Russo, G., Vitale, E., Zucchini, A., 2018, Chemo-mineralogical evolution and microstructural modifications of a lime treated pyroclastic soil, *Eng. Geol.* 2018, 245, 333–343.

Hossain, K.M., Mol, A., 2011, Some engineering properties of stabilized clayey soils incorporating 869 natural pozzolans and industrial wastes. *Constr. and Buil. Mat.* (25), 3495–3501. 870 https://doi.org/10.1016/j.conbuildmat.2011.03.042.

Ingles, O. G., & Metcalf, J. B., 1972, *Soil stabilisation: principles and practice*. Butterworths, Sidney, Australia.

James, R., Kamruzzaman, A. H., Haque, A., & Wilkinson, A., 2008, Behaviour of lime-slag-treated clay. *Proceedings of the ICE -Ground Improv.*, 161 (4), 207–2016.

Kamseu E., Leonelli C., Perera D.S., Melo U.C., Lemougna P.N., 2014, Investigation of Volcanic Ash Based Geopolymers as Potential Building Materials, INTERCERAM. - ISSN 0020-5214.58(2009), pp. 136–140.

Lembo-Fazio, A., Manfredini, M., Ribacchi, R. & Sciotti, M., 1984. Slope failure and cliff instability in the Orvieto hill, in *4 th Int. Symp. on Landslide*, Toronto, 2: 115–120.

Lemougna PN, MacKenzie KJD, Melo UFC, 2011, Synthesis and thermal properties of inorganic polymers (geopolymers) for structural and refractory applications from volcanic ash. *Ceram. Int.* 37: 3011–3018. doi:10.1016/j.ceramint.2011.05.002.

Lemougna P. N., Nzeukou A., Aziwo B., Tchamba A., Wang K.-t., Melo U. C., and Cui X., 2020, Effect of slag on the improvement of setting time and compressive strength of low reactive volcanic ash geopolymers synthetized at room temperature, *Mater. Chem. Phys* 2020, 239: 122077.

Little, D. N., 1996, Fundamentals of the stabilization of soil with lime: National Lime Association. Bulletin, 332, 1–20. Arlington, USA.

Locat, J., Tremblay, H., & Leroueil, S., 1996, Mechanical and hydraulic behaviour of a soft inorganic clay treated with lime. *Canadian Geotech. Journal* 33, 4, 654–669.

Martínez-Ramírez, S., Blanco-Varela, M., Ereña, I., & Gener, M., 2006, Pozzolanic reactivity of zeolitic rocks from two different Cuban deposits: characterisation of reaction products. *Applied Clay Science*, 32, 40–52.

Mertens, G., Snellings, R., Van Balen, K., Bicer-Simsir, B., Verlooy, P., & Elsen, J., 2009, Pozzolanic reactions of common natural zeolites with lime and parameters affecting their reactivity. *Cement and Concrete Research*, 39, 233–240.

Nappi G., Capaccioni B., Mattioli M., Mancini E., Valentini L., 1994, Plinian fall deposits from Vulsini Volcanic District (Central Italy), *Bull. of Volcanology* 56 (6-7), 502–515.

383

Nicotera M.V., Papa R., 2007. Comportamento idraulico e meccanico della serie piroclastica di Monteforte Irpino (AV), in: *C. Nunziata (ed.) Piattaforme Evolute di Telecomunicazioni e di Information Technology per l'Offerta di Servizi al settore Ambiente Petit-Osa*: 272–280. Rome: Aracne.

Papa R., 2007, *Indagine sperimentale sulla coltre piroclastica di un versante della Campania*. PhD Thesis, University of Napoli Federico II. Napoli, Italy.

Rao, S. M., & Shivananda, P., 2005, Compressibility behaviour of lime stabilized clay. *Geot.and Geol. Engineering*, 23, 309–319.

Rogers, C. D., & Glendinning, S., 1996, Modification of clay soils using lime, in S.G. CDF Rogers, *Lime stabilisation* (pp. 99–112). London: Thomas Telford.

Rotonda, T. & Tommasi, Paolo & Ribacchi, R., 2002, Physical and mechanical characterisation of the soft pyroclastic rocks forming the Orvieto cliff, in: *ISRM International Symposium-EUROCK 2002*, November 25-27, 2002, Madeira, Portugal. 137–146.

Russo G., 2019, Microstructural investigations as a key for understanding the chemo-mechanical response of lime-treated soils. *Rivista Italiana di Geotecnica*, 1, 2019, 100–114.

Russo G., Vitale E., Cecconi M., Pane V., Deneele D., Cambi C., Guidobaldi G., 2016, Microstructure insights in mechanical improvement of a lime-stabilised pyroclastic soil, in: *Int. Workshop on Volcanic Rocks and Soils*, Rotonda et al. Eds, Lacco Ameno, Ischia Island, Italy, 24-25 sept 2015.

Santi, P., 1991, New geochronological data of the Vulsini Volcanic district, (Central Italy), *Plinius*, 4, 91–92.

SETRA/LCPC, 2000, Traitement des sols à la chaux et/ou aux liants hydrauliques. Application à la realisation des remblais et des couches de forme. Guide Technique SETRA LCPC, Paris.

Sciotti M., 1981, Some problems of environment protection in Italy, in: *Proc. X Int. Conf. of Int. Soc. for Soil Mechanics and Foundation Engineering*, Stockholm. 2, 375–378.

Sivapullaiah, P. V., Sridharan, A., & Ramesh, A. N., 2000, Strength behaviour of lime treated soils in the presence of sulphate, *Canadian Geotech. Journal*, 37, 1358–1367.

Takeda H, Hashimoto S, Kanie H et al., 2014, Fabrication and characterization of hardened bodies from Japanese volcanic ash using geopolymerization. *Ceram. Int.* 40:4071–4076. doi:10.1016/j. ceramint.2013.08.061.

Tigue A.A.S., Dungca J.R., Hinode H., Kurniawan W., Promentilla M.A.B., 2018, Synthesis of a one-part geopolymer system for soil stabilizer using fly ash and volcanic ash, *MATEC Web of Conferences* 156, 05017 (2018)

Tommasi, P. and Ribacchi, R., 1998, Mechanical behaviour of the Orvieto tuff, in: *2nd Int. Symp., Hard Soils-Soft Rocks*, Napoli, 2, 901–909 Balkema, Rotterdam: Evangelista & Picarelli Eds.

Tommasi, P., Verrucci, L., Rotonda, T., 2015, Mechanical properties of a weak pyroclastic rock and their relationship with microstructure. *Canadian Geotech. Journal* 52 (2):211–223. DOI: 10.1139/cgj-2014-0149.

Tremblay, H., Leroueil, S., & Locat, J., 2001, Mechanical improvement and vertical yield stress prediction of clayey soil from eastern Canada treated with lime or cement. *Canadian Geotech. Journal*, 38, 567–579.

Verdolotti L., Iannace S., Lavorgna M., Lamanna R., 2008, Geopolymerization reaction to consolidate incoherent pozzolanic soil. *Journ. Mater. Sci.* 43:865–873. doi:10.1007/s10853-007-2201-x.

Verrucci L., Lanzo G., Tommasi P., Rotonda T., 2015, Cyclic and dynamic behaviour of a soft pyroclastic rock, *Géotechnique*, 65, 359–373.DOI: 10.1680/geot.SIP.15.P.012.

Viggiani G.MB., Cecconi M., 2007, Pyroclastic flow deposits from the Colli Albani. The example of the Pozzolana Nera, in: *Second International Workshop on Characterisation and Engineering Properties of Natural Soils*, vol. 2, p. 2411–2447, Tan T.S., Phoon K.K. and Hight D.W. and Leroueil Editors.

Vitale E., Deneele D., Paris M., Russo G., 2017. Multi-scale analysis and time evolution of pozzolanic activity of lime treated clays. *Applied Clay Science*, 141, 1 June 2017, Pages 36–45; 10.1016/j. clay.2017.02.013.

Vitale, E., Deneele, D., Russo, G., 2016. Multiscale Analysis on the Behaviour of a Lime Treated Bentonite, in: *6th Italian Conference of Researchers in Geotechnical Engineering, CNRIG 2016*. Procedia Engineering, 158, 2016, Pages 87–91. 10.1016/j.proeng.2016.08.409.

Vitale, E., Deneele, D., Russo, G., & Ouvrard, G., 2016, Short term effects on physical properties of lime treated kaolin. *Applied Clay Science*, 132, 223–231.

Zhou S., Yang Z., Zhang R., Li F., 2021, Preparation, characterization and rheological analysis of eco-friendly road geopolymer grouting materials based on volcanic ash and metakaolin., *Journal of Cleaner Production* 312 (2021) 127822.

Rock Mechanics and Engineering Geology in Volcanic Fields – Ohta, Ito & Osada (eds)
© 2023 copyright the Author(s), ISBN 978-1-032-27657-1

An experimental study on the shallow underground openings in rock mass with hexagonal discontinuity pattern

Michio Tamashiro*
Graduate School of Engineering and Science, University of the Ryukyus, Japan

Ömer Aydan, Takashi Ito & Naohiko Tokashiki
Department of Civil Engineering, University of the Ryukyus, Japan

ABSTRACT: Various discontinuity patterns are observed in rock masses as a result of their geological past. One of the common discontinuity patterns is the hexagonal pattern. This pattern is mostly observed in all extrusive volcanic rocks such as basalt, andesite, rhyolite and welded tuffs. In this study, the authors investigate the stability of shallow underground openings in rock mass with hexagonal discontinuity pattern through model tests. Furthermore, some considerations are given to the failure mechanism, which could be utilized for developing limiting equilibrium methods to assess their stability and displacement responses. The outcomes of these experimental studies are presented and their implications in practice are discussed.

Keywords: Hexagonal Discontinuity Pattern, Shallow, Underground Opening, Stability, Collapse

1 INTRODUCTION

Rock masses geologically have various type discontinuities. One of the common discontinuity patterns is hexagonal discontinuity pattern. Furthermore, there are some recent studies on the stability of underground and slopes. For example, Baihetan hydro-electric power complex has been built in basaltic rock mass. The construction of the hydro-electric power complex initiated various studies on hydro-mechanical properties of rock masses as well as the stability of underground caverns and diversion tunnels and the deformation response of the dam foundation, e.g., (Jiang et al. 2014; Jin et al. 2015).

Underground openings may be subjected to dynamic loading resulting from earthquakes and/or blasting. There is almost no damage to underground openings at depth unless the causative earthquake fault crosses the openings. Tanna tunnel, Inatori tunnel in Shizuoka prefecture and mine tunnels in Tangshan are known cases that the tunnels were damaged by earthquake faults at the locations. Moreover, when the overburden is shallow, the overall stability of the underground openings may be of great concern. Aydan et al. (2010) reported the effects of earthquakes on underground openings and gave some examples of collapses and damages to shallow underground openings. Furthermore, there are very few studies on the static and dynamic stability of shallow underground openings (Aydan et al. 1994; Genis and Aydan, 2002; Aydan et al. 2011; Ohta et al. 2014).

*Corresponding author: k208472@eve.u-ryukyu.ac.jp

DOI: 10.1201/9781003293590-47

In this study, the static and dynamic response and stability of shallow underground openings excavated in rock masses having hexagonal discontinuity pattern are evaluated through model tests at the rock mechanics laboratory of the University of the Ryukyus. Two different model testing devices were utilized. This article describes the model tests and the results obtained and discuss the static responses and stability of underground openings.

2 UNDERGROUND OPENINGS IN ROCK MASSES WITH HEXAGONAL DISCONTINUITY PATTERN AND INSTABILITY PROBLEMS

Columnar jointing in nature constitutes spectacular touristic spots worldwide. Figure 1 shows various examples of natural caves in USA, Azerbaijan, Japan and Taiwan. The discontinuity set number involved in columnar jointing is generally 4 or more. In plan, the columns generally have a hexagonal pattern and pentagonal pattern sometimes. The joints develop perpendicular to the flow direction of lava. The vertical joints, which will be called longitudinal joints, have generally rough undulations (Aydin and DeGraff, 1988). The orientation of joints perpendicular to flow direction are generally vertical. However, longitudinal joints may be horizontal or inclined in relation to magma or dyke intrusion direction. In addition, some secondary joints develop perpendicular to longitudinal joints. Depending upon the shape of openings, distribution of joints, gravity, seepage of ground water, sea wave actions and seismic forces from time to time, underground openings may become unstable as seen in Figure 2. As noted from Figure 2, instability modes may be categorized as rockfall, sliding, toppling. Rockfalls are expected to be due to the loss of arching action, particularly as a result of seismic forces and high seepage forces.

Figure 1. Examples of stable natural underground openings.

There are many examples of excavations of tunnels through rock masses having columnar jointing. Figure 3 shows three examples of tunnels excavated through andesite and basalt. The support systems of these tunnels consist of shotcrete and rockbolt. The

Figure 2. Examples of unstable natural underground openings.

main purpose of support is generally restrain the fall, sliding and/toppling of rockblocks. However, the portals could be vulnerable to fail in relation to heavy rainfall or the tunnel face may collapse when they cross fault, fracture or weakness zones especially at shallow depths (Figure 4). Jiang et al. (2014) also reported that there may be some stress-induced rock slabbing as observed during the excavation of Baihetan Undeground Power House.

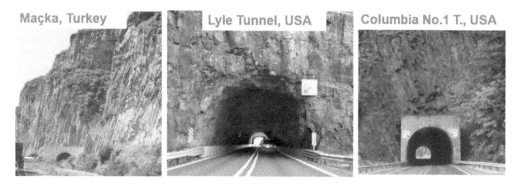

Figure 3. Views of tunnels excavated in rockmass with columnar jointing.

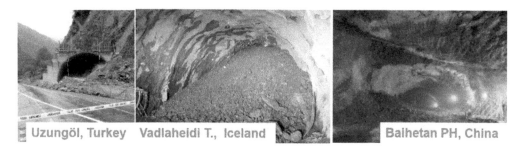

Figure 4. Views of failures of tunnels and underground power house in rock mass with columnar jointing.

3 MODEL MATERIALS AND TEST SET-UP

The authors have been performing model experiments on underground openings for some time (Aydan et al., 1994, 2011; Genis and Aydan, 2002; Ohta et al. 2014). The first series of experiments on shallow undergound openings in discontinuous rock mass using non-breakable blocks were reported by Aydan et al. (1994, 2011), in which a limit equilibirum method was developed for assessing their stability.These experiments have been repeated using different model materials.

Table 1. Physical and elastic constants of Steel and Aluminum.

Material Properties	Steel	Aluminum (Aydan et al., 2021)
Elastic Modulus (GPa)	210	68×10^4
Poisson's ratio	0.3	0.33
Unit weight (kN/mm^3)	56.0	26.89
Block interface friction angle	15-16	15-15.9

3.1 Model materials and their frictional properties

Steel and aluminum hexagonal blocks were selected to create underground opening models with hexagonal discontinuity pattern. Steel hexagonal blocks had a side-length of 10 mm and 8 mm thick. These blocks were utilized in model tests on the base-friction model experiment device. Aluminum blocks were 50 mm long and a side length of 6 mm (Aydan et al., 2021). The physical and mechanical properties of steel and aluminum are given in Table 1. In these particular experiments, the blocks were selected such that they will remain elastic while the movements can take place in the form of separation and sliding along block interfaces.

3.2 Discontinuity pattern of model rock masses

Fundamentally, the discontinuity patterns of model rock masses tests are denoted as Pattern A and Pattern B as illustrated in Figure 5. The 90 degrees rotation of Pattern-A results in Pattern-B. In other words, the variety of model rock masses is quite restricted to two patterns. If the friction angle of block interfaces is less than 30 degrees, the slope angle of the model rock mass cannot be greater than 60 degrees under gravitational conditional. Therefore, The Pattern A model rock mass with underground openings is possible with side restraints.

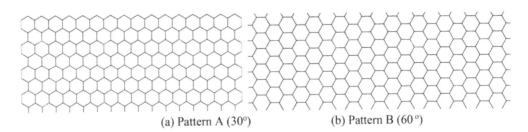

(a) Pattern A (30°) (b) Pattern B (60°)

Figure 5. Selected discontinuity Patterns.

4 MODEL TESTS

4.1 Model tests on underground openings with discontinuity pattern A

4.1.1 Model tests utilizing base-friction model testing apparatus

Some model experiments on underground openings using steel blocks with hexagonal shape were carried. The base-friction model testing device, which can simulate gravity in a horizontal plane, is used for this purpose. As the motions of blocks are slow, it is possible to observe failure process, which is generally difficult to be observed on model tests under gravitational field. Figure 6 shows the behavior of model underground openings at various stages. The failure process is initiated by the vertical motion (rock fall) of the block in the center and then two blocks on both sides start to slide into the opening. Following the motion of these unstable blocks, the failure progress upward. Each time, rockfall domain increases in width while the sliding of the blocks at both sides remains the same. Finally, the blocks cave into the opening as seen in Figure 6. In addition to this test, one side of the model rock mass was set to 60 degrees and the failure process was investigated. Nevertheless, those results would be given elsewhere due to the lack of space in this manuscript.

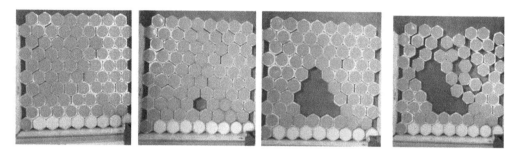

Figure 6.   Failure process during the model experiment utilizing the base-friction model apparatus.

4.1.2 *Model tests under vertical gravitational field*

The same experiment was repeated using aluminum blocks under true gravitational field. Figure 7 shows several views of the model experiment. It is interesting to notice that the overall response is almost the same. The failure process remained the same. However, an arching action was noticed just above the opening and that arching action was quite vulnerable to slight vibration.

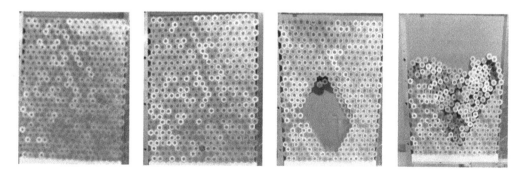

Figure 7.   Failure process during the model experiment under true gravitational field.

4.2   *Model tests on underground openings with discontinuity pattern B*

4.2.1 *Model tests utilizing base-friction model testing apparatus*

Next experiments were carried out on model underground openings with Discontinuity Pattern B. Figure 8 shows the behavior of model underground openings at various stages. The failure process is initiated by the vertical motion (rock fall) of the block in the center and then two blocks on both sides start to slide into the opening. Following the motion of these unstable blocks, the failure progress upward. Each time, rockfall and sliding blocks remain to be the same. Compared to the models with Discontinuity Pattern A, the failure process is different and its size remains to be the same. As the experiments were carried out using the base friction-model testing apparatus, the failure process could be easily seen.

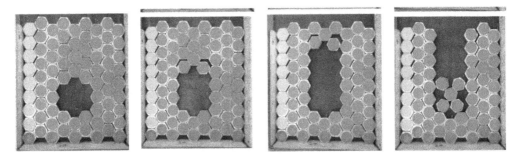

Figure 8.   Failure process during the model experiment utilizing the base-friction model apparatus.

4.2.2 *Model tests under vertical gravitational field*

The same experiment was repeated using aluminum blocks under true gravitational field. Figure 9 shows several views of the model experiment. It is interesting to notice that the overall response is almost the same. The failure process remained the same. However, an arching action was noticed just above the opening and that arching action was quite vulnerable to slight vibration.

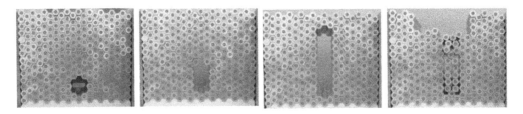

Figure 9.   Failure process during the model experiment under true gravitational field.

5   POSSIBLE FAILURE MODES FOR STABILITY ANALYSES

When the motions of the blocks around underground opening are considered, the motion patterns are entirely different for discontinuity patterns A and B as illustrated in Figure 10. While the motions of the blocks involve only rockfall and sliding over 60-degree planes for Discontinuity Pattern B, the rockfall domain becomes larger in relation to apparent stepped failure planes with an overall inclination of 60 degrees for Discontinuity Pattern A. It may be stated that the stabilization of underground openings in rock masses with Discontinuity Pattern B would be much simple and less costly.

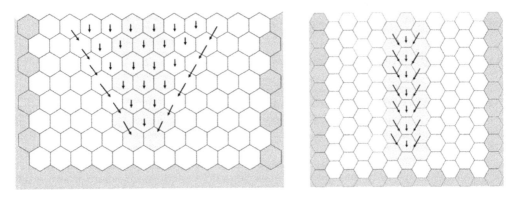

Figure 10.   Inferred motions of blocks for two different discontinuity patterns.

6 CONCLUSIONS

In this study, we investigated the stability of shallow underground openings in discontinuous rock mass with hexagonal discontinuity pattern and shallow overburden through model experiments. The experimental results clearly indicate the motion of blocks during the failure process are different and the size of failure is also different. Whether or not it collapses in this experiment is thought to be related to the effect of the arching action. Because arching action is noted during experiments. However, this arching action disappears when there is a shaking. Therefore, it is necessary to carry out similar model experiments using shaking tables.

REFERENCES

Aydin, A. and DeGraff, J.M., 1988, Evoluton of Polygonal Fracture Patterns in Lava Flows, *Science*, 239(4839), 471–476.

Aydan, Ö., Y. Shimizu, M. Karaca, 1994. The dynamic and static stability of shallow underground openings in jointed rock masses. *The 3rd Int. Symp. on Mine Planning and Equipment Selection*, Istanbul, October, 851–858.

Aydan, Ö., Ohta, Y., Geniş, M., Tokashiki, N. and Ohkubo, K., 2010, Response and stability of underground structures in rock mass during earthquakes. *Rock Mechanics and Rock Engineering*, 43(6), 857–875.

Aydan, Ö., Ohta, Y., Daido, M., Kumsar, H. Genis, M., Tokashiki, N., Ito, T. and Amini, M., 2011, Chapter 15: Earthquakes as a rock dynamic problem and their effects on rock engineering structures. *Advances in Rock Dynamics and Applications*, Editors Y. Zhou and J. Zhao, CRC Press, Taylor and Francis Group, 341–422.

Aydan, Ö., Tokashiki, N., Ito, T. and Murayama, Y. 2021, Dynamic stability of rock slopes with hexagonal discontinuity pattern. 15th Japan Rock Mechanics Symposium, 453–458.

Geniş, M., Aydan, Ö. 2002 Evaluation of dynamic response and stability of shallow underground openings in discontinuous rock masses using model tests. Korea-Japan Joint Symposium on Rock Engineering, Seoul, Korea, July, 787–794.

Jiang, Q. Xia-ting Feng, X.T., Hatzor, Y.H, Hao, X.J. and Li, S.J. 2014. Mechanical anisotropy of columnar jointed basalts: An example from the Baihetan hydropower station, China. Engineering Geology 175, 35–45.

Jin C, Yang C, Fang D, Xu S, 2015, Study on the failure mechanism of basalts with columnar joints in the unloading process on the basis of an experimental cavity. Rock Mech Rock Eng 48:1275–1288.

Ohta, Y., Aydan, Ö., Yagi, M. (2014): Laboratory model experiments and case history surveys on response and failure process of rock engineering structures subjected to earthquake. Proc. of the 8th Asian Rock Mechanics Symposium, Sapporo, 843–852.

Rock Mechanics and Engineering Geology in Volcanic Fields – Ohta, Ito & Osada (eds)
© 2023 copyright the Author(s), ISBN 978-1-032-27657-1

Development of a numerical simulation method for the fluid-mechanical coupled behavior of geothermal reservoir rocks within a fault zone

SangSun Jeong*, Adam K. Schwartzkopff & Atsushi Sainoki
Kumamoto University, Kumamoto, Japan

ABSTRACT: In this study, numerical simulations considering fault slip due to fluid injection were conducted using the coupled Extended Finite Element Method (X-FEM), which considers flow rate along the fault as well as into rock. A comparison is made between the results from a model that considers the heterogeneous permeabilities, in the fault damage zone, derived from the Discrete Fracture Network (DFN), and a model that uses the equivalent homogeneous permeability, in the damage zone, with the aim of quantifying the influence of fractured fault rocks with anisotropic characteristics. It has been shown that the two models produced the similar magnitude of seismic fault slip, however the heterogeneous permeability model produced the higher maximum seismic fault slip locally, attributed to the higher pressure regions along the fault. This resulted in the seismic to total slip ratio (the average dynamic shear movement divided by the average total shear movement) in the equivalent homogeneous permeability model of 0.27 % and 0.57 % in the heterogeneous permeability model. Interestingly, the seismic fault slip mainly produced on the edge of the pressurized zone in both models, which aligns well with field observations. These results imply the applicability of this coupled X-FEM approach to predict fault slip due to fluid injection and indicate a certain degree of influence of the damage zone with anisotropic characteristic on fluid-induced fault slip.

Keywords: Numerical simulation, Fluid injection/extraction, Fault slip, Discrete Fracture Network (DFN), Heterogeneous permeabilities

1 INTRODUCTION

As Japan has many volcanic zones, geothermal power generation is being developed as abundant renewable energy resource. However, fluid injection/extraction near faults from geothermal power generation can cause seismic events and significantly affect the mechanical and hydraulic properties of the rock mass. These seismic events can be caused by fault slip due to fluid pressure change. For this reason, it is important to efficiently predict and distinguish aseismic and seismic fault slip caused by fluid injection/extraction.

In this study, numerical simulations considering fault slip due to fluid injection were conducted using the coupled Extended Finite Element Method (X-FEM), which considers flow rate along the fault as well as into rock. A comparison is made between the results from a model that considers heterogeneous permeabilities, in the fault damage zone, derived from the Discrete Fracture Network (DFN), and a model that uses an equivalent homogeneous permeability, in the damage zone, with the aim of quantifying the influence of fractured fault rocks with anisotropic characteristics.

*Corresponding author: pro7954@gmail.com

DOI: 10.1201/9781003293590-48

2 DISCRETE FRACTURE NETWORK (DFN) MODEL

The Discrete Fracture Network (DFN) is a stochastic method of fracture simulation that can produce stochastically simulated fractures. DFN can produce realistic and stochastically similar discrete models based on limited field data. In addition, DFN represents a computational model that explicitly represents the geometric properties of each individual fracture (e.g. orientation, size, intensity) and the topological relationship between individual fractures and sets of fractures (Lei et al., 2017).

In this study, the one-dimensional fracture frequency P_{10} was converted into a two-dimensional fracture frequency P_{21} by applying Eq. (1) (Hirose et al., 2013).

$$P_{21} = P_{10}/0.3 \tag{1}$$

The function of fracture size distribution n is shown in Eq. (2).

$$n(l) = \alpha \cdot l^{-a} \tag{2}$$

where, α is fracture density, a is scaling index (the power law exponent is always negative, equal to $-a$, so that the scaling index is defined positive). The distribution is defined between l_{min} and l_{max}, the lower and upper bounds of the fracture sizes, respectively.

The function of probability density of the fracture length l is shown in Eq. (3).

$$F(l) = \frac{l_{min}^{1-a} - l^{1-a}}{l_{min}^{1-a} - l_{max}^{1-a}} \tag{3}$$

where, l_{min} is the minimum fracture length, and l_{max} is the maximum fracture length. The scaling index is an index in which the larger value results in the more short-length fractures and the smaller value results in the more long-length fractures (Gutierrez, M. and Youn, D.J., 2015).

The function of probability distribution of the fracture angle θ is shown in Eq. (4).

$$\frac{\partial F}{\partial \theta} = \frac{k \sin \theta e^{k \cos \theta}}{e^k - e^{-k}} \tag{4}$$

where, θ is the angular deviation from the mean vector, kappa k is the Fisher constant or dispersion factor. The Fisher distribution constant is a constant in which the larger value results in the larger the deviation of the angle, and the smaller the value results in the smaller the deviation of the angle (Gutierrez, M. and Youn, D.J., 2015).

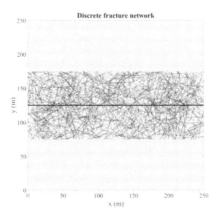

Figure 1.   DFN model (P_{10} = 0.3/m).

Table 1. DFN model parameters.

Parameter	Value
Number of fractures per m P_{10} (/m)	0.3
Scaling index (-)	3.5
α (-)	1
Minimum length (m)	1
Maximum length (m)	50
Kappa (Fisher distribution constant)	0.03
Aperture for X-FEM permeability tensor (m)	2.0×10^{-5}
Initial aperture for X-FEM permeability tensor (m)	2.0×10^{-5}

The DFN model produced using the above equation and the DFN model parameters are shown in Figure 1 and Table 1, respectively.

3 NUMERICAL SIMULATION METHOD

The coupled Extended Finite Element Method (X-FEM) modeling was conducted, because it is particularly computationally efficient when accommodating discontinuities. Compared to the conventional Finite Element Method (FEM), X-FEM can represent individual fractures implicitly without complicated meshing, thereby reducing the calculation time. In particular, the X-FEM approach enhances the FEM model by providing additional Degrees of Freedom (DOF) to the nodes of elements intersected by discontinuities. Therefore, a single mesh can be used for discontinuities in any length and orientation.

In order to represent the heterogeneous distribution of the components of the fractured rock in the numerical simulation, the mechanical properties of the fracture are represented in the form of a tensor based on the fracture tensor theory proposed by Oda (1986). This theory is based on two basic assumptions. First, individual fractures are represented by tiny flaws in the elastic continuum. Second, fractures are represented by twin parallel fracture walls connected by springs in both shear and normal deformation (Gan and Elsworth, 2016).

Considering that randomly distributed fractures can be intersected by multiple elements, the directional fracture permeability is defined as the permeability tensor k_{ij}, which can represent the orientation of the fracture and the explicit fracture volume intersecting any element block. The function of permeability tensor is shown in Eq. (5).

$$k_{ij} = \sum^{fracnum} \frac{1}{12}\left(P_{kk}\delta_{ij} - P_{ij}\right) = \sum^{fracnum} \frac{1}{12}\left(\frac{V_{ratio}}{b_{ini}}b^3 n_k^2 \delta_{ij} - \frac{V_{ratio}}{b_{ini}}b^3 n_i n_i\right) \tag{5}$$

where, *fracnum* is the number of fractures truncated in an element block, P is the basic fracture tensor, δ is the Kronecker's delta, V_{ratio} is the volumetric ratio of the truncated fracture over the element volume, b_{ini} is the initial aperture of fracture, b is the aperture of the fracture, n is the unit normal to each fracture (Gan and Elsworth, 2016).

The schematic of fault and numerical model is shown in Figure 2. The grid size was 50×51 elements in the x and y directions, respectively, with more elements in the y direction than the x direction so that the strong discontinuity crossed through the center of the elements. The model dimensions are 250 m in both x and y directions. The in-situ stresses are $\sigma_{xx} = -5.649$ MPa, $\sigma_{yy} = -3.351$ MPa and $\sigma_{xy} = 0.964$ MPa (negative values are compressive). The boundaries are fixed because it is remote from the fluid injection source. The model is run with a time increment of 5 s, with dynamic analysis time increment of 1×10^{-4} s. This time increase is the largest that produces stable and consistent results. The flow rate was set at 3×10^{-4} m^3/s (18 liters per minute).

The properties of rock parameters and fault parameters are shown in Tables 2 and 3, using the ranges given in Schwartzkopff et al. (2020).

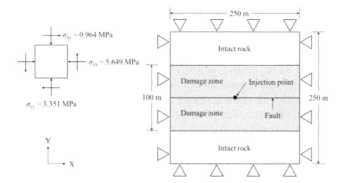

Figure 2. Schematic of fault and numerical model (Schwartzkopff et al., 2020)

Table 2. Rock parameters.

Parameter	Value
Intact elastic modulus E (GPa)	29.18
Poisson's ratio v	0.33
Density ρ_s (kg/m^3)	2364
Porosity n (%)	14.25
Biot poroelastic constant α_{Biot}	0.77

Table 3. Fault parameters.

Parameter	Value
Fault damage zone permeability for the homogeneous model k_f (m^2)	1.34×10^{-15}
Fault damage zone width (m)	50
Initial hydraulic aperture $2h_0$ (m)	4.71×10^{-5}
κ parameter	1.22
Apparent normal stiffness $\bar{k}_N$ (GPa/m)	27.74
Apparent tangential stiffness $\bar{k}_T$ (GPa/m)	11.5
Dilation angle φ_d (°)	19
Frictional coefficient μ_0	0.95 to 0.80 over 1 mm
a parameter	2.70×10^{-2}
b parameter	3.70×10^{-2}
Standard deviation of non-logarithmized fracture asperity heights σ_h (m)	2.00×10^{-4}

4 RESULTS AND DISCUSSION

The total displacement field for two models is shown in Figure 3 (a) and (b). It has been shown that the two models produced the similar magnitude of seismic fault slip, however the heterogeneous permeability model produced higher maximum seismic fault slip locally, attributed to the higher pressure regions along the fault. This resulted in the seismic to total slip ratio (the average dynamic shear movement divided by the average total shear movement) in the equivalent homogeneous permeability model of 0.27 % and 0.57 % in the heterogeneous permeability model. The reason for the different seismic to total slip ratio of the two models is that the higher pore pressure in the damage zone tends to produced more seismic fault slip. The maximum pore pressure at about 1,200 s the equivalent homogeneous permeability model was 3.15 MPa and the maximum pore pressure of the heterogeneous permeability model at approximately 1,200 s was 3.22 MPa, so that the pore pressure of the heterogeneous permeability model was higher locally. Therefore, this higher pore pressure along the fault produced more seismic fault slip. This high local fault slip is attributed to the permeability variability in the heterogeneous permeability model, where small changes in the permeabilities produce sections of the fault with higher (and lower) pressures compared with the equivalent homogeneous permeability model. Interestingly, the seismic fault slip mainly produced on the edge of the pressurized zone in both models, which aligns well with field observations (Figure 4 (a) and (b)). In addition, the heterogeneous permeability model illustrated that the seismic fault slip are produced in particular regions rather than evenly distributed. The maximum seismic fault slip movement of the equivalent homogeneous permeability model was 0.019 mm and for the heterogeneous permeability model it was higher at 0.024 mm. Therefore, the heterogeneous permeability model produced a more intense local seismic fault slip. However, the average seismic fault slip magnitude was almost the same for the two models. The average total fault slip for the homogeneous permeability model was 0.236 mm, whereas the heterogeneous permeability model produced 0.096 mm, because the slipping area is smaller in the heterogeneous permeability model. This produced a higher seismic to total fault slip ratio for the heterogeneous permeability model. These results imply the applicability of this coupled X-FEM approach to predict fault slip due to fluid injection and indicate a certain degree of influence of the damage zone with anisotropic characteristic on fluid-induced fault slip.

The relationship between P_{10} and seismic to total slip ratio is shown in Figure 5 (a). It was shown that the seismic to total slip ratio increased proportionally as the P_{10} increased. As P_{10} increased, the average total fault slip movement decreases proportionally due to the increase in permeability, but the average seismic fault slip movement is almost constant. This produced the seismic to total slip ratio, which increased proportionally with P_{10}. The trend line equation of relationship between P_{10} and the seismic to total slip ratio is shown in Eq. (6).

$$Seismic\ to\ total\ slip\ ratio(\%) = 0.336P_{10} + 0.227, R^2 = 0.32 \tag{6}$$

The relationship between maximum length and seismic to total slip ratio is shown in Figure 5 (b). It was shown that the seismic to total slip ratio was almost constant as the maximum length increased. As a result, it was confirmed that the maximum length did not significantly affect the seismic to total slip ratio. In addition, it was confirmed that the deviation of the seismic to total slip ratio occurred due to the influence of the DFN, which is assumed that because fractures are produced at random locations and random orientations around the injection point of the damage zone. Therefore, it is assumed that even the same P_{10} affects the seismic to total slip ratio.

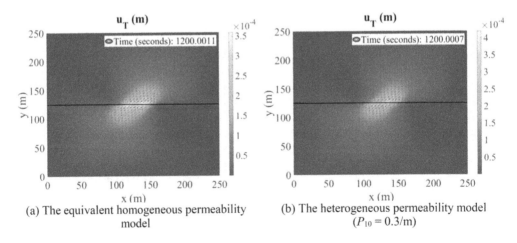

(a) The equivalent homogeneous permeability model

(b) The heterogeneous permeability model ($P_{10} = 0.3$/m)

Figure 3. The total displacement field for two models.

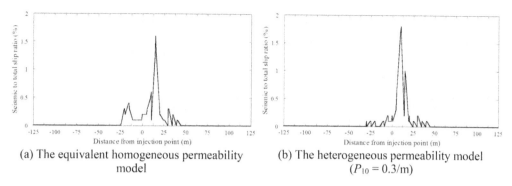

(a) The equivalent homogeneous permeability model

(b) The heterogeneous permeability model (P_{10} = 0.3/m)

Figure 4. Relationship between distance from injection point and the seismic to total slip ratio.

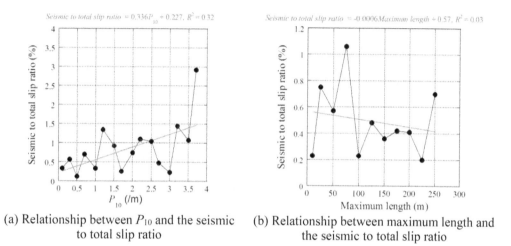

(a) Relationship between P_{10} and the seismic to total slip ratio

(b) Relationship between maximum length and the seismic to total slip ratio

Figure 5. Relationship between P_{10} and maximum length and the seismic to total slip ratio.

5   CONCLUSIONS

In this study, numerical simulations considering fault slip due to fluid injection were conducted using the coupled Extended Finite Element Method (X-FEM), which considers flow rate along the fault as well as into rock. It has been shown that the equivalent homogeneous and the heterogeneous permeability model produced the similar magnitude of the seismic fault slip however; the heterogeneous permeability model produced higher maximum seismic fault slip locally, attributed to higher pressure regions along the fault. The equivalent homogeneous permeability model produced lower maximum seismic fault slip but because it is over a larger seismic fault slip area than the heterogeneous permeability model it on average produced more total seismic fault slip. This produced the seismic to total slip ratio that is higher for the heterogeneous permeability model compared with the equivalent homogeneous permeability model. In addition, the seismic fault slip mainly produced on the edge of the pressurized zone in both models. These results imply the applicability of this coupled X-FEM approach to predict fault slip due to fluid injection and indicate a certain degree of influence of the damage zone with anisotropic characteristic on fluid-induced fault slip.

ACKNOWLEDGEMENT

This study was support by JSPS KAKENHI Grant Number 19KK0109.

REFERENCES

Gutierrez, M. and Youn, D.J., 2015, Effects of fracture distribution and length scale on the equivalent continuum elastic compliance of fractured rock masses, *J. of Rock Mech. and Geol. Eng.*, 7, 626–637.

Gan, Q. and Elsworth, D., 2016, Production optimization in fractured geothermal reservoirs by coupled discrete fracture network modeling, *Geothermics*, 62, 131–142.

Hirose, K., Edo, T. and Matsubara, H., 2013, Numerical mass-transfer simulation in widespread area considering its small area structures in fractured rock mass, *J. of Japan Soc. of Civil Eng., Ser. C*, 69, 367–377.

Lei, Q., Latham, J.P. and Tsang, C.F., 2017, The use of discrete fracture networks for modeling coupled geomechanical and hydrological behavior of fractured rock, *Computers and Geotechnics*, 85, 151–176.

Oda, M., 1986, An equivalent continuum model for coupled stress and fluid flow analysis in jointed rock mass, *Water Resources Research*, 22, 1845–1856.

Schwartzkopff, A.K., Sainoki, A. and Elsworth, D., 2020, Numerical simulation of an in-situ fluid injection experiment into a fault using coupled X-FEM analysis, *Rock Mech. and Rock Eng.*, 54, 1027–1053.

Rock Mechanics and Engineering Geology in Volcanic Fields – Ohta, Ito & Osada (eds)
© 2023 copyright the Author(s), ISBN 978-1-032-27657-1

Influence of Miocene igneous complexes on landslides in the southern part of the Kii Peninsula, southwestern Japan

Teruyuki Kikuchi*
J-POWER Design Co., Ltd., Tokyo, Japan

Katsumi Kimura
Fukada Geological Institute, Tokyo, Japan

Takato Takemura & Hinako Hosono
Nihon University, Tokyo, Japan

Chihaya Onda
Electric Power Development Co., Ltd., Tokyo, Japan

ABSTRACT: The aim of this study is to determine the relationship between intrusion of granite bodies, thermal metamorphism, and landslides. Omine granites and Kumano acid igneous rocks are widely distributed in the central to southern parts of the Kii Peninsula. These basic rock bodies have been active between 14 to 15 Ma. Research on the thermal history of hot springs and mines has progressed, and the estimation of large-scale high-temperature fluids in underground areas is often used in analysis.

Many landslides and collapses have been found in the Shimanto belt, which was deposited in the Cretaceous of the Mesozoic Era. It has been suggested that the location where landslides are concentrated is related to the location of granite bodies, but this has not been comprehensively investigated.

In this study, we focused on the illite crystallinity and physical characteristics of mudstone in the Ryujin and Miyama formations in the Shimanto belt and the adjacent Kumano formation, and verified the range of the thermal influence of Miocene igneous complexes and the location of landslides. Based on the results, it was clarified that there is a clear relationship between the distribution of illite crystallinity, and the location of landslides.

Keywords: Illite crystallinity, Landslide, Thermal metamorphism, Kumano acid igneous rocks

1 INTRODUCTION

A large number of floods have occurred in the Kumano-gawa (kawa or gawa means "river" in Japanese) system, which flow through the southern part of the Kii Peninsula in central Japan (Chigira et al, 2013). This flooding is typically accompanied by mass-wasting and landslides, which affect the environment, security, and social infrastructure of the surrounding area. The topography and geological features of this area have geologically advanced characteristics, such as complex geologic structures, numerous landslide topographies, and many hot springs and mines. However, as of date, no discussions are available regarding the relationships between the topographical and geological features and the landslide topography distribution. Thus, this study

*Corresponding author: kikuchi-t@jpde.co.jp

DOI: 10.1201/9781003293590-49

focuses on the region in the granitic rocks where thermal metamorphism occurred, which was active during the Miocene, and attempts to clarify its relationship with the occurrence of turbid water.

The Kumano-gawa branches off into the Kitayama-gawa and the Totsugawa 23 km from the river mouth at Shingu, in the Wakayama Prefecture. The catchment areas of the Totsugawa river and the Kitayama-gawa are 1,288 km^2 and 782 km^2. Most of the Kitayama-gawa basin is a contact thermal metamorphic zone of Omine Igneous rocks in a Kumano formation (Hashimoto and Kimura, 1999). In this basin, the production of anthracite and a high heated temperature due to vitrinite have been observed in the Kishu mine (Suzuki et al, 1982). However, such a thermal metamorphic zone cannot be found within the Totsugawa basin. The actual situation regarding the spatial distribution of such thermal metamorphism is unclear. In addition, there are few data relating to the degree of the thermal metamorphism and the weathering tolerance and rock strength. Thus, in this study, we investigate the following question to elucidate the relationship between the landslide topography distribution and the geological features: can the spatial distribution of the degree of thermal metamorphism perceived be presumed to be its heat source? The characteristics of the degree of thermal metamorphism in each basin are presented on a geological map, and their relationship with the landslide topography distribution is examined. The crystallinity of illite in the mudstone samples collected during a fieldwork study is called the IC (illite crystallinity) value and is used as the degree of thermal metamorphism to answer the above question.

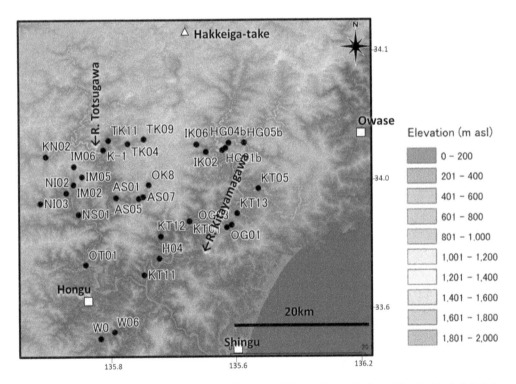

Figure 1.   Elevation map of the Kii peninsula in Japan. IC samples are indicated by the black dots (•).

2   STUDY AREA

Mt. Hakkeiga-take (elevation: 1,915 m) is the highest peak in the Kii Mountains as shown in Figure 1, and mountains with similar elevations form the ridge. A paleo-surface can be found in the southern and the western parts of the Kumano-gawa basin. This paleosurface is accompanied by the formation of a broad-based knick line; deep-seated landslides have occurred at many locations (Hiraishi and Chigira, 2009). The most prominent geological feature of the Kii Mountains is that it is located in the outer zone of southwest Japan, south of the Median Tectonic Line (Tsou et al, 2017). The Shimanto belt is separated into the Hidaka River zone, comprising the accretionary complex of the Cretaceous and the accretionary complex of the Paleogene by the Gobo–Hagi tectonic line and the Itaya-fault (Kimura, 1986). The Omine graphitic rocks that are distributed for a long distance in the north–south direction in the eastern part of the central Kii Peninsula pierce through the Cretaceous and the Paleogene formations (Hara and Hisada, 2007). Since the sedimentary rocks of the accretionary complex come into contact with granitic rocks, they may be affected by thermal metamorphism (Aramaki and Hada, 1965; Kawasaki, 1980). A gravity survey predicted the distribution of a granitic body in the deep underground (Umeda et al, 2003), and, in particular, it is thought that a high-temperature fluid present in the deep underground contributes to the low-temperature alteration observed in the Hongu and Totsugawa districts and acts as a heat source for hot springs (Hanamuro et al, 2008). Although there are no Quaternary volcanoes in the southern Kii Peninsula, high-temperature hot springs gush out in multiple locations around Hongu (Kanehara,1992). A relationship has been identified between hot springs and the distribution of groups of dykes of igneous rock from the Miocene.

A contrasting density is seen in the distribution of landslides in the Shimanto belt, and there is a high distribution frequency in the Totsugawa river basin. This frequency tends to be slightly higher in the northern part of the Kumano formation of the Paleogene (Figure 2, National Research Institute for Earth Science and Disaster Resilience,2015). As shown in Table 1, the location of the landslide topography distribution is sorted in the left and right banks of the Kitayama and Totsugawa rivers, which flow in the north–south direction. This result shows that the landslide topography number per unit area decreases toward the left bank of the Kitayama-gawa (western side) from the right bank of the Totsugawa (eastern side). In particular, the number per unit area and the number of locations where the landslide topography occurs in the right bank of the Totsugawa river increase by 34% and 5.8 times, respectively, compared with those on the left bank of the Kitayama-gawa.

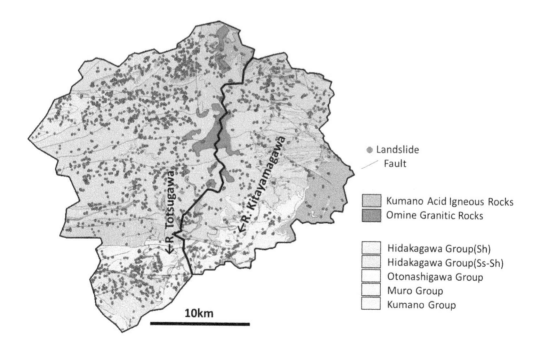

Figure 2. Simplified geological map of the Kii peninsula made from Landslide distribution map of National Research Institute for Earth Science and Disaster Resilience,2015 and Catchment area.

Table 1. Number of landslide topography distributions by Catchment area A: Catchment area, N: Number of landslides.

Catchment area	River side	A (km^2)	N	N/A (/km^2)
Totsu-gawa	right	796.0	1402	1.76
	left	491.8	799	1.62
Kitayama-gawa	right	317.0	381	1.20
	left	465.4	240	0.52

3 METHODOLOGY

Smectite in mudstone in the accretionary complex deposit forms illite through thermal metamorphism, and the crystallinity of illite can be used in measuring the degree of the thermal metamorphism (Otsuka and Watanabe, 1982; Tanabe and Kano,1996). The illite crystallinity can be quantified with the half-value width of the peak of the (001) surface of illite (lattice spacing: 10 Å), and it is expressed as an illite crystallinity (IC) value. In this study, the measurement of the IC value was measured using an X-ray diffractometer (Miniflex, Rigaku Corporation). The measurement conditions were as follows: voltage, 30 kV; current, 15 mA; anticathode element, CuKα; step angle, 0.01°; measurement time, 1.5 sec; and measurement angle (2θ), 3° to 14°. The making of the test samples for the IC value measurement was in accordance with Hara and Kimura (2000), and liquid paste with concentrated clay mineral was applied in a region on a slide glass with a constant area (5.5 cm^2). The results obtained are expressed as the half-value width of the primary peak of the (001) surface (Δ2°θ). The higher the crystallinity, the more uniformly the lattice arrangement of clay mineral, the sharper the peak, and the smaller the half-value width. The international standards of the degree of metamorphism of the IC value (0.2–0.3, 0.3–0.4, and 0.4 and above) correspond to epizone (pumpellyite phase and actinolite phase), anthizone (prehnite phase and pumpellyite phase), and diagenetic zone, respectively. In terms of the maximum heated temperature, 0.2–0.4 corresponds to approximately 300–200 degrees. Since the strength and tolerance to weathering of the rock relate closely to the degree of the thermal metamorphism of the sedimentary rock, these can be an important index in identifying the relationship with the distribution of landslide.

4 RESULT AND DISCUSSION

Shale samples for the IC value determination were collected in the Shimanto belt as follows: 24 samples from the Miyama formation and the Ryujin layer; one sample from the Muro formation and the Otonashi-gawa formation; and 5 samples from the Kumano formation (Figure 3). These sampling locations cover the western part of the Kii Peninsula where there is no effect of thermal metamorphism on abyssal rocks. According to Awan and Kimura (1996), the IC values of the sedimentary rocks in the Shimanto belt (0.48–0.60) correspond to the diagenetic zone, except in the Hanazono layer where schistosity has been developed; this zone is an area in which no effects of the abyssal rocks, such as the Omine granite, can be observed. The IC values indicate the degree of thermal metamorphism of average sedimentary rocks in the Shimanto belt in this area. Therefore, the effect of thermal metamorphism owing to intrusive rocks is evaluated by comparison with these IC values (0.48–0.60). The IC values in Section A of the samples collected from the area along the Kanno-gawa over the Taki-gawa where the Miyama layer is distributed were 0.54 at KN04, 0.52 at K01, and 0.44 at TK04, showing a change that they gradually decrease eastward [Figure 4(A)]. This reflects a change in the thermal metamorphism caused by the intruding rocks. Meanwhile, the thermal metamorphism was reflected at AS07, OK8, and KT12 near the intruding rocks (where the IC values were less than 0.35) in the Ryujin layer shown in Section B. Meanwhile, the IC values at AS01, NS01, and IM05, which are located away from the area where the intruding rocks are distributed, indicate a relatively high degree of crystallization [Figure 4(B)].

5 CONCLUSIONS

Spatial distribution of the degree of thermal metamorphism is displayed as a contour diagram of the illite crystallinity (IC value). The distribution landslide topography is also depicted in Figure 3 to show the relationship with the IC value. Since the zone with the IC value of 0.4 and lower is the zone equal to the granite contact thermal metamorphic zone, it is referred to as a high-temperature tectonic belt. A high-temperature tectonic belt can be found in the metamorphic zone of the Omine graphitic rocks that are distributed along the Omine mountains (a mountainous region of the Kitayama-gawa basin), with the IC value contour running parallel to it. Furthermore, the high-temperature tectonic belt parallel to the Omine mountains is present in the east. In this high-temperature tectonic belt, the sedimentary rocks are solidified and silicified due to the contact thermal metamorphism of granite and are changed into firm hornfels. Shear strength, tensile strength, and weathering resistance are all of high. Since the original bedding surfaces and the stripped surfaces are healed together and are not mechanically weak, the vertical joints act as weak surfaces from the viewpoint of geological structure. Instead, the vertically oriented developed joint becomes the weak surface. A sheer sharp peak found on the opposite shore from IK06 on the Ikego-gawa and a vertical rock face, such as "*Dorohaccho* "(KT12) in the lower Kitayama-gawa, are formed. Since such slopes are steep, and because the rocks have strong weathering resistance, forming the landslide topography is difficult, and little distribution is present.

Meanwhile, in the tributary basin that joins the right bank of the Totsugawa, areas with an IC value of 0.4 or more are predominant, which indicates that a developed high-temperature tectonic belt does not exist in the Kitayama-gawa basin. As shown Figure 3 in the area near K01, the actual exposure shows a clear sedimentary structure, and it is a stripped surface of black shale. There is no trace of undergoing the thermal metamorphism. Therefore, no contact thermal metamorphism is caused by granite; the shear

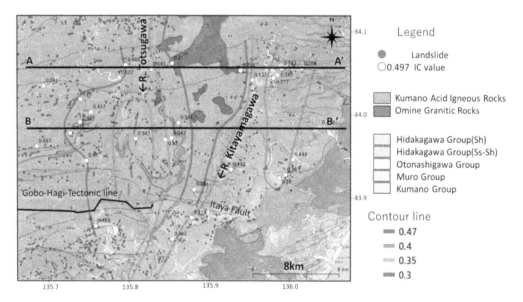

Figure 3. Illite crystallinity (IC value) and contour diagram of the Kumano River basin. A-A' and B-B': Position of cross section.

and tensile strengths of the rock are weaker than those of the rock in the high-temperature tectonic belt, and the weathering tolerance is weak. In particular, when pressure is released when the rocks are exposed in the subsurface, the bedding surface and the stripped surface (the weak surfaces of the sedimentary rocks) are loosened. This enables surface water to easily penetrate the rocks, which causes weathering and weakening to progress selectively. This selective weathering and weakening caused by phenomena such as earthquakes or heavy rains result in landslides and gravity transformation of mountain bodies. In this fashion, a difference in the distribution of the landslide topography bordering at the IC value of 0.4 is seen. In particular, the tendency for IC values to be higher and a greater number of landslide areas was most noticeable in the right bank of Totsugawa; the effect of thermal metamorphism and the distribution of landslide topography almost matched in this area.

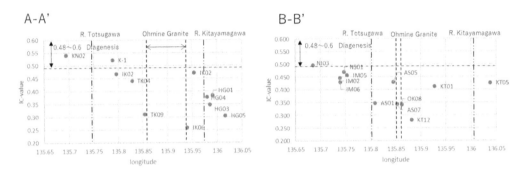

Figure 4.   Cross section of illite crystallinity (IC value).

ACKNOWLEDGEMENTS

Discussions with Masahiro Chigira of the Fukada Geological Institute, Tokyo and Teruyoshi Hatano of Nakanihon air co., ltd. were very helpful.

REFERENCES

Aramaki, S. and Hada, S., 1965, Geology of the central and southern parts of the acid igneous complex (Kumano Acidic Rocks) in southeastern Kii Peninsula, *J Geol. Soc. Japan*, 71, 494–512 (in Japanese with English abstract).

Awan, M.A. and Kimura, K., 1996, Thermal structure and uplift of the Cretaceous Shimanto Belt, Kii Peninsula, Southwest Japan: An illite crystallinity and illite bo lattice spacing study, *Island arc*, 5, 1, 69–88.

Chigira, M., Tsou, C.-Y., Matsushi, Y., Hiraishi, N. and Matsuzawa, M., 2013, Topographic precursors and geological structures of deep-seated catastrophic landslides caused by Typhoon Talas, *Geomorphology*, 201, 1, 479–493.

Hanamuro, T., Umeda, K., Takashima, I., Negishi, Y., 2008, The high temperature hot springs such as in Yunomine and Tosenji and the hydrothermal alteration zones of the Hongu area are distributed in the southern part of Kii peninsula, *Japanese Magazine of Mineralogical and Petrological Sciences*, 37, 2, 27–38 (in Japanese with English abstract).

Hara, H., Hisada. K., 2007, Tectono-metamorphic evolution of the Cretaceous Shimanto accretionary complex, central Japan: Constraints from a fluid inclusion analysis of syn-tectonic veins. *The Island Arc*. 2007, 16, 1, 57.

Hara, H., Kimura K., 2000, Estimation of errors in measurement of illite crystallinity: the limits and problems of application to accretionary complexes. *J Geol Soc Japan*, 106, 4, 264–279. (in Japanese with English abstract) 10.5575/geosoc.106.264

Hashimoto, Y., Kimura, G., 1999, Underplating process from melange formation to duplexing: example from the Cretaceous Shimanto Belt, Kii Peninsula, southwest Japan. *Tectonics* 18, 92–107.

Hiraishi, N., Chigira, M., 2009, Topographic evolution indicated by the distributions of knickpoints and slope breaks in the tectonically active Kii Mountains, southwestern Japan, *EGU General Assembly 2009, held 19-24 April, 2009 in Vienna, Austria*, http://meetings.copernicus.org/egu2009. 6722

Kanehara, K., 1992, Distribution Map and catalogue of hot and mineral springs in Japan, *National Institute of Advanced Industrial Science and Technology (eds.)*, 394

Kawasaki, M., 1980, Omine acid rocks, Kii peninsula —Geology and major element chemistry. J. *Japan Assoc. Min. Petr. Econ. Geol.*, 75. 86–102

Kimura, K.,1986. Stratigraphy and paleogeography of the Hidakagawa Group of the Northern Shimanto Belt in the southern part of Totsugawa village, Nara Prefecture, southwest Japan. *Journal Geological Society of Japan*, 92, 3, 185–203 (in Japanese with English abstract).

National Research Institute for Earth Science and Disaster Resilience,2015, Landslide distribution map (1:50000).

Otsuka,T.,and Watanabe, K., 1982, Illite crystallinity and low-grade metamorphism of pelitic rocks in the Mino Terrane, central Japan, *Mining Geology*, Vol.32, No.171, pp.55–65, DOI: https://doi.org/10.11456/shigenchishitsu1951.32.55 (in Japanese)

Suzuki, S., Oda, Y., Nambu, M., 1982, Thermal alteration of vitrinite in the Miocene sediments of the Kishu mine area, Kii Peninsula,Japan, *Mining Geology*, 32, 55–65 (in Japanese with English abstract).

Tanabe, H., Kano, K., 1996, Illite crystallinity study of the Cretaceous Shimanto Belt in the Akaishi Mountains, eastern Southwest Japan, *Island arc*, 5, 1, 56–68.

Tsou, C.Y., Chigira, M., Matsushi, Y., Hiraishi, N., Arai, N., 2017, Coupling fluvial processes and landslide distribution toward geomorphological hazard assessment: a case study in a transient landscape in Japan, *Landslide*, 14, 1901–1914.

Umeda, K., Uehara, D., Ogawa, Y., Kudo, K., Kakuda, C., 2003, Deep Structure of the Miocene Igneous Complex in the Kii Peninsula, SouthwestJapan, Inferred from Wide-band Magnetotelluric Soundings, *Volcanological Society of Japan*, 48, 461–470 (in Japanese with English abstract).

Rock Mechanics and Engineering Geology in Volcanic Fields – Ohta, Ito & Osada (eds)
© 2023 copyright the Author(s), ISBN 978-1-032-27657-1

Applicability of SBAS-DInSAR for monitoring displacements of a local steep slope

Kanta Nagasaki*, Mitsuo Kameyama, Myuji Yoshimoto, Shinichiro Nakashima & Norikazu Shimizu
Yamaguchi University, Ube, Japan

ABSTRACT: This paper investigates the applicability of DInSAR for monitoring the displacements of a steep slope along a roadway in a small area. Since DInSAR is usually applied for detecting displacements in extensive areas (e.g., more than 100 square kilometers), it is important to find a way to apply it to smaller targets (e.g., less than a square kilometer), which are often faced in Rock and Geotechnical Engineering. The SBAS-DInSAR method was employed as a multi-temporal analysis and applied to monitor a steep slope along a roadway. ALOS-2 SAR data, observed from 2014 to 2020, were used. Comparing the displacements obtained by DInSAR and GPS in the same area, the results showed a good agreement with a discrepancy of around 10-20 mm. It was found that SBAS-DInSAR has the potential to be applied for monitoring the displacements of local steep slopes.

Keywords: displacement monitoring, DInSAR, SBAS, steep slope

1 INTRODUCTION

Displacement monitoring is important for assessing the stability of slopes and for predicting the future behavior of the slopes. Various types of instruments are available for monitoring displacements in Rock and Geotechnical Engineering, such as extensometers, inclinometers, laser distance meters, etc. In addition, satellite technology, i.e., GPS (Global Positioning System) and SAR (Synthetic-Aperture Radar), has begun to be used for the monitoring of extensive areas. In particular, GPS has often been utilized as a monitoring system (Shimizu and Nakashima, 2017). It can continuously monitor three-dimensional displacements in real time with mm accuracy (Shimizu et al., 2014).

SAR is a high-resolution radar device that is mounted on an artificial satellite. It transmits pulse waves to the Earth's surface and receives the reflections. Interferometric SAR (InSAR) is a method for taking the signal phase difference from two scenes of SAR data to produce an interferogram, the two scenes being observed in the same area at different times. DInSAR (Differential InSAR) is a technique for detecting the displacements of the Earth's surface by removing the contribution of the topography and the distance between two satellite positions from the interferogram obtained in the InSAR process. It can provide the displacement distribution over huge areas, which are more than hundreds of square kilometers in size, without the use of any sensors on the ground surface (Ferretti, 2014).

Many applications of DInSAR have already been studied for detecting the distribution of ground surface displacements. The authors have also applied DInSAR to monitor ground subsidence (Yastika et al., 2019; Parwata et al., 2020b), landslides (Yastika et al., 2019; Yamaguchi et al., 2021), and a mining slope (Parwata et al., 2020a). However, there are still

*Corresponding author: a009veu@yamaguchi-u.jp

DOI: 10.1201/9781003293590-50

difficulties that must be overcome in order to use DInSAR as an engineering monitoring tool in practice, i.e., phase decorrelation depending on the ground surface conditions, evaluation of the reliability of the results, etc. Further studies are required for improving the method and confirming its reliability.

This study investigates the applicability of DInSAR for monitoring the displacements of a steep slope along a national roadway in Japan. As the target area is less than a square kilometer, it is challenging to reliably detect the displacements of a steep slope in such a small area with dense vegetation using DInSAR. The ALOS-2 SAR satellite data, observed from 2014 to 2020, are used. The results are compared with the displacements measured by a GPS continuous monitoring system in the same area, and the reliability of the DInSAR results is discussed.

2 OUTLINE OF DInSAR

SAR is a high-resolution radar device that is mounted on an artificial satellite. It can be used efficiently even at night and under any weather conditions. InSAR is a method for taking the signal phase difference from two scenes of SAR data, which are observed in the same area at different times. DInSAR is a technique for providing the displacement distribution of the Earth's surface by removing the contribution of the topography and the distance between two satellite positions from the interferogram obtained in the InSAR process (Ferretti, 2014).

The SAR satellite travels on ascending (northward) and descending (southward) orbits (Figure 1), and DInSAR produces the displacement in the Line of Sight (LOS) direction along the radar beam direction (Figure 2).

The Small Baseline Subset (SBAS) DInSAR (Berardino et al., 2002) is employed in this study to obtain the spatial distribution and temporal transition of the surface displacements. SBAS-DInSAR utilizes a series of SAR images in a target area and conducts multiple DInSAR processing to produce the time-series displacement distribution.

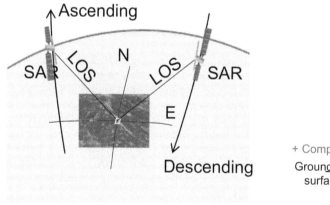

Figure 1.   Ascending and descending orbits.

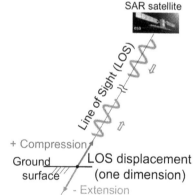

Figure 2.   Line of sight (LOS) displacement.

3 STUDY AREA AND SAR DATA

3.1 *Study area*

The study area is a steep slope, 150 m wide, 300 m long, and 120 m high, along a Japanese national roadway (Figure 3). The slope is composed primarily of rhyolite and granite which formed in the Cretaceous period, and its surface is partially covered with a colluvial deposit. The vegetation extends over the slope, which may create an undesirable condition in terms of the reflection of the SAR microwave. Since the 1970s, the slope has sometimes collapsed, deposits have flowed, and the road has been damaged. To reduce the risk to traffic along the road, a check dam was constructed to collect colluvial deposits, and a rock shed was built to protect the road.

Figure 4 shows the ranges in ALOS-2 SAR data for the ascending and descending orbits including the study area.

Figure 3. Local steep slope.

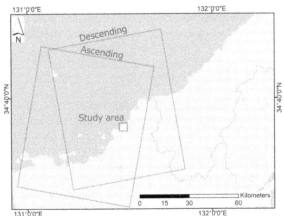

Figure 4. Study area and ranges in SAR data (ALOS-2).

3.2 *SAR data set*

In this research, the SAR dataset from the ALOS-2 satellites (1.2GHz L-band radar), operated by JAXA (Japan Aerospace Exploration Agency), is employed to monitor the displacements of the target slope. The numbers of SAR images used in this research are 16 in the ascending orbit (November 6, 2014 to March 4, 2021) and 26 (February 9, 2015 to March 1, 2021) in the descending one, respectively (Table 1).

The number of possible combinations of SAR pairs for the multiple DInSAR processing is based on the thresholds of the spatial and temporal baselines. In this research, the temporal baseline threshold is 360 days, and the spatial baseline thresholds are 450 and 600 meters for the ascending and descending orbits (20% of the critical baseline length), respectively. The temporal baseline threshold is quite a bit larger than that in a standard SBAS-DInSAR analysis, because many pairs of SAR data are needed to produce the interferograms even for such a small number of images. Figure 5 shows the connection graphs representing the pairs of SAR data. The total numbers of pairs to obtain the time-series analysis are 29 and 78 for the ascending and descending orbits, respectively. All the SAR data processing is conducted using Envi-SARScape 5.6. (Harris Geospatial Solutions, Inc.).

4 RESULTS AND DISCUSSION

4.1 *Spatial distribution of LOS displacement*

Figures 6 and 7 show the spatial distributions of the LOS displacements for about 5 years in the target area obtained by SBAS-DInSAR using the ascending and descending data, respectively. The resolution (unit of pixels) of the ground surface is 10m x 10m. The displacements are represented by the range in colors from red to blue denoting mm units from −200 mm to 200 mm. The red and blue colors mean negative (extension) and positive (compression) LOS displacements (see Figure 2), respectively.

Table 1. List of SAR data.

Ascending				Descending			
No	Data	No	Data	No	Data	No	Data
1	2014-11-06	14	2020-03-05	1	2015-02-09	14	2017-11-13
2	2015-04-09	15	2020-06-11	2	2015-02-23	15	2018-03-05
3	2015-07-02	16	2021-03-04	3	2015-09-07	16	2018-06-11
4	2015-12-17			4	2015-09-21	17	2018-08-20
5	2016-03-24			5	2015-11-30	18	2018-11-12
6	2016-06-30			6	2016-03-07	19	2019-03-04
7	2016-12-01			7	2016-04-18	20	2019-08-19
8	2017-03-09			8	2016-05-02	21	2019-11-11
9	2017-06-15			9	2016-06-13	22	2020-03-02
10	2018-04-05			10	2016-11-14	23	2020-06-08
11	2019-03-07			11	2017-03-06	24	2020-08-17
12	2019-06-13			12	2017-06-12	25	2020-11-09
13	2019-11-28			13	2017-08-21	26	2021-03-01

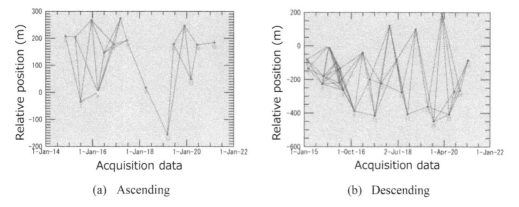

(a) Ascending   (b) Descending

Figure 5. Connection graphs of ascending and descending data for SBAS-DInSAR.

In the area with low coherence below the threshold value (0.3), the results of the LOS displacement have been removed from the map. The coherence is an index, taken as 0 to 1, to represent whether a good interferogram has been produced. A higher value indicates a better condition for obtaining reliable results. It is found that the area not showing results extends to the north-east of the target area due to low coherence. It may have been caused by both the dense vegetation and the unfavorable relationship between the slope direction and the direction of the microwave transmission.

This paper focuses on investigating the validity of the displacements obtained by SBAS-DInSAR at the target slope.

4.2 *Comparison between displacements by SBAS-DInSAR and GPS*

In order to investigate the validity of the results of SBAS-DInSAR, the LOS displacements are compared with those measured by the GPS displacement monitoring system. Since the system continuously provides three-dimensional displacements, they are converted to the LOS direction. Figures 8 and 9 provide comparisons between SBAS-DInSAR and GPS displacements at G-3 and G-4 as examples in the ascending and descending LOS directions, respectively.

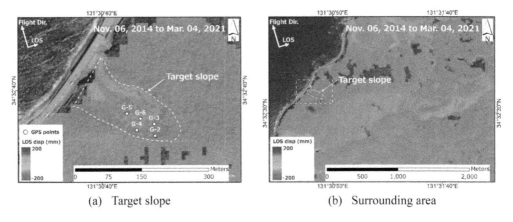

(a) Target slope   (b) Surrounding area

Figure 6. LOS displacement distribution by SBAS-DInSAR (Nov. 6, 2014 to Mar. 4, 2021) (Ascending).

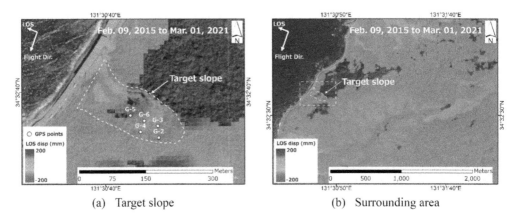

(a) Target slope   (b) Surrounding area

Figure 7. LOS displacement distribution by SBAS-DInSAR (Feb. 9, 2015 to Mar. 1, 4) (Descending).

The discrepancies between the two sets of displacements are within 10 mm and 20 mm for ascending and descending results, respectively. Figure 10 shows the correlation between SBAS-DInSAR and GPS at all GPS monitoring points, namely, G-2, G-3, G-4, G-5, and G-6. Although a few results are largely scattered, it is found that the SBAS-DInSAR results agree well with the GPS results. This means that DInSAR can be applied for monitoring the displacements at local steep slopes.

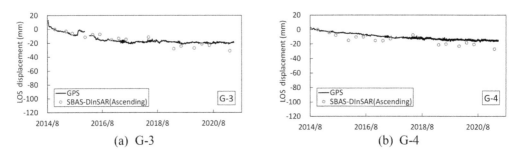

(a) G-3   (b) G-4

Figure 8. Time-transition of LOS displacements by SBAS-DInSAR (ascending) and GPS.

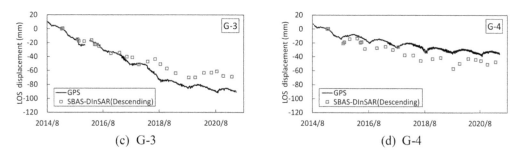

Figure 9. Time-transition of LOS displacements by SBAS-DInSAR (descending) and GPS.

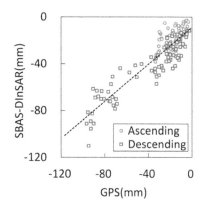

Figure 10. Correlation of LOS displacements obtained by SBAS-DInSAR and GPS.

Figure 11 shows the time-transition of the LOS displacements at G-3 and G-4 in both ascending and descending directions. At G-4, both descending and ascending LOS displacements increase toward the negative side (extension) with time and the descending LOS displacement is similar to or slightly larger than the ascending one (Figures 11(b)). This indicates that the actual displacement moves slightly west and downward.

At G-3, both descending and ascending LOS displacements also increase toward the negative side (extension) with time and the descending LOS displacement is quite a bit larger than the ascending one (Figures 11(a)). This indicates that the actual displacement moves west and downward.

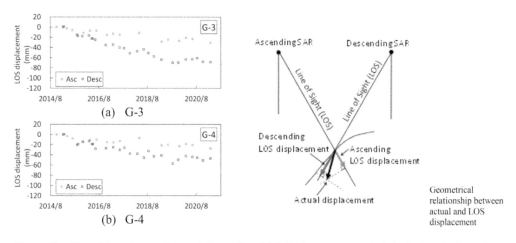

Figure 11. Transition of ascending and descending LOS displacements at pixels including GPS monitoring points (G-3 and G-4).

411

From the above discussion, it is found that the target slope shows typical sliding landslide behavior downward in the slope direction. Therefore, SBAS-DInSAR could be an effective method for monitoring the displacements of smaller areas of less than a square kilometer and for assessing the stability of slopes.

On the other hand, from Figures 8(b) and 9(b), the displacements increase along the mountain ridges and small valleys (yellow and red areas). Such behavior does not seem natural. The coherence in those areas is lower than that in other areas, because it may be influenced by vegetation and errors in the digital elevation model of the topography, which was used in this analysis. Further discussions will be needed to evaluate the reliability of the results in this area.

5 CONCLUSIONS

In this study, the applicability of DInSAR for monitoring the displacements at a local steep slope, which extends over a rather small area (150m x 300m), has been investigated. SBAS-DInSAR, using ALOS-2 data, was applied to obtain the time-transition of the distribution of spatial displacements.

Comparing the results with the displacements continuously monitored for five years by GPS, the discrepancy was within 10-20 mm. Therefore, SBAS-DInSAR was found to have the potential to be an effective method for monitoring the displacements of local steep slopes.

On the other hand, non-negligible amounts of displacement were found along mountain ridges and small valleys. They seem to have been influenced by dense vegetation and errors in the digital elevation model. Further discussions are needed to evaluate the reliability of the results.

ACKNOWLEDGEMENTS

This research was partially supported by JSPS KAKENHI (Grant-in-Aid for Scientific Research, Japan Society for the Promotion of Science) Grant Number 16H03153. The authors express their gratitude to the Japan Aerospace Exploration Agency (JAXA) for providing the ALOS-2 SAR data based on the collaborative relationship between JAXA and Yamaguchi University. They also extend their appreciation to Dr. I. Nyoman Sudi Parwata, University of Udayana, Indonesia, a former student of Yamaguchi University, for his support in conducting the SBAS-DInSAR analysis, and to Ms. H. Griswold for proofreading the manuscript.

REFERENCES

Berardino, P., Fornaro, G., Lanari, R. and Sansosti, E., 2002, A new algorithm for surface deformation monitoring based on small baseline differential SAR interferograms, *IEEE Trans Geosci Remote Sens*, 40:2375–2383.

Ferretti, A., 2014, Satellite InSAR data: reservoir monitoring from space, EAGE Publication, The Netherlands.

Parwata, I N. S., Nakashima, S., Shimizu, N. and Osawa, T., 2020a, Effect of digital elevation models on monitoring slope displacements in open-pit mine by Differential Interferometry Synthetic Aperture Radar, *Journal of Rock Mechanics and Geotechnical Engineering*, 12(5):1001–1013.

Parwata, I N. S., Shimizu, N., Grujić, B., Zekan, S., Čeliković, R. and Vrkljan, I., 2020b, Monitoring the Subsidence Induced by Salt Mining in Tuzla, Bosnia and Herzegovina by SBAS-DInSAR Method, *Rock Mechanics and Rock Engineering*, 53(11):5155–5175.

Shimizu, N. and Nakashima, S., 2017, Review of GPS displacement monitoring in rock engineering, *Rock Mechanics and Engineering*, Volume 4, ed. Xia-Ting Feng, CRC Press, Chapter 19:593–626.

Shimizu, N., Nakashima, S. and Masunari, T., 2014, ISRM suggested method for monitoring rock displacements using the Global Positioning System, *Rock Mech Rock Eng*, 47:313–328.

Yamaguchi, M., Yastika, P. E., Shimizu, N., Milev, N. and Vrkljan, I., 2021, Application of SBAS-DInSAR to monitoring landslides along the northern Black Sea coast in Bulgaria, *EUROCK-Torino 2021* (in press)

Yastika, P. E., Shimizu, N. and Abidin, H. Z., 2019, Monitoring of long-term land subsidence from 2003 to 2017 in coastal area of Semarang, Indonesia by SBAS DInSAR analyses using Envisat-ASAR, ALOS-PALSAR, and Sentinel-1A SAR data, *Advances in Space Research*, 63:1719–1736.

Yastika, P. E., Shimizu, N. and Verbovšek, T., 2019, A case study on landslide displacement monitoring by SBAS-DInSAR in the Vipava River Valley, Slovenia, *Proceedings of ISRM Specialized Conference on 5th ISRM Young Scholars' Symposium on Rock Mechanics and International Symposium on Rock Engineering for Innovative Future*, 406–411.

Rock Mechanics and Engineering Geology in Volcanic Fields – Ohta, Ito & Osada (eds)

Monitoring three-dimensional displacements by GPS - case study of a large steep slope

Takeshi Sato*
Yamaguchi University, Ube, Japan

Nguyen Trung Kien
Thuyloi University, Hanoi, Vietnam

Shinichiro Nakashima & Norikazu Shimizu
Yamaguchi University, Ube, Japan

ABSTRACT: The monitoring of ground surface displacements plays an important role in the risk assessment of various types of slopes. The Global Positioning System (GPS) is one of the useful tools for monitoring slope behavior because it realizes the automatic and continuous measurements of three-dimensional displacements over extensive areas. However, in order to obtain highly reliable displacements under any environmental conditions, appropriate error correction/reduction methods should be applied to enhance the monitoring results. Tropospheric delays and overhead obstacles cause large measurement errors especially in steep slopes when there is a large height difference between the reference point and a monitoring point.

In this research, the GPS monitoring system, applied to monitor the displacements of a large steep slope, is examined. The monitoring has been continuously conducted for eight years. By applying error correction/reduction methods, the three-dimensional movements of the landslide could be clearly observed. This paper presents the transitions of the three-dimensional displacement components and vectors from 2014 to 2021.

Keywords: GPS, steep slope, three-dimensional displacement, displacement vectors, error correction

1 INTRODUCTION

Monitoring displacements is important for assessing the stability of slopes. The Global Positioning System (GPS) is one of the useful tools for monitoring slope behavior because it realizes automatic and continuous measurements of three-dimensional displacements over extensive areas. However, in order to obtain highly reliable displacements under any monitoring environmental conditions, appropriate error correction/reduction methods should be applied. Actually, a large height difference between the reference point's sensor and a monitoring point's sensor can cause non-negligible errors due to the tropospheric delays of the microwaves transmitted from the satellites. In addition, overhead obstacles above a sensor can disturb the signals and cause a scatter in the measurement results.

*Corresponding author: a007veu@yamaguchi-u.ac.jp

DOI: 10.1201/9781003293590-51

The authors proposed the "ISRM Suggested Method for Monitoring Rock Displacements Using the Global Positioning System" (Shimizu et al., 2014), which describes the error correction/reduction methods for improving the monitoring accuracy.

In this research, GPS displacement monitoring has been conducted to assess the stability of an unstable steep slope along a national road in Japan. The above suggested method was applied, and monitoring has continued for almost eight years. Since the target slope has experienced repeated local failures several times over the last 50 years, displacement monitoring has been conducted by borehole inclinometers and surface extensometers. However, some of the instruments have occasionally not worked due to large deformations, and it has become difficult to perform the monitoring continuously. In order to overcome such trouble, the GPS displacement monitoring system (Masunari et al., 2003) has been in place since 2012 (Furuyama et al., 2014; Kien et al., 2017).

The system has been automatically monitoring the displacements every hour and has provided the results on the Internet's web site in real time. It has detected a small increase in displacement and then found that the displacement was about to exceed the safety criteria for road management due to heavy rains (Furuyama et al., 2014). The road was temporally closed to traffic based on the monitoring results. The GPS monitoring contributed to the safe road traffic management. Ever since carrying out countermeasure works by drainage boring from 2013-2014, the displacement has been stable (Kien et al., 2017; Nakashima et al., 2018). However, it still continues to increase at a constant rate.

In this paper, the recent monitoring results and the time transitions of the displacements and their vectors are demonstrated.

2 GPS DISPLACEMENT MONITORING SYSTEM AND ERROR CORRECTION METHODS

The system developed by the authors (Masunari et al., 2003) is shown in Figure 1. Sensors, composed of an antenna and a terminal box, are set on measurement points and a reference point. The sensors are connected to a control box into which a computer, a data memory, and a network device are installed. The data, emitted from the satellites, are received by the sensors and transferred to the control box through cables. The server computer, which is located at an office away from the measurement area, automatically controls the entire system to acquire the data from the control box and to analyze it. Then, the three-dimensional displacements at all the measurement points are obtained. The monitoring results are provided to users on the web through the Internet in real time.

GPS measurement results generally include random errors and bias errors. Random errors are brought about by random fluctuations in the measurements. These errors are caused by receiver noise, the operating limit of the receiver (resolution), and other accidental errors. Typical bias errors, on the other hand, are tropospheric delays and other signal disturbances caused by obstructions above the antennas. Since both random and bias errors influence the monitoring quality, they should be corrected by appropriate methods which can reduce or eliminate them.

Measurement value y is assumed to be composed of exact value u_0, bias errors ε_P and ε_T, and random error ε_R, as shown in Equation (1).

$$y = u_0 + \varepsilon_p + \varepsilon_T + \varepsilon_R \qquad (1)$$

where ε_P and ε_T are signal disturbances caused by obstructions above the antennas and tropospheric delays affecting the signal traveling through the troposphere from the satellite, respectively.

In the ISRM suggested method, the authors proposed that signal disturbances due to obstructions above the antennas can be reduced by the mask procedure, and that errors due to tropospheric delays can be corrected by a tropospheric model using meteorological data (Shimizu et al., 2014). After removing the bias errors from the measurement value, the "trend model" is applied to estimate the exact displacement (Shimizu et al., 2014).

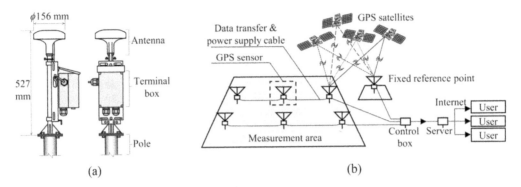

(a)

(b)

Figure 1. Schematic diagram of GPS displacement monitoring system: (a) Sensor and (b) Monitoring system.

3 MONITORING SITE

The monitoring site is a large steep slope along a national road in Japan (see Figure 2). The slope is composed primarily of rhyolite and granite which formed in the Cretaceous period of the Mesozoic era, and its surface is partially covered with a colluvial deposit. The vegetation extends over the slope and creates an obstruction above the antennas.

In July 1972, the slope collapsed in large areas and deposits flowed, causing damage to a railway and a national road. The railway was later replaced by a new track with a tunnel running through the mountain. A check dam was then constructed to collect deposits from the upper part of the slope, and a rock shed was built to protect the road. In 2010, a failure with a volume of 1600 m^3 was recorded due to heavy rainfall. Other small surface collapses and runoffs occurred in 2011 after heavy rainfall.

GPS sensors were set at reference point K1 and monitoring points G1 and G2 in 2013. Since a small slope collapse occurred just beneath G1, the sensor at G1 was removed in August 2014. In the same period, two new monitoring points, G3 and G4, were installed on the right-hand

Figure 2. Monitoring site and locations of GPS sensors (dates below GPS points indicate monitoring period of equivalent monitoring points).

side (southwest side) of the slope (see Figure 2). All the sensors were firmly fixed at the monitoring points on the ground surface by a tripod with anchors (see Figure 3). The height differences between monitoring points G1, G2, G3, and G4 and reference point K1 were 103 m, 112 m, 103 m, and 94 m, respectively. Figure 4 shows the sky views above the sensors. It is found that the vegetation and the slope behind the sensors cover the sky area from the northeast to south.

The system provides the three-dimensional displacement every hour at each point.

(a)   (b)   (c)   (d)   (e)

Figure 3. GPS sensors at reference point (a) and at monitoring points (b) G1, (c) G2, (d) G3, and (e) G4.

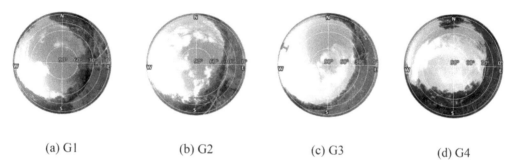

(a) G1   (b) G2   (c) G3   (d) G4

Figure 4. Sky views above GPS antennas at.

4 RESULTS AND DISCUSSION

4.1 *Time-transition of three-dimensional displacements*

The monitoring results from March 2013 to August 2014 were already shown and discussed in previous papers (Furuyama et al., 2014; Kien et al., 2017). They are summarized as follows. The displacement at G-1 increased at a rate of 2-3 mm/month in the beginning, and then gradually increased every time it rained. Finally, it reached 355 mm in the north direction and 234 mm in the west direction, and the settlement reached 137 mm. The displacement velocity exceeded the safety criteria (Level 2: 20 mm/day) two times during the rainy season, and the road was closed at those times. It was proven that the GPS displacement monitoring system was useful and effective for assessing the slope stability and conducting the safety road management.

Figure 5 shows the monitoring displacements at G2, G3, and G4 from 12 August 2014 to 1 May 2021 together with the amounts of rainfall (mm/day). The standard deviations in the directions of latitude, longitude, and height are 1.6 mm, 1.4 mm, and 3.6 mm at G2, 1.5 mm, 1.6 mm, and 3.7 mm at G3, and 1.8 mm, 1.1 mm, and 3.2 mm at G4, respectively. Continuous three-dimensional displacement monitoring was performed for almost seven years without missing any data, except during the maintenance period.

Since August 2014, there have been no heavy rainfall events with more than 200 mm of continuous rainfall, just like before 2013, and the average velocity of the displacements at G2,

G3, and G4 were around 5-25 mm/year, which is quite a bit smaller than that before 2013. It has become particularly small in the last few years.

Although it can be seen that the slope is stable at present, the GPS displacement monitoring system has detected that displacements are still continuously increasing at a slow rate of less than 10 mm/year.

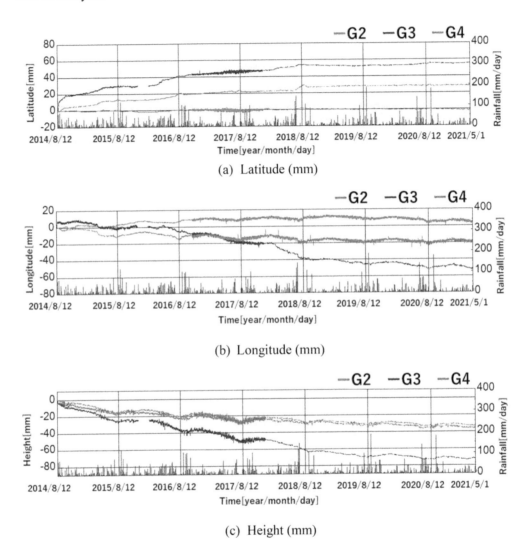

(a) Latitude (mm)

(b) Longitude (mm)

(c) Height (mm)

Figure 5. Time-transition of three-dimensional displacements.

4.2 *Transition of displacement vectors*

Figure 6 shows the transition of the displacement vectors in the plan view and in the vertical sections of the slope from August 2014 to December 2018. During this period, the displacement at G3 headed to the northeast to north colluvium slopes until 9 May 2016 and then turned to the northwest (downward movement). The displacement at G4 constantly increased in the direction of the slope, and G2 subsided. Therefore, the block of the slope that includes G2, G3, and G4 seems to show a rotational slide.

On the other hand, the slope behavior has changed since January 2019, as shown in Figure 7. The displacements at G3 and G4 have been heading in the same direction as that at G3 since 9 May 2016. In addition, G2 has also turned and is moving in the same direction as that at G3 and G4 in the plan view and as that at G3 in the vertical section. This block shows translational movement along a sliding plane. The GPS displacement monitoring system can detect the detailed three-dimensional displacements of the slope.

5 CONCLUSIONS

A GPS displacement monitoring system and error correction/reduction methods have been applied for monitoring the behavior of a steep slope along a national roadway since 2012. The following conclusions can be drawn:

- Three-dimensional displacements have been continuously monitored every hour, without missing any data, for a period of 8 years. This paper presented the results from 2014 to 2021.
- The GPS displacement monitoring system has been able to detect small gradual increases in displacement with average velocities of less than 10 mm per year. The standard deviation of the measured displacement is 1-2 mm in the horizontal direction and 3-4 mm in the vertical direction.
- Changes in the direction of the displacement vectors could be clearly found. This was useful information for discussing the mechanism of the behavior and for assessing the stability of the slope.

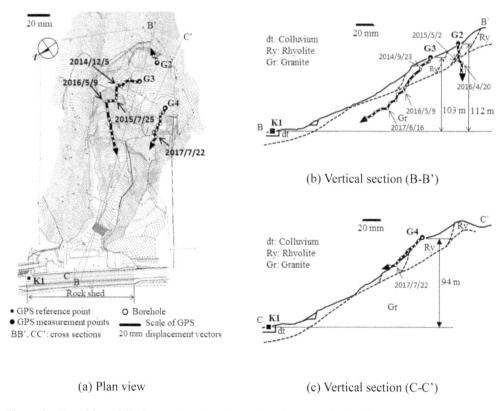

(a) Plan view

(b) Vertical section (B-B')

(c) Vertical section (C-C')

Figure 6.   Transition of displacement vectors (August 2014 – December 2018).

419

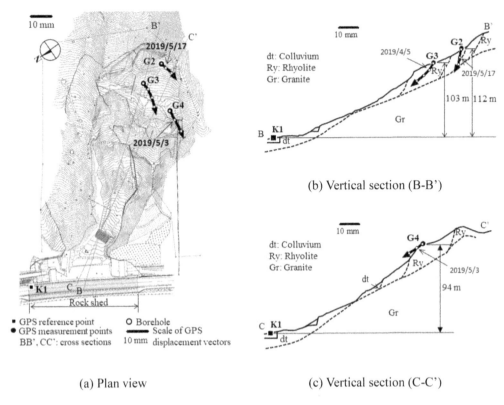

(a) Plan view

(b) Vertical section (B-B')

(c) Vertical section (C-C')

Figure 7. Transition of displacement vectors (January 2019 – May 2021).

ACKNOWLEDGEMENTS

This research has been supported by JSPS KAKENHI (Grant-in-Aid for Scientific Research, Japan Society for the Promotion of Science) Grant Number 16H03153. The authors wish to express their sincere appreciation to the branch office of the Ministry of Land, Infrastructure Transport and Tourism for supporting the field monitoring, and to Ms. H. Griswold for proofreading this paper.

REFERENCES

Furuyama, Y., Nakashima, S. & Shimizu, N. 2014. Displacement monitoring using GPS for assessing stability of unstable steep slope by means of ISRM suggested method, *Proceedings of the 2014 ISRM International Symposium - 8th Asian Rock Mechanics Symposium (ARMS8)*, Sapporo, 1897–1904.

Masunari, T., Tanaka, K., Okubo, N, Oikawa, H., Takechi, K., Iwasaki, T. & Shimizu, N. 2003. GPS-based continuous displacement monitoring system. *Proceedings of International Symposium on Field Measurements in Geomechanics (FMGM03)*, Oslo, 537–543.

Kien, N. T., Hayashi, Y., Nakashima, S. and Shimizu, N. 2017. Long-term displacement monitoring using GPS for assessing the stability of a steep slope, *Proceedings of Young Scholars' Symposium on Rock Mechanics (YSRM 2017) & International Conference on New Development in Rock Mechanics and Geotechnical Engineering (NDRMGE 2017)*, Jeju, 165–168.

Shimizu, N., Nakashima, S. & Masunari T. 2014. ISRM Suggested Method for Monitoring Rock Displacements Using the Global Positioning System, *Rock Mech. Rock Eng.*, 47: 313–328. DOI 10.1007/s00603-013-0521-5.

Nakashima, S., Furuyama, Y., Hayashi, Y., Kien, N. T., Shimizu, N. & Hirokawa, S. 2018. Accuracy enhancement of GPS displacements measured on a large steep slope and results of long-tern continuous monitoring, *Journal of the Japan Landslide Society*, 55(1):13–24.

Rock Mechanics and Engineering Geology in Volcanic Fields – Ohta, Ito & Osada (eds)
© 2023 copyright the Author(s), ISBN 978-1-032-27657-1

Development of rock-identification technique for cuttings obtained by advancing boring for tunneling using hyperspectral camera

Minato Tobita*, Suguru Shirasagi, Yasuyuki Miyajima, Hayato Tobe &
Kazuhiko Masumoto
Kajima Corporation, Tokyo, Japan

Youhei Kawamura & Narihiro Owada
Akita University, Akita, Japan

ABSTRACT: In the excavation of rock tunnels, rocks behind the face have to be known in advance, especially in the case that rocks like serpentinite, which would cause squeezing while tunneling, might exist. A long advancing boring from the working face gives such information with cuttings, but recognizing its rock type is difficult even for geologists. Accurate, simple rock-identification technique of cuttings helps to draw out a rational scheme beforehand and improve productivity of tunnel excavation. Recently, Convolutional Neural Network (CNN) using hyperspectral images of samples established a rock-identification technique where a rock is identified only with a picture taken by a special camera called "hyperspectral camera." Here, we apply this technique to cuttings obtained by advancing boring for tunneling.

This study employed four kinds of cuttings samples obtained from a tunnel ground: greenstone 100% (G100), serpentinite 100% (S100), greenstone 90% + basalt 10%, and greenstone 90% + serpentinite 10%. Appearances of these samples are quite similar and are difficult to discriminate visually. Using hyperspectral images of the samples, two types of CNN models were trained; one is for G100 and S100 (Model A), and the other is for all of the four cuttings samples (Model B).

The validation accuracy of Model A was 99.40%, whereas that of Model B decreased to 73.10%. A mislabeling for a mixed cuttings sample by geologist would account for the low accuracy of Model B. The CNN model for mixed cutting samples should better be trained using hyperspectral images of artificially calibrated mixed samples. On the other hand, the high accuracy of Model A shows that this technique is already practical to cuttings purely composed of a single rock type. The CNN model developed in this study gives a correct, quick discrimination of serpentinite and greenstone with no professional technique, which will give a great help to avoid troubles and contribute to safety and profitability in tunnel excavation of the ground including serpentinite.

Keywords: Advancing boring, Cuttings, Hyperspectral camera, Rock, Convolutional neural network (CNN)

1 INTRODUCTION

In the excavation of rock tunnels, rocks behind the face have to be known in advance, especially in the case that rocks like serpentinite, which would cause squeezing while tunneling,

*Corresponding author: tobitam@kajima.com

DOI: 10.1201/9781003293590-52

might exist. Recently, a long advancing boring (>1000 m) from the working face plays an important role in order to predict appearance of such kind of rocks as soon as possible and to avoid problems at the face. However, a long advancing boring does not produce rock cores but cuttings, the grain size of which are too small (up to tens of millimeters in diameter) to recognize its rock type correctly. Generally, rock-identification of cuttings requires costly and time-consuming procedures such as X-ray diffraction measurements and professional observation by geologists, reducing productivity of the excavation of rock tunnels.

In this study, we employed cuttings obtained from a 1000 m class advancing boring in a tunnel ground including serpentinite. Due to its swelling characteristics, it is necessary to obtain information as soon as possible where and when serpentinite appears during the excavation. The tunnel ground also includes greenstone and basalt, which require no special care, as well as serpentinite. Thus, correct and fast identification of cuttings obtained from the long advancing boring is essential for rational design and construction of the rock tunnel.

Recently, Sinaice et al. (2017) developed a rock-identification technique for igneous rocks in which a rock is identified only with a picture taken by a special camera called "hyperspectral camera" (hereinafter referred to as "HS") based on Convolutional Neural Network (CNN) of hyperspectral images of samples. Here, we apply this technique to rock-identification of cuttings (not a mass but small-grained sample) containing a metamorphic rock, serpentinite.

2 OBTAINING HS DATA

2.1 Samples

We employed four kinds of cuttings samples obtained from a non-core boring in a tunnel ground consisting mainly of greenstone, serpentine and basalt. In the surrounding area, greenstone, containing green-colored minerals such as glauconite and chlorite, is considered to have formed with alteration and metamorphism of basalt. All of the cuttings samples have a grain size of 5 to 20 mm, and have a similar dark green color (Figure 1), making it difficult for even geologist to distinguish them visually.

Figure 1. Cuttings samples.

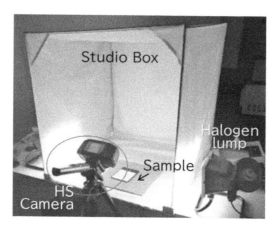

Figure 2.   The setting of HS camera.

The total number of samples obtained from the non-core boring was 84 bags. According to professional observation by a geologist, 78 bags contain only greenstone (hereinafter referred to as "G100"), a bag contains only serpentine (hereinafter referred to as "S100"), 4 bags were a mixture of greenstone 90% and basalt 10% (hereinafter referred to as "G90B10"), and a bag was a mixture of greenstone 90% and serpentine 10% (hereinafter referred to as "G90S10").

2.2   *HS camera*

Whereas general digital (RGB) camera produces two-dimensional image each of which has wavelength information with 3 bands (red, green, and blue), HS camera can acquire that with more than 100 spectral bands. Therefore, a picture taken by HS camera has not only two-dimensional planar data in the vertical and horizontal directions (i.e., x-y grid charts) but also a depth data in the wavelength direction (i.e., z grid chart). The HS camera employed in this study is Specim IQ (SPECIM, SPECTRAL IMAGING LTD.), which acquires wavelength from 400 nm (almost the lowest wavelength in the visible radiation) to 1000 nm (a low part of the near-infrared radiation) with 204 spectral bands.

2.3   *Shooting method*

The shooting method of HS camera in this study was followed by Tobita et al. (2021). In a studio box in dark room, a cuttings sample was spread flatly in a 20cm square on white paper. Two halogen lamps illuminated the sample from outside of the studio box to reduce heterogeneity of the illumination at the surface of the sample (Figure 2). The halogen lamps yield light covering the wavelength range that the HS camera acquires (from 400nm to 1000nm). The brightness of the light source was adjusted to 3000-3500 lux at the surface of the sample. The entire frame of the HS camera was filled with the spreading cuttings sample, and each sample was taken three times.

2.4   *Data processing*

Before training the CNN model, HS data was preprocessed as shown in Figure 3 (Tobita et al., 2021). The obtained HS image data has 512 pixels in length and breadth, and 204 bands in depth. This image was divided into 256 equal square parts (16×16 for length and breadth, respectively), each of which has 1024 pixels (32 pixel × 32 pixel for length and breadth, respectively). Then, each of 204 bands was averaged over the 1024 pixels, resulting in 256 datasets of 1 pixel in the length and breadth and 204 bands in the depth. This data processing produces a dataset of 256 csv files of HS data per a picture. The HS data were then input with 1 × 204 dimension into the CNN consisting of three of two convolution layers plus a pooling layer.

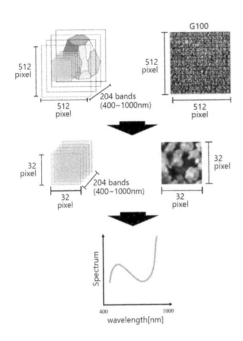

Figure 3.   Data processing of HS data.

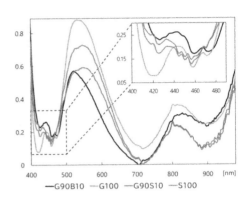

Figure 4.   The HS of four cuttings samples.

Output \ Input	G100 (77)	S100 (77)
G100	98.7% (76)	0.0% (0)
S100	1.3% (1)	100% (77)

Figure 5.   The confusion matrix of model A.

3   HS DATA OF THE CUTTINGS SAMPLES

The HS data for each cuttings sample is shown in Figure 4, which is averaged for each sample and is normalized using the maximum and minimum values of each spectrum. The HS signature of each sample is distinct in the range of wavelength from 400 nm to 500 nm, whereas that greater than 500 nm shows no clear difference. The peak positions of G100 and S100 are distinct, while those of S100 and G90S10, and those of G90B10 and G100 are in good agreements. In this study, two types of CNN models were trained to classify the samples; one is for G100 and S100 (Model A), and the other is for all of the four cuttings samples (Model B).

4   RESULTS OF CNN MODELS

4.1   Model A

The number of HS data obtained from the non-core boring is 768 and 59904 csv files for S100 and G100, respectively. For training Model A, all of HS data from S100 and randomly extracted 768 data from G100 were employed in order to equalize the number of training data for each. The training data, validation data, and test data were set to be a ratio of 8:1:1, respectively. The mini-batch size and the number of epochs was set to 256 and 200, respectively.

As a result of the training, Model A was able to classify the validation data with 100% accuracy and the test data with 99.40% accuracy (Table 1). Figure 5 shows the confusion matrix of Model A with the predicted classification (Output Class) in the columns and the true classification (Input Class) in the rows. Although there was a case in which greenstone was classified as serpentine, there was no other error and the accuracy was very high.

Table 1.   Result of the CNN models.

	For validation data	For test data
Model A	100%	99.40%
Model B	77.27%	73.10%

4.2 Model B

The number of HS data obtained from the non-core boring is 768, 59904, 3072, and 768 csv files for S100, G100, G90B10, and G90S10, respectively. For training Model B, all of HS data from S100 and G90S10 and randomly extracted 768 data from each of G100 and G90B10 were employed in order to equalize the number of data and the training quality for all of the samples. The training data, validation data, and test data were set to be a ratio of 8:1:1, respectively. The mini-batch size and the number of epochs was set to 256 and 200, respectively.

As a result of training of Model B, the accuracy was 77.27% and 73.10% for the validation data and the test data, respectively (Table 1). Figure 6 shows the confusion matrix with the predicted classification (Output Class) in the columns and the true classification (Input Class) in the rows. The accuracy for G100 was the highest (87%), and those for S100 and G90B10 were still high as well (73% and 78%, respectively). The lowest accuracy was 55% for G90S10, many of which was derived from the combination of G100 and G90 B10. In particular, a large number of G90S10 were misidentified as S100 (45%, Figure 6).

In / Out	S 100	G 100	G90 S10	G90 B10
S 100	73%	0%	45%	0%
G 100	0%	87%	0%	21%
G90 S10	27%	12%	55%	1%
G90 B10	0%	1%	0%	78%

Figure 6.   The confusion matrix of model B.

5   EVALUATION OF THE CNN MODELS

The accuracy of Model A was very high (99.4%), which is likely because there was a clear difference in the HS signature between G100 and S100 in the range of wavelength from 400 nm to 500 nm. However, the accuracy of Model B, classification of four samples, decreased by more than 20% compared to the classification of two samples (Model A, Table 1). This decreasing might be due to the similar HS signatures between G90S10 and S100, and between G90B10 and G100. Particularly, the HS trends of G90S10 and S100 produced almost the same peak positions and heights, strongly indicating that few unique characteristics of each sample was trained into the CNN models.

Considering that the "G90S10" might have been mislabeled by the geologist due to the unlikely distinguishable appearances of cuttings samples, a spectrum was theoretically calculated where those of G100 and S100 were mixed in a ratio of 9:1 (Figure 7). The theoretical spectrum of greenstone 90% and serpentinite 10% (i.e., G90S10(fake) in Figure 7) is similar to that of G100 but is dissimilar to that of S100. If G90S10 had consisted 90% of greenstone as observed by the geologist, the spectrum would be alike that of G100. The inconsistency implies that the cuttings sample, G90S10, contained serpentinite much more abundant than expected. Therefore, in order to improve the accuracy of classification for cuttings samples with multiple kinds of rocks, the CNN model should be trained with HS data of cuttings sample consisting of multiple kinds of rocks that are artificially prepared to a certified mixture rate.

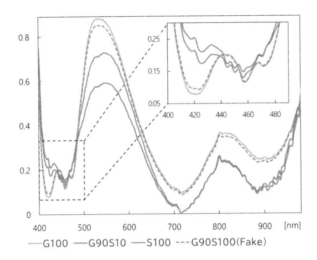

Figure 7. The HS of cuttings samples.

6 CONCLUSIONS AND FUTURE PLAN

From the deep learning of HS, the Model A was able to identify rock-type of cuttings consisting purely of greenstone and serpentinite with a high accuracy of 99.40%. By acquiring appropriate HS data shown as the method in section 2.3, the Model A developed in this study provides a correct, instant discrimination of serpentinite and greenstone for cuttings with no professional technique, leading to a quick, cost-free feed back to the optimization of excavation in the tunnel ground including serpentinite.

On the other hand, for the four cuttings samples including multiple kinds of rocks (Model B), the accuracy of classification decreased by more than 20% compared to the classification of two single-rock samples (Model A). The low accuracy was likely due to an improper labeling to a mixed-rock sample, which will be improved with HS data of artificial mixed-rock samples prepared with a certified mixture ratio. Until the accuracy of Model B is sufficiently improved, observing HS data by a field worker would be useful for rock-identification of cuttings; if a HS trend of sample is similar to that of S100 obtained in this study, the sample can contain significant amount of serpentinite.

This study developed a new rock-identification technique with HS camera for cuttings obtained from a long advancing boring. Using this technique, only taking a picture provides meaningful information where and when kinds of rocks requiring special cares appear during excavation of rock tunnel. In the future, we are planning to improve the accuracy of classification for mixed-rock samples, and to apply this technique to other rocks than serpentinite and further to minerals requiring special cares (e.g., asbestos) as well.

REFERENCES

Sinaice, B., Kawamura, Y., Shibuya, T., Sasaki, J., Yoshimoto, H., Ito, Y., and Utsuki, Y., 2017, Development of a differentiation and identification system for igneous rocks using hyper-spectral images and a convolutional neural network (CNN) system., *Sapporo Journal of MMIJ*, Sapporo, Japan, 26–28, September 2017, Vol. 4, No.2.

Tobita, M, Shirasagi, S., Miyajima, Y., Tobe, H., Masumoto K., Kawamura, Y., and Owada, N., 2021, Identification of rocks behind the face with cuttings obtained by advancing boring using hyperspectral camera. *15th Japan Symposium on Rock Mechanics*, Osaka, Japan, 14-15, January 2021, RS–134.

Rock Mechanics and Engineering Geology in Volcanic Fields – Ohta, Ito & Osada (eds)
© 2023 copyright the Author(s), ISBN 978-1-032-27657-1

Employing NCA as a band reduction tool in rock identification from hyperspectral processing

Brian B. Sinaice*
Graduate school of International Resource Sciences, Akita University

Narihiro Owada
Faculty of International Resource Sciences, Akita University

Shinji Utsuki
UGS-Utsuki Geo Solution

Zibisani B. Bagai
Department of Geology, University of Botswana

Elisha Shemang
Department of Earth and Environmental Science, Botswana International University of Science and Technology

Youhei Kawamura
Graduate school of International Resource Sciences, Akita University

ABSTRACT: Hyperspectral imaging is a highly viable method in which rock delineation can be performed with minimal classification error. Though previous studies based on this method yield high accuracy results, it is often reported that the computational requirements necessary to process data of this magnitude are high due to the hundreds of spectral bands, hence making field applicability rather challenging. To counter this problem with field applicability in mind, this study proposes a method of employing Neighbourhood Component Analysis (NCA) as a dimensional reduction tool, which, based on feature weights, determines the most important spectral bands which can in turn be used in delineating rocks based on their spectral signatures. From our hyperspectral images database, we attempted to reduce each spectral anomaly from their original resolution of 204 bands captured using our hyperspectral camera; which has a spatial range within the Visible-Near-Infrared Range (VNIR) of 400-1000nm. To achieve this, we performed NCA which assigns feature weights to each one of the 204 spectral bands, thereby allowing the user to select the number of spectral bands they would subsequently prefer to employ in field applicable classification and identification problems, which in our case was 5 bands. Tied into one task, our algorithm then automatically performs classification of these rocks based on the 5 chosen spectral bands with the highest feature weights. Our model was able to achieve a validation accuracy of 70.9%, and an average per-class precision of 72%, hence, proving that high accuracy rock classifications can indeed be performed on dimensionally reduced spectral bands. It can therefore be said that advantages of using NCA include; low computational requirements, shorter classification times, and most importantly, the ability to reduce the dimensions of hyperspectral layers, thereby making it possible to accurately classify rocks with fewer spectral bands.

Keywords: Neighbourhood Component Analysis, Hyperspectral Imaging, Feature Selection, Dimensional Reduction, Machine Learning

*Corresponding author: bsinaice@rocketmail.com

DOI: 10.1201/9781003293590-53

1 INTRODUCTION

The integration of 'smart' technics into various industries has become the norm as current practices have a growing need to optimize every aspect of their systems design. The mining, rock and mineral engineering industries have been no stranger to this trend as researchers (Jain et al., 2021; Saha and Annamalia., 2021) have demonstrated various benefits of employing these smart Artificially Intelligent (AI) technics in rock related problems as they consistently yield better accuracy results, enhance safety in the workplace, and offer flexibility with regards to handling of data pre, during and post-analysis. Motivated by this trend and the potential it holds in optimizing various aspects of the rock, mining and geotechnical engineering works, our paper proposes the fission of hyperspectral imaging converted to multispectral imaging via a dimensionality reducing Neighbourhood Component Analysis (NCA) algorithm, together with an AI based Machine Learning (ML) method into a system by which specific rock identification can be based (Figure 1).

1.1 *Conversion to multispectral imaging from hyperspectral imaging*

Though our rock hyperspectral imaging database contains high resolution information from the hundreds of spectral bands it retains pertaining to the reflectance of rocks (Meer, 2006), it however possesses what researchers (Zuniga et al., 2021 and Goldberger et al., 2004) refer to as the 'dimensionality curse'. This term stems from the difficulty in visualizing such depth retentive data (Koren and Carmel, 2004). In addition to the high cost of purchasing a hyperspectral camera such as the one used in the development of our 204 band rock hyperspectral database, this data is computationally costly and often contains redundant information.

To counter this, we propose the conversion from high resolution rock hyperspectral imaging, to a lower though specialized resolution, multispectral rock identification imaging as it has the potential to address the previously mentioned disadvantages. As multispectral imaging contains a few spectral bands (Zhang and Li, 2014), adopting this method in rock identification problems requires a strategic method by which specific rock identification specialized bands may be extracted. To achieve this, dimensionality reduction is a key technique in data analysis, aimed at revealing expressive structures and unexpected relationships in multivariate data (Koren and Carmel, 2004). This makes it highly useful in the in the visualization of such multivariate data, and moreover allowing for feature selection, thereby reducing the footprint of the data in successive data processing stages such as clustering and classification (Saha and Annamalia, 2021). In principle, success in dimensionality reduction means that low-dimensional presentation of original data should provide enough information for classification (Goldberger et al., 2004).

There are several dimensionality reduction methods, distinguishable on the basis of whether they are supervised on unsupervised, the latter possessing a slight handicap in that they only use the global structure of the sample while ignoring the equally important local structures (Zuniga et al., 2021). Having learnt this, we opted for a supervised dimensionality reduction algorithm as they take into account the local structures and their data labels. One of such highly regarded algorithms is NCA, which performs better in terms of learning from a training set distance matric, storage size reduction, and an overall advancement in classification or class separation via assignment of relevant feature weights for each dimension (Goldberger et al., 2004), hence our decision to employ this method.

1.2 *ML based verification post dimensionality reduction*

Though NCA provides the opportunity to ignore redundant data heavy bands, it does not provide any information related to the retained classification accuracies post dimensionality reduction. To assess the capabilities of our potential specialized rock identification multispectral bands, ML models, a subdomain if AI, are capable of deriving useful information from data and utilize that information in self-learning, which in turn aids in making reliable and good accuracy predictions (Jain et al., 2018). We, in this instance employed several supervised ML models as such models require labeled training data such as our labelled rock database, that in turn aids in making predictions about future data (Saha and Annamalia, 2021). These models require labeled data to guide the machine in looking for exact patterns, which is highly advantageous as we are faced with a rock identification problem based on multispectral anomalies. Figure 1 is an overview of our proposed system design.

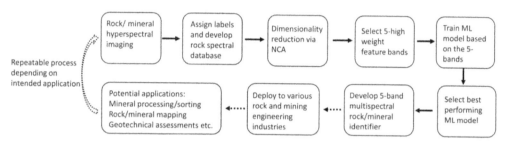

Figure 1.   Proposed rock and/or mineral identification system design based on dimensionality reduction from hyperspectral imaging to specialized multispectral imaging via NCA, combined with ML models to prove the viability of selected multispectral bands.

2   METHODOLOGY

2.1   *Attaining rock hyperspectral spectral image signatures*

In building this proposed integrated system, we were faced with the challenge of proving its viability before potential deployment into various rock engineering works. Hence, we employed 8 rocks namely Granite, Diorite, Gabbro, Granodiorite, Rhyolite, Andesite, Basalt and Dacite; these being specimens of Akita Mining Museum, each with several representative samples as our test subjects. To capture their hyperspectral signatures, a *Specim IQ* hyperspectral camera capable of capturing 204 bands within the VNIR of 400-1000nm was employed. The main components of this experimental setup are shown in Figure 2. As it has been pointed out by Meer (2006); Zhang and Li (2014), capturing spectral data requires the standardization of the capturing environment, hence maintenance of the right exposure was verified via a white board (supplied by the manufacturer, Specim), whose role is to act as the base reflectance, allowing for the extraction of true rock spectra. Thereafter, this depth possessing data is converted into quantitative data via a hyperspectral analyzing software such as *Python Spider ver. 3.6*. This data undergoes a pre-processing stage where each of the individual spectral signature data are assigned labels of the rocks they belong to, doing so aids in tractability of data, and is moreover a prerequisite in employing supervised methods; hence a rock hyperspectral database is developed.

2.2   *Band selection based on NCA*

It is well documented that supervised methods generally outperform unsupervised methods (Saha and Annamalia, 2021), as such, we employed the NCA dimensionality reduction technique. The primary purpose of employing NCA is to reduce the number of dimensions in data and assign the appropriate feature weights to the most important spectral bands. This is however not without its challenges as selecting the appropriate number of features to employ is

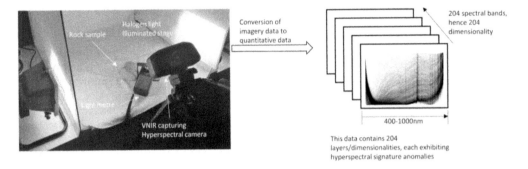

Figure 2. Rock hyperspectral images captured using a *Specim IQ* hyperspectral camera with a spatial resolution of 400-1000nm, and a spectral resolution of 204 bands under a controlled environment to extract the true spectral signatures of rocks.

often dictated by the intended use of such features. Therefore, based on the current industry trend of 5-band multispectral imaging devices, our target was to perform a dimensionality reduction from 204-bands to 5-bands. We should note however that as a way of validating our system, we initiated our assessment with all 204 bands, followed by our targeted 5-bands so as to appreciate the difference in attainable classification capabilities given the magnitude of our dimensionality reduction. From these 5-bands, we are in essence able to identify and confirm the most appropriate multispectral bands suitable in the identification of rocks in or related to our database, which in principle means redundant data has been successfully eliminated.

2.3 Classification post NCA

Having extracted the 5-high-feature-weight possessing spectral bands as dictated by NCA which in principle means eliminating redundant data, several supervised ML models are employed in the training, validation and classification task with spectral data as input variables. This task is performed in two parts, first, we commence with a training and classification task that acts as a control undertaking using 100% of the 204 rock spectral signature bands to note the ability of each model. Thereafter, we initiate the post NCA classification task using 5-bands, this data undergoes the same training, validation and testing task via various ML models, results from which demonstrate the capabilities of those particular ML models in the identification of rocks based on the 5 multispectral spectral bands.

3 EXPERIMENTAL AND ANALYTICAL RESULTS

3.1 Feature selection via NCA

With NCA, large feature weights are assigned to the most relevant dimensions which coincide with their specific spectral bands, while lower feature weights are assigned to the least relevant or redundant feature bands as it seeks to identify and down-scale the global unwanted variability within data. Chunklets, refereeing to a subset of points that belong to the same though unknown class, are what the algorithm uses to obtain the most relevant dimensions required for future rock identification (Goldberger et al., 2004). Based on the NCA output results on Figure 3, one is able to appreciate the ease in identifying spectral bands with the highest feature weights as it shows the performance of each of the 204 hyperspectral bands in terms of being the most relevant rock discrimination bands. As our goal is to identify the most relevant 5-multispectral bands, the NCA output dictates band number 14, 46, 116, 133 and 169 from the 204 hyperspectral bands are the most viable spectral bands to employ in future field and/or laboratory rock identification of employed and/or related rocks. Converting these spectral band positions into electromagnetic wavelength bands, we get 441nm, 535nm, 741nm, 791nm and 897nm as the most discriminatory multispectral bands. A point to note is that the hyperspectral camera we employed for our investigation has a spectral resolution of approximately 3nm, implying each of the 5 selected spectral bands inherently possess a +/-3nm error; hence, for example, for band 14 (441nm), the range of rock identification is 438-444nm. Another point to note in that, based on the intended application of a multispectral rock, mineral, soil or environmental phenomenon identifier, a specific number of spectral bands can be selected based on their post NCA feature weights.

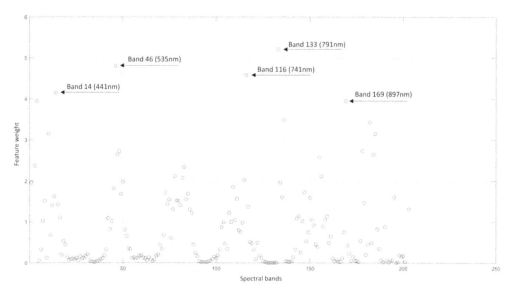

Figure 3. NCA feature selection via the assignment of high feature weights to spectral bands with the most significance in the identification of rocks, thereby reducing redundancy. The best 5-multispectral-bands are 441nm, 535nm, 741nm, 791nm and 897nm.

3.2 Post NCA classification capabilities via ML

Having discarded 97.5% (199 spectral bands) of the primary hyperspectral data (204 bands) as dictated by NCA, we employed 2.5% (5-bands) of the existing now multispectral data in our ML classification task. ML models attempt to acquire familiarity pertaining to the input data during the training process, thereafter, learnt patterns aid in the successive testing process, hence we are able to judge the prediction capabilities of the model. Using '*MATLAB R2020b classification learner Machine Learning toolbox*', we employed several models in the attempt to find the best performing models for our 5-bands multispectral rock identification problem. Outputs from the top 5 performing models have been compiled and summarized in Table 1. Our control task entailing employment of all 204 spectral bands (hyperspectral imaging) has the Cubic SVM (support vector machine) model as the best performing classifier with a global accuracy of 90.7%, this being an expected output given the resolution of hyperspectral imaging. With regards to the post NCA 5-multispectral-bands, Table 1 ML output results dictate that the best performing model in classification of rocks in our database is again Cubic SVM, with a global accuracy of 70.9% and a per-class average precision of 72%. We consider these output results viable bearing in mind the magnitude of our dimensionality reduction from 204-hyperspectral bands to 5-multispectral bands. From this dimensionality reduction, we can safely say we have achieved a reduction in computational costs, data storage, easier data management, simplified data application, and most importantly, viability in rapid field applicability as we have determined the best spectral bands along with the best classification model for our rock classification problem.

Table 1. Classification capabilities of the 5 best performing ML models pre-dimensionality reduction from 204-bands, versus post dimensionality reduction (via NCA) for 5-bands.

Number of spectral bands	Machine learning algorithm	Output capabilities (%)		Training time (seconds)
		Global accuracy	Average per-class precision	
204-bands (pre-NCA)	SVM (Cubic SVM)	90.7	90	28.7
	SVM (Quadratic SVM)	87	86	27.3
	SVM (Linear SVM)	79.1	78.5	13.8
	Linear discriminant	80.4	78.4	4.6
	Ensemble (Subspace discriminant)	81.2	79.3	41.5
5-bands (post NCA)	SVM (Cubic SVM)	70.9	72	182.1
	Ensemble(Bagged trees)	69.6	67	12.3
	SVM (Quadratic SVM)	68.4	65.8	76.7
	SVM (Fine Guassian SVM)	68.4	66.6	7.5
	KNN (Fine KNN)	67.3	66	5.7

As the training, validation and classification process has proved Cubic SVM as the best performing ML model in the delineation of our rocks post NCA; using the Figure 4, we are able to appreciate the in-depth capabilities of this particular classification model. Based on Figure 4 output results, the average per-class precision for all 8 rocks is 72%, with granite rock classification being where the model performed the best at a prediction precision of 88.9%, and basalt rock classification being where the model performed the least compared to all other rock classifications at a prediction precision of 61%.

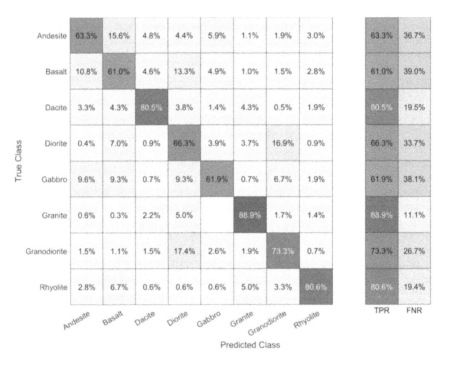

Figure 4. Confusion matrix illustrating the prediction capabilities of the Cubic SVM model in the classification of rocks based on 5 post NCA selected multispectral bands.

Given our findings, we can therefore confirm that our proposed system, which consists of dimensionality reduction of rock hyperspectral data and subsequently employing specific discriminant features for our rock database, performs as well as hypothesized. This in principle means for geological, geotechnical and/or mining engineering problems requiring discrimination of rocks, minerals, soils and other environmental phenomena based on their spectral signatures are indeed able to employ this system. By setting desired attributes founded on pre-knowledge of a site such as types of rocks present within a geologic site, mine site, rocks transported via a conveyer belt, or the general mapping of the environment, it makes sense to exploit data collection based on precise multispectral imaging without having to incur unnecessary storage, processing or classification costs associated with massive data.

4 CONCLUSIONS

With this paper, we have proposed the dimensionality reduction of hyperspectral imaging via NCA to multispectral imaging, integrated with ML as a method by which subsequent spectral features of rocks, minerals and the environment can be performed without involuntary processing of redundant spectral bands. With our NCA algorithm, we were able to achieve dimensionality reduction without altering the intrinsic variability that exists in spectral data, as well as assigning relevant feature weights to hyperspectral bands, thereby converting from 204-bands, to the current industry trend 5 multispectral bands. From NCA, we can consequently conclude that the most sustainable discriminative 5 multispectral bands viable in the identification of Granite, Diorite, Gabbro, Granodiorite, Rhyolite, Andesite, Basalt and Dacite, are bands with the following wavelengths: 441nm, 535nm, 741nm, 791nm and 897nm.

Our proposed method moreover faultlessly combines with numerous ML models, consequently providing quantitative outputs pertaining to the classification abilities of each ML model, thus we can say for our rock identification problem, the Cubic SVM model is the most feasible classification model as it achieved higher output values with a global accuracy of 70.9% as well as a per-class average precision of 72%, this being impressive given the transition from the initial 204-hyperspectral bands to a considerable dimensionality reduced magnitude of 5-multispectral-bands.

ACKNOWLEDGEMENTS

This work was supported by JSPS 'Establishment of Research and Education Hub on Smart Mining for Sustainable Resource Development in Southern African Countries'; Grant number: JPJSCCB2018005.

REFERENCES

Goldberger, J., Roweis, S., Hinton, G.E., & Salakhutdinov, R. (2004). Neighbourhood Components Analysis. NIPS.

Jain, U., Nathani, K., Ruban, N., Joseph Raj, A. N., Zhuang, Z., & G.V. Mahesh, V. (2018). Cubic SVM Classifier Based Feature Extraction and Emotion Detection from Speech Signals. 2018 International Conference on Sensor Networks and Signal Processing (SNSP), 386–391. https://doi.org/10.1109/SNSP.2018.00081

Koren, Y., & Carmel, L. (2004). Robust linear dimensionality reduction. IEEE Transactions on Visualization and Computer Graphics, 10(4),459–470. https://doi.org/10.1109/TVCG.2004.17

Meer, F. (2006). The effectiveness of spectral similarity measures for the analysis of hyperspectral imagery. International Journal of Applied Earth Observation and Geoinformation, 8(1), 3–17. https://doi.org/10.1016/j.jag.2005.06.001

Saha, D., & Annamalai, M. (2021). Machine learning techniques for analysis of hyperspectral images to determine quality of food products: A review. Current Research in Food Science, S2665927121000034. https://doi.org/10.1016/j.crfs.2021.01.002

Zhang, X., & Li, P. (2014). Lithological mapping from hyperspectral data by improved use of spectral angle mapper. International Journal of Applied Earth Observation and Geoinformation, 31, 95–109. https://doi.org/10.1016/j.jag.2014.03.007

Zuniga, M. M., Murangira, A., & Perdrizet, T. (2021). Structural reliability assessment through surrogate based importance sampling with dimension reduction. Reliability Engineering & System Safety, 207, 107289. https://doi.org/10.1016/j.ress.2020.107289

Rock Mechanics and Engineering Geology in Volcanic Fields – Ohta, Ito & Osada (eds)
© 2023 copyright the Author(s), ISBN 978-1-032-27657-1

Lithology identification during rotary percussion drilling based on acceleration waveform 1D convolutional neural network

Lesego Senjoba* & Yoshino Kosugi
Graduate School of International Resource Sciences, Akita University, Akita, Japan

Masaya Hisada
MMC Ryotec Corporation, Gifu, Japan

Youhei Kawamura
Graduate School of International Resource Sciences, Akita University, Akita, Japan
North China Institute of Science and Technology, Hebei, China

ABSTRACT: Identifying the lithology of underground formations is of great importance in the mining industry as it provides useful information about the geometry of rocks under the surface. Current methods are based on core logging and deploying an array of down-the-hole instruments to determine the lithology and structure of the subsurface. Although these methods are highly dependable and accurate, they result in serious time lag. In recent years, various methods have been applied to lithology identification based on acoustic data, vibration data, and two-dimensional convolutional neural networks (CNN). This study presents a system that uses one-dimensional CNN with time acceleration as input data to identify the lithology of an area in real-time. 18 m^3 rocks of granite and marble were drilled horizontally with similar drilling parameters, using a rock drill and an intact tungsten carbide drill bit. Time acceleration of drill vibration was measured using acceleration sensors mounted on the guide cell of the rock drill. Model accuracy verification and prediction were carried out. The lithology identification model achieved a verification and classification accuracy of 98.55% and 98.9%. The proposed model was compared to state-of-the-art (SOTA) deep learning neural networks. The model outperformed SOTA methods in terms of validation and classification accuracy. The proposed model gave satisfactory results when evaluated with time acceleration data from defective and abrasion drill bits. It was therefore proven that using drill vibration as input data to a 1D CNN algorithm, provides an efficient method for obtaining the lithology of the subsurface.

Keywords: Lithology identification, Rotary percussion drilling, Drill vibration, Time acceleration, Convolutional neural network

1 INTRODUCTION

In the mining industry, it is imperative to know the type and properties of underground formations. Current methods involve core observation and logging. The cores are obtained from the ground by drilling, it is often troublesome to obtain the core samples from underground. The cores obtained are subjected to numerous lab tests. Although these methods are reliable, they result in high latency. In today's world, obtaining real-time data has become crucial,

*Corresponding author: senjobal@gmail.com

DOI: 10.1201/9781003293590-54

hence it is necessary to have a system that would result in low latency. This will allow engineers to better plan blast designs and support systems in mine sites. Rotary percussion drilling is a method that is often used in underground mines. It can be employed in both hard and soft rocks of confined compressive strength 80-200 MPa. In rotary percussion drilling, the rock is made to fail using a piston that delivers a rapid impact to the drill stem thereby transferring energy to the drill bit. In recent years, many companies and researchers have been adopting machine learning techniques to improve the efficiency of operations such as drilling. Several methods have been developed for finding lithology of an area during or after drilling. Shreedharan et al. (2014) suggested a method to monitor and evaluate the sounds produced at the drill bit and rock interface to predict the type of rock being drilled. A laboratory rotary drill was used to vertically drill 8 types of rock samples. A microphone was used to capture the sounds produced. Afterward, FFT analysis was carried out, the results obtained were promising, it proved that it was possible to identify the rocks based on FFT analysis. Gang et al. (2020) proposed a method for lithology classification based on 2D CNN by utilizing drill string vibration data. The vibration data sets were from a drill string data of a 5000-6000 m deep well, vibration time-frequency characteristics were extracted into time-frequency images and used as input for the CNN. The model was tested using fine gravel sandstone, fine sandstone, and mudstone. The lithology model had a good accuracy of 89.7%.

Most previous approaches utilize 2D CNN for lithology identification and require heavy data pre-processing which is tedious. This paper proposes a new and reliable method to identify the lithology of an area by employing deep learning:1D CNN and time-acceleration as input data. The application of 1D CNN allows for minimal data pre-processing and low computational complexity that allows real-time monitoring. In this study, a simple and compact 1D CNN is presented and evaluated. The proposed model was then compared to two state-of-the-art (SOTA) deep learning neural network models.

2 METHODOLOGY

2.1 *Experimental design*

18m^3 rock blocks of granite and marble were drilled horizontally using an underground rotary percussion drill jumbo. A drill jumbo has a rock drill, which is mounted on the boom. A rock drill uses a top hammer drilling style, which consists of a drifter, shank, rod, and drill bit. A tungsten carbide button drill bit was used for drilling. A TEAC'S piezoelectric acceleration transducer 600 series acceleration sensor was utilized to measure the acceleration of the vibrations. A Graphtec data platform DM 3300 data logger was used to monitor, store and transfer the data to a PC. Figure 1 shows the setup of the experiment.

Figure 1. Experimental setup.

2.2 Experimental procedure

An acceleration sensor was mounted on the guide cell of the rock drill, facing the direction of drilling. Drilling was performed on granite and marble rock. 7 holes were drilled in each rock with a length of 1 m. The number of hits per minute was set at 3120 for all holes. Table 1 and Table 2 illustrate experimental conditions and data acquisition parameters used in the experiment. The impact pressure for each hole was set at 13.5-7 MPa. The striking frequency, the number of revolutions, and the drilling speed were determined by the striking pressure, the rotating pressure, and the feed pressure. The rotation pressure was 4-6 MPa and the feed pressure was 4 MPa. The sampling frequency was set at 50 kHz. The maximum acceleration used by the accelerometer was ± 10,000 m/s². The acceleration sensor output a voltage which was then transmitted to the data logger which had a built-in amplifier. The acceleration was written out as CSV data on the PC by inputting the calibration coefficient of the acceleration sensor into the data logger.

Table 1. Experimental conditions.

Rock Type	Type of bit	Number of holes	Number of hits per minute	Drill length (m)	Time (s)
Granite	Normal	7	3120	1	60
Marble	Normal	7	3120	1	60

Table 2. Parameters of the experiment.

Impact pressure (MPa)	Striking Frequency (Hz)	Rotary Pressure (MPa)	Feed pressure (MPa)	Sampling Frequency (kHz)
13.5-7	52	4-6	4	50

2.3 Data pre-processing

Time acceleration data was used as input for the 1D CNN. The sampling frequency, sampling time, and sampling number were taken into consideration for data augmentation. The sampling frequency was set at 50 kHz at a sampling time of 60 s, hence each test drill hole had approximately 3 000 000 data points. Data augmentation was performed on raw time acceleration data before feeding the waveform into the network. Each test drill hole signal $x[N]$, was divided into approximately 1000 segments each with 3000 data points $x_n[N]$ with no overlapping. Figure 2 depicts the data augmentation process. Each segmentation had a fixed length of 0.06 s, which guaranteed that at least 3 drill bit hits were represented within one segment. After augmentation, each rock type had 7200 test samples to be used as training data for the 1D CNN model.

Figure 2. Data augmentation.

2.4   1D Convolutional neural network

CNNs were developed with the idea of local connectivity, each node is connected only to a local region in the input. The local connectivity is achieved by replacing the weighted sums from the neural network with convolutions. Convolution can be seen as applying and sliding a filter over the time series. Unlike images, the filters exhibit only one dimension (time) instead of two dimensions (width and height). The filter can also be seen as a generic non-linear transformation of a time series. A general form of applying the convolution for a centered time stamp t is given in the following equation:

$$C_t = f(\omega * X_{t-l/2} : t+l/2 + b)| \forall \ t \in [1, \ T]$$ (1)

where C denotes the result of a convolution (dot product *) applied on a univariate time series X of length T with a filter ω of length l, a bias parameter b, and a final non-linear function f such as the Rectified Linear Unit (ReLU).

The result of a convolution (one filter) on an input time series X can be considered as another univariate time series C that underwent a filtering process. Thus, applying several filters on a time series will result in a multivariate time series whose dimensions are equal to the number of filters used. An intuition behind applying several filters on an input time series would be to learn multiple discriminative features useful for the classification task. The values of the filter ω are learned automatically since they depend highly on the targeted dataset. Local pooling such as average or max pooling takes an input time series and reduces its length T by aggregating over a sliding window of the time series. A global aggregation is adopted to reduce drastically the number of parameters in a model thus decreasing the risk of overfitting. In addition to pooling layers, some deep learning architectures include normalization layers to help the network converge quickly. For time-series data, the batch normalization operation is performed over each channel, therefore, preventing the internal covariate shift across one mini-batch training of time series. The final discriminative layer takes the representation of the input time series (the result of the convolutions) and gives a probability distribution over the class variables in the dataset. Finally, to train and learn the parameters of a deep CNN: a feed-forward pass followed by backpropagation is carried out (Fawaz et al., 2019).

2.5   Lithology identification model's architecture

The lithology identification model is made up of two pairs of convolution layers, pooling layers, two fully connected layers, and a softmax layer. Figure 3 depicts the architecture of the lithology identification model. The first convolutional layer had 128 filters with a kernel size of 1024, the stride of the first and preceding layers was set at 2. The second convolutional layer had 128 filters with a kernel size of 256. Batch normalization and ReLU were applied after each convolution to speed up the convergence and to help improve generalization.

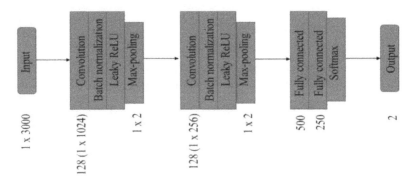

Figure 3.   1D CNN lithology identification model.

To run all experiments, cross validation was used to sufficiently train and evaluate the performance of the model. The data was split into 6500 for training, 200 for validation, and 500 for testing. A max epoch and mini batch size of 50 and 100 were used to estimate the error gradient. Adaptive moment estimation (Adam) was used as an optimizer at a learning rate of 0.001.

3 RESULTS AND DISCUSSION

3.1 *Evaluation of the lithology identification model*

A two-class classification learning model was built with acceleration as input data. Figure 4 illustrates the learning curve and the confusion matrix obtained after training the model. The learning curve was reviewed while training and afterward, to diagnose any problems associated with the learning. From the learning curve, it was observed that the model was a good fit as the training and validation loss decreased to a point of stability with a minimal gap between the two final loss values. The training and validation datasets were sufficiently represented as there was no noise in the validation loss curve. The model had a training accuracy of 99.69% and a validation accuracy of 98.5%. The model's generalization performance was assessed on unseen test data. The confusion matrix in Figure 4 shows the summary of the prediction results made by the lithology identification model. It indicates the true label and false label made by the model for each class. The prediction accuracy for granite and marble were 98.4% and 96.9% respectively. Out of 500 test datasets from each rock, the model could accurately classify 484 and 492 as granite and marble. The overall classification accuracy of the model was 98.9%. The results indicate that it is possible to distinguish the two rocks based on vibrations produced during the drilling process.

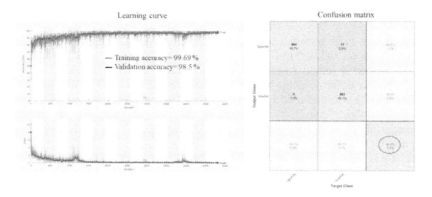

Figure 4. Lithology identification model's learning curve and confusion matrix.

3.2 *Comparison with SOTA deep neural networks*

The performance of the proposed lithology identification model was compared with two commonly used deep learning neural networks (DNNs) for time series classification: multilayer perceptrons (MLP) and fully convolutional networks (FCN). MLPs are the most traditional forms of DNNs. They are described as a baseline model for time series classification. The network has four layers, each is fully connected to the output of its previous layer. FCNs are convolutional networks that do not contain any local pooling layers which means that the length of the time series is kept unchanged throughout the convolutions. The basic block is a convolutional layer followed by a batch normalization layer and a ReLU activation layer. FCNs use a shorter kernel size of 8,5, and 3 which is different from the proposed model's kernel size of 1024 and 256. The architecture of the two models is described in full depth by Wang et al. (2017). The two models were trained using similar training conditions as the former model.

Table 3 indicates the validation and classification accuracies of the three models. The proposed model outperformed the SOTA models in terms of validation and classification accuracy. The lithology identification model had the highest validation accuracy of 98.55% and classification accuracy of 98.9% hence proving that the use of a longer kernel size and local pooling works well for our data. FCN performed relatively well with a validation and classification accuracy of 95.10 % and 95.8%. MLP had the lowest performance because it often disregards spatial information contained in the data.

The lithology identification model had the longest training time because it has the highest number of parameters compared to the other two models. MLP had the shortest training time because it has three fully connected layers, each with 500 neurons hence, forward and backpropagation can be carried out swiftly. The results validate that the proposed 1D CNN's architecture is effective in identifying the lithology of an area.

Table 3. Comparison of the proposed model with FCN and MLP.

Model	Validation Accuracy (%)	Classification Accuracy (%)	Training time (s)	Number of parameters
Lithology identification	98.55	98.9	4784	31 936 950
FCN	95.10	95.8	4064	167558
MLP	77.75	80.7	3730	2003002

3.3 *Effect of drill bit wear on the lithology identification model*

During drilling it is quite common for drill bits to get worn out, hence this study assessed the impact of bit wear on the performance of the pre-trained lithology identification model. The experiment described in section 2 of this paper was repeated with similar drilling parameters and equipment, however defective and abrasion drill bits were used as shown in Figure 5. The pre-trained lithology identification model was tested using datasets from the two bits. For a defective bit, the model had 78.2% and 77.5% prediction accuracy for granite and marble, whilst for abrasion, the prediction for granite and marble were 31.5% and 86.9%. The model had a poor prediction accuracy for abrasion-granite as depicted in Figure 5, because abrasion-granite displayed similar waveform characteristics as abrasion-marble. Even though the lithology identification model was not trained using defective and abrasion's time acceleration data, it yielded satisfactory prediction capabilities for worn-out bits except for abrasion-granite.

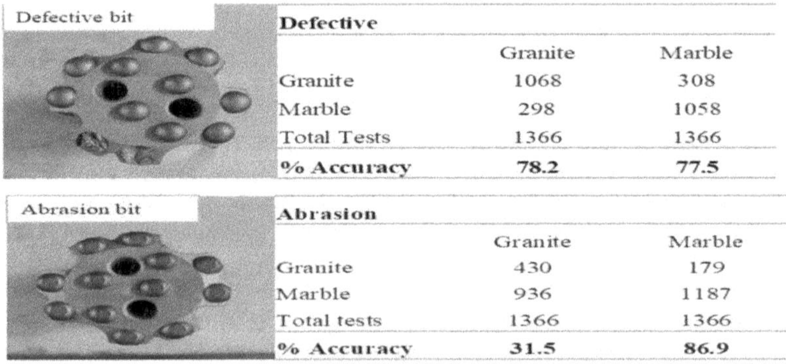

Defective	Granite	Marble
Granite	1068	308
Marble	298	1058
Total Tests	1366	1366
% Accuracy	78.2	77.5

Abrasion	Granite	Marble
Granite	430	179
Marble	936	1187
Total tests	1366	1366
% Accuracy	31.5	86.9

Figure 5. Effect of drill bit wear (abrasion and defective) on the lithology identification model.

4 CONCLUSION

This research illustrates the capabilities of utilizing drill vibration 1D CNN during rotary percussion drilling to identify the lithology of underground formations. 1D CNN is employed because of its high feature extraction capabilities and minimum data preprocessing. A two-layered CNN model with 128 filters, a stride of 2, and a kernel size of (1,1024) and (1,256) was utilized to classify two types of rocks: granite and marble. The model had a validation and classification accuracy of 98.55% and 98.9%. It was able to successfully classify each rock with few incorrect predictions. The proposed model performed better than the SOTA models. Furthermore, the model was evaluated using time acceleration from defective and abrasion drill bits. The model yielded satisfactory results as it resulted in prediction accuracy higher than 75% expect for abrasion granite. It was therefore proven that the lithology identification model using drill vibration as input data, provides an efficient method for obtaining the lithology of the subsurface.

Although this study is based in the mining industry, we believe the same approach can be applied in volcanic fields as core samples are often obtained by drilling for classification and further investigation by volcanologists.

ACKNOWLEDGMENTS

The authors would like to show gratitude to MMC Ryotec Corporation for data collection assistance and financial support.

REFERENCES

Chen G., Chen M., Hong G., Lu Y., Zhou B., & Gao Y. (2020). A New Method of Lithology Classification Based on Convolutional Neural Network Algorithm by Utilizing Drilling String Vibration Data. *Energies*, 13(4), 888. https://doi.org/10.3390/en13040888

Fawaz H. I., Forestier G., Weber J., Idoumghar L., & Muller P.-A. (2019). Deep learning for time series classification: A review. Data Mining and Knowledge Discovery, 33(4), 917–963. https://doi.org/10.1007/s10618-019-00619-1

Shreedharan S., Hegde C., Sharma S., & Vardhan H. (2014). Acoustic fingerprinting for rock identification during drilling. *International Journal of Mining and Mineral Engineering*, 5(2), 89. https://doi.org/10.1504/IJMME.2014.060193

Wang Z., Yan W., & Oates T. (2016). Time Series Classification from Scratch with Deep Neural Networks: A Strong Baseline. ArXiv:1611.06455 [Cs, Stat]. http://arxiv.org/abs/1611.06455

Rock Mechanics and Engineering Geology in Volcanic Fields – Ohta, Ito & Osada (eds)

Development of hyperspectral database and web based classifying system for rock type identification

Narihiro Owada* & Brian Bino Sinaice
Akita University, Akita, Japan

Shinji Utsuki
Utsuki Geo Solution, Tokyo, Japan

Hisatoshi Toriya & Youhei Kawamura
Akita University, Akita, Japan

ABSTRACT: In geological surveys, it is necessary to employ geological experts in determining rock samples correctly, this is however very costly and time consuming. Sinaice et al. (2020) demonstrated how employing hyperspectral data of igneous rocks and machine learning can be used in classifying rocks without any knowledge or background in geology. Although machine learning is able to achieve high accuracy with a sufficient amount of data (teaching data), there are no existing big-data-sets of rock hyperspectral data. In order to improve the accuracy and robustness of our machine learning model, we collected a large amount of spectral data of various types of rocks as an attempt to solve the machine learning bottleneck by creating web application that is able to share hyperspectral data among users. We created a web application that allows users to upload hyperspectral image data of rocks taken by various users, determine the rock type using our trained machine learning model, and subsequently browse the spectral database of the Mining Museum of Akita University. The machine learning model is capable of automatically improving its accuracy as data is uploaded by various users. Hence, each user is able to use the database and the determination function. By using this web application, it is possible to collect spectral data from a wide variety of rocks from various users of the web application, thereby improving the accuracy and robustness of rock type determination using hyperspectral data and machine learning, hence solving the aforementioned difficulties borne by researchers in previous studies.

Keywords: Hyperspectral Imaging, Machine Learning, Web Application

1 INTRODUCTION

Hyperspectral imaging is one of the ways to get physical properties of rocks. Basically, only three wavelengths of reflectance can be captured with an RGB camera. However, with a hyperspectral camera, it is able to capture hundreds of wavelengths of reflectance. Conventionally, hyperspectral imagery has been used for geological and mineralogical mapping in the remote sensing field. Because hyperspectral data can be identified materials on the basis of their characteristic spectral curves.

Classifying materials based on the spectral feature is one of the most beneficial applications of hyperspectral imaging. There are various algorithms to determine materials using acquired

*Corresponding author: owada@gipc.akita-u.ac.jp

DOI: 10.1201/9781003293590-55

data with hyperspectral cameras. Spectral Angle Mapper (SAM) is commonly used for mineralogical maps (Freek, 2006). Recently, other methods are developed, for example, Support Vector Machine (SVM) and Gaussian Process based method (GP-OAD) (Giorgos et al, 2011), (Sven et al, 2014). More recently, machine learning algorithms, such as Convolutional Neural Network (CNN) and Generative Adversarial Network (GAN) have been employed successfully. These approaches require reference spectra (Sinaice et al, 2019), (Lin et al, 2018). Especially, machine learning methods need an enormous amount of data for accurate classification.

The USGS spectral library and the ASTER spectral library provide comprehensive collections of spectra covering a wide range of wavelengths including minerals and rocks (Kokaly et al, 2017), (Baldridge et al, 2009). Measured samples usually are in powder form to purify and homogenize the sample. This type of measurement provides single distinctive spectra per sample. These spectral data are appropriate conventional classification approaches, such as SAM but not optimal for machine learning based methods.

The shortage of hyperspectral data will be an obstacle to develop a classification system of the rock types with machine learning and hyperspectral data. Machine learning extracts spectral character and learns distinctive feature from an enormous number of hyperspectral data. Then machine learning will get the capability to identify rock types based on the extracted unique spectral features. To get robustness for prediction, learning various types of rock spectra is essential because rocks in the real world are not only a single type. Despite the same rock name, the signature of rocks is different with some factors such as region and weathering. A spectral library containing a large amount of spectrum data with a wide variety is necessary for machine learning based classification methods to identify rock types comprehensibly and correctly.

In order to collect a wide variety of hyperspectral data of rocks, we created a web application that is able to share hyperspectral data among users. This web application allows users to upload, share and compare hyperspectral data in the browser and gives classification results using the created prediction model. This web application promotes improving the prediction accuracy of machine learning based methods.

2 THEORY

2.1 Hyperspectral imaging

In this experiment, we used SpecimIQ to measure spectra. SpecimIQ is a hyperspectral camera that can capture the 204 bands in the range 400 to 1,000 nm containing visible light and near infrared. Unlike image data taken by RGB camera, hyperspectral data has not only locational information but also spectral information that represent absorption and reflection feature of the material. It is able to classify material using some algorithms on the assumption each material has distinctive spectral curves.

In this experiment, for data upload test, some rock samples are selected from Akita University Mining Museum and take spectral image under the condition of halogen light (3,000 to 3,500lx) directed to sample from both sides in a dark room as shown in Figure 1.

2.2 Machine learning

Machine learning is defined as "A computer program is said to learn from experience E with respect to some class of tasks T and performance measure P if its performance at tasks in T, as measured by P, improves with experience E." (Tom, 1997). In this experiment, Convolutional Neural Network (CNN) is employed to classify rock type (task: T) and improve classification accuracy (performance measure: P) with spectral data (experience: E). A CNN can automatically extract features of the input data such as peaks and slopes peculiar to each rock of input spectral data and learn that feature. Based on the learned feature, trained CNN can classify the rock type.

As shown in Figure 2, CNN is structured as repeating the convolution and pooling operation. Convolution is a sum of products with input matrix and small matrix called kernel. This operation extracts the feature of input data. After convolution, pooling is to be applied. With pooling, convoluted data compress keeping extracted their own feature. As a result of

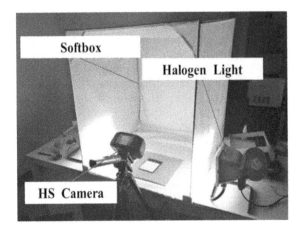 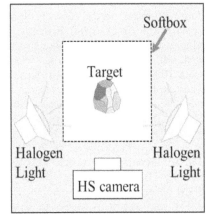

Figure 1.   Set-up of experiment.

repeating operation, the characteristics of the input data are emphasized so that machine learning would be discriminate the rock type based on input spectral data. In addition, enough data is needed to get a spectral feature.

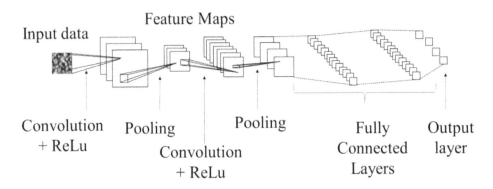

Figure 2.   Structure of the CNN.

2.3   *Web application*

Some specialized software that must be installed computer allows to see the detail of spectral data. However, it is hard to share spectral data among the same camera user because this type of software run on own computer. Also, limitation of reference spectral data exists due to the number of the open-source library.

Web based applications specialized in spectra would be a solution. Using the web application, spectral data can be accessed anywhere having an internet connection with an interactive graphical interface so that spectral signature is shown and compared. Hyperspectral camera users can share their spectral data with conditional information of the rock and photographing.

Figure 3. shows the structure of the web application. Clients mean someone who accesses the web application server. Nginx, Gunicorn, and Django are employed for web server, WSGI and web application framework on the server, respectively. WSGI is a connector with a web server and web application framework.

Django is one of the libraries for making web applications in python. Python is a computer language that is broadly used in machine learning fields. Also, python has capability to make web applications. Because of that, it is easy to create a web application containing a machine learning prediction system.

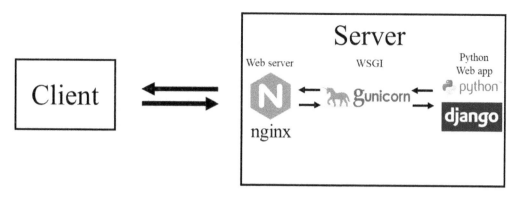

Figure 3.   Structure of the web application.

In this Django web application project, two types of functions are written. They are functions to render the HTML pages and process the spectral data. On the browser, a dat file containing raw spectral data and a metadata file describing photographing date and settings such as integration time must be uploaded to record the raw spectral data and metadata into the database. Data processing is needed to read spectral data correctly because of the file extension of raw data.

Figure 4. is shown the interface of the web application. Web application provides interactive spectral graph and classification result by machine learning trained model. Red rectangular on the upper left of Figure 4. shows the selected area. After selecting the area, the average spectrum is computed and returns as a graph in the middle of Figure 4. At the same time, the extracted spectrum is inputted into a prediction model trained by machine learning. The classification result return on the left side of the graph.

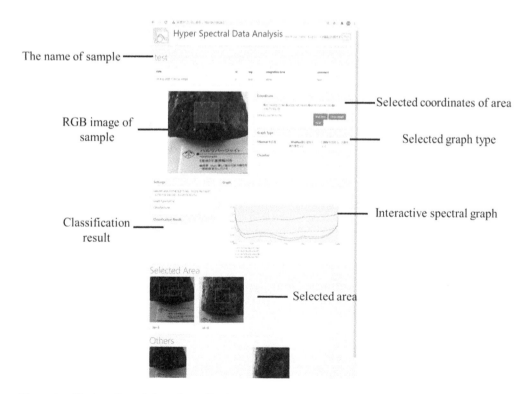

Figure 4.   The interface of the web application.

3 DISCUSSION

Hyperspectral imaging and machine learning have colossal potential to distinguish rock types. Despite that potential, no library exists to provide enough spectral data for machine learning. Aiming to collect a wide variety of spectral data taken in various conditions to solve the lack of data, the web application has been created. The web application allows to share spectral data uploaded by users and provides an interface to compare each sample's spectral curves easily. In addition, a learned prediction model by machine learning is applied as a classifier for rocks. The more user works with this application, the higher the number of spectral data is. Using collected data, the prediction model would be updated. However, this application does not have CSV export function so that users can analyze spectral data as they want. Furthermore, this application is only adapted for SpecimIQ. Other spectral camera data can't process properly. As long as the web application is not adapted to all spectral cameras, it is not to be said that this web application is practical.

4 CONCLUSION

Classification is one of the main applications of hyperspectral imaging. Recently, machine learning based classification method performs successfully. However, machine learning cannot achieve high accuracy prediction without a sufficient amount of learning data. In order to collect a large quantity of data, we have developed the web application to share spectral data among users. This application will facilitate the development of a prediction model for rock types.

REFERENCES

Baldridge, A. M., Hook, S. J., Grove, C. I., and Rivera, G., 2009, The ASTER spectral library version 2.0, Remote Sensing of Environment, 113, 711–715.

Meer, F., 2006, The effectiveness of spectral similarity measures for the analysis of hyperspectral imagery, International Journal of Applied Earth Observation and Geoinformation, 8, 3–17.

Mountrakis, G., Im, Jungho., and Ogole, C., 2011, Support vector machines in remote sensing: A review, ISPRS Journal of Photogrammetry and Remote Sensing, 66, 247–259.

Kokaly, R.F., Clark, R.N., Swayze, G.A., Livo, K.E., Hoefen, T.M., Pearson, N.C., Wise, R.A., Benzel, W.M., Lowers, H.A., Driscoll, R.L., and Klein, A.J., 2017, USGS Spectral Library Version 7: U.S. Geological Survey Data Series 1035, 61.

Zhu, L., Chen, Y., Ghamisi, P., and Benediktsson, A. J., 2018, Generative Adversarial Networks for Hyperspectral Image Classification, IEEE Transactions on Geoscience and Remote sensing, 56, 5046–5063.

Sinaice, B.B., Kawamura, Y., Kim, J., Okada, N., Kitahara, I., and Jang, H., 2020, Application of Deep Learning Approaches in Igneous Rock Hyperspectral Imaging, Proceedings of the 28th International Symposium on Mine Planning and Equipment Selection - MPES 2019, 228–235.

Schneider, S., Murphy, R. J., and Melkumyan, A., 2014, Evaluating the performance of a new classifier – the GP-OAD: A comparison with existing methods for classifying rock type and mineralogy from hyperspectral imagery, ISPRS Journal of Photogrammetry and Remote Sensing, 98, 145–156.

Mitchell, T., 1997, Machine Learning, McGraw-Hill Science Engineering.

Rock Mechanics and Engineering Geology in Volcanic Fields – Ohta, Ito & Osada (eds)

Application of machine learning technique for X-ray CT image segmentation of cement grouted rock

Chihiro Tagawa* & Shinichiro Nakashima
Yamaguchi University, Yamaguchi, Japan

Yoichi Yoshizu
NEWJEC Inc, Osaka, Japan

Hirotaka Iseki
Kanasai Electric Power Co., Osaka, Japan

Kiyoshi Kishida
Kyoto University, Kyoto, Japan

ABSTRACT: X-ray CT is a useful tool for observing internal structure of rock samples in non-destructive manner. It has been employed by many researchers in rock engineering and geological fields to see void structure, crack distribution, crack apertures, inhomogeneous material distribution and so on. This study deals automatic image segmentation using deep learning algorithm for X-ray CT imaging. 15 core samples were collected from check hole (depth = 10~20 m) after grouting in a gravity dam construction site. The samples were scanned with X-ray CT. Two cores are collected from mudstone layers to analyze, namely, Cores A and B. Core A is at the depth of 18.00 meter with a length of 130 mm, which has planar open joint with cement grout. Core B is at the depth of 27.5 meter with a length of 100 mm, which looks crushed and filled with cement grout. The CT images were automatically segmented into rock part, grout part and void part using a deep learning algorithm. The automatic segmentation method successfully identified rock part and grout part for the mudstone specimen with simple planar fracture, although the CT value range of rock part and grout part are overlapping each other. It also identified the difference of cement milk mixtures. However, for the specimen having intricate fracture geometry that was extracted from sheared stratum, narrow ungrouted cracks (less than 1.0 mm width) were sometimes misidentified. This indicates that automatic segmentation requires more training data for intricate fracture geometry.

Keywords: Rock mechanics, X-ray CT, Deep learning

1 INTRODUCTION

X-ray CT is often utilized in geo-science field as a means of observing geo-materials in a non-invasive manner. Through CT image processing and analysis, various types of useful information can be obtained on the internal structure, fracture geometry, mineral composition, and so on. Researchers have visualized the heterogeneous micro-structure of rock and micro-crack propagation (Verhelst et al. 1995, Sugawara 1997), the fluid flow of sedimentary rocks (Sato

*Corresponding author: a008veu@yamaguchi-u.ac.jp

DOI: 10.1201/9781003293590-56

et al. 2002), the tracer diffusion and migration into the rock matrix and fractures (Nakashima et al. 2004; Sato et al. 2007), and the fluid flow within deformed rock (Hirono et al. 2007). X-ray CT has also been used to measure fracture apertures and to detect contact areas (Yoshino et al. 2003; Nakashima et al. 2010, 2017; Yoshida et al. 2018). In a CT image analysis, one of the most basic but challenging processes is material recognition and segmentation based on the voxel CT values. To treat thousands of slices of CT images, some automatic algorithm for image segmentation is needed. This study deals with the image segmentation method using a deep learning algorithm for the X-ray CT images of rock samples. Mudstone cores with joints and cement grout were imaged with a X-ray CT scan and the images were then analyzed for material segmentation. Based on the results, the results, the efficacy of the automatic segmentation method is discussed.

2 X-RAY CT SCAN OF MUDSTONE CORES WITH JOINTS AND CEMENT GROUT

The scanned specimens are boring cores extracted from the foundation of a concrete gravity dam (Yoshizu et al. 2020). Cement grout was injected into the dam foundation as curtain grouting using cement milk at water/cement ratios of 1.5, 0.8, and 0.6. The coring was done after the curtain grouting, with diameter of 1.86 inches, to check the post-grout effects. As shown in Figure 1, sandstone and mudstone alternately appear in this foundation. The cement grout penetrated into some of the fractures.

Two cores were chosen for this study from the mudstone layers, namely, Cores A and B, as shown in Figure 2. Core A is at a depth of 18.00 m, with a length of 130 mm, and has a planar open joint with cement grout. Core B is at a depth of 27.5 m, with a length of 100 mm, looks crushed, and is filled with cement grout.

The specimens were scanned with a medical helical X-ray CT scanner under the conditions given in Table 1. The spatial resolution of the scanned images is 0.0568 mm in cross section and the slice thickness is 0.25 mm. The images were output in 16-bit signed values.

The scanned images are shown in Figure 3 in volumetric and cross-sectional views. From the figure, it is found that the cement grout appears as brighter zones in the images. In Core A, the cement grout fills filling the planar joint in a thin stratum. On the other hand, in Core B, the ragged continuous apertures are filled with cement grout, and the small blind voids are empty.

For further image analysis, small cuboid elements were clipped from the whole core images. The cuboid elements are shown in Figure 4. Their dimensions are 14.53 mm × 14.53 mm × 15.33 mm. They were clipped as the joint surface is almost parallel to one of the cuboid surfaces.

Their CT value histograms are shown in Figure 5. The CT histogram for Element A has three distinct peaks, namely, CT = 2000, 1700, and -1000, which correspond to the grout part, rock part, and void part, respectively. On the other hand, the histogram for Element B is rather gentle and has only one peak around CT = 800. From these histograms, it can be said that each material varies in its CT values and that the histograms overlap each other. It is difficult to determine the threshold of the CT value for each material and to identify the material based on only the CT value of each voxel. Therefore, this study employed a deep learning algorithm for the CT image segmentation.

3 SEGMENTATION METHOD WITH DEEP LEARNING

Deep learning has been attracting attention in image recognition in recent years. FusionNet is a learning model designed for medical image analysis. By introducing a Fully Convolutional Network (FCN), end-to-end learning is performed on a pixel-to-pixel basis. FusionNet is built based on the architecture of a convolutional autoencoder, which consists of an encoding path to retrieve the features of interest and a symmetric decoding path that enables the prediction of a synthesis (Quan et al. 2016, Long 2015).

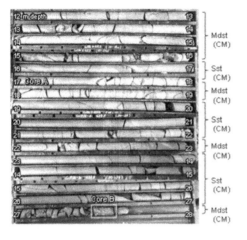

Figure 1. Boring core sample after the curtain grouting.

Figure 2. Mudstone cores scanned with X-ray CT. Core A has a planar open joint with cement grout. Core B is sheared mudstone.

Table 1. X-ray CT scanning conditions.

X-ray tube voltage	140kV
X-ray tube current	200mA
Exposure	300mA•s
Slice thickness	0.25mm
Voxel size	0.0568×0.0568×0.25 mm

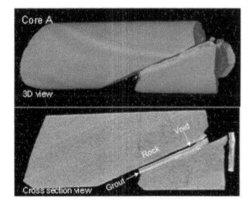

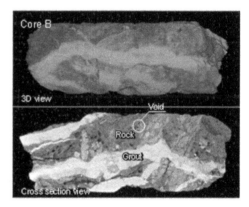

Figure 3. X-ray CT images of specimen in volumetric and cross-sectional views. The cement grout can be recognized as brighter (higher CT values) zones.

In this study, 200 datasets were manually labeled as shown in Figure 6. Of the 200 datasets, 160 are training data used for adjusting the model itself, and 40 are validation data used for evaluating the learning results (Aoshima et al. 2018). The network was trained using pairs of images, namely, the CT image and its corresponding labelled image. The output image was compared with the manually segmented image. The mean square error loss function was used to back-propagate in order to adjust the weights of network. The learning parameters are shown in Table 2. During the training process, the number of epochs and the batch size of Element B were set to be twice those of Element A because the loss function did not decrease.

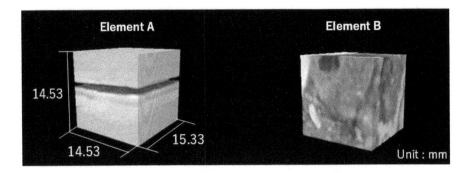

Figure 4. Cuboid elements from CT images. Element A and B are clipped from Cores A and B.

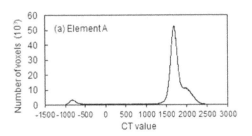
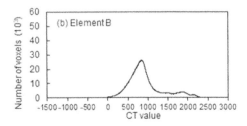

Figure 5. CT value histograms for Elements A and B.

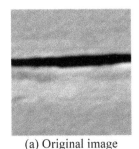

(a) Original image   (b) Annotation image

Figure 6. Example of labelling CT images. The red, green, and purple colors in (b) indicate the rock part, grout part, and void part, respectively.

Table 2. Learning parameters in Fusion Net.

	Element A	Element B
Image size	256×256	256×256
Input channel	3	3
Batch size	8	16
Epoch	200	400
LR	0.0002	0.0002

4 SEGMENTATION RESULTS

Automatic material segmentation with FusionNet was applied to Element A and B. The segmentation results for Element A are shown in volumetric view in Figure 7(a). From the figure, it is found that the rock part, grout part, and void part have been successfully separated with

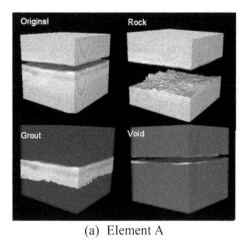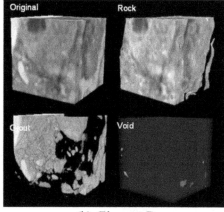

(a) Element A   (b) Element B

Figure 7. Results of segmentation for Elements A and B with FusionNet.

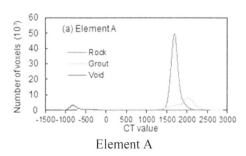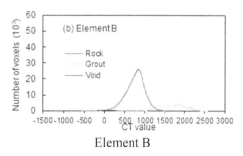

Element A   Element B

Figure 8. CT value histogram of Elements for A and B after segmentation with FusionNet.

Table 3. Mode and standard deviation of.

Element	Element A	Element B
Depth	18.00 m	27.50 m
Mode	1684	850
Standard deviation	148	227

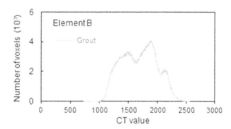

Figure 9. CT value of grout in Element B. The cement grout has three peaks in the histogram, corresponding to the three mixtures of the cement milk injected into the curtaingrouting.

automatic segmentation. From Figure 7, it is seen that grout is uniformly filled into the fracture. FCN performs convolution processing on the image and extracts feature points (Fujiyoshi et al. 2017). According to the segmentation results, the feature points were correctly extracted for Element A that probably contained consecutive fracture. This is due to the similar shapes of the annotation images and that of the test images. From the results, (a) the average fracture aperture width, (b) the average filling grout thickness, and (c) the grout filling rate inside the fracture were calculated as (a) 4.71mm, (b)3.53mm, and (c)75.9%

The segmentation results for Element B are shown in volumetric view in Figure 7(b). Compared with Element A, the geometry of the grout part in Element B is clearly intricate.

Figure 8 shows the CT value histograms for Element A and B after the segmentation. The overlapping of the ranges in CT values of the rock part (red line) and the grout part (green line) can be seen for both elements. By comparing the histograms of the elements, it is easily found that the CT values of the rock part are different between Elements A and B. As shown in Table 3, the peak CT value of the rock part in Element B is clearly lower than that in Element A and has a wider variation. One possible reason is that Element B existed in a severely sheared zone which leads to great porosity or density.

The CT value histogram of the grout part in Element B is magnified in Figure 9. From the figure, the grout CT value in Element B is seen to have three peaks, corresponding to the three mixtures of the used cement milk W/C = 1.5, 0.8, and 0.6, as previously described. It is interesting that the X-ray CT can sensitively distinguish the mixtures of cement milk.

From the viewpoint of automatic segmentation with machine learning the fractures and voids contained in Element B are much more irregular compared to those in Element A. Therefore, it is possible that the learning accuracy has deteriorated. Since the shape of each material in the cross-sectional of the crushed specimen is unpredictable, it is necessary to increase the teacher images.

5 EVALUATING IMAGE SEGMENTATION MODELS

A correct image was manually created to confirm the discrimination accuracy of deep learning. The Table 4 shows the accuracy rate of the class given to each material (rock, grout and void). From the results, it can be said that the accuracy rate is high because both of them exceed 94%. Then, Intersection over union (IoU) was used to compare the accuracy. IoU is known to be a good metric for measuring overlap between two bounding boxes or masks. The IoU metric measures the number of pixels common between the target and prediction masks divided by the total number of pixels present across both masks. The Table 5 shows the calculation results of IoU. From the results of IoU, there are many mis-segmentations in the void.

Table 4. Pixel-based comparison and accuracy rate of the test images and the correct images of Elements A and B.

Element A	Correct(px)	Incorrect(px)	Accuracy(%)	Element B	Correct(px)	Incorrect(px)	Accuracy(%)
Rock	10842025	212805	98.1	Rock	12731562	399882	97.0
Grout	3901004	36895	99.1	Grout	2624566	454562	85.2
Void	1155531	170204	87.2	Void	91265	16627	84.6
Total	15898560	419904	97.4	Total	15447393	871071	94.7

Table 5. IoU calculation results of Elements A and B.

Element A	Rock	Grout	Void	Mean	Element B	Rock	Grout	Void	Mean
IoU	0.969	0.924	0.872	0.921	IoU	0.936	0.826	0.223	0.662

Element A   Element B

Figure 10.   The misidentification parts for Elements in volumetric view.

The misidentifications parts for Elements are shown in volumetric view in Figure 10. In the Element A, there are many misidentifications at the boundary between substances. In the Element B, there in no feature of the misidentified.

6   CONCLUSIONS

This study imaged cement grouted mudstone cores with medical X-ray CT, and the images were processed with automatic segmentation with a deep learning algorithm. The knowledge obtained in this study is summarized below.

In the case of Element A, having a simple planar fracture, the rock part, grout part, and void part were successfully identified with automatic segmentation. IoU exceeded 0.9, but there is a misidentification at the boundary of the substance. On the other hand, in the case of Element B having intricate fracture geometry, the accuracy of identification was lower than that of Element A, and there was no feature in the misidentified part. More training data are required in the case intricate fracture geometry.

The difference of grouting material mixtures was sensitively identified with X-ray CT.

REFERENCES

Aoshima, K., Kawamura, S., Nakano, S., and Nakamura, H. 2018. Study on variant extraction of concrete structures using image recognition by deep learning. *Journal of Japan Society of Civil Engineers, Ser. E2*. 74(4). 293–305.

Fujiyoshi, H., and Yamashita, T. 2017. Image Recognition by Deep Learning. *Journal of the Robotics Society of Japan*. 35(3). 180–185.

Hirono, T., M. Takahashi and S. Nakashima. 2003. In situ visualization of fluid flow image within deformed rock by X-ray CT. *Engineering Geology*. 70: 37–46.

Long, J. Shelhamer, E., and Darrell, T. 2015. Fully Convolutional Networks for Semantic Segmentation. arXiv: 1411.4038.

Nakashima, S., D. Hasegawa, K. Kishida and H. Yasuhara. 2010. Measurement of fracture aperture in granite core using microfocus X-ray CT. In *Proceedings of the 44th US Rock Mechanics Symposium and 5th U.S.-Canada Rock Mechanics Symposium*, CD-ROM Paper No. ARMA 10-205

Nakashima, S., T. Sakamoto, H. Yasuhara and K. Kishida. 2017. Observation and quantification of fracture aperture in granite core using X-ray tomography and edge detection technique. In *Proceedings of the 51st US Rock Mechanics/Geomechanics Symposium*, CD-ROM Paper No. ARMA 17-247.

Nakashima, Y., T. Nakano, K. Nakamura, K. Uesugi, A. Tsuchiyama and S. Ikeda. 2004. Three-dimensional diffusion of non-sorbing species in porous sandstone: computer simulation based on X-ray microtomography using synchrotron radiation, *Journal of Contaminant Hydrology*. 74: 253–264.

Quan, T.M, Hildebrand, D.G.C., and Jeong. W. 2016. FusionNet: A deep fully residual convolutional neural network for image segmentation in connectomics. arXiv:1612.05360

Re, F. and C. Scavia. 1999. Determination of contact areas in rock joint by X-ray computer tomography. *International Journal of Rock Mechanics and Mining Science*. 36 (7): 883–890.

Sato, A, S. Kubota, T. Kawaguchi, K. Ezoe and K. Sugawara. 2002. Evaluation of permeability in rock by means of X-ray CT. In *Proceedings of the 3rd Korea-Japan Joint Symposium on Rock Engineering*. Seoul. 327–334.

Sato, A. and A. Sawada. 2007. Analysis of tracer migration process in the crack by means of X-ray CT. In *Proceedings of the 11th International Congress on Rock Mechanics*, Lisbon. 15–18.

Sugawara. K, Y. Obara, K. Kaneko, K. Koike, M. Ohmi and T. Aoi. 1997. Visualization of three-dimensional structure of rocks using X-ray CT method. In *Proceedings of the 1st Asian Rock Mechanics Symposium*, Rotterdam. 769–774.

Verhelst, F., A. Vervoort, P.H. De Bosscher and G. Marchal. 1995. X-ray computerized tomography: Determination of heterogeneities in rock samples. In *Proceedings of the 8th International Congress on Rock Mechanics*, Tokyo. 105–108.

Yoshida, R, Nakashima, S., Yoshizu, Y., Song, C., and Kishida. K. 2018. Co-registration of T images by SIFT method and observation of temporal changes in granite fractured aperture under long-term loading, In *Proceedings of the 52nd US Rock Mechanics Symposium*, Paper No. ARMA 18-1253.

Yoshino, N., A. Sawada and H. Satou. 2003. An examination of aperture estimation on fractured rock. In *Proceedings of the 32nd Symposium on Rock Mechanics*, Tokyo. 347–352.

Yoshizu, Y., Iseki, K., Iseki, H., Nakashima, S., Kishida, K. 2020. Image analysis and evaluation on grouting conditions employed modified gin grouting method through μ-focus X ray CT. *Journal of Japan Society of Civil Engineers*, Ser. C, 76(4), 394–404.

Rock Mechanics and Engineering Geology in Volcanic Fields – Ohta, Ito & Osada (eds)
© 2023 copyright the Author(s), ISBN 978-1-032-27657-1

Long-term displacement monitoring by GPS - case study of a large rockfill dam

Kaito Hiromitsu*, Shinichiro Nakashima & Norikazu Shimizu
Yamaguchi University, Ube, Japan

Morimasa Tsuda & Shigeki Ichikawa
Japan Water Agency, Saitama, Japan

ABSTRACT: The monitoring of ground displacement plays an important role in the risk assessment of volcanoes and slopes; and therefore, stable observations are required over a long period of time. This paper reports the displacement of a dam body surface observed with the GPS displacement monitoring system. The site was a central soil core type of rockfill dam with a height of 161 meters. The GPS measurement of this dam has continued for more than 14 years, including the first impoundment. The tropospheric delay error in the GPS measurement data was corrected using the modified Hopfield model and local weather data. The corrected GPS displacement at the dam crest in the vertical direction showed a good agreement with the settlement of the dam core observed by means of a settlement meter system.

Keywords: Rockfill dam, Exterior displacement, GPS, Settlement meter system

1   INTRODUCTION

In the safety management of rockfill dams after their construction, the displacement of the dam body is one of the most important items to measure in addition to seepage. While the displacement has generally been measured by conventional survey techniques, namely, total station surveys and leveling, more rapid and sophisticated measurement methods are required these days from the viewpoints of saving labor, reducing costs, and ensuring safety in the measurement works after unusual events.

Global Positioning System (GPS) technology has been utilized in various geotechnical engineering fields as a means to measure the displacement of the ground surface (Kondo et al. 1996; Masunari et al. 2003). This GPS-based displacement measurement system has several advantages in that it can automatically, rapidly, and continuously measure the three-dimensional displacements of multiple measurement points. Moreover, it is available for use regardless of the weather and does not require a line-of-sight between the measurement points, which is not the case with conventional survey techniques. The GPS-based displacement measurement system has been examined for its applicability to rockfill dam management (Yamaguchi et al. 2009; Nakashima et al. 2011; Kobori et al. 2014, 2015; Yamato et al. 2019), and it is now in operation at many dam sites in Japan.

This paper reports the GPS displacement measurement at a large rockfill dam. It presents the observational results and verifies the effectiveness of the GPS system through

*Corresponding author: a011veu@yamaguchi-u.ac.jp

DOI: 10.1201/9781003293590-57

a comparison of the displacement data with the results of settlement meter sensors. In addition, the difference in displacement behavior, depending on the part of the dam body, will be discussed based on the GPS displacement data.

2 OUTLINE OF THE GPS MEASUREMENT AT A ROCKFILL DAM

2.1 *GPS displacement monitoring system*

Figure 1 shows a schematic of the GPS displacement system typically used at rockfill dam sites. This system consists of L1-band GPS receivers and a control box. Using the static positioning technique, the three-dimensional coordinates of the measurement points are measured every hour with their origin at a fixed reference point.

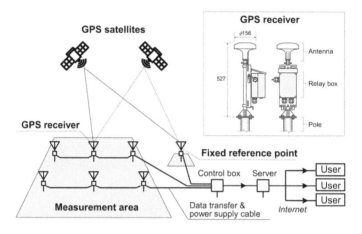

Figure 1.   GPS displacement monitoring system.

The raw GPS displacement data contain some random noise. The accuracy of displacement measurement by GPS depends on the number of GPS satellites used, the overhead obstructions, the multipath effects, the atmospheric conditions, and so on. The general accuracy is 5 - 10 mm in the horizontal direction and 10 - 20 mm in the vertical direction even with the precise static relative positioning method. Therefore, statistical treatment is needed to estimate the true displacements. In our monitoring system, the trend model, one of the smoothing models based on a polynomial regression model improved by introducing a probability structure (Kitagawa and Gersch 1984; Shimizu 1999) has been used.

2.2 *Measurement conditions at the rockfill dam*

This study targets a rockfill dam in Japan with a height of 161.0 m and a length of 427.1 m. The plan and cross-sectional views are shown in Figure 2. There are seven GPS receivers on the embankment surface as measurement points on the crest (GD3, GD4, and GD5) and the downstream face (GD11, GD18, GD25, and GD30), as indicated by red circles in the figure. Three fixed reference points (K1, K2, and K3) are set on the right bank, as indicated by light blue squares in the figure.

Figure 3 shows the two types of GPS antennas used at this dam. Pole-mounted antennas, like that shown in Figure 3(a), are usually used in the GPS displacement monitoring system. In snowy areas, however, the antenna poles, which have a cantilever structure, may be largely deformed due to the lateral load brought about by deep snow piled up on the embankment slope. To avoid such harmful lateral loading, low antennas, like that shown in Figure 3(b),

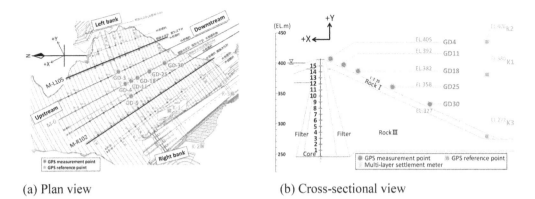

(a) Plan view (b) Cross-sectional view

Figure 2.   Schematic of the rockfill dam and layout of GPS receivers.

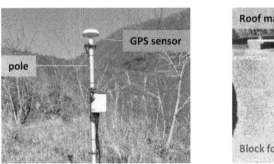
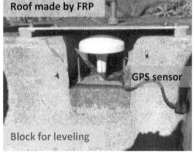

(a) GPS sensor at reference point (b) GPS sensor at measurement point

Figure 3.   Two types of GPS antennas.

were employed at this dam. With this type of antenna, the GPS sensors are recessed into concrete blocks for leveling surveys. In addition, a snow visor, made of fiber-reinforced plastic board, is set over each antenna to avoid snow accretion (Kobori et al. 2014). At the reference points, namely, K1, K2, and K3, on the other hand, pole-mounted antennas, with 2.5-m-high poles, were used because they were located on almost the level ground.

The GPS measurement at this dam was started on November 18, 2006, two months after the first impoundment was initiated, and it has been continued up to the present, namely, for more than 14 years.

3   GPS MEASUREMENT RESULTS AT LARGE ROCKFILL DAM

3.1   *Measurement error correction*

As mentioned previously, GPS measurement contains some errors and noises. Figure 4 shows the three-dimensional GPS displacement results for the measurement points on the dam crest (GD3, GD4, and GD5) from November 2006 to June 2018. Random noise in the measurement was filtered using the trend model (Kitagawa and Gersch 1984; Shimizu 1999). From this figure, it is seen that the measurement results show a spike-like fluctuation every year in winter, especially in the vertical direction (Z displacement) at all the measurement points. The duration of the fluctuation corresponds to the snowy season. This phenomenon is known as the snow effect on the GPS measurement. The attenuation and refraction of the GPS satellite signals in travelling through the snow layer on the GPS antennas are the reasons for the fluctuation, and it is difficult to correct this snow effect (Kobori et al. 2014).

Another characteristic of GPS measurement is the tropospheric delay. The effect of the troposphere on the GPS signals appears as an extra delay in the measurement of the signals traveling from the satellite to the receiver. This delay depends on the atmospheric conditions, namely, the temperature, pressure, and humidity. When the elevations of a measurement point and the reference point are almost the same, the effect of the tropospheric delay is canceled out in the process of calculating the phase difference. However, when there is a large difference in height between the two points, the extra delay remains in the displacement measurement in proportion to the height difference. Figure 5 shows the GPS vertical displacement results (Z direction) for 2011 and 2012. From Figure 5(a), a yearly periodic fluctuation in the results can be seen in the vertical direction and the amplitudes are proportional to the height difference between the reference point and the measurement point, which are typical characteristics of the tropospheric delay error. To correct the tropospheric delay error, this study employed the modified Hopfield model, an estimation model for the tropospheric delay that uses meteorological observations. Figure 5(b) presents the results of the correction. The yearly periodic fluctuation is seen to have been largely reduced by the error correction.

Figure 6 shows the GPS displacement results for five measurement points on the downstream slope after the tropospheric delay correction. The displacement data during the snowy season is hidden in this figure. From this figure, the long-term displacement behavior of the dam body can be grasped. The monotonic motion toward the downstream in the horizontal direction and the settlement in the vertical direction can be found.

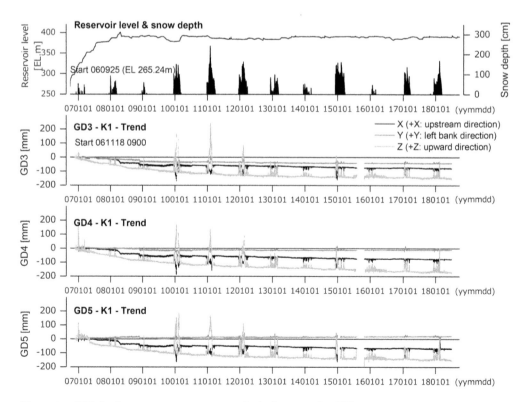

Figure 4.   GPS displacement measurement results (reference point: K1).

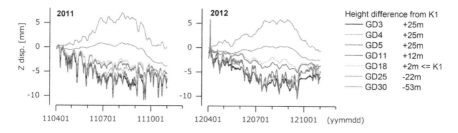

(a) GPS vertical displacement before correction

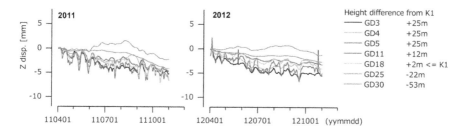

(b) GPS vertical displacement after tropospheric delay correction with modified Hopfield model

Figure 5. GPS vertical displacement before and after tropospheric delay correction.

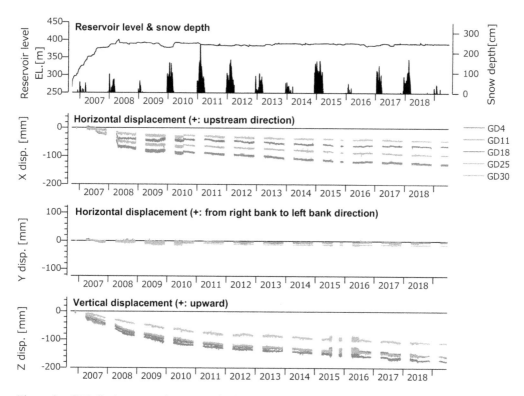

Figure 6. GPS displacement after tropospheric delay correction (reference point: K1).

3.2 Comparison of vertical displacement results by GPS and settlement meter system

As illustrated in Figure 2(b), multi-layer settlement meter sensors have been installed in the soil core of the dam body. The timeline chart of the settlement meter system since November 2006 (the start of the first impoundment) is shown in Figure 7(a). Unfortunately, some of the settlement sensors went down during the embankment construction and impoundment. Therefore, this figure shows only the reliable measurements.

The vertical displacement by the GPS measurement at the dam crest (GD4) was compared with the measurement by the settlement meter system. The profile of the settlement by the settlement meter system is shown in Figure 7(b) with the vertical displacement by the GPS (GD4). From this figure, it is found that the GPS measurement exists almost as an extension of the profile by the settlement meter system until the end of 2008. This means that the GPS measurement agrees well with the direct measurement of the settlement. On the other hand, nine years later, in 2017, the GPS measurement is seen to be much larger than the extension line of the settlement meter profile. The self-weight consolidation of the dam core or the collapse settlement of the coarse rockfill material might have progressed during those years around the part of the dam body at the higher elevation, although no evidence can confirm it due to the failure of the settlement meter system at the higher elevation.

4 CONCLUSIONS

The exterior displacement of a large rockfill dam body has been observed for 14 years using the GPS displacement monitoring system. This paper presented the measurement results, the error correction method, and a comparison with the settlement meter measurement results.

- The tropospheric delay error in the GPS measurement was effectively corrected using the modified Hopfield model. This kind of error is common in the GPS displacement monitoring of large dams, and the presented correction will serve as a good case study.

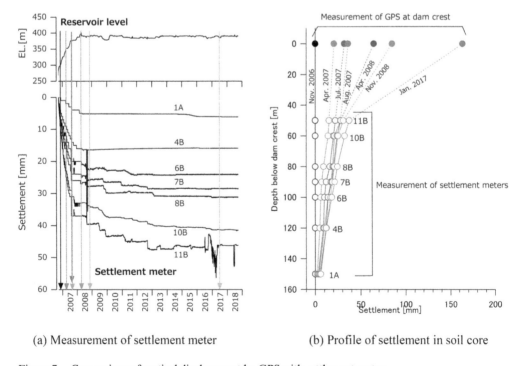

(a) Measurement of settlement meter   (b) Profile of settlement in soil core

Figure 7.   Comparison of vertical displacement by GPS with settlement meter.

460

- The GPS displacement at the dam crest agreed well with the settlement of the dam core directly observed with the multi-settlement meter system. This means that the GPS system is a reliable monitoring tool for the safety management of dam bodies.

REFERENCES

Goad, C. C. and Goodman, L. L., 1974, A modified Hopfield tropospheric refraction correction model, Presented at the fall annual meeting, *American Geophysical Union*, 1–16.

Kitagawa, G. and Gersch, W., 1984, A smoothness priors-state space modeling of time series with trend and seasonality, *J. American Statistical Association*, 79(386),378–389.

Kobori, T., Yamaguchi, Y., Iwasaki, T., Nakashima, S. and Shimizu, N., 2014, Application of GPS to monitoring of exterior deformation of fill dams using antennas installed inside the manhole at the dam crest, *J. Japan Society of Civil Engineers* Ser. F3, 70 (1), 18–28.

Kobori, T., Yamaguchi, Y., Nakashima, S., and Shimizu, N., 2015, Continuous displacement monitoring of rockfill dam during earthquakes by using Global Positioning System, *J. Japan Society of Dam Engineers*, 25(1),6–15.

Kondo, H., Cannon, M., Shimizu, N. & Nakagawa, K., 1996, Developments of a ground displacement monitoring system by using the Global Positioning System. *J. Construction Management and Engineering*, Japan Society of Civil Engineers, 546/6-32: 157–168.

Masunari, T., Tanaka, K., Ookubo, N., Oikawa, H., Takechi, K., Iwasaki, T. and Shimizu, N., 2003 GPS continuous displacement monitoring system. *Proc. Int. Symp. Field Measurements in Geomechanics*, FMGM03, 537–543.

Nakashima, S., Kawasaki, H., Shimizu, S., Kubota, S., Nakano, T., 2011, Measurement of exterior deformation of an earth-fill dam using GPS displacement monitoring system, *Proc. of the 12th ISRM Cong.*, pp.1069–1072.

Shimizu, N., 1999, Displacement monitoring by using Global Positioning System of assessment of slopes, *Proc. 9th Int. Congr. Rock Mech.*, 1435–1438.

Yamaguchi, Y., Kobori, T., Yoshida, H., Sakamoto, T., Itaya, H. & Iwasaki, T. (2009): Real-time monitoring of exterior deformation of embankment dams using GPS. *Proc. 23rd ICOLD Congr.*: Q.91–R.1.

Yamato, K., Nakashima, S., Shimizu, N., Kubota, T., Sato, N., & Tsuda, M., 2019 Tropospheric delay correction in GPS displacement monitoring: case study of Tokuyama Dam in Japan, *Proc. 5th ISRM Young Scholars' Symposium on Rock Mechanics*, ISRM-YSRM-2019-050.

Rock Mechanics and Engineering Geology in Volcanic Fields – Ohta, Ito & Osada (eds)
© 2023 copyright the Author(s), ISBN 978-1-032-27657-1

The evaluation of rockfalls at Ayvalı in Cappadocia Region due to differential weathering of tuff layers

Nazli Tunar Özcan* & Resat Ulusay
Department of Geological Engineering, Hacettepe University, Ankara, Turkey

Ömer Aydan
Department of Civil Engineering, University of the Ryukyus, Okinawa, Japan

Naoki Iwata & Kazuki Kanose
Chuden Consultants Co. Hiroshima, Japan

ABSTRACT: Differential weathering of tuffs results in overhanging configurations so that rockfall events take place. This paper presents some observations in-situ and characteristics of tuffs, some studies on numerical simulations on determining the rockfall travel distances to investigate the rockfall hazard problem at Ayvalı in Cappadocia (Turkey). Laboratory tests on rocks indicated that the rock sampled from the layer just below the top formation is quite vulnerable to water content variation and its strength decreases drastically with an increase in water content. The rockfall hazard is an important issue in Cappadocia (Turkey) and this study may cast some insights to this issue.

Keywords: Differential weathering, tuff, rockfall, travel distance, Cappadocia, laboratory tests

1 INTRODUCTION

Rock fall is one of the instability forms of rock slopes. It may result in loss of life and property, closing main transportation roads and railways and major economic losses. It is important to estimate the rockfalls and their trajectories in order to protect structures and facilities from rockfall events. When rocks have different strength and weathering characteristics against natural agents such as winds, rainfall and snowfall. Tuffs of Cappadocia Region have layered structure deposited at different time episodes associated volcanic activities in the past. As a result, differential weathering of tuffs occurs and they result in overhanging configurations so that rockfall events take place.

The authors have been involved with the differential weathering issue near Ayvali village and did some site investigations on the characteristics of tuffs, rock falls, their travel distances. Some observations in-situ and characteristics of tuffs, some studies on material properties and numerical simulations of the rockfall travel distances with different slope angles are investigated. The rockfall hazard is an important issue in Cappadocia Region, which was included in the World Heritage List and is a well-known internationally as the most popular region in Turkey due to its to historical heritages and open air museums, and this study may cast some insights to this issue.

*Corresponding author: ntunar@hacettepe.edu.tr

DOI: 10.1201/9781003293590-58

2 GEOGRAPHY, GEOLOGY AND OBSERVATIONS

Ayvali Village is located almost in the center of the modern Cappadocia Region as shown in Figure 1. Although most of tuff formations are denoted as ignimbrites by geologists (Figure 1), the authors would prefer the word "tuff" for most of rocks seen in the region.

A view of intense rockfalls is shown in Figure 2a. There is a soft tuff layer beneath a hard tuff and the rock undergoes differential weathering (Figure 2b). Furthermore, there are some vertical and/or sub-vertical cooling or faulting induced joints in hard rock layer. Rockfalls are associated with this hard rock layer and notch depth may be 5 to 6 m. The soft rock having vertical desiccation cracks crumbles and the erosion depth increases during the rainy season as well as dry seasons. When the erosion depth reaches to a certain depth, the overhanging rock blocks topples and tumble down.

The measurements on the site indicated that the travel distance could be less than 60 m and the estimated "fahrböschung angle (angle defined by the highest scarp to maximum horizontal travel distance)" ranges between 11.3 and 17.8 degrees with an average of 14.2 degrees. The road between Ayvalı village and Mustafa Paşa town passes near the site. The observations on the site indicated that the rockfalls do not present any danger to the roadway. Figure 3 shows a typical cross-section.

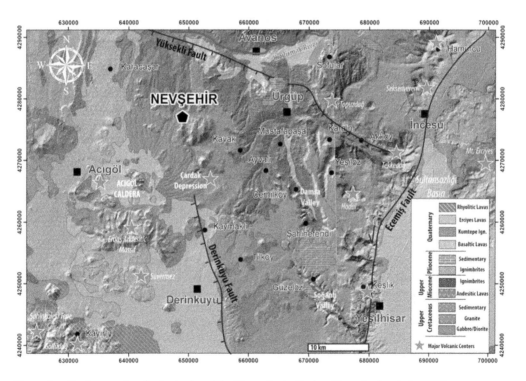

Figure 1. Geology and major tectonics of the Cappadocia Region (from Aydar et al., 2012).

3 ROCK MECHANICS INVESTIGATIONS

Some rock blocks were collected from the site of underground differential weathering and samples were prepared for rock mechanics testing. Figure 4 compares the strain-stress responses of white color and beige color tuffs during uniaxial compression tests. As noted from Figure 4., the strength of beige color tuff is drastically reduced upon

| (a) A general view of the location | (b) tuff layers |

Figure 2.   Views of the study location and tuff layers involved in differential weathering. The degree of weathering of the beige color tuff is W3, white color tuff is W2 and overhanging layer is W1.

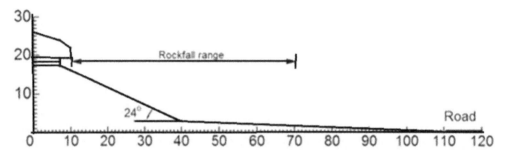

Figure 3.   A simplified cross-section of Ayvalı rock-fall location.

saturation and desiccation cracks observed in this tuff during site investigations while no strength reduction was observed in white color tuff. Table 1 summarizes physico-mechanical properties of tuffs.

4   ROCKFALL INITIATION MECHANISM AND SOME CONSIDERATIONS

As seen in Figure 2b, soft tuff layers below the harder top tuff layers eroded due to percolation of groundwater, wind and cyclic freezing-thawing and wetting-drying actions. Aydan (1989), Tokashiki and Aydan (2010) studied this issue in relation to the cliff failures due to toe erosion. This failure is classified as "flexural toppling". The following relation was derived for the highest tensile stress of a trapezoidal shaped cantilever beam (see Aydan 1989, Aydan and Kawamoto (1993) Tokashiki and Aydan (2010) for details) (Figure 5a).

$$\sigma_b = k_h \gamma h_b L \left(\frac{1+\alpha}{2} \right) + 6 \frac{M}{h^2} \tag{1}$$

Where

$\alpha = \frac{h_s}{h_b}; \ V_o = (1 + k_v) \gamma h_b L \left(1 - \frac{(1-\alpha)}{2} \right), \ M_o = -(1 + k_v) \gamma h_b L^2 \left(\frac{1}{2} - \frac{(1-\alpha)}{3} \right)$

h_b, h_s, γ and L are beam height at the base and at the far end, unit weight of rock mass and erosion depth, respectively. k_h and k_v are horizontal and vertical seismic coefficients.

Equation (1) is utilized to estimate the maximum erosion depth for a given thickness for different tensile strength values as shown in Figure 6. The estimations imply that tensile

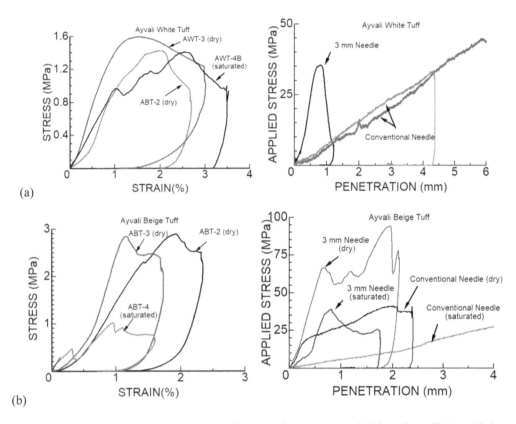

(a)

(b)

Figure 4. Responses of strain-stress and needle penetration responses of white color tuff (a) and beige color tuff (b).

Table 1. Physico-mechanical properties of the investigated tuffs (dry).

Tuff	Unit weight (kN/m^3)	P-wave velocity (km/s)	S-wave velocity (km/s)	UCS (MPa)	NPI N/mm
White	13.9-15.1	1.16-1.36	0.81-0.92	1.43-1.77	6.5-13.8
Beige	14.6-16.7	1.14-1.44	0.80-0.97	2.8-2.9	24.9-30.0

UCS: Uniaxial Compressive Strength   NPI: Needle Penetration Index

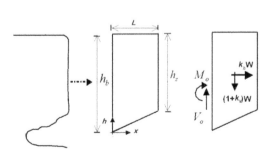

Figure 5. Modeling of overhanging cliffs.

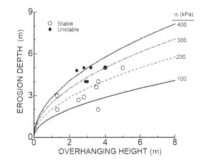

Figure 6. Comparison of observations with estimations.

465

strength of the fallen rock blocks should be greater than 200 kPa. The maximum erosion (notch) depth is generally less than 5 m.

5 ROCKFALL SIMULATIONS

A number of simulation programs have been developed and applied in practical rock-fall prediction recently. The Colorado Rockfall Simulation Program (CRSP) (developed at Colorado School of Mines; Pfeiffer and Bowen, 1989) is the most well-known computer code and most of following programs essentially similar to CRSP (e.g. Rocfall 2D by Rocscience, 2019). These programs use simplified parameters to simulate rock-fall behavior based on rigid body concept and calculate trajectories and provide useful statistics for the design of mitigation measures. In order to predict of rockfall trajectories accurately by using rockfall simulation codes, the parameters such as the size and shape of the rock blocks, the coefficients of friction, the coefficients of restitution, characteristics slope material and slope geometry should be defined clearly. In this study, 2D statistical analysis program RocFall 2D is utilized for the simulation of the model experiments. RocFall 2D allows the user to perform both lumped mass and rigid body rockfall analyses with different block shapes (Rocscience, 2019). As the block shape used in the model experiments and the parameters of path material are known, rigid body rockfall analyses method should provide some insight to physical model tests on rockfalls.

A series of rockfall analyses were carried out near Ayvalı village of the Cappadocia Region as shown in Figure 7 (see Özcan et al. 2021 for details). Analysis were performed for square, rectangular and hexagonal shaped blocks with a size range between 1 m^3 to 10 m^3 with the consideration various shapes observed during site-investigations. Figure 7 shows the trajectories for 50 rock falls using the rigid body simulation for 10 m^3 square, rectangular and hexagonal blocks and their distribution of rockfall travel locations. The highest reach (fahrböschung) angles were 23, 20 and 14 degrees for square, rectangular and hexagonal blocks, respectively. These results are quite close to the observations mentioned in Section 2. Furthermore, none of the blocks reached to the roadway.

6 SAR INVESTIGATIONS -LOS ANALYSES

Satellite images of the area of investigations are shown in Figure 8. As noted from the images, the vegetation is quite poor so that it constitutes best conditions for SAR-based analyses for the images from Sentinel I. The authors utilized the 109 Synthetic Aperture Radar (SAR) images taken by Sentinel I between the period of 2014 Dec. 10 and 2019 October with an interval of two weeks. Sentinel-I provides images of C-Band with a range of 37.5 and 75 mm. Figure 8a shows a satellite view of the region. The close-up view of the domain is shown in Figure 8b. Figure 9a shows the image obtained from LOS (Line of Sight) time-series analyses. Although the coherence was quite poor, some movements were noted in some areas and time-series data of selected three points are shown in Figure 9b. Although these analyses are of first-kind to be utilized in the region, it is expected that it may provide some quantitative data on the movements and rockfalls in the area for long-period of time.

7 CONCLUSIONS

This paper presents some observations in-situ and characteristics of tuffs, some studies on numerical simulations on determining the rockfall travel distances to investigate the rockfall hazard problem at a location in Ayvalı of Cappadocia (Turkey). The laboratory tests on rock

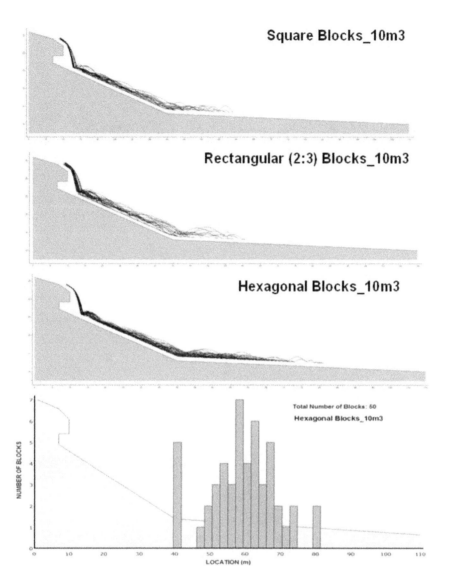

Figure 7. Rockfall analysis results for Ayvalı site.

samples indicated that the rock layers are quite vulnerable to moisture content variation and subsequently to differential weathering. The numerical simulation results indicated that the travel distance of rock falls depend upon slope angle, properties of rock block and also the surface morphology and frictional characteristics besides some dynamic properties such as restitution coefficient. The rockfall hazard is an important issue in Cappadocia and this study may cast some insights to this issue.

Rockfall simulations for Ayvalı village in Cappadocia are in accordance with in-situ observations and rockfalls could not reach to the roadway. However, these are preliminary analyses and evaluation of the rockfall hazards in Ayvalı village in Cappadocia, we need the exact geometry of rockfall paths, the identification of fallen rock blocks and their geometry and potential blocks along the cliff. For this purpose, remotely operated unmanned drones or helicopters with laser equipment capable of penetrating beneath the vegetation are necessary for further detailed analyses.

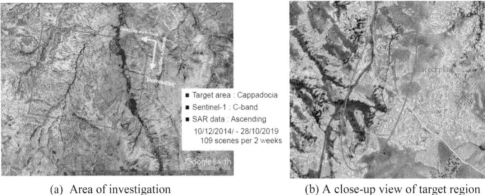

(a) Area of investigation   (b) A close-up view of target region

Figure 8. Satellite views of the area of investigation.

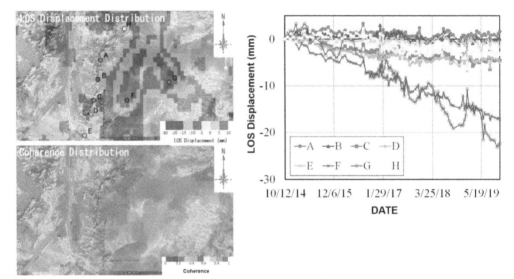

(a) Areal distribution of estimated deformation   (b) time-series of selected points

Figure 9. Results of LOS time-series analyses.

REFERENCES

Aydar E., Schmitt A.K., Çubukcu H.E., Akın L., Ersoy O., Şen E., Duncan R.A. and Atıcı G., 2012, Correlation of ignimbrites in the central Anatolian volcanic province using zircon and plagioclase ages and zircon compositions. J. Volcanol. Geotherm. Res., 213-214, 83–97

Aydan Ö., 1989. The stabilization of rock engineering structures by rockbolts. Doctorate Thesis, 204p, Nagoya University.

Aydan Ö. and Kawamoto T., 1992, The stability of slopes and underground openings against flexural toppling and their stabilisation. *Rock Mechanics and Rock Engineering*, 25(3), 143–165.

Aydan Ö., Shimizu Y. and Kawamoto T., 1992, The reach of slope failures. *The 6th Int. Symp. Landslides*, Christchurch, *ISL 92*(1), 301–306.

Aydan Ö. and Shimizu Y., 1993, Post-failure motions of rock slopes. *Int. Symp. Assessment and Prevention of Failure Pheomena in Rock Engineering*, Istanbul, 409–415.

Aydan Ö., Ohta Y., Tokashiki N. and Akagi T., 2005, Prediction of post-failure motions of rock slopes induced by earthquakes. *35th Japan Rock Mechanics Symposium*, 103–108.

Pfeiffer T.J. and Bowen T. D., 1989, Computer simulation of Rockfalls. Bulletin of the Association of Engineering Geologists, 26 (1), 135146.

Özcan Tunar N., Aydan Ö., Murayama Y., Horiuchi K. and Ulusay R., 2021, A Study on Model Experiments and Numerical Simulations on Rockfalls and Its Utilization to Assess the Rockfall Hazards in Miyagi Island (Japan) and Cappadocia (Turkey), *IACMAG*, 482–490.

Rocscience, 2019, Rocfall 2D software, Rocscience Inc.,Toronto, Canada, (accessed on Dec. 2019).

Tokashiki N. & Aydan Ö., 2010, The stability assessment of overhanging Ryukyu limestone cliffs with an emphasis on the evaluation of tensile strength of Rock Mass. *Journal of Geotechnical Engineering*, JSCE, 66(2), 397–406.

Rock Mechanics and Engineering Geology in Volcanic Fields – Ohta, Ito & Osada (eds)
© 2023 copyright the Author(s), ISBN 978-1-032-27657-1

Evolution of permeability in a single granite fracture under coupled conditions

Hideaki Yasuhara* & Naoki Kinoshita
Ehime University, Ehime, Japan

ABSTRACT: Waste liquid with high radioactivity generated when reprocessing spent nuclear fuel produced in nuclear power generation is called high level radioactive waste. When geological disposal of high-level radioactive waste is carried out, it is assumed that heat generation from the waste and high overburden earth pressure will affect the permeability characteristics of the surrounding rock mass. Therefore, it is indispensable to evaluate the long-term permeability change of underground rock mass. In this study, a suite of permeability experiments was conducted under various temperature and confining pressure conditions assuming underground environment, and the effects of temperature, pH, confining pressure and fracture surface roughness on permeability characteristics of underground rock mass were evaluated. From the test results, it was confirmed that the higher the pH, the lower the permeability, and that the precipitation phenomenon significantly affects the permeability change, when the permeated water is different. It was also confirmed that when deionized water and high-pH water were used, the permeability more decreased in the high-temperature condition, and when mineral-saturated water and simulated seawater were used, the permeability more decreased in the room-temperature condition. When the confining pressure and fracture surface roughness were different, no clear difference was observed in the permeability change. Moreover, it was confirmed that the elemental concentration of permeated water after permeability was higher in the high temperature condition than in the room temperature condition. This may be due to the promotion of free-face dissolution phenomena by high temperature conditions. From SEM-EDX observations, the precipitation that may be salt crystals was observed under the room temperature condition on the fracture surface after permeability experiments using simulated sea water for permeation water, but it was not confirmed under the high temperature condition. When simulated seawater was used, the precipitation of salt crystals seemed to decrease the permeability in the room temperature condition, because the permeability decreased more than in the high temperature condition. The fracture surface measurements concluded that the roughness of the fracture surface became smoother by carrying out the permeability experiments.

Keywords: Rock fracture, Permeability, Dissolution, Precipitation

1 INTRODUCTION

In nuclear power generation, high-level radioactive waste is generated in the process of reprocessing spent nuclear fuel produced during power generation. High-level radioactive wastes are geologically disposed to a depth of 300 m underground and are conserved by a multiple barrier structure consisting of an artificial barrier and a natural barrier. Then, it is assumed that a disposal facility will be installed in the coastal area from the viewpoint of marine transportation of high-level radioactive wastes, and that it becomes a high-temperature state by heat generation

*Corresponding author: yasuhara.hideaki.me@ehime-u.ac.jp

DOI: 10.1201/9781003293590-59

from a waste body, and that groundwater contains salt derived from sea water, so that it is considered to have a large effect on permeability characteristics of rock mass. Moreover, long-term performance evaluation of underground rock mass is indispensable, because high-level radioactive waste requires management in units of several tens of thousands of years. In this study, in order to simulate groundwater containing salt derived from sea water, continuous permeability test is carried out using simulated sea water as permeation water for granite specimens under temperature and confining pressure controlled. The effects of temperature, permeation water, confining pressure, pH, etc. on permeability characteristics are evaluated from the test results. Then, the permeated solution is collected, and the concentrations of the elements are measured by inductively coupled plasma atomic emission spectroscopy (ICP-AES), thereby confirming the chemical reaction such as mineral dissolution and precipitation. Microstructural observations using scanning electron microscope-energy dispersive X-ray spectrometry (SEM-EDX) are also conducted on the specimens before and after the permeability test to observe secondary minerals precipitated on the fracture surfaces.

2 METHODOLOGY

In this study, continuous permeability tests using granite were carried out. The mechanical and physical properties of granite are shown in Table 1. In the permeability test, permeated solution after water flow was taken every 24 hours. Cylindrical specimens with smooth fracture were prepared for the tests (Figure 1). A method for producing a test piece is as follows. Firstly, the coring of Φ50 mm was carried out for the granite block. Then, the fracture surface was produced using a diamond cutter. Thereafter, using an indoor boring machine after the fracture surface fabrication, coring of Φ30 mm was conducted. Then, the end faces are shaped using a polishing machine, and the test piece height was molded to be about 60 mm. In the continuous permeability test, the boundary conditions given to each specimen are shown in Table 2. Since actual seawater contains many organic substances such as plankton, and long-term preservation is difficult, simulated seawater (manufactured by Yasu Chemical Co., Ltd.) was produced and used as permeated water. The elemental concentration of simulated seawater is shown in Table 3. In addition, we compared the results among E-4, a de-ionized water condition that was previously studied, E-8 that uses oil as permeant that may refrain chemical reaction, and E-16, E-18 that are a mineral saturated water conditions in which the silica (SiO_2) content, a main component of granite, was made to be a supersaturated condition, to evaluate the effects of differences in permeation water and temperature on the evolution of permeability. The experimental setup for continuous permeability tests is shown in Figure 2.

Figure 1.   Rock sample with a single smooth fracture.

Table 1.   Mechanical and physical properties of granite used in this work.

Density [kg m⁻³]	Uniaxial compressive strength [MPa]	Young's modulus [GPa]	Poisson's ratio [-]	Effective porosity [%]
2.58×10^3	171	50.7	0.28	1.0

Table 2. Experimental conditions (Kinoshita and Yasuhara, 2012).

Specimen	fracture surface	Height [mm]	Diameter [mm]	Temperature [°C]	Confining pressure [MPa]	Permeant	pH
E-4	Saw-cut	57.44	29.79	25-90	5.0	Deionized water	6.0
E-8	Saw-cut	62.30	28.95	25-90	5.0	Oil	-
E-16	Saw-cut	59.18	29.68	20	5.0	Mineral saturated water	7.61
E-18	Saw-cut	60.70	29.70	90	5.0	Mineral saturated water	7.84
E-22	Saw-cut	60.36	29.25	20	5.0	Simulated seawater	8.26
E-23	Saw-cut	59.95	30.81	90	5.0	Simulated seawater	8.26

Table 3. Elemental concentrations of simulated seawater.

	Si	Al	K	Fe	Ca	Na	Mg
Concentration [mol L^{-1}]	1.85×10^{-5}	1.42×10^{-4}	1.25×10^{-2}	1.86×10^{-5}	2.25×10^{-2}	4.58×10^{-1}	5.99×10^{-2}

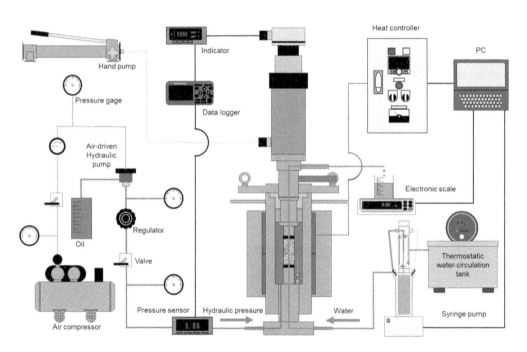

Figure 2. Experimental setup for continuous permeability test.

3 RESULTS

Hydraulic aperture, b, can be obtained from the continuous permeability tests by Equation (1). Then, fracture permeability, k, can be estimated by using Equation (2).

$$b = \left\{ \frac{12\mu Q l}{w(P_0 - \rho_w g l)} \right\}^{-3} \tag{1}$$

$$k = \frac{b^2}{12} \tag{2}$$

Where μ is the dynamic viscosity [Pa s], Q is the flow rate [m³ s⁻¹], l is the sample height [m], w is the sample size [m], P_0 is the differential pressure [Pa], ρ_w is the fluid density [kg m⁻³], and g is the gravity acceleration [m s⁻²]. The differential pressure prescribed in this work ranges from 50 to 1200 kPa, and the resulting Reynolds numbers should be less than 1, which secures Darcy's law within the fractures.

Obtained test results are shown in Figures 3 and 4. As shown in Figure 3a, the permeability of E-22 (simulated seawater) decreased by about two orders of magnitude throughout the test period, indicating the greatest reduction compared with the other conditions. This may be due in part to the precipitation of salt minerals derived from simulated seawater on the fracture surface. It is also expected that the higher the pH, the greater the permeability decreases, and that the evolution of permeability depends on the pH. From Figure 3b, the permeability reduction is within one order in all the conditions, and the change behavior is also very similar, and no clear difference between the conditions is found under high temperature conditions. Figure 4 shows the comparison of permeability at 20 and 90 °C using simulated seawater. The permeability reduction at 20 °C is greater than that at 90 °C. This may be attributed in part to the increase in solubility of salt in water, and the precipitation of salts might be suppressed.

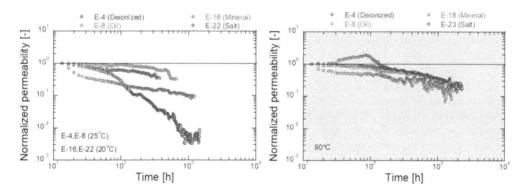

Figure 3. Evolution of permeability obtained from continuous permeability tests (Kinoshita and Yasuhara, 2012).

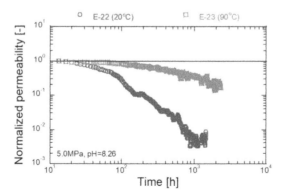

Figure 4. Comparison of permeability change between 20 and 90 °C in case of simulated seawater.

473

The concentrations of effluent solutions for each element in E-22 and E-23 obtained by ICP-AES are shown in Figure 5. In E-22, the Na concentration shows negative values at almost all the time during the water permeability test. It indicates that secondary minerals containing Na were precipitated on the fracture surfaces. No clear change was observed for other elements. In E-23, Ca always shows a negative value, and Na and Mg, which are relatively much contained in simulated seawater, traveled in the positive and negative directions from the beginning of the permeability test to about 800 hours, and thereafter show large negative values. No clear change was found for the other elements. Therefore, it is considered that precipitation phenomena were predominant in Ca. In both permeability tests of E-22 and E-23, the variation of Na content is greatest compared with other elements, and it is considered that the effect on fracture topography should be most apparent.

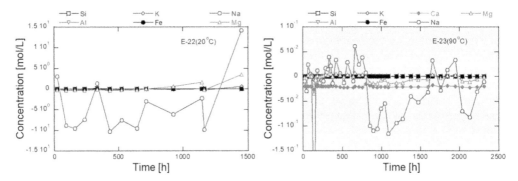

Figure 5.   Element concentrations obtained from effluent solutions in E-22 and E-23.

In order to observe secondary minerals produced by permeability tests, SEM-EDX observations were carried out on specimens E-22 and E-23 after the permeability tests. We compare SEM-EDX observations of E-22 and E-23 to examine the effect of temperature-difference on the formation of secondary minerals on fracture surfaces. E-22 at 20 °C, sedimentation of secondary minerals derived from simulated seawater was confirmed (Figure 6a). Especially, the secondary minerals of cubic type which seemed to be salt crystals could be confirmed. The EDX analysis revealed that the precipitated minerals are salt crystals because large values for Na and Cl were confirmed at the points. In contrast, in E-23 at 90 °C, cubic-shaped crystals were not observed and other secondary minerals were apparent (Figure 6b). The significant reduction of permeability in E-22 at 20 °C may be attributed to the precipitation of salt crystals, but further investigations are required.

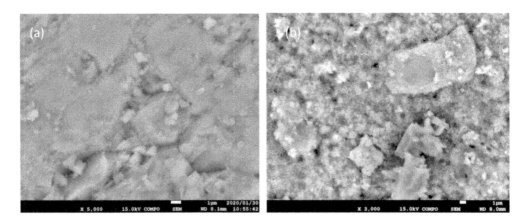

Figure 6.   Microstructural observation by SEM in (a) E-22 and (b) E-23.

4 CONCLUSIONS

In this study, continuous permeability tests were conducted on a granite with a single fracture under high temperature, high pressure and salt water conditions to simulate the underground environment. The effect of simulated seawater on the permeability of the granite fracture was extensively examined. In addition, ICP-AES and SEM were conducted to investigate the cause of the change in permeability with time. The findings of this study are described below.

(1) In the continuous permeability test, the permeability decreased monotonically from the beginning of the permeability test, and the final permeability was about one order and two orders of magnitude lower than the initial permeability in E-22 and E-23, respectively. The normalized permeability was lower at 20°C than at 90°C.

(2) For the evaluation of the concentration of substances, we focused on the concentration of Na in the simulated seawater. It was confirmed that precipitation was more dominant at 20°C than at 90°C in simulated seawater.

(3) For the observation of microstructure, the room temperature experiment (E-22) and the high temperature experiment (E-23) were compared. The cubic precipitates, which were observed in the room temperature experiment (E-22), but not in the high temperature experiment (E-23), were considered to be salt crystals. Elemental mapping analysis also showed that Na was not scattered in the high temperature condition, which was observed in the room temperature condition, and the existence of salt crystals could not be confirmed.

ACKNOWLEDGEMENTS

This work was partly supported by JSPS KAKENHI Grant Number 19H02237. This support is gratefully acknowledged. The authors thank Fuminori Ohnishi and Daichi Sako for their contribution to experiments conducted in this work.

REFERENCES

Kinoshita, N. and Yasuhara, H., 2012, Evolution of fracture permeability in granite under high temperature and high confining pressure conditions, *Journal of MMIJ*, 128, 72–78.

Author Index

Milton Keynes UK
Ingram Content Group UK Ltd.
UKHW032232151223
434481UK00021B/460

9 781032 276571